MELVIN D. JOESTEN
Professor of Chemistry
Vanderbilt University
Nashville, Tennessee

JOHN T. NETTERVILLE
Guest Professor, Retired
David Lipscomb University
Nashville, Tennessee

JAMES L. WOOD
Resource Consultants, Inc.
Brentwood, Tennessee

The World of Chemistry Boxed features in each chapter provided by:
Isidore Adler
Late of the University of Maryland

Nava Ben-Zvi
Hebrew University of Jerusalem

Text Typeface: Times Roman
Compositor: Progressive Typographers
Acquisitions Editor: John Vondeling
Developmental Editor: Jennifer Bortel
Managing Editor: Carol Field
Project Editor: Laura Maier
Copy Editor: Donna Walker
Manager of Art and Design: Carol Bleistine
Text Designer: Rosemary Murphy
Cover Designer: Lawrence R. Didona
Text Artwork: J & R Art Services, Inc.
Layout Artist: Dorothy Chattin
Director of EDP: Tim Frelick
Production Manager: Charlene Squibb
Marketing Manager: Marjorie Waldron

Cover Credit: Computer Graphic © Orion/Westlight. Title page photo is a polarized light micrograph of a polyethylene net. (Dr. Harold Rose/Science Photo Library)

Printed in the United States of America

WORLD OF CHEMISTRY—ESSENTIALS

0-03-076114-X

Library of Congress Catalog Card Number: 92-053779

2345 069 987654321

THIS BOOK IS PRINTED ON **ACID-FREE, RECYCLED** PAPER

THE ELEMENTS*

Name	Symbol	Atomic Number	Atomic Weight	Name	Symbol	Atomic Number	Atomic Weight
Actinium	Ac	89	(227)	Neon	Ne	10	20.1797
Aluminum	Al	13	26.981539	Neptunium	Np	93	(237)
Americium	Am	95	(243)	Nickel	Ni	28	58.69
Antimony	Sb	51	121.75	Niobium	Nb	41	92.90638
Argon	Ar	18	39.948	Nitrogen	N	7	14.00674
Arsenic	As	33	74.92159	Nobelium	No	102	(259)
Astatine	At	85	(210)	Osmium	Os	76	190.2
Barium	Ba	56	137.327	Oxygen	O	8	15.9994
Berkelium	Bk	97	(247)	Palladium	Pd	46	106.42
Beryllium	Be	4	9.012182	Phosphorus	P	15	30.973762
Bismuth	Bi	83	208.98037	Platinum	Pt	78	195.08
Boron	B	5	10.811	Plutonium	Pu	94	(244)
Bromine	Br	35	79.904	Polonium	Po	84	(209)
Cadmium	Cd	48	112.411	Potassium	K	19	39.0983
Calcium	Ca	20	40.078	Praseodymium	Pr	59	140.90765
Californium	Cf	98	(251)	Promethium	Pm	61	(145)
Carbon	C	6	12.011	Protactinium	Pa	91	231.03588
Cerium	Ce	58	140.115	Radium	Ra	88	(226)
Cesium	Cs	55	132.90543	Radon	Rn	86	(222)
Chlorine	Cl	17	35.4527	Rhenium	Re	75	186.207
Chromium	Cr	24	51.9961	Rhodium	Rh	45	102.90550
Cobalt	Co	27	58.93320	Rubidium	Rb	37	85.4678
Copper	Cu	29	63.546	Ruthenium	Ru	44	101.07
Curium	Cm	96	(247)	Samarium	Sm	62	150.36
Dysprosium	Dy	66	162.50	Scandium	Sc	21	44.955910
Einsteinium	Es	99	(252)	Selenium	Se	34	78.96
Erbium	Er	68	167.26	Silicon	Si	14	28.0855
Europium	Eu	63	151.965	Silver	Ag	47	107.8682
Fermium	Fm	100	(257)	Sodium	Na	11	22.989768
Fluorine	F	9	18.9984032	Strontium	Sr	38	87.62
Francium	Fr	87	(223)	Sulfur	S	16	32.066
Gadolinium	Gd	64	157.25	Tantalum	Ta	73	180.9479
Gallium	Ga	31	69.723	Technetium	Tc	43	(98)
Germanium	Ge	32	72.61	Tellurium	Te	52	127.60
Gold	Au	79	196.96654	Terbium	Tb	65	158.92534
Hafnium	Hf	72	178.49	Thallium	Tl	81	204.3833
Helium	He	2	4.002602	Thorium	Th	90	232.0381
Holmium	Ho	67	164.93032	Thulium	Tm	69	168.93421
Hydrogen	H	1	1.00794	Tin	Sn	50	118.710
Indium	In	49	114.82	Titanium	Ti	22	47.88
Iodine	I	53	126.90447	Tungsten	W	74	183.85
Iridium	Ir	77	192.22	Unnilennium	Une	109	(266)
Iron	Fe	26	55.847	Unnilhexium	Unh	106	(263)
Krypton	Kr	36	83.80	Unniloctium	Uno	108	(265)
Lanthanum	La	57	138.9055	Unnilpentium	Unp	105	(262)
Lawrencium	Lr	103	(260)	Unnilquadium	Unq	104	(261)
Lead	Pb	82	207.2	Unnilseptium	Uns	107	(262)
Lithium	Li	3	6.941	Uranium	U	92	238.0289
Lutetium	Lu	71	174.967	Vanadium	V	23	50.9415
Magnesium	Mg	12	24.3050	Xenon	Xe	54	131.29
Manganese	Mn	25	54.93805	Ytterbium	Yb	70	173.04
Mendelevium	Md	101	(258)	Yttrium	Y	39	88.90585
Mercury	Hg	80	200.59	Zinc	Zn	30	65.39
Molybdenum	Mo	42	95.94	Zirconium	Zr	40	91.224
Neodymium	Nd	60	144.24				

* Values listed for atomic weights are those reported by IUPAC in 1987. The atomic weights given here apply to elements as they exist naturally on earth. Values in parentheses are used for radioactive elements whose atomic weights cannot be determined accurately. In these cases, the number given in parentheses is the atomic mass number for the isotope with the longest half-life. Refer to Chapter 3 for an explanation of isotopes, atomic masses, and atomic weights.

World of Chemistry

Saunders Golden Sunburst Series

SAUNDERS COLLEGE PUBLISHING
A Harcourt Brace Jovanovich College Publisher

*Fort Worth Philadelphia San Diego New York
Orlando Austin San Antonio Toronto
Montreal London Sydney Tokyo*

Preface

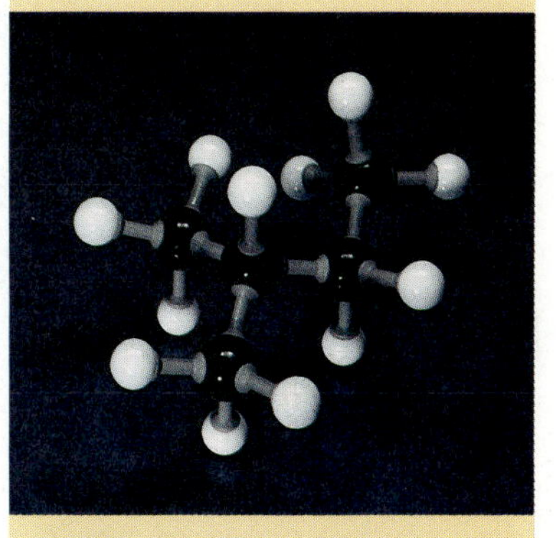

Ball-and-stick model of one of the isomeric pentanes C_5H_{12}. (Charles Steele)

Approach and Scope

World of Chemistry — Essentials has been written for those students wishing a one-semester course in college chemistry. This text is based on the successful approach used in the text and video series entitled *World of Chemistry.* The focus of this shorter text is on the applications and consequences of chemical change and the modification of the total human experience. While less emphasis is given to the theoretical base of the science, the fundamental approach remains. Physical and chemical discoveries are presented along with the impact of these discoveries on our way of life. These investigations of nature and the consequent human-provoked changes have produced the modern chemical world. Throughout the text, the student encounters applications of chemical knowledge that dramatically affect the quality of life.

No previous knowledge of chemistry is assumed or required in this presentation. However, the approach in this text is sufficiently different to challenge and interest the student with a background in high school chemistry.

To the beginning student chemistry may be a mystery, but to leave the workings of the chemist unexplained argues that the liberally educated person must be dependent upon the chemist for those chemical decisions that affect society as a whole. *World of Chemistry — Essentials* is based on the belief that the nonscience major can see and appreciate the chain of events leading from chemical fact to chemical theory as well as the ingenious manipulation of materials based on these chemical theories. Thoughtful students will see that the intellectual struggles in chemistry are closely akin to their own intellectual pursuits and will feel that each educated individual can and should have a say in how the applications of chemical knowledge affect our quality of life.

The topics covered in this book were selected based on what we have observed to be student concerns. As a team of authors, and as individual teachers of chemistry at the collegiate level for many years, we have observed the following intense interests of our students:

1. Feeling the satisfaction of understanding the cause of natural phenomena.
2. Understanding the scientific basis for making the important personal choices demanded in using of chemicals and chemical products.
3. Participating on a rational basis in the societal choices that will affect the quality of human life.
4. Helping to preserve and restore the quality of the environment and to sensibly approach the recycling of natural resources.
5. Developing an insight into the perplexing problem of chemical dependency.
6. Sensing the balance involved in population control, the chemical control of disease, and the ability of the world to produce food.
7. Choosing personal habits in exercise programs and in nutritional selections that are compatible with healthful living.
8. Using present energy reserves at a sensible rate while new energy sources are developed for the long term.

All of these paramount interests, as well as many of lesser note, are featured in this chemistry text on material substances and their uses.

World of Chemistry—Essentials uses a common sense approach that is too often lost when the chemical community presents itself to the educated public at large. As in the total human experience, there is in chemical studies a fundamental relationship between cause and effect—structure causes function, chemical periodicity, and consequent material properties. We have carefully selected that thread of chemical history that shows chemistry to be the human endeavor it is. The text has been written consistent with the essential chemical story to be told, the effective communications with college students that we have employed over the years, and with the critical reviews received from our peers. The philosophical setting for the presentation is made in Chapter 1, allowing the text and teacher to whet the students' appetite to understand what they may have previously thought to belong only to the scientific elite. Chapters 2, 3, 4, 6, and 9 lay the necessary chemical groundwork to consider chemistry's impact on society; these chapters are replete with interesting applications to which the liberal arts students readily relate. The remaining chapters of the text address problems of intense interest to the general public including energy available for human use, synthetic materials that dramatically alter the human environment, the nutritional basis of healthy living, medicines and drugs, pollution and the conservation of natural resources, consumer chemistry, and the agricultural production of food for a hungry world population.

A concerted effort has been made to inform the reader about certain vital and/or interesting matters that have their basis in chemistry. Some examples are designer drugs; viral diseases in general, and AIDS in particular; chemical treatment for major diseases; hair growth; and the almost unbelievable applications being developed using genetic engineering.

Enhancements

The use of full color throughout the text greatly enhances the effectiveness of the teaching aids that help the book communicate. Objectives at the beginning of each chapter help the student to clearly define study goals for the particular unit of study. Self-tests help the student to measure retention and comprehension. Questions at the end of the chapters provide an additional measure of study and opportunities for extended research. Boldface type for new terms and concepts as well as marginal notes add emphasis to focus the reader's attention. Numerous interesting features such as personal notes about scientists and information about new commercial products have been added throughout the text. The many illustrations, which are a logical extension of the text, often communicate better than words. Boxed features in each chapter from *The World of Chemistry* videos further develop interesting concepts that are only mentioned in the video presentation.

The World of Chemistry Video Package

This text is presented either as a stand alone course in chemistry for nonscience majors or as an integral part of a comprehensive telecourse package including a series of 26 thirty-minute video programs, a telecourse study guide, telecourse faculty manual, and a telecourse laboratory manual. The video series, entitled *The World of Chemistry,* was developed by the late Dr. Isadore Adler of the University of Maryland, and Dr. Nava Ben-Zvi of Hebrew University of Jerusalem and was sponsored by the Annenberg/CPB Project and corporate sponsors. The video programs feature Nobel laureate and Priestly medalist Roald Hoffmann and provide a comprehensive survey of the field of chemistry and its impact on modern society. The series was produced jointly by The University of Maryland and The Educational Film Center. The individual videos are listed on the back cover of this book. For information on ordering these videocassettes, call 1-800-LEARNER.

In addition to the videotapes, JCE: SOFTWARE has produced "The World of Chemistry: Selected Demonstrations and Animations I," a videodisc that is a selection of demonstrations and animations taken from *The World of Chemistry* videotapes. It is a 12 in, double-sided, 60 min, CAV-type videodisc; NTSC standard. For information on this issue of JCE: SOFTWARE, contact: JCE: Software, John W. Moore and Jon L. Holmes, University of Wisconsin–Madison, Madison, WI 53706 (ISSN-1050-6942, *Journal of Chemical Education: Software* # SP-3).

The Text Package

The Instructors Manual/Test Bank contains teaching suggestions, solutions and test questions, as well as a correlation guide to *The World of Chemistry* videotapes and the JCE: SOFTWARE videodisc.

The EXAMaster Computerized Test Bank is available in both IBM and Macintosh versions. Overhead transparencies show 119 full-color images reproduced from both *The World of Chemistry* and *The World of Chemistry—Essentials.*

A laboratory manual by Jones, Johnston, Netterville, Wood, and Joesten presents 45 experiments that are easily incorporated into the course.

The Student Study Guide carefully and intelligently directs the student through the text.

Acknowledgments

We are deeply grateful to all who have contributed to the improvement of the manuscript and teaching aids for this book. We sincerely appreciate the help of our fellow teachers who reviewed our manuscript and offered suggestions for improvement: Earl C. Alexander, San Diego Mesa College; Erwin Boschmann, Indiana University–Purdue University at Indianapolis; Robert C. Byrne, Illinois Valley Community College; Richard Conway, Shoreline Community College; Jack Cummins, Metropolitan State College; Howard D. Dewald, Ohio University; Ronald Distefano, Northampton Community College; Alton Hassell, Baylor University; Chu-Ngi Ho, East Tennessee State University; Stanley N. Johnson, Orange Coast College; Jerry L. Mills, Texas Tech University; Tom Mines, Florissant Valley Community College; David S. Newman, Bowling Green State University; Marie Nguyen, Indiana University–Purdue University at Indianapolis; Robert J. Palma Sr., Midwestern State University; James Schreck, University of Northern Colorado; Berton C. Weberg, Mankato State University; Donald H. Williams, Hope College; Bruck Winkler, University of Tampa; and Robert Yolles, DeAnza College. We are thankful for the numerous other users of our texts who have helped with corrections. Jerry Mills was especially helpful with the final manuscript draft as we sought to reduce our presentation of classical chemistry to only the segments necessary to understand the current chemical problems facing our society.

We appreciate the help of the entire staff at Saunders College Publishing, professionals every one; they know how to get the job done! Jennifer Bortel, Developmental Editor, facilitated the flow of information and ideas, prompted the necessary decisions, pushed for meeting deadlines, and worked for excellence in every facet of editorial control. Special thanks is given to Laura Maier, Project Editor, and to Donna Walker, Copy Editor, for their excellent work in the production of this text. We also wish to express appreciation to Margie Waldron, Senior Marketing Manager; Charlene Squibb, Production Manager; Carol Bleistine, Manager of Art and Design; and to intern Andrew Beckwith who conceived of and collected the opening art for each of the chapters.

We give our special thanks to John Vondeling, Publisher, who has placed confidence in our series of chemistry texts for more than twenty years. We appreciate his leadership in the world of collegiate chemistry.

Much help has come our way, but of course, the responsibility for the contents of the text rests entirely on us.

As in all of our previous works, we dedicate this effort to our spouses and gratefully acknowledge their support and understanding during the preparation of this manuscript.

Melvin D. Joesten
John T. Netterville
James L. Wood

June 1992

Contents Overview

A rocket upon takeoff requires much energy in a short time, that is, much power. (NASA)

Table of Contents

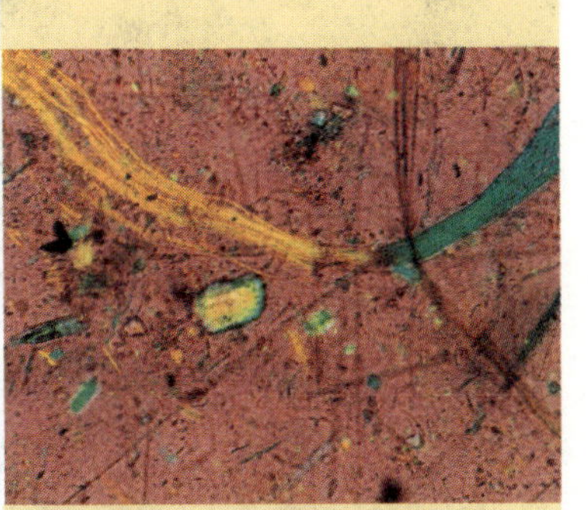

Asbestos sample as seen under a polarizing light microscope. This sample contains 60% chrysotile asbestos. (Courtesy of Particle Data Laboratories)

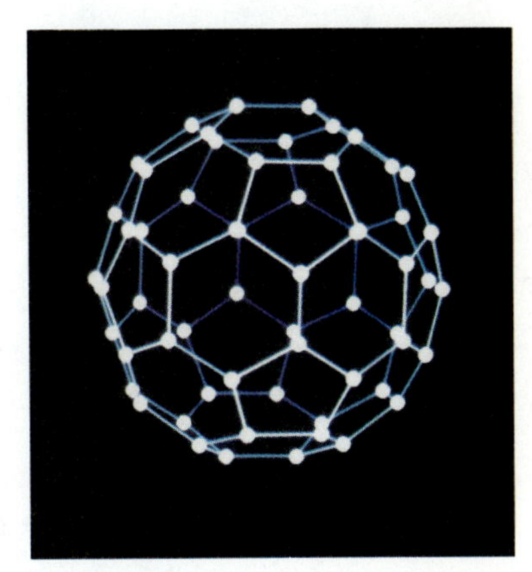

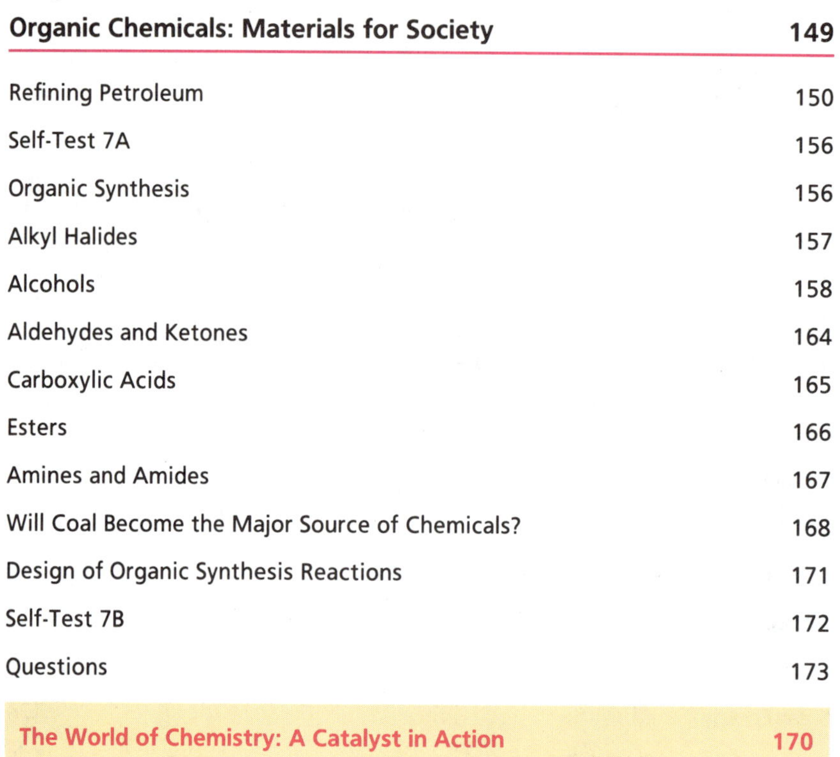

Chapter 9
Chemistry of Life 201

Chapter 10
Toxic Substances 235

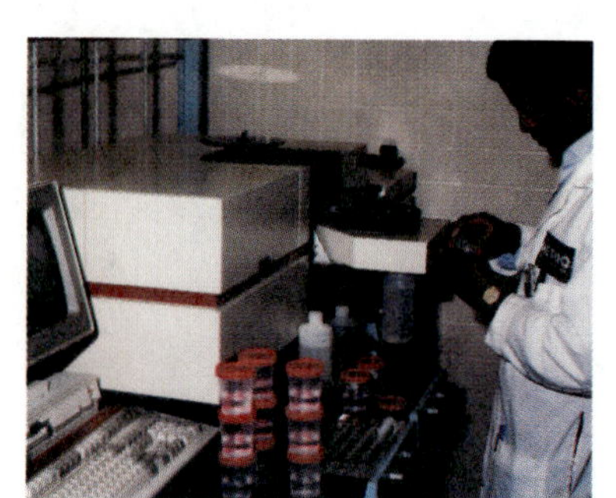

Multiple Vitamins

The montage of city scenes in Robert Rauschenberg's *Estate* represents some of the consequences of chemical manipulations by humans. (The Philadelphia Museum of Art: Given by the Friends of the Philadelphia Museum of Art)

CHAPTER

1

Impact of Science and Technology on Society

Chemistry is an important science with subdisciplines and applications that cover many technologies. Chemistry studies matter and its changes. Technology uses chemistry to solve practical problems.

1. What is the difference between science and technology?
2. How is science done?
3. What are facts, laws, and theories?
4. What are the experimental methods of chemistry?
5. How do scientists communicate?
6. How has technology affected the Industrial Revolution?
7. What are the risks of technology?
8. What are the causes of scientific controversy?

Science can be defined in a number of ways. Perhaps the place to start is with the derivation of the word *science,* which comes from the Latin *scientia,* meaning "knowledge." Science is a human activity involved in the accumulation of knowledge about the universe around us. Pursuit of knowledge is common to all scholarly endeavors in the humanities, social sciences, and natural sciences. Historically, the natural sciences have been closely associated with our observations of nature—our physical and biological environment. Knowledge in the sciences is more than a collection of facts; it involves comprehension, correlation, and an ability to explain established facts, usually in terms of a physical cause for an observed effect.

Figure 1-1 is a classification for the natural sciences. There is no sharp distinction between the physical and biological groups of sciences or among members within a group because new disciplines emerge that bridge areas at different levels. In the physical sciences, for example, there are biophysicists, geochemists, bioinorganic chemists, and chemical physicists. Some of these names define broad interdisciplinary fields, and others refer to more specialized subfields. The dynamic character of science is illustrated by the emergence of new disciplines. Current chemical research is particularly active in those areas that either link subdisciplines of chemistry, such as organometallic chemistry, or that link chemistry with other sciences, such as bioinorganic chemistry or chemical physics.

WHAT IS CHEMISTRY?

There are many different subdivisions of science because there are many different ways to focus on the world around us. In this book we shall study the science of chemistry, which is one of the physical sciences. Chemistry is concerned with the study of **matter** and its changes. Since matter is the material of the universe, every object we see or use is part of the chemical story. Our body is a sophisticated chemical factory with hundreds of chemical reactions occurring even as you read this page. Because chemistry is so intimately involved in every aspect of our contact with the material world, chemistry can be regarded as the central science, an integral part of our culture, having an influence on almost every aspect of our lives.

Matter is anything that has mass and occupies space.

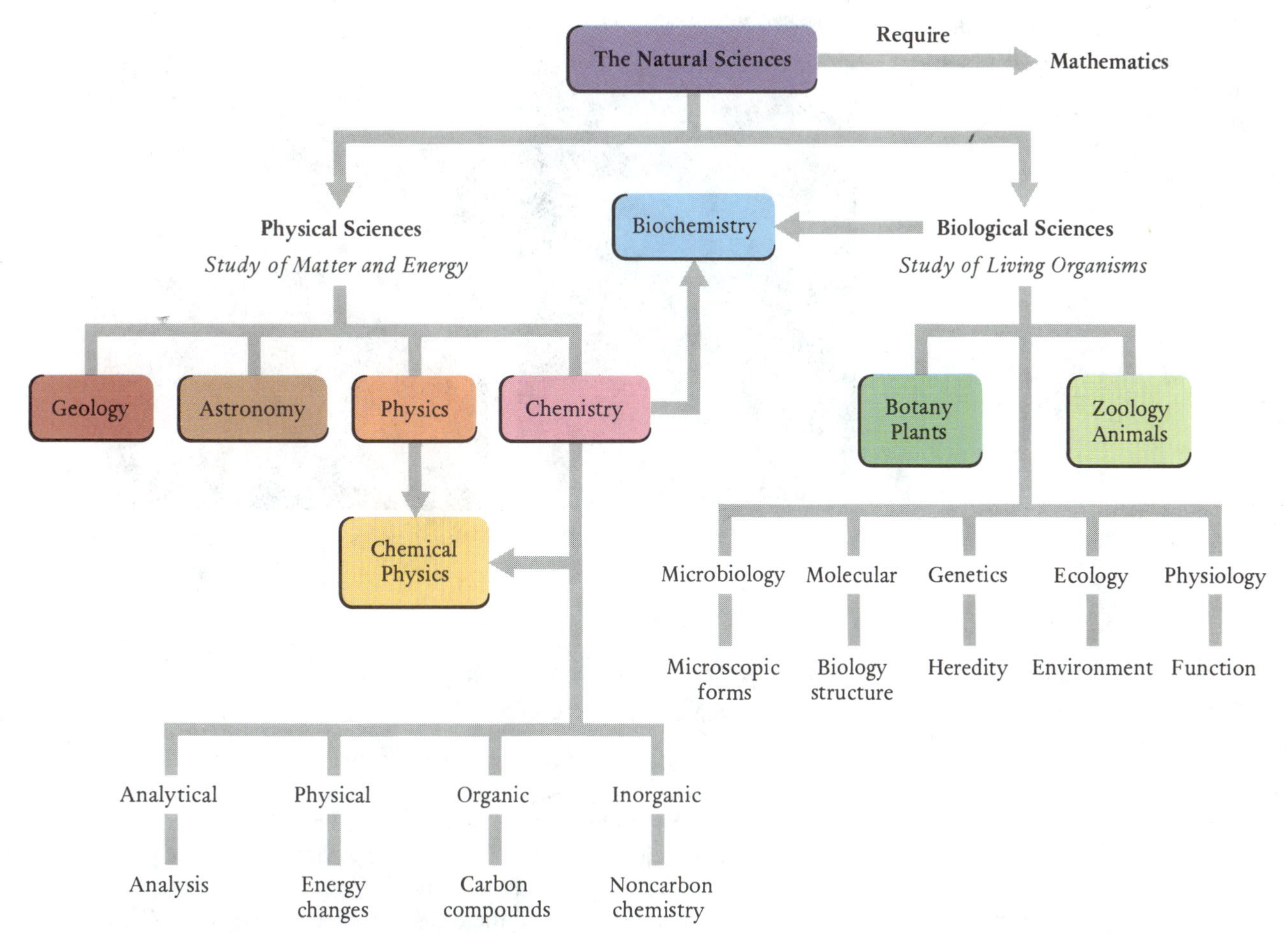

Figure 1–1 Organizational chart for the natural sciences, with emphasis on chemistry.

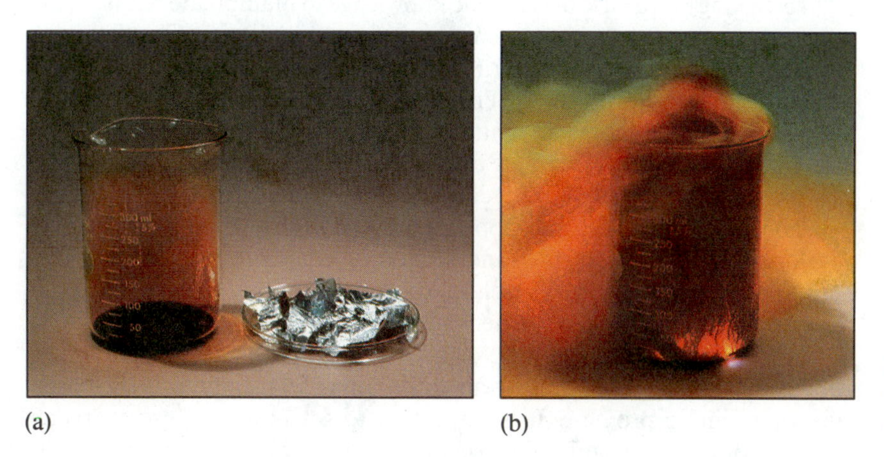

(a) (b)

Reaction of aluminum with bromine to give aluminum bromide. (Charles D. Winters)

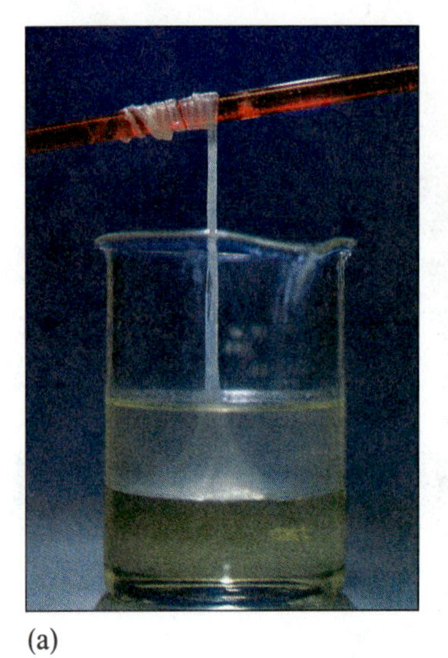

(a) (b)

(a) Basic research led to the discovery of nylon by Dr. Wallace Carothers in 1935. (Charles D. Winters) (b) Applied research and technology led to today's applications of the use of nylon in rugs. (Courtesy of Du Pont de Nemours & Company)

WHAT IS THE DIFFERENCE BETWEEN SCIENCE AND TECHNOLOGY?

A deeper understanding of scientific knowledge comes when we distinguish among basic science, applied science, and technology. The difference between basic and applied science is determined by the motivation for doing the work. **Basic science** is the pursuit of knowledge about the universe with no short-term practical objectives for application. An example of **basic research** is seeking the answer to the question "What is penicillin?" by determining the molecular structure of penicillin. **Applied science** has well-defined, short-term goals related to solving a specific problem. For example, after the antibacterial action of penicillin was discovered, scientists conducted **applied research** on the effectiveness of penicillin against different types of bacterial infections. Thus, both basic and applied scientific research produce new knowledge.

Technology is the use of scientific knowledge to manipulate nature. This may involve production of (1) new drugs, (2) better plastics, (3) safer automobiles, (4) nuclear weapons, or (5) chicken feed that causes eggs to have lower cholesterol. For example, engineers developed economical methods for large-scale production of penicillin from a knowledge of the results of basic and applied research on the chemistry and biochemistry of penicillin.

The important point is that technology, like science, is a human activity. Decisions about technological applications and priorities for technological developments are made by men and women; whether scientific knowledge is used to promote good or bad technological applications depends on those persons in industry and government with authority to make such decisions.

The image most people have of science is strongly influenced by their familiarity with technological advances and, in most instances, is their only view of scientific progress. It is important to get beyond these everyday images in order to recognize the **symbiotic** relationship of science and technology. Modern science depends on technological advances, especially in

Symbiosis is the close association of two dissimilar things in a mutually beneficial relationship.

TABLE 1–1 How Long It Has Taken Some Fruitful Ideas to Be Technologically Realized

Innovation	Conception	Realization	Incubation Interval (Years)
Antibiotics	1910	1940	30
Cellophane	1900	1912	12
Cisplatin, anticancer drug	1964	1972	8
Heart pacemaker	1928	1960	32
Hybrid corn	1908	1933	25
Instant coffee	1934	1956	22
Nuclear energy	1919	1945	26
Nylon	1927	1939	12
Photography	1782	1838	56
Radar	1907	1939	32
Recombinant DNA drug synthesis	1972	1982	10
Roll-on deodorant	1948	1955	7
Xerox copying	1935	1950	15
X rays in medicine	Dec. 1895	Jan. 1896	<1
Zipper	1883	1913	30

the development of more sophisticated instrumentation, to examine in greater depth the unanswered questions about the universe. Examples of this interrelationship are given throughout this book.

Perhaps more easily seen are the advances in technology that have occurred whenever new scientific discoveries are made. Regardless of the type of scientific discovery, there is a delay between a discovery and its technological application. The incubation period depends on (1) the rate of information transmittal, (2) the recognition of the applicability of the discovery, (3) the invention of a technological application for the new science, and (4) the large-scale manufacture of the new invention. The incubation times for several ideas or scientific discoveries are given in Table 1–1. Although there are exceptions, innovations based on applied research tend to happen faster than those developed from basic research.

HOW IS SCIENCE DONE?

Scientific Method

The methodology of science is often summarized by the term *scientific method.* A scientific method is a logical approach to solving scientific problems that may include (1) **observation,** or facts gathered by experiment, (2) **inductive reasoning** to interpret and classify facts by a general statement **(law),** (3) **hypothesis,** or speculation about how to explain facts or observations, (4) **deductive reasoning** to test a hypothesis with carefully designed experiments, and (5) **theory,** or a tested hypothesis or model to explain laws. Variations of this approach are often practiced in scientific research, and the imagination, creativity, and mental attitude of the scientist are often more important than the actual procedure.

Inductive reasoning moves from specific facts to generalization. Deductive reasoning moves from generalization to specific facts.

Another outline of the scientific method: observe, generalize, theorize, test, and retest.

The strictest intellectual honesty is required in the collection of observable facts and in the effort to arrange these facts into a pattern that reveals the underlying cause of the *observed* behavior. The data normally must be collected under conditions that can be reproduced anywhere in the world. New data can then be obtained to confirm or to refute the correctness of the suggested pattern. The results represent a unique type of objective truth that is ideally independent of differences in the language, culture, religion, or economic status of the various observers. Such established truth is appropriately referred to as **scientific fact.** A scientific fact is an observation about nature that usually can be reproduced at will.

A scientific fact can usually be verified by any independent observer.

Often a large number of related scientific facts can be summarized into broad, sweeping statements called **natural,** or **scientific, laws.** The law of gravity is a classic example of a natural law. This law—all bodies in the universe have an attraction for all other bodies that is directly proportional to the product of their masses and inversely related to the square of their separation distance—summarizes in one sweeping statement an enormous number of facts. It implies that any object more dense than air that is lifted a short distance from the surface of the earth will fall back if released. Such a natural law can be established in our minds only by inductive reasoning; that is, you conclude that the law applies to all possible cases, since it applies in all of the cases studied or observed. A well-established law allows us to predict future events. When convinced of the generality of a scientific law, we may reason deductively, based on our belief that if the law holds for all observed situations, it will surely hold for any new related event.

A scientific law summarizes a large number of related facts. A scientific law predicts what *will* happen. A governmental law describes what people *should* or *should not* do.

The same procedure is used in the establishment of chemical laws, as can be seen from the following example. Suppose an experimenter carried out hundreds of different chemical changes in closed, leakproof containers, and suppose further that the containers and their contents were weighed before and after each of the chemical changes. Also, suppose that in every case the container and its contents weighed exactly the same before and after the chemical change had occurred. Finally, suppose that the same experiments were repeated over and over again, the same results being obtained each time, until the experimenter was absolutely sure that the facts were reproducible. The experimenter could reasonably conclude that:

All chemical changes occur without any detectable loss or gain in weight.

This is indeed a basic chemical law that finds application in everything from chemical analysis to waste recycling.

After a natural law has been established, its explanation will be sought because chemists are usually not satisfied until they have explained chemical laws logically in terms of the structure of matter. This is a difficult process, and until recently its progress had been painfully slow because of the lack of direct access into the submicroscopic structure of matter. In the past decade, scientists have built microscopes that can actually "see" the images of atoms using a device called a scanning tunneling microscope (see Chapter 3).

Atoms are discussed in Chapter 3.

Submicroscopic: too small to be seen with an ordinary microscope.

Consider again the chemical law concerning the conservation of weight in chemical changes. What is a possible *theoretical model* that could explain this law? If we assume that matter is made up of atoms that are grouped in a particular way in a given pure substance, we can reason that a chemical change is simply the rearrangement of these atoms into new groupings

Some chemical facts	A chemical law	A model or theory to explain the law
(a) 2 units of hydrogen by weight react with 16 units of oxygen. Result: 18 units of water (b) 3 units of hydrogen by weight react with 14 units of nitrogen. Result: 17 units of ammonia (c) 14 units of nitrogen by weight react with 16 units of oxygen. Result: 30 units of nitric oxide	All chemical reactions occur without any detectable loss or gain in weight.	H atoms—one unit of mass O atoms—16 units of mass N atoms—14 units of mass (a) $H + H + O \longrightarrow H_2O$ water (b) $H + H + H + N \longrightarrow NH_3$ ammonia (c) $N + O \longrightarrow NO$ nitric oxide

Figure 1–2 Example of relationship between facts, chemical laws, and theories.

without the loss or destruction of those atoms and, consequently, rearrangement into new substances. If the same atoms are still there, they should have the same individual characteristic weight, and hence the law of conservation of weight is explained. The set of boxes in Figure 1–2 summarizes the relationship among facts, a chemical law, and a theory or model. Note that the chemical law, known as the **law of conservation of matter** for chemical reactions, summarizes the facts shown in the first set of boxes. In the last box is a version of the atomic theory model that explains the chemical law.

Theories are ideas or models used to explain facts and laws.

For a scientific theory to have lasting value, it must not only explain the pertinent facts and laws at hand but also be able to explain or accommodate new facts and laws that are obviously related. If the theory cannot consistently perform in this manner, it must be revised until it is consistent, or, if this is not possible, it must be discarded completely. You must not allow yourself to think that this process of trying to understand nature's secrets is nearing completion. The process is a continuing one.

The word *theory* is often used in a different sense from the one discussed previously. If a student is absent from the chemistry class, his neighbor may say, "I do not know why he is absent, but my theory is that he is sick and unable to come to class." The speculative guess of the student about his absent friend is what scientists call a **hypothesis** and is vastly different from the broad theoretical picture used to explain a number of laws. The reader should be alert for the considerable amount of confusion that has resulted from the different meanings associated with this word. In this book the word *hypothesis* is used when speaking of a speculation about a particular event or set of data, and the word *theory* is reserved for the detailed imaginative concepts that have gained wide acceptance by withstanding scrutiny, and by their ability to explain facts and laws over a long period.

Experimental Methods

Discoveries come about through the observation of nature or by experimentation that can be categorized as trial and error, planned research, or accidental discoveries (serendipity).

The use of the term *serendipity* for accidental discoveries was first proposed in 1754 by Horace Walpole after he read a fairy tale titled "The Three Princes of Serendip." Serendip was the ancient name of Ceylon (now Sri Lanka) and the princes, according to Walpole, "were always making discoveries by accident, of things they were not in quest of."

Discovery by trial and error begins when one has a problem to solve and does various experiments in the hope that something desirable will emerge. The next set of experiments then depends on the results obtained in the first set. The discovery of the Edison battery by Thomas Edison's group is an example of discovery by trial and error. Edison's group performed more than 2000 experiments, each guided by the previous one, before settling on the composition of Edison's battery.

Discovery by planned research comes from carrying out specific experiments to test a well-defined hypothesis. The carcinogenic nature of some compounds is determined by progressing through a set pattern of experimental tests.

Carcinogens are substances that cause cancer.

Discovery by accident is really a misnomer. The investigator is usually actively involved in investigating nature through experimentation but "accidentally" finds some phenomenon not originally imagined or conceived. Thus, the accident has an element of serendipity and is not likely seen unless the investigator is a trained observer. As Louis Pasteur said, "Chance favors the prepared mind."

Figure 1–3 Dr. Barnett Rosenberg holds in his left hand a mouse that will die of cancer in a few days. In his right hand is a mouse that has been infected with cancer but will survive because it has received cisplatin. (Barros Research Institute)

The discovery of one of the leading anticancer drugs, cisplatin, is an example of such an accidental discovery. In 1964, Barnett Rosenberg (Fig. 1–3) and his co-workers at Michigan State University were studying the effects of an electrical current on bacterial growth. They were using an electrical apparatus with platinum electrodes to pass a small alternating current through a live culture of *Escherichia coli* bacteria. After an hour, they examined the bacterial culture under a microscope and observed that cell division was no longer taking place. After thorough analysis of the culture medium and additional experimentation, they determined that traces of several different platinum compounds were produced during electrolysis from the reaction of the platinum electrodes with chemicals in the culture medium.

Careful observation was essential because platinum electrodes are commonly regarded as inert or unreactive, and only a few parts of platinum compounds per million parts of culture medium were present. Additional testing indicated a compound known as cisplatin was responsible for inhibiting cell division in *E. coli.* Approximately two years after its initial discov-

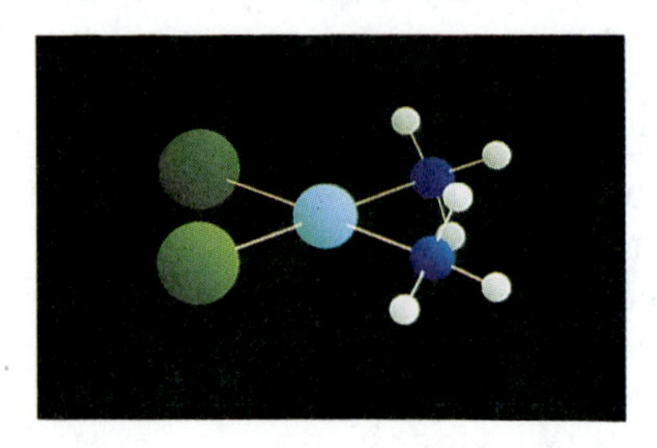

Model of the cisplatin molecule. The green spheres represent chlorine atoms; the white spheres, hydrogen atoms; the dark blue, nitrogen atoms; and the light blue, the platinum atom. The formula is $Pt(NH_3)_2Cl_2$. See Chapter 2. (Courtesy of Dr. George Parks, Phillips Petroleum)

THE WORLD OF CHEMISTRY

Serendipity

The history of science is replete with examples of serendipity, simply described as a fortuitous and happy discovery or observation from which many important future developments have flowed. There is a special quality to serendipity. It is significant only when the discoverer or observer recognizes through a burst of intuition that there is something that needs further exploration and then proceeds to devote a serious effort to exploit the discovery. Much more often than not, of course, is the requirement of a prepared mind.

A classic example of serendipity is to be found in the story of W. H. Perkin as told by science historian John K. Smith of Lehigh University.

He was a brilliant young chemist, who, while working in his home laboratory in 1856 in an effort to synthesize badly needed quinine succeeded instead in creating the dye "mauve." Perkin recognized the importance of his accidental discovery and as a consequence of his efforts succeeded in establishing the beginning of the dye industry. The spinoffs were enormous. There are today as a consequence a large variety of materials such as drugs, explosives, fertilizers which play such an important role in the affairs of society. One of the most important consequences, for example, is aspirin, easily one of the most useful drugs in the history of pharmaceutical chemistry.

The World of Chemistry (Program 2) "Color."

W. H. Perkin. (*The World of Chemistry*, Program 2)

ery, the Rosenberg group had the answer to the question "What caused inhibition of cell division in *E. coli* bacteria?" At this point they hypothesized that cisplatin might inhibit cell division in rapidly growing cancer cells. The compound was tested as an anticancer drug, and in 1979 the U.S. Food and Drug Administration approved its use as such. The drug has now been proved to be effective alone or in combination with other drugs for the treatment of a variety of cancers.

Another interesting aspect of the story is that cisplatin, a compound that contains two chloride ions, two ammonia molecules, and platinum, was first prepared in 1845. Although its chemistry had been studied thoroughly since then, the biological effects of cisplatin and its inhibition of cell division were not discovered until "the accident" 120 years later.

An ion is an electrically charged atom or group of atoms.

HOW DO SCIENTISTS COMMUNICATE?

Scientific knowledge is cumulative, and progress in science and technology depends on access to this body of knowledge. Since the earliest beginnings of science, this knowledge has been transmitted primarily by the written word. The invention of the printing press led to the development of

Figure 1–4 The 11th Collective Index of Chemical Abstracts is over 17 feet tall. *Chemical Abstracts* have asked the *Guinness Book of World Records* to consider whether the 11th Collective Index is the world's largest index. (Courtesy of *Chemical Abstracts*)

scientific journals and other publications collectively known as the scientific literature. The explosive expansion of the scientific literature since the 1940s makes information management an essential part of modern science and technology. So many scientific journals are published that abstracting services like *Chemical Abstracts* are used to keep up with the published articles. Even the indices of *Chemicals Abstracts* are large by any comparison (Fig. 1–4).

TECHNOLOGY AND THE INDUSTRIAL REVOLUTION

Historical

Over the last 200 years, accumulated scientific and technical knowledge has been put to use on an extensive scale in Europe and in those areas of the world that had the means and the will to follow the examples set during the Industrial Revolution. The result is the development of a society largely dependent upon and supported by a constantly changing technology. The first consequence of this technology was to increase the rate at which things can be produced. This, in turn, continually changed the occupational patterns of millions of human beings and brings forcefully to mind the persistence of change in our pattern of life.

These changes profoundly influenced the way people think about their material wants and the ways those desires can be satisfied. Although we are dependent on technology, people are today beginning to doubt its ability to solve both personal and social problems in the long run. It is obvious that confusion exists on this point because the cries about the curses of technology come from people who are highly dependent on it and who are even asking for more from technology.

Almost as soon as the Industrial Revolution began in England, the public realized that technological progress brought with it a series of problems. For example, the increased use of fuels of all sorts, especially the introduction of coal and coke into metallurgical plants and then the use of coal to fuel engines, led to widespread problems with air pollution that were recognized and discussed over 200 years ago.

Automobile assembly line. (*The World of Chemistry*, Program 22)

The Industrial Revolution caused people to move from farms to cities. As life became easier, populations grew, which caused strains on agriculture's ability to provide adequate food. By the late 1800s scientists recognized that the world's future food supply would be determined by the amount of nitrogen compounds made available for fertilizers. The source of these nitrogen supplies was then limited to rapidly depleting supplies of guano (bird droppings) in Peru and to sodium nitrate in Chile. It was realized that when these were exhausted, widespread famine would result unless an alternative supply could be developed. This problem was recognized first by English scientists as a potentially acute one, because by the 1890s England had become very dependent on imported food supplies. The Industrial Revolution allowed the population to grow rapidly, so the number of hungry people soon outstripped the domestic food supply.

Widespread interest in this problem led to research on a number of chemical reactions to obtain nitrogen from the relatively inexhaustible supply present in air. Air is 21% oxygen and 79% nitrogen. The nitrogen in the air is present as the unreactive molecule N_2, and in this form it can be used as a source of other nitrogen compounds by only a few kinds of bacteria. Some is also transformed into nitrogen oxide by lightning, and when this is washed into the soil by rain, the nitrogen can be utilized by plants. The amount of nitrogen transformed by these processes into chemical compounds useful to plants is quite limited and cannot be increased easily.

Fritz Haber (1868–1934).

Several chemical reactions were developed to form useful compounds from atmospheric nitrogen, but the best known and most widely used one has an ironic history. While England was interested in nitrogen for fertilizers, Germany was interested in nitrogen for explosives. The German General Staff realized that the British Navy could blockade German ports and cut them off from the sources of nitrogen compounds in South America. As a consequence, when the German chemist Fritz Haber showed the potential of an industrial process in which nitrogen reacts with hydrogen in the presence of a suitable catalyst to form ammonia (an essential ingredient in the preparation of explosives), the German General Staff was quite interested and furnished support through the German chemical industry for the study of the reaction and the development of industrial plants based on it. The first such plant was in operation by 1911, and by 1914 such plants were being built very rapidly.

When World War I broke out in August of 1914, many people thought that a shortage of explosives based on nitrogen compounds would force the war to end within a year. Unfortunately, by this time the ammonia industry in Germany was capable of supplying the needed compounds in large amounts. This process thus prolonged the war considerably and resulted in an enormous increase in mortality. Subsequently, the ammonia process (Haber process) has been used on a huge scale to prepare fertilizers and is now largely responsible for the fact that Earth can support a population of more than 5 billion. Ammonia production by this process exceeds 40,000 tons per day in the United States alone.

The Chip and the Splice

Applications of ammonia to a field. (Courtesy Farmland Industries, Inc.)

We currently are in the midst of two vastly important technological revolutions. The first is the microelectronic revolution, characterized by the computer chip—a device fabricated on a wafer of highly purified silicon. The second is the biotechnological revolution started by the discovery of recombinant DNA. Silicon-based microelectronic circuits were first produced in 1968, and the recombinant DNA gene-cloning experiments began in the early 1970s.

The **chip,** a nickname for the integrated circuit, is a small slice of silicon that contains an intricate pattern of electronic switches (transistors) joined by "wires" etched from thin films of metal. Some chips are information storers called memory chips; others combine memory with logic function to produce computer or microprocessor chips. These two applications make

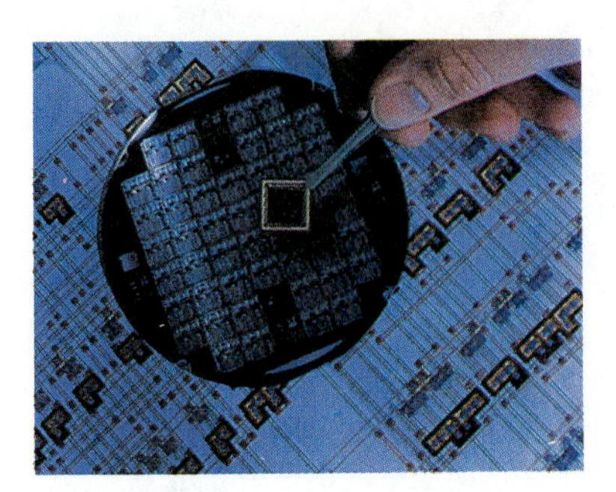

Figure 1–5 A typical computer chip. (AT&T Bell Laboratories)

the chip capable of almost infinite application. A microprocessor chip, for example, can provide a machine with decision-making ability, memory for instructions, and self-adjusting controls.

In everyday life we see many examples of the influence of the chip: digital watches; microwave oven controls; new cars with their carefully metered fuel-air mixtures; hand calculators; cash registers that total bills, post sales, and update inventories; and computers of a variety of sizes and capacity—all of these make use of the chip (Fig. 1–5).

One of the earliest benefits of recombinant DNA was the biosynthesis of **human insulin** in 1978. Millions of diabetics depend on the availability of insulin, but many are allergic to animal insulin, which was the only previous source. Biosynthesized human insulin is now being marketed. Biotechnology firms are also producing **human growth hormone,** which is used in treating youth dwarfism, and **interferon,** which is a potential anticancer agent. The FDA approved in fall, 1987, the marketing of **tissue-plasminogen-activator** (TPA), which dissolves the blood clots that cause heart attacks. If treatment with TPA begins soon enough, it can not only save lives but also reduce the damage caused by an attack.

Scientists have created mice that produce human TPA in their milk. These mice can produce grams of TPA per liter of milk, compared with a typical output from a bioreactor of *E. coli* of milligrams per liter—a thousandfold increase. Although the gene for making human TPA is present in all of the cells of the genetically altered (transgenic) mice, the protein is synthesized only in the mammary gland. The researchers accomplished this by fusing genes that normally control the production of mouse milk proteins to the gene for TPA. Since extracting milk from mice would not be simple, research is underway to produce transgenic cows, sheep, or goats that secrete TPA. Biogenetic engineers estimate that the world's supply of TPA could be obtained by milking a herd of 100 transgenic cows.

THE RISKS OF TECHNOLOGY

We have described some of the potential benefits of technology, but we also need to examine its risks. A decade ago, at the beginning of the recombinant DNA era, many people, including the scientists working in the area, saw danger in biotechnology. Since *E. coli* is an intestinal bacterium, what if some of the genetically engineered *E. coli* escaped and found its way into people's intestines? These fears led to an 18-month moratorium on recombinant DNA research. However, the evidence to date shows that the *E. coli* used in recombinant DNA technology is too delicate to survive outside its environment. In addition, strict regulations are being followed in the experiments with genetically engineered bacteria to ensure against such problems.

The public is aware of the dangers of chemicals in the environment. Many persons have developed **chemophobia** (an unreasonable fear of chemicals) because of careless industrial practices such as improper disposal of hazardous wastes; environmental pollution of air, water, and earth; and catastrophic accidents. The names Bhopal, Chernobyl, and *Challenger* remind us of the influence of human error in increasing the risks associated with technology.

"We fear things in proportion to our ignorance of them." Livy, Roman historian (64-59 B.C.–A.D. 17)

The chemical-plant accident that occurred in Bhopal, India, on December 3, 1984, was the worst in history. Methyl isocyanate, a deadly gas used in the preparation of pesticides, escaped from a storage tank at the Union Carbide plant, killing over 2000 people and injuring tens of thousands. Numerous violations of safety procedures contributed to the disastrous leak. The explosion of the space shuttle *Challenger* on January 28, 1986, was caused by defective plastic O-rings between casing sections in the booster rocket. The explosion at the Chernobyl nuclear plant in the Soviet Union on April 26, 1986, which released large amounts of radioactive material into the atmosphere, was the result of a number of violations of operating regulations by the workers at the plant.

The Chernobyl nuclear power plant after the accident. (Photography by V. Zufarov; courtesy of Fotokhronika Tass)

Improper disposal of hazardous wastes has been the cause of many problems. Love Canal in Niagara Falls, New York, the Times Beach community in Missouri, and the Minamata Bay and Jinzu River in Japan are just a few locations where serious problems have resulted from the disposal of hazardous wastes.

Love Canal, the neighborhood that in 1977 discovered it was built on a toxic chemical dump, was the first publicized example of the problems of chemical waste dumps. In the mid-1970s heavy rains and snows seeped into the dump and pushed an oily black liquid to the surface. The liquid contained at least 82 chemicals, 12 of which were suspected carcinogens.

The entire community of Times Beach, Missouri, was bought by the U.S. Environmental Protection Agency (EPA) in 1983, and the 2200 resi-

Discarded waste barrels. (*The World of Chemistry,* Program 25)

dents were relocated because dioxins, a group of very toxic chemicals produced in small amounts during the synthesis of a herbicide, were found in the soil at concentrations as high as 1100 times the acceptable level. It is interesting to note that in 1991 the EPA issued a statement that they had overestimated the degree of the dioxin hazard in Times Beach. Whether or not the former residents of this small town will ever go back there is an issue that has not been settled.

In the 1950s, tons of waste mercury were dumped into the bay at Minamata, Japan. In the next few years thousands of persons in the Minamata area suffered paralysis and mental disorders, and over 200 people died. Several years passed before it was determined that these people had been poisoned by methyl mercury compounds. Anaerobic bacteria in the sea bottom converted mercury to methyl mercury compounds, which were eaten by plankton. The methyl mercury compounds were carried up the food chain and eventually accumulated in the fatty tissue of fish. Since fish are a major part of the Japanese diet, intake of methyl mercury compounds reached levels that caused the sickness now known as **Minamata disease.**

What Is an Acceptable Risk?

Risk assessment for individuals involves a consideration of the likelihood or probability of harm and the severity of the hazard. Assessment of societal risks combines probability and severity with the number of persons affected. The science of risk assessment is still evolving, but it is clear that the importance of public perception of risks needs to be recognized before risk assessment can be quantified. Often there is little correlation between the actual statistics of risk and the perception of risk by the public or by individuals. For example, we are all aware that the risk of injury or death is much lower from traveling in a commercial airplane than from traveling in an automobile, yet all of us know persons who avoid airplane flights because of their fear of a crash.

What factors influence public perception of risk? Catastrophic accidents such as those at Bhopal and Chernobyl obviously affect public perception of risk. In addition, people tend to judge involuntary exposure to activities or technologies (such as living near a hazardous dump site) as riskier than voluntary exposure (such as smoking). In other words, persons rate risks they can control lower than those they cannot control.

The living must accept risk. The question is how much.

No absolute answer can be provided to the question "how safe is safe enough?" The determination of acceptable levels of risk requires value judgments that are difficult and complex, involving the consideration of scientific, social, and political factors. Over the years a number of laws designed to protect human health and the environment have been enacted to provide a basic framework for making decisions. The fact that three types of laws exist in this area adds to public confusion about risk assessment and its meaning.

Chapter 14 discusses food additives.

Risk-based laws are zero-risk laws that allow no balancing of health risks against possible benefits. The Delaney Clause of the Federal Food, Drug, and Cosmetic Act is such a law. It specifically bans the use of any intentional food additive that is shown to be a carcinogen in humans or animals, regardless of any potential benefits. The rationale for this law is the nonthreshold theory of carcinogenesis, which assumes that there is no safe level of exposure to any cancer-causing agent.

The Safe Drinking Water Act, the Toxic Substances Control Act, and the Clean Air Act are **balancing laws;** they balance risks against benefits. The EPA is required to balance regulatory costs and benefits in its decision-making activities. Risk assessments are used here. Chemicals are regulated or banned when they pose "unreasonable risks" to or have "adverse effects" on human health or the environment.

Technology-based laws impose technological controls to set standards. For example, parts of the Clean Air Act and the Clean Water Act impose pollution controls based on the best economically available technology or the best practical technology. Such laws assume that complete elimination of the discharge of human and industrial wastes into water or air is not feasible. Controls are imposed to reduce exposure, but true balancing is not attempted; the goal is to provide an "ample margin of safety" to protect public health and safety.

Clean air is discussed in Chapter 12. Clean water is discussed in Chapter 11.

SCIENTIFIC CONTROVERSY

It may seem surprising that scientists do not always agree on some aspects of science. Certainly, experiments can be performed and recorded so that others can reproduce the results, but the interpretation of those results and even the basic design of the experiments can often be very controversial. Animal carcinogen testing is a good example. Many chemicals have been found to be carcinogenic to laboratory animals, mostly white rats. These chemicals are usually administered to the animal in the maximum tolerated dose (MTD, see Chapter 10). Many scientists think that this is a biased way of doing science because many of the chemicals administered in this way lead to a cancer in the laboratory animals. They claim that if the doses were lower, that is, more in line with the doses humans might receive on exposure to these chemicals, fewer of them would be found to be carcinogenic. In fact, the MTD, they say, causes cell proliferation (mitogenesis), which is the cause of cancer.

Scientists at agencies like the EPA, which regulate chemical carcinogens, differ with this view. They claim that the chemicals tested are suspected to be carcinogens because of their molecular structure (see Chapter 6) or other characteristics and that any bias lies in the choice of the chemicals rather than the method of dosing the animal. For example, vinyl chloride, used to make many plastics (see Chapter 8), causes a rare liver cancer in rats and mice. It causes the same cancer in humans and at the same doses.

What is the correct view? Probably a mixture of both. Certainly some chemicals can cause cancer in humans. Many cause cancer in rats and other animals. If the doses delivered to these animals are too high, then perhaps more experimentation should be done using lower doses and different methods. That is what science is all about.

WHAT IS YOUR ATTITUDE TOWARD CHEMISTRY?

Before beginning this study of chemistry and its relationship to our culture, each of us needs to examine our prejudices (if any) and attitudes about chemistry, science, and technology. Many nonscientists regard science and its various branches as a mystery and have the attitude that they cannot possibly comprehend the basic concepts and consequent societal issues. Many also have chemophobia and a feeling of hopelessness about the environment. Many of these attitudes are the result of reading about the harmful effects of technology. Some of these harmful effects are indeed tragic. However, what is needed is a full realization of both the benefits and the harmful effects that can be attributed to science and technology. In the analysis of these pluses and minuses, we need to determine why the harmful effects occurred and whether the risk can be reduced for future generations. This book will give you the basics in chemistry, which we hope will afford you a healthier and more satisfying life by allowing you to make wise decisions about personal problems and problems that concern our world.

In order to use chemistry and the other sciences, all citizens need to be able to evaluate scientific and technological advances. This requires a basic knowledge about matter and what matter does, whereas more sophisticated chemical problems require a deeper understanding of the workings (facts and theories) of chemistry. You should be involved. As an educated person, this is your responsibility and privilege.

SELF-TEST 1*

1. The ultimate test of a scientific theory is its agreement with ______________.
2. Different workers, in different countries, who carry out a particular laboratory experiment in exactly the same way should get ______________ result.
3. Another name for accidental discovery is ______________.
4. The "chip" is fabricated on a wafer of highly purified ______________.
5. An unreasonable fear of chemicals is called ______________.
6. An example of a technology-based law is ______________________.
7. A genetically engineered product called TPA is used for the treatment of ________________________.

* Use these self-tests as a measure of how well you understand the material. Take a test only after careful reading of the material preceding it. Do not return to the text during the self-test, but reread entire sections carefully if you do poorly on the self-test on those sections. The answers to the self-tests are in Answers to Self-Test Questions at the end of the text.

QUESTIONS

1. Distinguish between theory and law in chemistry.
2. Give an example of a chemical fact.
3. Give an example of a chemical law.
4. How many times do you think a given experiment should yield the same result before a scientific fact is considered to have been established?
5. Distinguish between basic science, applied science, and technology.
6. Persons often confuse science with scientism. Look up the definition of *scientism* in a dictionary and discuss why it is important to society that science not be confused with scientism.
7. What is the difference between a scientific fact and an historical fact?
8. Discuss the risks we face living in today's world. Include some caused by technology and some reduced by technology.
9. Compare a scientific law with a governmental law, and describe how each came to be and which is more likely to change.

The melange in Jackson Pollack's *One (Number 31, 1950)* reflects nature because most natural samples of matter are mixtures. (Collection, The Museum of Modern Art, New York: Sidney and Harriet Janis Collection Fund (by exchange))

CHAPTER

2 The Language of Chemistry

Chemistry is concerned primarily with chemical change, the disappearance of one or more pure substances with the concurrent appearance of new ones. The science of chemical change was rather late in developing because nearly all of the natural pure substances are hidden in the complex mixtures that make up our environment.

1. What is a pure substance?
2. What are the three fundamental types of changes in matter?
3. Why are purification techniques so important in chemistry?
4. How is a solution different from a pure substance?
5. What are the two major classes of pure substances?
6. What are the most abundant elements?
7. Is chemical shorthand necessary or just a convenience?
8. What factors determine the rate of chemical reactions?
9. What is chemical equilibrium?
10. How many measurement units do we really need?
11. Is it possible to destroy either matter or energy in chemical change?

Do you enjoy the material things around you? Sometimes yes and sometimes no, right? Think beyond your immediate setting: How many materials and things are in this universe of ours? Too many to count? These things — all of them — that we can see, touch, and weigh are made of **matter.** Although an uncountable number of materials and objects exist, is there any order and simplicity in the make-up of matter? Can we reasonably hope to control matter to make life more pleasant for the human race? The science of **chemistry** addresses these fundamental questions.

Matter occupies space and has weight.

Most of the things we use in our daily lives are very different from the materials found in nature. Practically everything we use has been changed from a natural state of little or no utility to one of very different appearance and much greater utility. The processes by which the materials found in nature can be changed and a detailed description of such changes are highly intriguing. This is a basic dimension of the science of chemistry: the **changes** in matter.

Matter can undergo three basic kinds of change, and our attention in chemistry will be focused on one of these — **chemical change** (Fig. 2-1). In any chemical change, the starting material is changed into a different kind of matter. In **physical changes,** new forms of the same material are produced, and in **nuclear changes,** some matter is changed into energy while producing new substances. More complete definitions of these types of changes will follow. Also, as you study further about physical and chemical changes, you will come to realize that the categorization of material change is not as clear-cut as it first appears.

Chemical change alters the kind of matter without changing the amount of matter. Examples: a. chemical change — burning, rusting, souring; b. Physical change — melting, boiling, cutting; c. Nuclear change — producing nuclear energy and radioisotopes.

What causes changes to occur in matter? It is **energy!** Examples of energy are heat, light, sound, and electricity. Energy and matter are not the same even though they are closely related. Energy has the ability to move matter (engines, eardrums, and motors, for example). Matter is converted into energy in nuclear reactors and nuclear bombs. Energy can infiltrate matter and manifest itself through the actions of the matter. A sample of hot water contains more energy than the same sample when it is cold. Some forms of energy can exist apart from matter; examples are light and radiant heat. It appears that all forms of energy are generated by changes in matter, and that matter, in turn, can absorb energy to produce other physical and chemical changes. Indeed, energy by definition is that which can produce change in matter. It follows, then, that a study of chemistry involves still another dimension: the *energy* associated with chemical changes.

In physical terms, energy is the ability to do work and work is a physical force exerted through a distance.

It is difficult, in a few words, to establish the exact bounds of chemistry. Even so, it will be helpful to think of chemistry as

the study of the kinds of matter and the changes of one kind of matter into another with the associated energy changes.

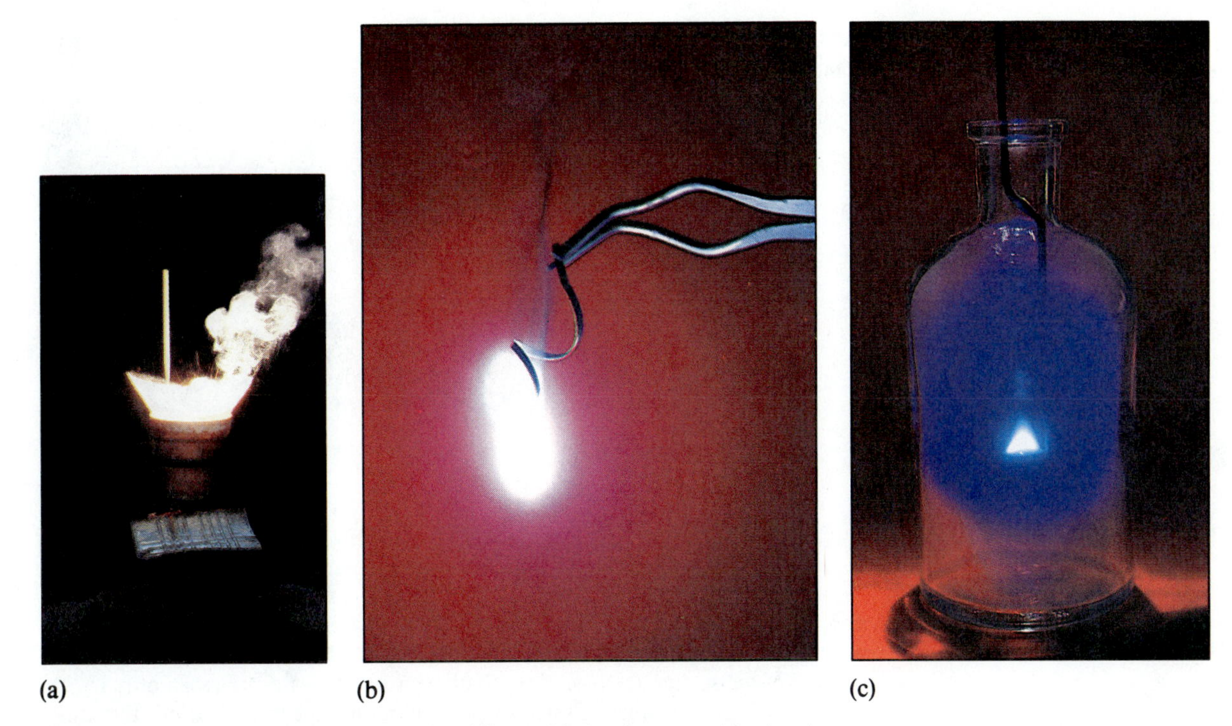

(a) (b) (c)

Figure 2–1 Chemical change results in different kinds of matter. (a) Aluminium powder reacts with iron oxide in the thermite reaction, a reaction that railroaders used to weld rails with the resulting molten iron. (b) Metallic magnesium burns in air to produce a white solid, magnesium oxide. (c) Yellow sulfur, a solid, reacts with oxygen, a gas, to produce another gas, poisonous sulfur dioxide. Note that a chemical change, in contrast to a physical mixing, produces a new substance. (Charles D. Winters)

Because the feature used to recognize a chemical change is the production of a different kind of matter, recognition of a chemical change requires a recognition of different kinds of matter. In a natural state the kinds of matter are usually mixed together, and the separation of such mixtures has to precede their systematic classification. After an examination of the methods of separating such mixtures into their components, we can appreciate some of the problems involved in an accurate definition of the terms *kinds of matter* and *chemical change.*

MIXTURES AND PURE SUBSTANCES

Most natural samples of matter are mixtures. Often, it is easy to see the various ingredients in a mixture (Fig. 2–2). Some mixtures are obviously heterogeneous, as the uneven texture of the material is clearly visible. Some mixtures appear to be homogeneous when actually they are not. For example, the air in your room appears homogeneous until a beam of light enters

Homogeneous means smooth texture, the same throughout.
Heterogeneous means nonuniform texture, not the same at every observed point.

Figure 2–2 This NASA photograph of a moon rock and many similar ones show that lunar materials, like the solid formation in the Earth's crust, tend to be mixtures of more basic substances. It is likely that this is characteristic of crust materials in the universe.

Figure 2–3 The Tyndall effect. A colloid in a clear liquid or gas scatters light because of the relatively large size of the dispersed particles in contrast to the much smaller sizes of atoms and small molecules. A solution, which is a dispersion at the molecular level, passes light with no scatter. (a) A laser light show works best in relatively "dirty" air, as the beams can be seen in every direction because of the scatter. (Fritz Goro.) (b) Sunbeams in a forest can be seen only if there is colloidal moisture or dust. In clear air neither the laser light nor the sunbeam can be observed at any angle of view, the light traveling straight through with no scatter. (H. Armstrong Roberts)

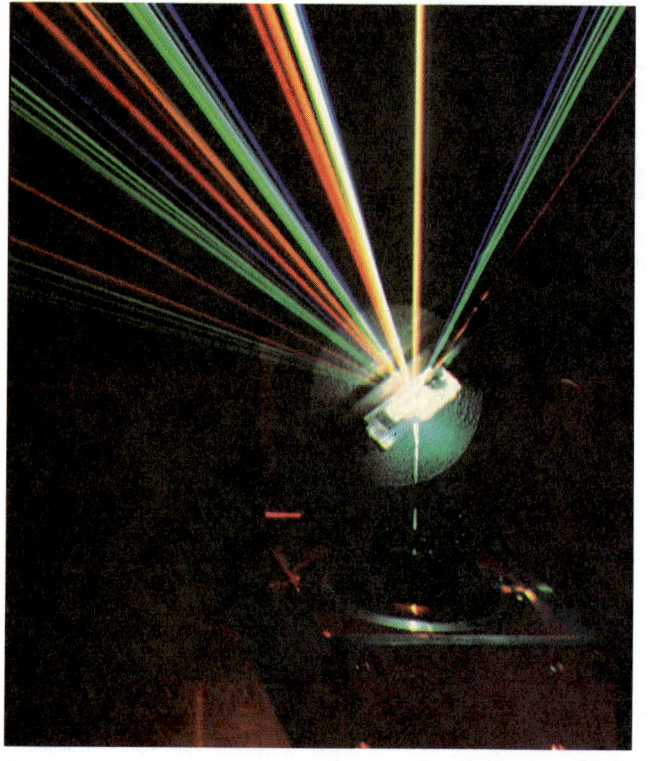

(a)

(b)

the room, revealing floating dust particles. Milk appears smooth in texture to the eye, whereas magnification reveals an uneven distribution of materials. **Colloids** are mixtures that appear to be homogeneous in normal lighting but actually are heterogeneous (Fig. 2–3). Homogeneous mixtures do exist; such mixtures are **solutions.** No amount of optical magnification reveals a solution to be heterogeneous, for heterogeneity in solutions exists only at atomic and molecular levels where the individual particles are too small to be seen with ordinary light. Examples of solutions are clean air (mostly nitrogen and oxygen), sugar-water, and some brass alloys (which are homogeneous mixtures of copper and zinc).

When a mixture is separated into its components, the components are said to be *purified.* However, most efforts at separation are incomplete in a single operation or step, and repetition of the process is necessary to produce a purer substance. Ultimately in such a procedure the experimenter may arrive at **pure substances,** samples of matter that cannot be purified further. For example, if sulfur and iron powder are ground together to form a mixture, the iron can be separated from the sulfur by repeated stirrings of the mixture with a magnet (Fig. 2–4). When the mixture is stirred the first time and the magnet removed, much of the iron is removed with it, leaving the sulfur in a higher state of purity. However, after just one stirring the sulfur may still have a dirty appearance due to a small amount of iron that remains. Repeated stirring with the magnet, or perhaps the use of a very strong magnet, will finally leave a bright yellow sample of sulfur that apparently cannot be purified further by this technique. In this purification process a property of the mixture, its color, is a measure of the extent of purification. After the bright yellow color is obtained, it could be assumed that the sulfur has been purified.

Purification separates the kinds of matter.

Drawing a conclusion based on one property of the mixture may be misleading because other methods of purification might change some other properties of the sample. It is safe to call the sulfur a pure substance only when all possible methods of purification fail to change its properties. This

Native gold. (© Gemological Institute of America)

Figure 2–4 Mixed powdered iron and sulfur illustrate heterogeneous mixtures. The magnetic property of iron allows a physical separation of the two pure substances. (Charles Steele)

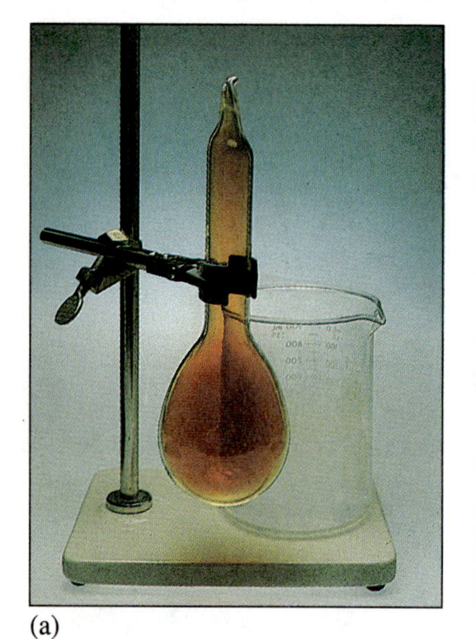
(a)

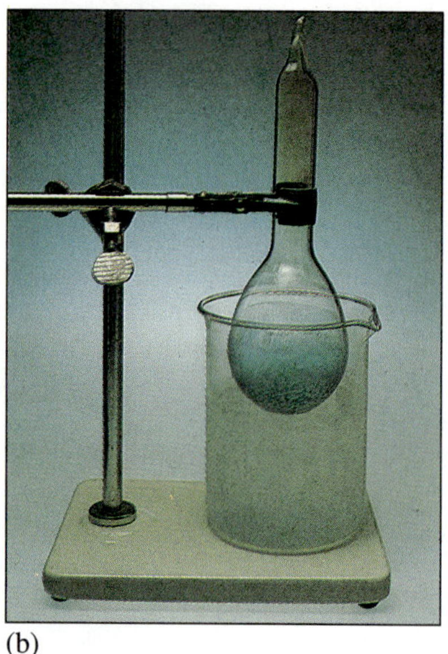
(b)

(c)

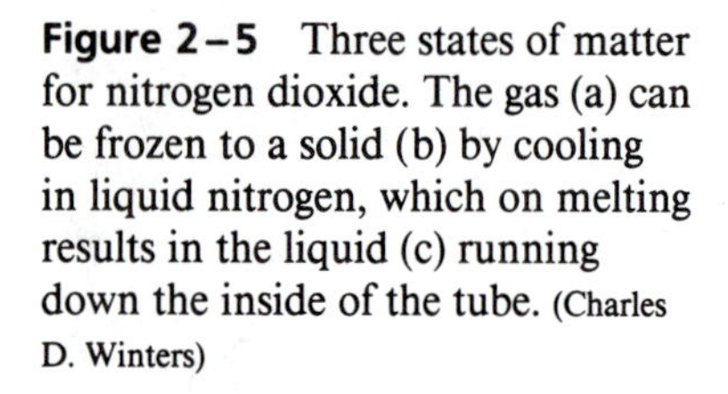
Figure 2–5 Three states of matter for nitrogen dioxide. The gas (a) can be frozen to a solid (b) by cooling in liquid nitrogen, which on melting results in the liquid (c) running down the inside of the tube. (Charles D. Winters)

assumes that all pure substances have a set of properties by which they can be recognized, just as a person can be recognized by a set of characteristics.

> **A pure substance is a kind of matter with properties that cannot be changed by further purification.**

A pure substance may exist in more than one state, as liquid water, solid ice, and gaseous steam illustrate. Figure 2–5 shows the three phases of nitrogen dioxide.

There are some naturally occurring pure substances. Rain is very nearly pure water, except for small amounts of dust, dissolved air, and various pollutants. Gold, diamond, and sulfur are also found in very pure form. These substances are special cases. The human, a complex assemblage of mixtures, lives in a world of mixtures—eating them, wearing them, living in houses made of them, and using tools made of them.

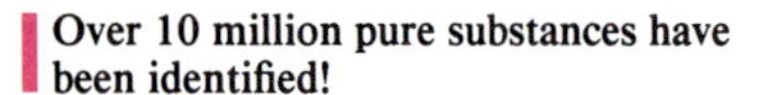
Over 10 million pure substances have been identified!

Although naturally occurring pure substances are not common, it is possible to produce many pure substances from natural mixtures. Relatively pure substances are now very common as a consequence of the development of modern purification techniques. Common examples are refined sugar, table salt (sodium chloride), copper, sodium bicarbonate, nitrogen, dextrose (glucose), ammonia, uranium, and carbon dioxide—to mention just a few. In all, over 10 million pure substances have been identified and cataloged.

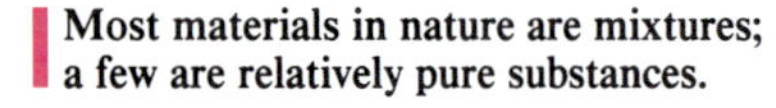
Most materials in nature are mixtures; a few are relatively pure substances.

The concept of a pure substance allows for a better definition of chemical change. A **chemical change** involves the disappearance of one or more pure substances and the appearance of one or more other pure substances. In contrast, the pure substance is preserved in a **physical change** even though it may have changed its physical state or the gross size and shape of its pieces. Shaping wood into furniture is a physical process, whereas burning wood for heat is chemical in nature because the compounds in the wood are changed into other compounds.

SEPARATION OF MIXTURES INTO PURE SUBSTANCES

The separation of mixtures is usually more difficult than the magnetic separation of iron and sulfur described previously. Most beginning chemistry students would find it bewildering to separate a piece of granite into pure substances; indeed, a trained chemist might find this difficult. Since each of the pure substances in granite has a set of properties unlike those of any other pure substance, it should be possible to use these properties to separate the pure substances, just as the attraction of iron to a magnet is used to separate iron from sulfur.

Refer to Figure 2–6 as you consider the definitions of the classifications of matter. A homogeneous sample of matter may be a pure substance or a mixture, as in the case of a solution. Note that a pure substance is considered *elemental* only if all attempts to reduce it to two or more pure substances fail, as illustrated in the following section. Also take note of whether a physical change or a chemical change is required to go from one type of matter to another.

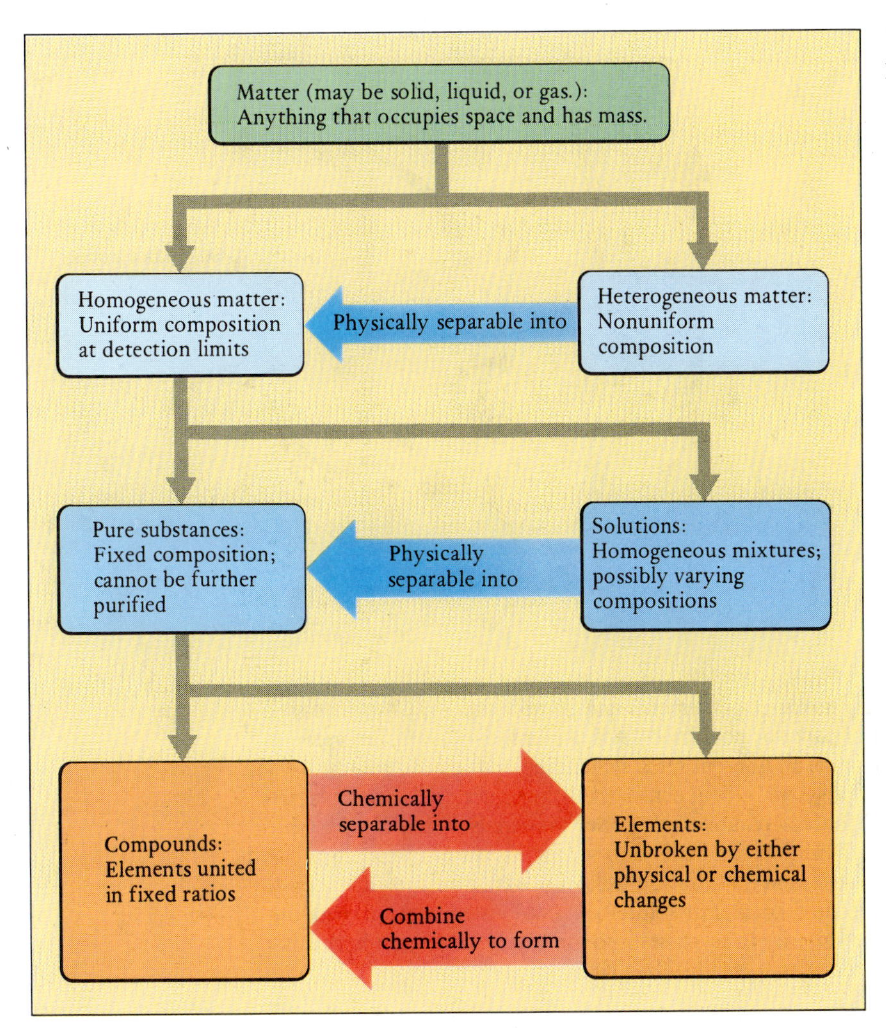

Figure 2–6 A chemical and physical classification of matter.

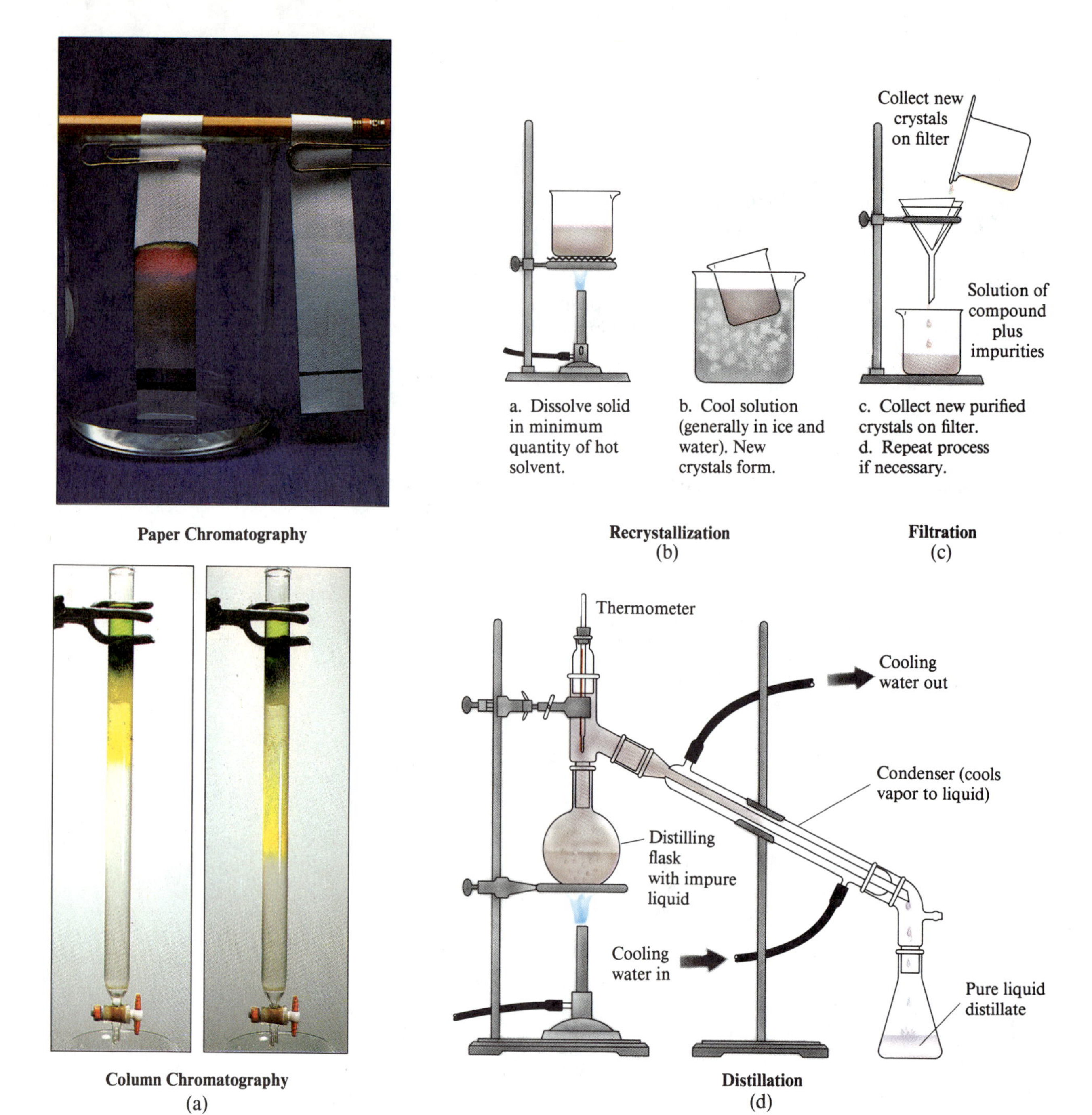

Figure 2–7 Four methods of purifying mixtures of elements and compounds. (a) Chromatography. Owing to the absorbent character of paper or powdered silica, water moves along the medium and carries the ink dyes or spinach pigments at different rates, depending on the different attractions of the dyes or pigments for the paper or powdered silica; hence, the colored materials are separated. Chromatography is now widely applied to the separation of colorless materials with different "seeing" techniques. (Charles Steele) (b) Recrystallization can be used to separate some solid mixtures based on different degrees of solubility in a solvent, and (c) filtration isolates solids from liquids. (d) Distillation. Sodium chloride dissolves in water to form a clear solution. When heated above the boiling point (indicated by thermometer), water vaporizes and passes into the cool condenser, where it liquefies as pure water. Sodium chloride remains in the distilling flask.

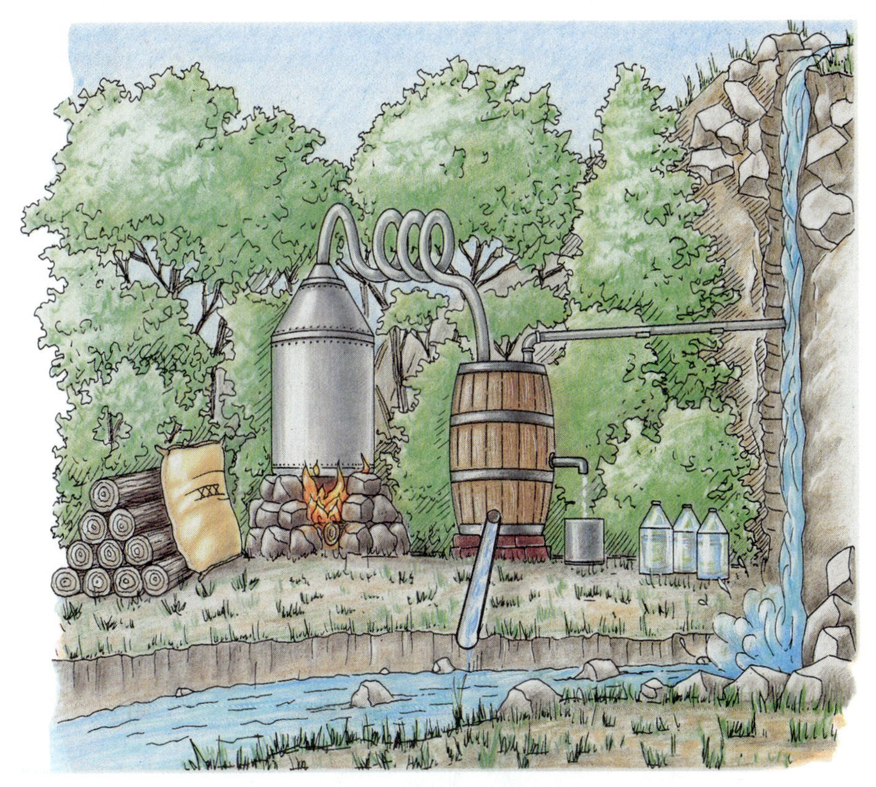

Figure 2–8 Distillation. Some of the most useful purification techniques copy processes in nature and date back to alchemical times. Distillation allows a more volatile substance to be separated from a less volatile one. In this case, alcohol is partially separated by evaporation from water and other ingredients. One distillation can produce a mixture that is 40% alcohol from one that is only 12%. Further distillations would produce an even better separation.

Many different methods have been devised to separate the pure substances in a mixture. In each case, differing properties of the pure substances are exploited to effect the separation. Figure 2–7 illustrates four commonly used methods: chromatography, distillation, recrystallization, and filtration. Figure 2–8 illustrates a separation by distillation that has been of interest to many.

ELEMENTS AND COMPOUNDS

Experimentally, pure substances can be classified into two categories: those that can be broken down by chemical change into simpler pure substances and those that cannot. Table sugar (sucrose), a pure substance, decomposes when heated in an oven, leaving carbon, another pure substance, and evolving water. No chemical operation has ever been devised that decomposes carbon into simpler pure substances. Obviously sucrose and carbon belong to two different categories of pure substances. Only 89 substances

Figure 2–9 displays the most common elements in the universe and on Earth.

found in nature cannot be reduced chemically to simpler substances; 20 others are available artificially. These 109 substances are called **elements.** Pure substances that can be decomposed into two or more different pure substances are referred to as **compounds.** Even though there are currently only 109 known elements, there appears to be no practical limit to the number of compounds that can be made from the 109 elements.

Elements are the basic building blocks of the universe and the world in which we live. A complete list of the elements is found inside the front cover of this text. Several elements are found as the elementary substance in nature; examples include gold, silver, oxygen, nitrogen, carbon (graphite and diamond), platinum, sulfur, and the noble gases (helium, neon, argon, krypton, xenon, and radon). Many more elements, however, are found chemically combined with other elements in the form of compounds.

Elements in compounds no longer show all of their original, characteristic properties, such as color, hardness, and melting point. Consider ordinary sugar as an example. It is made up of three elements: carbon (which is usually a black powder), hydrogen (the lightest gas known), and oxygen (a gas necessary for respiration). The compound sucrose is completely unlike any of the

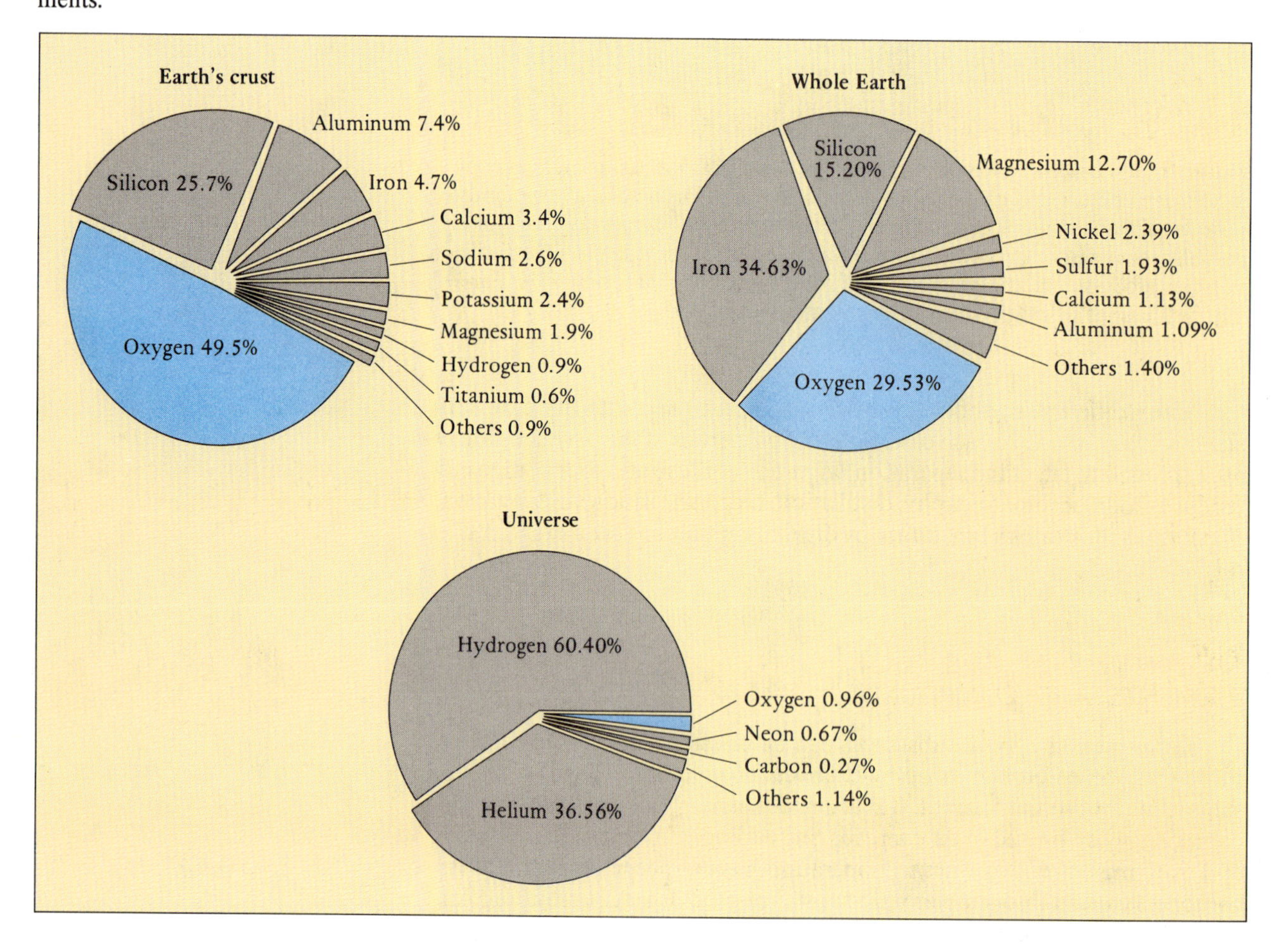

Figure 2–9 Relative abundance (by mass) of the most common elements in Earth's crust, the whole Earth, and the universe. Note that Earth's crust differs significantly from the cosmic array of the elements.

three elements of which it is composed; it is a white crystalline powder that, unlike carbon, is readily soluble in water.

A careful distinction should be made between a compound of two or more elements and a *mixture* of the same elements. The two gases hydrogen and oxygen can be mixed in all proportions. However, these two elements can and do react chemically to form the compound water. Not only does water exhibit properties peculiar to itself and different from those of hydrogen and oxygen, but it also has a definite percentage composition by weight (88.8% oxygen and 11.2% hydrogen). In addition to the distinctly different properties between compounds and their parent elements, there is this second distinct difference between compounds and mixtures:

Compounds have a fixed composition of the elements they contain.

Compounds have a definite percentage composition by weight of the combining elements.

SELF-TEST 2A

1. Four common materials that cannot be pure substances are:
   a. ______ **c.** ______
   b. ______ **d.** ______
2. Four common materials that are very nearly pure substances are:
   a. ______ **c.** ______
   b. ______ **d.** ______
3. All of the properties of two different pure substances could be identical.
   True ( ) False ( )
4. In a word, what causes changes to occur in matter? ______
5. A homogeneous mixture is a ______.
6. What always appears in a chemical change? ______
7. How many elements are currently known? ______
8. A compound has properties that are combinations of the elemental properties.
   True ( ) False ( )
9. List four purification processes.
   a. ______ **c.** ______
   b. ______ **d.** ______
10. What is constant for the compound iron sulfide that is not necessarily constant for a mixture of iron and sulfur? ______
11. Which element is most abundant:
    a. in the crust of the Earth? ______
    b. in the bulk of the Earth? ______
    c. in the universe? ______
12. List as many elements as you can recall without looking at a reference.

SYMBOLS, FORMULAS, AND EQUATIONS

Symbols, formulas, and equations are used in chemistry to convey ideas quickly and concisely. These shorthand notations are merely a convenience and contain no mysterious concepts that cannot be expressed in words.

Three-letter symbols have been proposed for elements 104 to 109:
Unnilquadium—Unq
Unnilpentium—Unp
Unnilhexium—Unh
Unnilseptium—Uns
Unniloctium—Uno
Unnilennium—Une

Certain characters are used often, and a general familiarity with them helps in reading a chemical text.

A **chemical symbol** for an element is composed of one, two, or three letters—the first letter a capital and the second and third are lowercase letters. The symbol represents three concepts. *First,* it stands for the element in general. H, O, N, Cl, Fe, and Pt are shorthand notations for the elements hydrogen, oxygen, nitrogen, chlorine, iron, and platinum, respectively. It is often useful and timesaving to substitute these symbols for the words themselves in describing chemical changes. Some symbols originate from Latin words (such as Fe, from *ferrum,* the Latin word for iron); others come from English, French, and German names. *Second,* the chemical symbol stands for a single atom of the element. The **atom** is the smallest particle of the element that can enter into chemical combinations. *Third,* the elemental symbol stands for a mole of the atoms of the element. The **mole** is a term in chemical usage (derived from the Latin for "a pile of" or "a quantity of") that means the quantity of substance that contains 602 sextillion identical particles (Fig. 2-10).

Chemical symbols are abbreviations for the different elements.

A mole contains 6.022×10^{23} particles.

Just as a dozen apples would be 12 apples, a mole of atoms would be 602,200,000,000,000,000,000,000 atoms or 6.022×10^{23} atoms. How big is this number? A mole of textbooks like this one would cover the entire surface of the continental United States to a height of 190 miles! To match the population density on Earth, a mole of people would require 150 trillion planets. It takes 134,000 years for a mole of water drops (0.05 mL [milliliter] each) to flow over Niagara Falls at a flow rate of 112,500,000 gallons per minute. A mole is a *very* large number. Yet, it turns out that a mole of very small atoms is usually a convenient amount for laboratory work. Thus, the symbol Ca can stand for the element calcium, or a single calcium atom, or a mole of calcium atoms. It will be evident from the context which of these meanings is implied.

A mole of water molecules in the liquid state occupies only about 4 teaspoonfuls. Are molecules small . . . or are they small?

Atoms can bond together to form molecules. A **molecule** is the smallest particle of an element or a compound that can have a stable existence in the

The formula for sulfuric acid is H_2SO_4 and the molecular weight is 98. One mole of H_2SO_4 weighs 98 g and contains 6.022×10^{23} molecules and 4.215×10^{24} atoms.
$S + O_2 \rightarrow SO_2$ $2SO_2 + O_2 \rightarrow 2SO_3$
$SO_3 + H_2O \rightarrow H_2SO_4$
?
"chemistry"

The shorthand used in chemical expressions is merely a time-saving device, a technique used in recording most human enterprises. Consider R, H, E, HR, SO, ERA, X, etc., used in baseball jargon.

(a)

(b)

Figure 2–10 Molar amounts, 1 mole, of (a, back row) bromine, aluminum, mercury, copper (front row) sulfur, zinc, and iron. (Charles Steele) (b) White sodium chloride, orange potassium dichromate, red cobalt chloride hexahydrate, green nickel chloride hexahydrate, and blue copper sulfate pentahydrate. (Richard Roese)

close presence of like molecules. One or more of the same kind of atom can make a molecule of an element. For example, two atoms of hydrogen bond together to form a molecule of ordinary hydrogen.

When unlike atoms combine, as in the case of water (H_2O) or sulfuric acid (H_2SO_4), the formulas tell what atoms and how many of each are present in a molecule of the compound. For example, an H_2SO_4 molecule is composed of two hydrogen atoms, one sulfur atom, and four oxygen atoms. A **formula** of a molecular substance can stand not only for the substance itself but also for one molecule of the substance or for a mole of such molecules, depending on the context.

H_2SO_4: 2 atoms of hydrogen, 1 atom of sulfur, 4 atoms of oxygen

Some compounds are composed of charged atoms or groups of atoms held together by the attraction between positive and negative ions. Such compounds contain no molecules at all. See Chapter 3.

When elements or compounds undergo a chemical change, the formulas, arranged in the form of a **chemical equation,** can present the information in a very concise fashion. For example, carbon (C) can react with oxygen (O_2) to form carbon monoxide (CO). Like most solid elements, carbon is written as though it had one atom per molecule; oxygen exists as diatomic (two-atom) molecules, and carbon monoxide molecules contain two atoms, one each of carbon and oxygen. Furthermore, one oxygen molecule combines with two carbon atoms to form two carbon monoxide molecules. All of this information is contained in the equation

Chemical equations summarize information on chemical reactions in a concise fashion.

$$2\,C + O_2 \longrightarrow 2\,CO$$

The arrow (→) is often read "yields"; the equation then states the following information:

1. Carbon plus oxygen yields carbon monoxide.
2. Two atoms of carbon plus one diatomic molecule of oxygen yield two molecules of carbon monoxide.
3. Two moles of carbon atoms plus one mole of diatomic oxygen molecules yield two moles of carbon monoxide molecules.

The number written before a formula, the **coefficient,** gives the amount of the substance involved, and the **subscript** is a part of the composition of the pure substance itself. Changing the coefficient only changes the amount of the element or compound involved, whereas changing the subscript necessarily involves changing from one substance to another. For example, 2 CO means either two molecules of carbon monoxide or two moles of these molecules, whereas CO_2 means a molecule or a mole of carbon dioxide, a very different substance.

When coefficients are properly selected so that the number of atoms of each element represented on the left (reactant side) is equal to the number of atoms of that same element on the right (product side), the equation is **balanced.** A balanced equation displays the conservation of matter in chemical change. Atoms are not created or destroyed as a chemical change occurs; they are simply rearranged as new substances are formed.

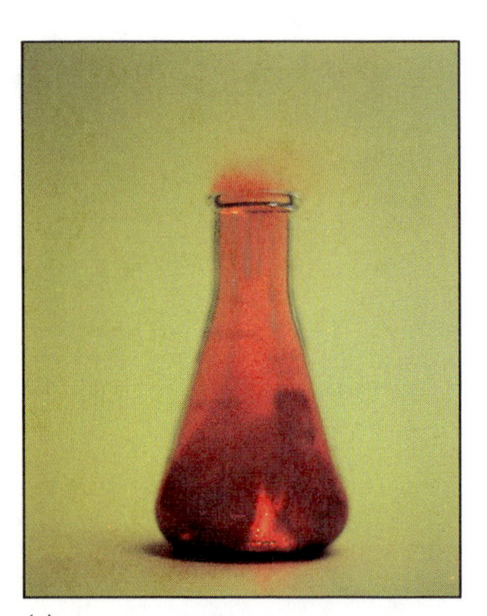

(a)

RATES OF CHEMICAL REACTIONS

Chemical reactions can proceed extremely fast, as in the case of explosives, or very slowly, as observed in the tarnishing of silver. The rate of a reaction is measured in terms of the amounts of chemical substances consumed or produced per unit of time as the reaction occurs. Consider the burning of sulfur (S) to produce sulfur dioxide (SO_2).

$$S + O_2 \longrightarrow SO_2$$

The rate of this reaction can be measured in terms of the amount of SO_2 formed per minute or the amount of S or O_2 consumed per minute. The rate

(b)

Figure 2–11 The elements antimony (Sb) and bromine (Br) react on contact with each other. The reaction at 75°C (b) is much faster than at 25°C (a). Relate this to what you know about the speed of cooking at different temperatures and the retarding of food decay when food is refrigerated. (J. Morgenthaler) of a particular chemical reaction depends on the temperature, the concentration of the chemicals involved, and the presence of catalysts.

Temperature Chemical reactions proceed faster as the temperature is increased. We make use of this principle in cooking foods (a roast cooks faster at a higher temperature) and in preserving foods (foods spoil less quickly if refrigerated). Figure 2-11 contrasts the reaction of antimony with bromine at 25°C with the same reaction at 75°C.

Raising the temperature speeds up chemical reactions.

For many reactions, a temperature rise of 10°C doubles the rate.

Concentration Increasing the concentration of the reacting chemicals speeds up the rate of a reaction. For example, in the reaction of sulfur and oxygen given above, if air replaces oxygen, the reaction proceeds at a slower rate because air is a mixture of about one part oxygen and four parts nitrogen. Smoking tobacco in air is dangerous to your health, but smoking tobacco in pure oxygen would pose the immediate risk of a severe facial burn, as the tobacco and paper would produce a flash fire in the higher concentration of oxygen.

An interesting and sometimes very dangerous aspect of the concentration effect on reaction rates is the state of subdivision of the chemical reactant. You would find it difficult to impossible to burn a sack of flour in an ordinary fireplace, as the flour would tend to smother the wood and block its contact with oxygen in the air. Would you be surprised to learn that the same flour as dust in the air of a flour mill forms an explosive mixture so powerful that it can literally blow a concrete building apart? Figure 2-12 illustrates this principle with the burning of lycopodium powder in bulk and in a dust dispersion in air.

Grain dust has caused explosions destroying entire grain elevators and storage facilities.

Catalysts Slow chemical reactions often proceed at a much faster rate in the presence of a third chemical or group of chemicals. For example, in the

Figure 2-12 Lycopodium powder is made up of the ground-up spores of a common moss. If a flame is directed on the powder (left photo), it burns with difficulty, in contrast to the explosive burning (right) when the powder is sprayed into the flame. (Charles D. Winters)

manufacture of sulfuric acid (H_2SO_4), the number one chemical of commerce, it is necessary to convert sulfur dioxide (SO_2) to sulfur trioxide (SO_3):

$$2\ SO_2 + O_2 \longrightarrow 2\ SO_3$$

If the pure chemicals are mixed, the reaction is very slow, much too slow for a profitable industrial process. However, if some oxides of nitrogen are introduced into the system, the desired reaction proceeds rapidly. Furthermore, the nitrogen compounds are not permanently changed in the process. Such a chemical is referred to as a catalyst.

A catalyst is a substance that increases the rate of a chemical reaction without being permanently consumed.

Biological catalysts are called enzymes.

Later, in the study of biochemistry, it will be amazing to note the ability of biological systems to produce catalysts on demand and then destroy them after a particular chemical need is met.

CHEMICAL EQUILIBRIUM

Chemical reactions are capable of going forward or backward.

Most chemical reactions can be reversed under suitable conditions. Figure 2–13(a) pictures the electrolysis of water as electrical energy is employed to decompose water into the elements hydrogen and oxygen.

$$2\ H_2O \longrightarrow 2\ H_2 + O_2$$

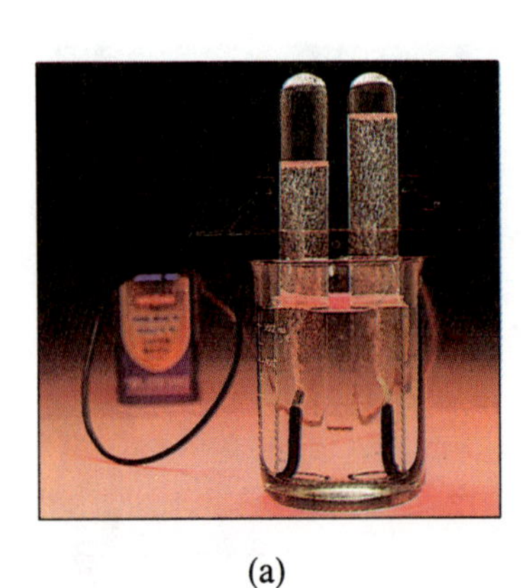

(a)

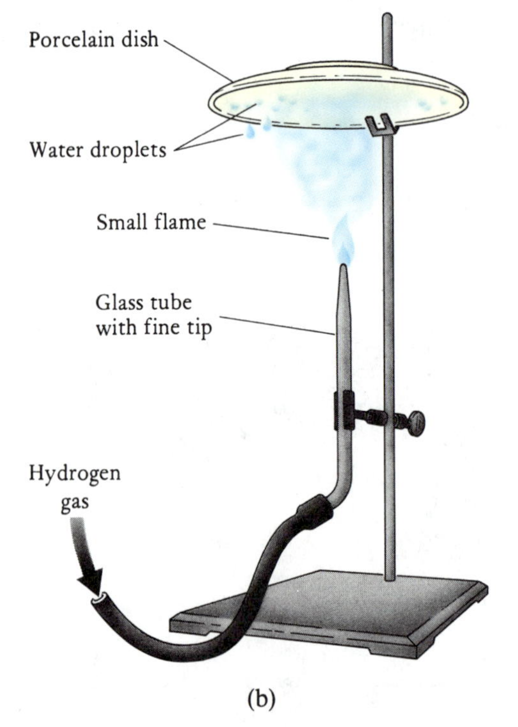

(b)

Figure 2–13 (a) Electrical energy is required to decompose water into hydrogen (left tube) and oxygen (right tube). Note that two volumes of H_2 are produced for each volume of O_2. (Charles D. Winters) (b) Hydrogen and oxygen burn to produce water in the gaseous state. The water is condensed on the cooler porcelain dish.

In part (b) of the illustration, water is shown to form when hydrogen is burned in air, the reverse of the electrolytic reaction.

$$2\ H_2 + O_2 \longrightarrow 2\ H_2O$$

Chemical reactions are generally reversible.

There are many reversible reactions important to human life. One of these involves the transport of atmospheric oxygen from the lungs to the various parts of the body. This task is carried out by hemoglobin, a complex compound found in the blood. This substance takes up oxygen while in the lungs to form oxyhemoglobin, only to release the oxygen in the various parts of the body for metabolic purposes (Fig. 2–14).

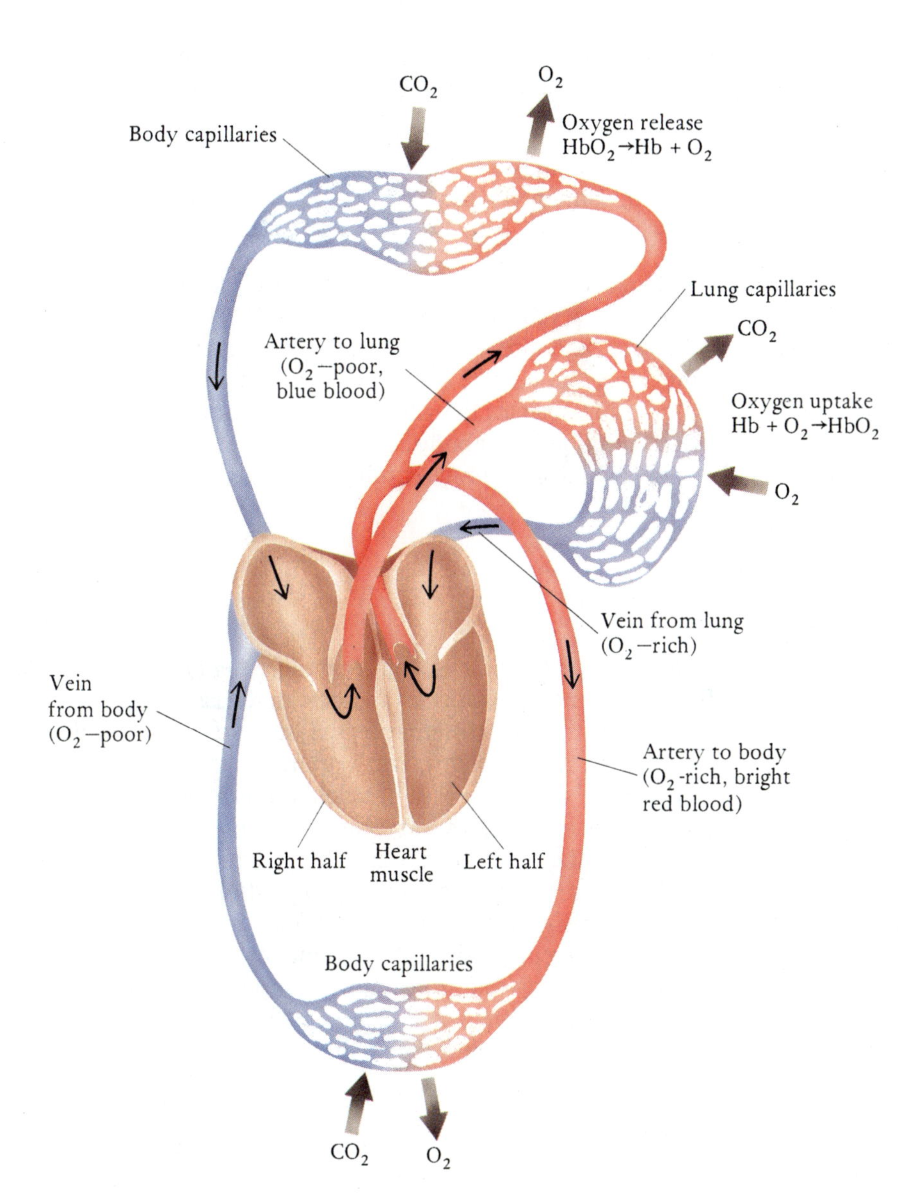

Figure 2–14 Chemical reversibility is illustrated in this simplified diagram of human circulation. The heart (shown in the front) is divided into two parallel halves. The right half pumps oxygen-poor blood to the lungs; the left half pumps oxygen-rich blood to the body. Hb = hemoglobin; HbO_2 = oxyhemoglobin. The oxygen is bound chemically in the lungs and, under different conditions, released in the body tissues.

$$\text{Hemoglobin} + O_2 \rightleftharpoons \text{Oxyhemoglobin}$$

The double arrow is used to show that the reaction can proceed in both the forward and reverse directions.

Chemicals do not always react to form products with the complete extinction of the reactants. We may get the idea that all chemical reactions go to completion when we watch a piece of wood "burn up." However, nature quite often displays a reaction in which both reactants and products are present in the reaction medium in constant, but not necessarily the same, concentration levels. When reversible reactions reach the point at which the forward reaction is proceeding at the same rate as the reverse reaction, the amounts of chemicals present remain constant because a particular chemical is produced as fast as it is consumed. At this point, we have **chemical equilibrium.**

A chemical change is at equilibrium when products are produced at the same rate that products are consumed in reproducing reactants.

Consider the equilibrium among limestone ($CaCO_3$), lime (CaO), and carbon dioxide (CO_2). If dry limestone is placed in a vacuum, carbon dioxide gas soon appears in the container with the limestone. The reaction is:

Double arrows ($\rightleftharpoons$) indicate that the reaction proceeds in both directions at one time and that it has reached equilibrium.

$$CaCO_3 \rightleftharpoons CaO + CO_2$$

After the CO_2 builds up to a certain concentration level, the system is at equilibrium and the CO_2 level does not rise further. We know that the reaction is proceeding in both directions when the system is at equilibrium because we can introduce radioactive carbon in either the reactant or the product and use a radiation detector to trace the radioactive carbon through the reaction in either direction. Your body controls the appropriate concentration levels for many chemicals through equilibrium systems in your body fluids.

MEASUREMENT

As we have moved from classical systems of measurement to the metric system (dating from the French Revolution) and now in the last few years to the *Système International d'Unites* (International System of Units, usually referred to in English as the SI system), two important goals have been pursued. First, there is the desire for ease of calculations which has been facilitated by defining units in multiples and submultiples of ten relative to a basic defined unit. A milliliter (mL), the thousandth part of a liter, and the kilometer (km), a thousand meters, are familiar illustrations of such relationships. This goal has largely been achieved. Second, there is the desire to define basic units in terms of reproducible phenomena of nature rather than in terms of the length of the king's foot or even some museum piece set aside to serve as a standard. For example, length is now defined in terms of the wavelength of a particular light source, a length that appears constant in the human frame of reference. This second objective has been only partially achieved.

It should be noted that one system of units is no more accurate or precise than another. Accuracy and precision depend on our ability to use and read

TABLE 2-1 Prefixes for Multiples and Submultiples

Multiple or Submultiple		Prefix	Symbol
10^9	1000 000 000	giga	G
10^6	1000 000	mega	M
10^3	1000	kilo	k
10^{-2}	0.01	centi	c
10^{-3}	0.001	milli	m
10^{-6}	0.000 001	micro	μ
10^{-9}	0.000 000 001	nano	n

divisions on instruments, not on the size and names of the defined units marked on the instrument.

All of the present quantitative knowledge about the universe can be described in terms of seven basic units (see Appendix A). These units define standard measurements for mass, length, time, temperature, electric current strength, light intensity, and the amount of a substance (the mole). Because it is not the purpose of this text to engage in a study of measurement, we will employ and define units as needed and will not limit ourselves to units in the SI system. For example, we will measure volumes in liters (L) and milliliters (mL) rather than the SI-approved cubic meter because the liter is very familiar to us (2-liter soft drink bottle). Furthermore, we will measure energy in terms of calories and kilocalories rather than in the SI-approved joules (1 calorie = 4.184 joules) because everyone is familiar with calorie-counting relative to food intake. (One food calorie (Cal) is a kilocalorie, or 1000 calories (cal)).

Exponential Notation In studying chemistry we encounter very large and very small numbers. When considering the number of atoms in a mole, it is simply easier to write 6.022×10^{23} than to count and write all of the zeros. The exponential notation here simply means to multiply by ten 23 times. It is similar with very small numbers. If the impurity in purified silicon is one part in a billion parts of the solid, the impurity is 0.000000001 of the whole. It is easier to write 10^{-9}, which means to divide one by ten nine times.

Prefixes are also useful in expressing large and small numbers. Most people would rather express distances between towns in terms of kilometers than thousands of meters because both the written and oral expressions are quicker without any loss in meaning. The prefix kilo (k) means 1000 times the measurement it modifies. Prefixes that will be useful in this text are given in Table 2-1.

WEIGHT RELATIONSHIPS IN CHEMICAL REACTIONS

A mole of the atoms (6.022×10^{23} atoms) of any of the elements weighs its **atomic weight** in grams (g). For example, the atomic weight of hydrogen is 1.0079, and a mole of hydrogen atoms weighs 1.0079 g. A list of the atomic

A. L. Lavoisier and wife by David. (Metropolitan Museum of Art) Lavoisier led in finding weight relationships in chemistry.

weights for all of the elements is given inside the front cover of this book. Find oxygen on this list and note that it has an atomic weight of 15.9994. Therefore, a mole of oxygen atoms weighs 15.9994 g. The weight of a mole of molecules is readily summed from the atomic weights. Consider two examples:

1. Molecular oxygen, O_2 (oxygen is composed of diatomic molecules). A mole of oxygen molecules contains 2 moles of oxygen atoms. Hence, a mole of oxygen *molecules* weighs two times the weight of a mole of oxygen *atoms* (2 times 15.9994 g, or about 32 g).
2. Sulfuric acid, H_2SO_4. A mole of sulfuric acid contains 2 moles of hydrogen atoms, 1 mole of sulfur atoms, and 4 moles of oxygen atoms. Rounding the atomic weights to whole numbers, the mole of sulfuric acid weighs

$$2 \times 1\text{ g} + 1 \times 32\text{ g} + 4 \times 16\text{ g} = 98\text{ g}$$

All calculations involving chemical equations are based on the law of conservation of matter.

As pointed out in Chapter 1, one of the first laws established in chemistry is the conservation of matter in chemical change. Pure substances are destroyed and created, but the matter involved is preserved. The weight relationships characteristic of any chemical change are readily ascertained from the balanced chemical equation and the list of atomic weights. Weight ratios between reactants, between products, or between reactants and products are the consequence of the weights of the number of moles stated in the balanced equation.

EXAMPLE

What weight of hydrogen, H_2, can be expected from the electrical decomposition of 36 g of water (Fig. 2-13)?

Write the equation.

$$H_2O \longrightarrow H_2 + O_2$$

The beginning student is not expected to know reactants and products and the formulas of the chemicals involved as new chemical changes are encountered. Such information is known as a result of experimental observations and measurements.

Balance the equation.

$$2\,H_2O \longrightarrow 2\,H_2 + O_2$$

The mole ratio from the balanced equation tells us that 2 moles of water produce 2 moles of hydrogen molecules; the ratio is 2:2 or 1:1.

Calculate the weight ratio from the mole ratio and atomic weights.

One mole of water weighs 2×1 g + 16 g, or 18 g.

One mole of hydrogen molecules weighs 2×1 g, or 2 g.

Therefore, 18 g of water yields 2 g of hydrogen.

Using this ratio, we can calculate for any amount of water used.

$$?\ \text{g}\ H_2 = 36\ \text{g}\ H_2O \times \frac{2\ \text{g}\ H_2}{18\ \text{g}\ H_2O} = 4\ \text{g}\ H_2$$

All weight relationships in chemical change are this simple in principle even though the numbers are not such that you can do the calculations in your head.

QUANTITATIVE ENERGY CHANGES IN CHEMICAL REACTIONS

Chemical reactions may produce heat or absorb it. Heat production is an **exothermic** process, and heat absorption is an **endothermic** one. Furthermore, the amount of heat energy involved in a chemical change is just as quantitative as the amounts of chemicals involved.

Consider the following reactions:

Reactants			Products		Heat Effects*†
CaO Calcium oxide (quicklime)	+ H_2O Water	$\longrightarrow$	$Ca(OH)_2$ Calcium hydroxide (slaked lime)	+	15.6 kcal/mole $Ca(OH)_2$ [65.3 kjoule/mole]
2 Na Sodium	+ $Cl_2(gas)$ Chlorine	$\longrightarrow$	2 NaCl Sodium chloride (table salt)	+	196.4 kcal (98.2 kcal/mole NaCl) [821.7 kjoule (410.9 kjoule/mole NaCl)]
$H_2(gas)$ Hydrogen	+ $I_2(gas)$ Iodine	$\longrightarrow$	2 HI(*gas*) Hydrogen iodide	−	12.4 kcal (−6.20 kcal/mole HI) [−51.9 kjoule (−25.9 kjoule/mole HI)]

* Heat energy is measured in calories (cal). A kilocalorie (kcal) is 1000 cal. A calorie is the amount of heat required to raise the temperature of 1 g of water by 1°C.

† +kcal means heat is liberated (exothermic); −kcal means heat is required (endothermic).

1 calorie = 4.184 joules
cal = calorie
kcal = kilocalorie = 1000 calories
kjoule = kilojoule = 1000 joules

In the first reaction, CaO (quicklime) reacts with H_2O to give $Ca(OH)_2$ (slaked lime) with the evolution of heat. In the second reaction, Na reacts with the greenish-yellow gas Cl_2 to give NaCl (table salt). If a piece of hot Na is put into a flask containing Cl_2, the Na burns quickly and liberates a great deal of heat and light. White crystals of NaCl are produced. In the last reaction, gaseous H_2 reacts with gaseous I_2 to produce gaseous HI with the absorption of heat. These facts, along with similar ones, lead to a second generalization about chemical reactions:

> **A given amount of a particular chemical change corresponds to a proportional amount of energy change.**

For example, the preparation of 1 mole of $Ca(OH)_2$ from CaO and H_2O releases 15.6 kcal. To prepare 2 moles of $Ca(OH)_2$, 2 × 15.6, or 31.2 kcal of heat is released.

Sometimes energy changes in reactions are difficult to observe because of the very slow rate of reaction. An example is the rusting of iron. The reaction involved is complicated, but we can represent it by the simplified equation:

$$\underset{\text{Iron}}{4\,Fe} + \underset{\text{Moist air}}{3\,O_2 + 6\,H_2O} \longrightarrow \underset{\text{Iron hydroxide (rust)}}{4\,Fe(OH)_3} + \underset{\text{Heat}}{788\text{ kcal}}\ [197\text{ kcal/mole }Fe(OH)_3]$$

Ordinarily, the rusting of iron occurs so slowly that the liberation of heat is perceptible only with the aid of special instruments. The total amount of heat evolved in rusting is considerable, but it typically takes place over a long period.

WHY STUDY PURE SUBSTANCES AND THEIR CHANGES? COMFORT, PROFIT, AND CURIOSITY!

Perhaps by now you are wondering why we should be interested in elements and compounds and their chemical properties. There are two basic reasons. The first is the belief that the knowledge of chemical substances and

Rolls of sheet aluminum obtained from the reduction of aluminum oxide. (Aluminum Association of America)

chemical changes will allow us to bring about desired changes in the nature of everyday life. Two hundred years ago most of the materials surrounding a normal person could be changed only by physical means. Only a few useful materials, such as iron and pottery, were the product of chemical change. By contrast, today's synthetic fibers, plastics, drugs, latex paints, detergents, new and better fuels, photographic films, and audio and video tapes are but a few of the materials produced by controlled chemical change. (We shall return to examine the chemistry of many of these later.) You will find it difficult to find more than a few objects in your home that have not been altered by a desirable chemical change. Not only is it important to bring about desirable changes, but also in the areas of toxicity and pollution it is important to avoid undesirable changes.

A second driving reason for the study of elements, compounds, and their properties is simple curiosity. Chemicals and chemical change are a part of nature that is open to investigation, and, like the mountain climber, we shall find this task both interesting and challenging simply because it is there. If we hope to understand matter through basic research, the first steps are to discover the simplest forms of matter and to study their interactions. Curiosity draws many chemists toward these basic research activities.

SELF-TEST 2B

1. Consider the chemical equation: $CH_4 + 2\ O_2 \rightarrow CO_2 + 2\ H_2O$. Explain what is meant by the symbols:
   a. O ______
   b. $2\ O_2$ ______
   c. CH_4 ______
   d. $\rightarrow$ ______
   e. H_2O ______
   f. $2\ H_2O$ ______
2. Name the three concepts that a chemical symbol can represent.
   ______, ______, and ______
3. A chemical formula gives what two pieces of information?
   ______ and ______
4. Which is proper to change when you balance a chemical equation: a coefficient or a subscript? ______
5. In photosynthesis, carbon dioxide is combined with water to form the simple sugar glucose and oxygen:
   a. Balance the equation.

   $$\_\_\_\ CO_2 + \_\_\_\ H_2O \longrightarrow \_\_\_\ C_6H_{12}O_6 + \_\_\_\ O_2$$

   b. How many molecules of CO_2 are necessary to produce one molecule of sugar? ______
   c. How many moles of CO_2 are necessary to produce one mole of sugar? ______
   d. What is the molecular weight of CO_2? ______, of $C_6H_{12}O_6$? ______
   e. How many grams of CO_2 are required to make 1 mole of sugar?
   ______

6. If 68 kilocalories of energy are released in the formation of 18 g (1 mole) of water by the combination of H_2 and O_2, how much energy would be released in the formation of 36 g of water? ________
7. Balance the following equations:
   a. ___ Mg + ___ $O_2 \rightarrow$ ___ MgO
   b. ___ Si + ___ $Cl_2 \rightarrow$ ___ $SiCl_4$
   c. ___ Al + ___ $O_2 \rightarrow Al_2O_3$
8. At least what two factors drive chemical research? ________ and ________
9. What is the primary advantage of the SI system of units over the classical English system of units for measurement? ________
10. How many ears would you expect on a dozen elephants? ________ How many hydrogen atoms would you expect in 1 mole of hydrogen molecules, H_2? ________
11. Is chemical equilibrium static or dynamic? ________ Explain. ________
12. What three factors are known to affect the rate of a chemical reaction? ________, ________, and ________

QUESTIONS

1. Name as many materials as you can that you have used during the past day which were not chemically changed from their natural occurrence.
2. Identify the following as physical or chemical changes.
   a. Formation of snowflakes
   b. Rusting of a piece of iron
   c. Ripening of fruit
   d. Fashioning a table leg from a piece of wood
   e. Fermenting grapes
   f. Boiling a potato in water
3. Would it be possible for two pure substances to have exactly the same set of properties? Give reasons for your answer.
4. Cite an example of a chemical change that is useful and one that is destructive relative to your interests.
5. Classify each of the following as a physical property or a chemical property.
   a. Density (the mass per unit volume for a given sample of matter)
   b. Melting temperature
   c. A substance that decomposes upon heating
   d. A conductor of electricity
   e. A substance that does not react with sulfur
   f. Ignition temperature of a piece of paper
6. Most natural materials are mixtures. Name five natural mixtures for which you can identify at least two ingredients in the mixture. For example, air contains oxygen and nitrogen.
7. Classify each of the following as an element, a compound, or a mixture. Justify your answers.
   a. Mercury
   b. Milk
   c. Pure water
   d. A tree
   e. Ink
   f. Iced tea
   g. Pure ice
   h. Carbon
8. Which of the materials listed in Question 7 can be pure substances?
9. Is it possible for the properties of iron to change? What about the properties of steel? Explain your answer.
10. Suggest a method for purifying water slightly contaminated with a dissolved solid.
11. A glass of liquid appears free of solids as you observe it. What simple test could you perform to tell if the liquid is really a colloidal suspension?
12. Define a solution as a particular kind of mixture.
13. Aspirin is a pure substance—a compound of carbon, hydrogen, and oxygen. If two manufacturers produce equally pure aspirin samples, what can be said of the relative worth of the two products?
14. Explain how the definition of a pure substance allows for the possibility that the pure substance may in fact be impure at the molecular level.
15. Except for air-borne particulates, rain water is free of solid impurities. What is the natural purification process that causes this high degree of purity?
16. Is it possible to have a mixture of two elements and also to have a compound of the same two elements? Explain. Can you think of an example?
17. What is the difference between a chemical change and a nuclear change?
18. A company has announced bottled "pure water" from

12,000-year-old glaciers. Explain how nature uses crystallization to purify water.

19. Given the following sentence, write a chemical reaction using chemical symbols that conveys the same information. "One nitrogen molecule containing two nitrogen atoms per molecule reacts with three hydrogen molecules, each containing two hydrogen atoms, to produce two ammonia molecules, each containing one nitrogen and three hydrogen atoms."

20. Describe in words the chemical process that is summarized in the following equation:

$$2\ Na + Cl_2 \longrightarrow 2\ NaCl$$

21. How many *atoms* are present in each of the following?
 a. One mole of He
 b. One mole of Cl_2
 c. One mole of O_3

22. If you had a mole of dogs, how many moles of dog feet would you expect to have? If you have a mole of O_3 molecules, how many moles of oxygen atoms would you have?

23. Balance the following equations:
 a. $SO_2 + O_2 \rightarrow SO_3$
 b. $K + Br_2 \rightarrow KBr$
 c. $CH_4 + O_2 \rightarrow CO_2 + H_2O$
 d. $O_3 \rightarrow O_2$

24. (a) Give a definition for the rate of a chemical reaction. (b) What is a chemical catalyst, and how would a catalyst affect a reaction rate? (c) Why is dust a safety factor in grain elevators and flour mills?

25. What is the weight in grams of 1 mole of:
 a. H_2O_2
 b. H_3BO_3
 c. $C_2H_4(OH)_2$ (*Note:* The subscript after the parentheses indicates a multiple for each symbol indicated within the parentheses.)
 d. Fe_2O_3

26. (a) How many moles of KCl can be made by using 1 mole of potassium and an excess of chlorine?

$$2\ K + Cl_2 \longrightarrow 2\ KCl$$

(b) How many grams of KCl can be made from 1 mole of potassium?

27. What is constant about a compound?
 a. The weight of a sample of the compound
 b. The weight of one of the elements in samples of the compound
 c. The ratio by weight of the elements in samples of the compound

28. Is the burning of hydrogen in oxygen an endothermic or exothermic process?

29. (a) Look up the term *spontaneous combustion* and relate this term to endothermic and exothermic chemical processes. (b) Fires have started by water seeping into bags in which quicklime (CaO) was stored. Why would this produce a fire?

30. Explain how nature can use recrystallization to form relatively pure chemicals in caves.

Just as Picabia uses the abstract nature of his painting *Catch as Catch Can* to represent intangibles, so do models of electrons in atoms and molecules portray a reality that thus far has escaped an exact representation. (Philadelphia Museum of Art: The Louis and Walter Arensberg Collection)

CHAPTER

3

The Chemical View of Matter

Topics essential to the understanding of chemistry include the structure of the atom, the periodic table and its use in predicting a wealth of chemical information, and the bonds that hold matter together.

1. What is the significance of Dalton's atomic theory?
2. What are the three basic particles of the atom, and where are they found?
3. What is the experimental evidence for Rutherford's model of the nucleus?
4. What are the different types of radiation given off by radioactive elements?
5. What are isotopes?
6. Where are electrons in atoms, and how are they arranged?
7. How was the periodic table developed?
8. Why is the periodic table so useful?
9. What kinds of bonds hold matter together?
10. What is hydrogen bonding, and why is it important?

Why does an element or compound have the properties it has? Why does one element or compound undergo a change that another element or compound does not undergo? Inanimate matter is the way it is because of the nature of its parts. The use of atoms to represent the "parts" dates back to about 400 B.C., when the Greek philosophers Leucippus and Democritus proposed the atom as the smallest unit of matter. However, it wasn't until John Dalton introduced his atomic theory in 1803 that the importance of using atomic theory to explain properties of matter was recognized. Dalton, drawing from his own quantitative experiments and those of earlier scientists, proposed that

John Dalton (1766–1844) is considered the father of chemical theory.

1. Matter is composed of indestructible* particles called atoms.
2. All atoms of a given element have the same properties such as size, shape, and weight,† which differ from the properties of atoms of other elements.
3. Elements and compounds are composed of definite arrangements of atoms, and chemical change occurs when the atomic arrays are rearranged.

Dalton's atomic theory and the development of the periodic table by Mendeleev in 1869 led to rapid growth of chemistry as a science. In this chapter we will use the current knowledge about the atom together with the periodic table to serve as the basis for our understanding of the chemical view of matter.

STRUCTURE OF THE ATOM

There is experimental evidence now for more than 60 subatomic particles. However, only three are important to our understanding of the chemical view of matter: **electrons, protons,** and **neutrons** (Table 3–1). The proton and neutron are found in the nucleus of the atom, and the electrons are distributed around the nucleus. Because the proton and neutron have nearly the same mass and are about 1840 times heavier than the electron, the nucleus contains essentially all the mass of the atom. Protons have a +1 charge, so a given nucleus has a positive charge equal to the number of protons it contains. In fact, each chemical element is defined by its **atomic number**—the number of protons in the nucleus of its atoms. Since atoms are electrically neutral, the number of protons and electrons must be equal,

* Radioactive atoms self-destruct, but natural radioactivity wasn't discovered until 1896, so Dalton had no knowledge of this phenomenon.

† We know that all of the atoms of the same element do not necessarily have the same weight, but isotopes were not known at the time of Dalton's work.

TABLE 3-1 Summary of Properties of Electrons, Protons, and Neutrons

	Relative Charge	Relative Mass	Location
Electron	−1	0.00055	Outside the nucleus
Proton	+1	1.00727	Nucleus
Neutron	0	1.00867	Nucleus

Figure 3-1 Marie and Pierre Curie. The Curies and H. A. Becquerel shared the Nobel Prize in physics for their research in radioactivity. Marie Curie received a second Nobel Prize in 1911 for the discovery of radium and polonium. She was the first person to receive two Nobel Prizes.

so the atomic number of an element also indicates the number of electrons outside the nucleus.

Look at the periodic table in Figure 3-7. Note that each element has a unique atomic number. For example, the atomic number of nitrogen is 7, so nitrogen has 7 protons in the nucleus and 7 electrons outside the nucleus. How do we know that the protons are inside the nucleus? Where are the electrons outside the nucleus? How big are atoms? How big is the nucleus inside the atom? Answers to these questions were obtained through experiments done in the early 1900s. Many of the early experiments on atomic structure used particles emitted from naturally radioactive elements.

Natural Radioactivity

In 1896 Henri Becquerel (1852-1908) discovered natural radioactivity in minerals containing natural uranium and radium. Marie Sklodowska Curie (1867-1934), a student of Becquerel, continued to do research on radioactivity. She and her husband, Pierre, discovered two radioactive elements, radium and polonium (Fig. 3-1). Their studies of radioactive elements laid the groundwork for Rutherford's later experiments on the atom. Radioactive elements commonly emit alpha, beta, and gamma rays (Fig. 3-2 and Table 3-2). Alpha and beta rays are composed of particles with

The word *radioactivity* was first used by Marie Curie to represent the penetrating rays emitted from elements such as radium.

Marie Curie named the element polonium after her native country, Poland.

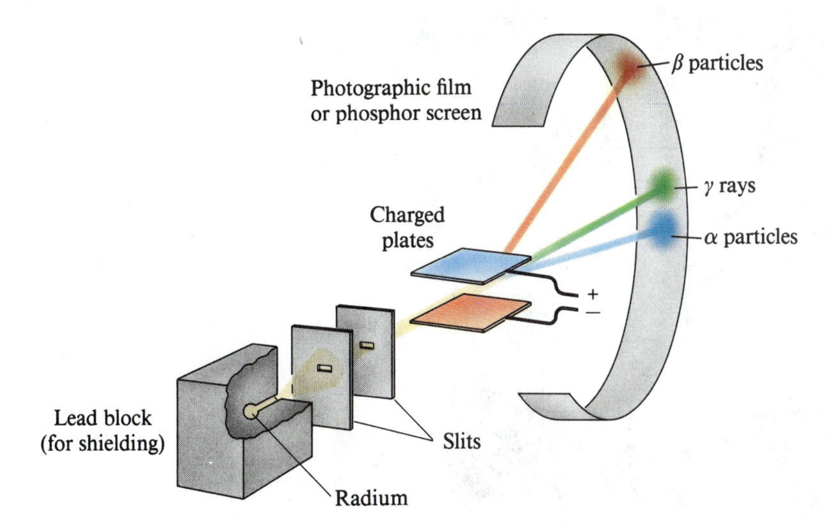

Figure 3-2 Separation of alpha and beta particles and gamma rays by an electrical field.

TABLE 3–2 Summary of Properties of Alpha Particles, Beta Particles, and Gamma Rays

	Charge	Relative Mass	Symbols
Alpha particle	Positive (+2)	4	α, ${}^{4}_{2}\alpha$, ${}^{4}_{2}He$
Beta particle	Negative (−1)	0.0005	β, ${}^{0}_{-1}\beta$, ${}^{0}_{-1}e$
Gamma ray	Neutral (0)	0	γ, ${}^{0}_{0}\gamma$

charges and masses, but gamma rays, like light and X rays, are part of the electromagnetic spectrum (Fig. 3–3). Alpha particles are high-energy helium nuclei and beta particles are high-energy electrons. It is important to keep in mind that these particles are emitted from radioactive nuclei and have much higher energy than regular helium nuclei or the electrons found outside the nucleus in an atom.

The Nucleus

Alpha particles are scattered by the nuclei of the gold atoms.

When Ernest Rutherford and his students directed alpha particles toward a very thin sheet of gold foil in 1909, they were amazed to find a totally unexpected result (Fig. 3–4). As they had expected, the paths of most of the alpha particles were only slightly changed as they passed through the gold foil. The extreme deflection of a few of the alpha particles was a surprise. Some even "bounced" back toward the source. Rutherford expressed his astonishment by stating that he would have been no more surprised if someone had fired a 15-inch artillery shell into tissue paper and then found it in flight back toward the cannon.

What allowed most of the alpha particles to pass through the gold foil in a rather straight path? According to Rutherford's interpretation, the atom is

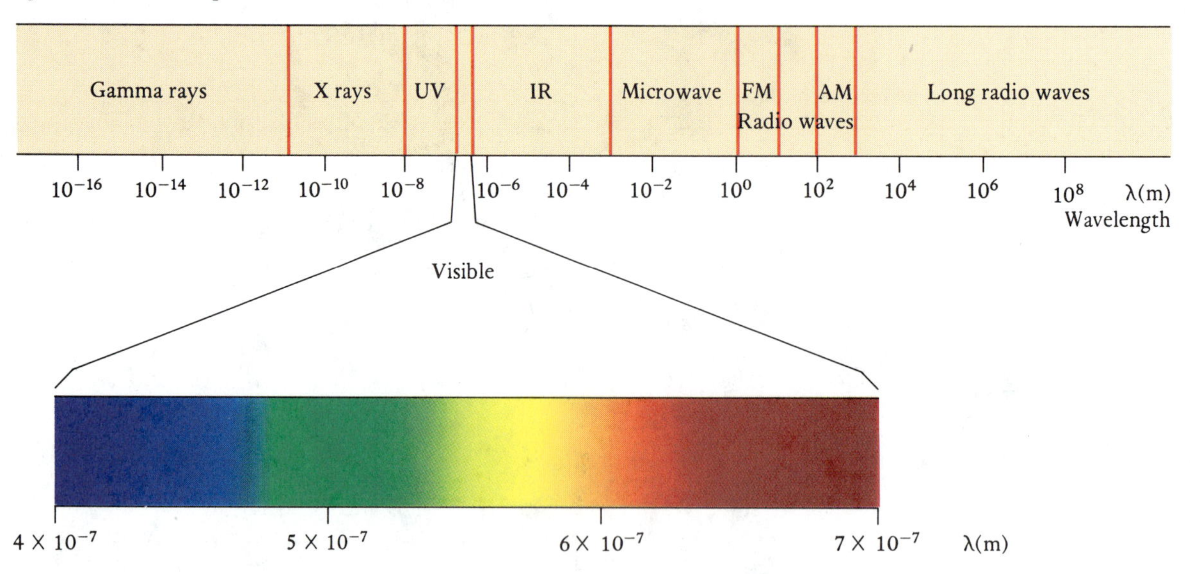

Figure 3–3 Electromagnetic spectrum. The visible spectrum is only a small part of the entire spectrum.

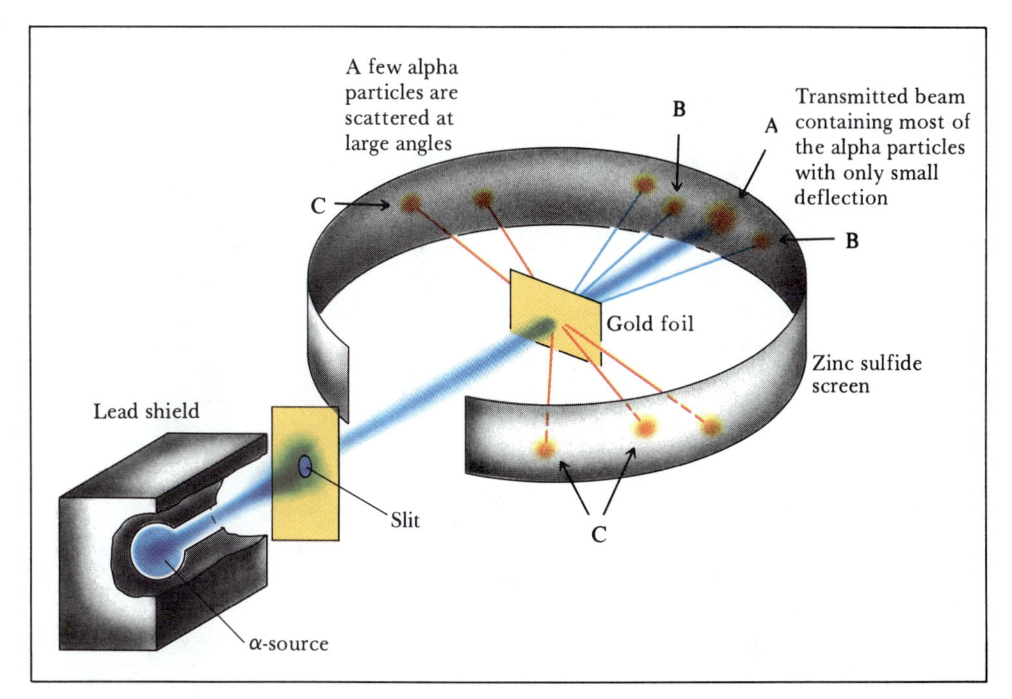

Figure 3–4 Rutherford's gold foil experiment. A cylindrical scintillation screen is shown for simplicity; actually, a movable screen was employed. Most of the alpha particles pass straight through the foil to strike the screen at point A. Some alpha particles are deflected to points B, and some are even "bounced" backward to points such as C.

mostly *empty space* and, therefore, offers little resistance to the alpha particles (Fig. 3–5).

What caused a few alpha particles to be deflected? According to Rutherford's interpretation, concentrated at the center of the atom is a **nucleus** containing most of the mass of the atom and all of the positive charge. When an alpha particle passes near the nucleus, the positive charge of the nucleus repels the positive charge of the alpha particle; the path of the smaller alpha particle is consequently deflected. The closer an alpha particle comes to a target nucleus, the more it is deflected. Those alpha particles that meet a nucleus head on bounce back toward the source as a result of the strong positive-positive repulsion, because the alpha particles do not have enough energy to penetrate the nucleus.

Like charges repel. Unlike charges attract.

Rutherford's calculations, based on the observed deflections, indicate that the nucleus is a very small part of an atom. An atom occupies about a million million times more space than does a nucleus; the radius of an atom is about 100,000 times greater than the radius of its nucleus. Thus, if a nucleus were the size of a baseball, then the edges of the atom would be over 2 miles away. And most of the space in between would be absolutely empty.

Alpha-particle scattering can be explained if the nucleus occupies a very small volume of the atom.

Since the nucleus contains most of the mass and all of the positive charge of an atom, the nucleus must be composed of the most massive atomic

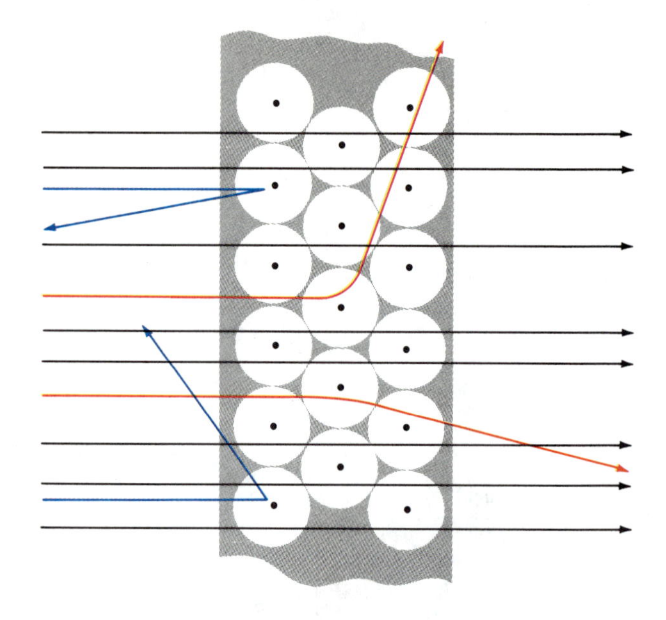

Figure 3–5 Rutherford's interpretation of how alpha particles interact with atoms in a thin gold foil. Actually, the gold foil was about 1000 atoms thick. For illustration purposes, points are used to represent the gold nuclei, and the path widths of the alpha particles are drawn much larger than scale.

particles, the protons and neutrons. The electrons are distributed in the near-emptiness outside the nucleus.

Truly, Rutherford's model of the atom was one of the most dramatic interpretations of experimental evidence to come out of this period of significant discoveries.

Ernest Rutherford (1871–1937) received the Nobel Prize in 1908. He also guided the research of ten future recipients of the Nobel Prize.

The Neutron

Although Rutherford's calculations showed that the nucleus contains most of the mass, he could account for only about half of the atomic masses of various elements by using the number of protons (atomic number). This led him to predict the existence of a neutral particle with about the same mass as the proton. The neutral particle proved hard to find because the usual methods of detecting small particles were based on detecting charged particles. In 1932, James Chadwick, one of Rutherford's former students, designed an experiment that produced neutrons by a nuclear reaction (see Chapter 5) and then detected them by having the neutrons knock hydrogen ions, a detectable species, out of paraffin.

Isotopes

Isotopes are atoms of the same element having different numbers of neutrons.

Experiments have shown that atoms of the same element may not weigh the same. Atoms of the same element with different masses are called **isotopes.** Isotopes have the same number of protons and electrons but a different number of neutrons. The different isotopes of an element have a different **mass number:** the number of protons plus neutrons. A notation frequently used to show atomic mass and atomic number is shown below:

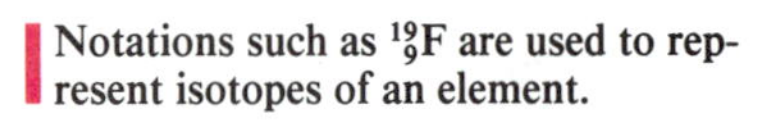

Notations such as $^{19}_{9}F$ are used to represent isotopes of an element.

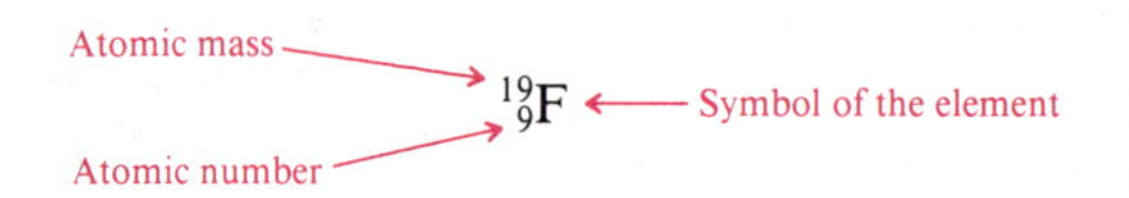

(For an atom of fluorine, $^{19}_{9}F$, the number of protons is 9, the number of electrons is also 9, and the number of neutrons is $19 - 9 = 10$.)

The number of neutrons in a nucleus can vary from zero (99.8% of hydrogen atoms have no neutrons) to more than 150 in heavy atoms such as curium, the element named after Marie Curie. There are more than 1000 known isotopes of the 109 elements, and many of them are produced artificially. The first element, hydrogen, is the only element that has different names for its isotopes: hydrogen (sometimes called protium), deuterium, and tritium. Tritium is radioactive.

Some really heavy atoms have over 155 neutrons in the nucleus.

$^{1}_{1}H$	$^{2}_{1}D$	$^{3}_{1}T$
Hydrogen	Deuterium	Tritium

Other elements use the name of the element followed by the mass number. For example, two important isotopes of uranium are uranium-235 and uranium-238.

Niels Bohr (1885–1962) received the Nobel Prize in 1922.

The **atomic weight*** of an element represents a weighted average of the isotopes found in a naturally occurring sample of the element. For example, a natural sample of neon gas is a mixture of three isotopes of neon: $^{20}_{10}Ne$, $^{21}_{10}Ne$, and $^{22}_{10}Ne$. Analysis of natural samples of neon gas gives the following relative abundance for the three isotopes: 90.92%, 0.26%, and 8.82%, respectively. The atomic weight value of 20.179 for neon is based on these percentages.

All atomic weights and isotopic masses are referenced to the weight of 1 mole of carbon-12 atoms. Carbon-12 is one of the carbon isotopes. In fact, by international agreement, the SI system defines the mole as the number of carbon-12 atoms in *exactly* 12 g of this isotope. The number of atoms in a mole has been determined in several different ways, and there is good agreement that there are 6.022×10^{23} atoms in exactly 12 g of the carbon-12 isotope.

Review the discussion of moles in Chapter 2.

The **atomic mass unit** (amu) is used when the masses of individual atoms, ions, molecules, or even subatomic particles are described. One carbon-12 atom has a mass of 12 amu, and one atom of deuterium has a mass of 2 amu. Just as 1 mole of carbon-12 atoms weighs 12 g, 1 mole of deuterium atoms weighs 2 g. One mole of atomic mass units equals 1 g (6.022×10^{23} amu = 1.000 g).

Max Planck (1858–1947) was awarded the Nobel Prize in 1918.

Where Are the Electrons in Atoms?

The Bohr theory of the atom, proposed in 1913 by Niels Bohr for the hydrogen atom, is still a useful model for representing the relative positions and energies of electrons in an atom. In Bohr's concept, electrons travel around the nucleus in definite orbits (energy levels), much as planets revolve around the sun. With brilliant imagination, Bohr applied a little algebra, some classical mathematical equations of physics, and Planck's **quantum theory** to his tiny solar-system model of the hydrogen atom. Bohr assumed that only a few allowable orbits are available in which electrons can move around the nucleus. He expressed the energy differences between any two

In 1900 Planck proposed that energy is not continuous but comes in discrete "bundles" or "packets" called quanta.

* Although *atomic mass* would be more correct, chemists usually use the term *atomic weight* to represent the weighted average of the isotopes found in a naturally occurring element. We will use *mass number* when referring to individual isotopes and *atomic weight* when referring to the weighted average "mass" for all of the naturally occurring isotopes of a particular element.

THE WORLD OF CHEMISTRY

The Scanning Tunneling Microscope

Chemists and physicists had more than ample evidence of the existence of atoms before the invention of the scanning tunneling microscope (STM). They were able to "see" atoms through a large variety of phenomena, but to say that they saw atoms had a special meaning. What they were seeing by such techniques as X-ray diffraction was a manifestation of many atoms and a composite picture due to the scattering of X rays from many planes of a crystal.

Yet chemists and physicists have always dreamed of being able to see individual atoms directly, that is, of being able to produce images with a direct correspondence to the atom's actual position in the sample.

These dreams began to be realized in the 1950s. An early and spectacular effort was reported by Erwin Mueller using a field ion microscope that he invented, which made it possible to image individual atoms on a crystal's surface. The even more remarkable STM not only makes it possible to see individual atoms and how they are arranged on a surface, but also permits the study of atom migration and atomic dislocations on surfaces. The development of the STM is considered an event of such magnitude that its developers, Gerd Binnig and Heinrich Rohrer of IBM's Zurich Research Laboratory in Switzerland, received the Nobel Prize in physics in 1986. The STM is an astonishing device because of its inherent simplicity. It consists of a tungsten needle, hardly more than a single atom wide at the end. When this needle is lowered to within a few atoms thickness of the surface to be imaged and a small voltage is applied, electrons tunnel, that is, they pass from the tungsten atom into the electron clouds of the atoms on the surface and produce a measurable current. By adjusting the up–down position of the tungsten needle as it moves across the surface, a constant tunneling current is maintained. As this takes place, however, the positions of the atoms are actually measured, giving a picture of the atomic landscape.

The potential for studying materials using the STM is great, particularly in the area of catalysts. Recently, STM studies have also been able to see actual amino acid molecules on the surfaces of crystals. Amino acids are the basic building blocks of all living matter.

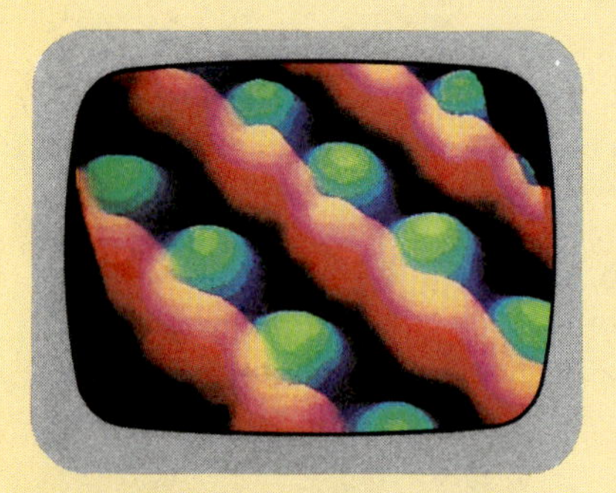

STM image of gallium arsenide. (*The World of Chemistry*, Program 6)

The World of Chemistry (Program 6) "The Atom."

Bohr assumed that atoms can exist only in certain energy states.

orbits in terms of quanta. Bohr was able to use his model to explain the spectrum of hydrogen, but it gave only approximate agreement with spectra of atoms with more than one electron. Later models of the electron have been more successful by considering electrons as having both particle and wave characteristics. This led to mathematical treatment of the locations of electrons as **probabilities** rather than as the precise locations envisioned by Bohr. In this text, we will use the Bohr model as an approximation of the energy levels for electrons in atoms.

Bohr used the term *orbits*, but they are now called *energy levels*.

The periodic table in Figure 3–7 gives the arrangement of electrons in Bohr energy levels for each element. The maximum number of electrons per energy level is $2n^2$, where n is the number of the energy level. This gives numbers of 2, 8, 18, 32, and 50 for the maximum number of electrons in energy levels 1 through 5, respectively. The last energy level listed for the elements in Figure 3–7 is the highest energy level, with electrons the greatest distance from the nucleus. These are the most important electrons in the study of chemistry because they are the ones that interact when atoms react with each other. For example, look at phosphorus (P). The arrangement of electrons is 2, 8, 5. This means that the stable state of the phosphorus atom has electrons in three energy levels. The one closest to the nucleus has two electrons; the second energy level has eight electrons; and the highest energy level has five electrons. The energy level with five electrons is farthest from the nucleus, and these are the electrons most available for interactions with outer electrons of other atoms in chemical reactions.

SELF-TEST 3A

1. Atoms were first proposed (a) by early British philosophers, (b) by early Greek philosophers, (c) in the early 1900s.
2. How many subatomic particles are necessary to describe the chemical and physical properties of atoms? (a) over 60, (b) 2, (c) only 1, (d) 3
3. What is the charge on the proton? (a) −1, (b) +1, (c) 3
4. If an atom has an atomic number of 10, then it has ____ protons and ____ electrons. If its mass number is 21, then it has ____ neutrons.
5. Henri Becquerel discovered (a) natural radioactivity, (b) isotopes, (c) radium.
6. A beta particle is a high-energy (a) neutron, (b) electron, (c) proton, (d) none of these.
7. Ernest Rutherford discovered (a) isotopes, (b) natural radioactivity, (c) the nucleus, (d) all of these.
8. The mass of an atom is concentrated in its (a) nucleus, (b) electrons, (c) neither of these.
9. The names of most isotopes are given by the element name followed by the mass number, like uranium-235. Use this naming method and name tritium ____________ and deuterium ____________.
10. In the symbol, $^{19}_{9}F$, the number 9 is the ______________________, and the number 19 is the ______________________.
11. The maximum number of electrons in the third (n = 3) energy level of an atom is (a) 3, (b) 9, (c) 18.

PERIODIC TABLE

In Chapter 2 we learned that atoms of elements are the building blocks of all matter. From Figure 2–6 we see that all matter contains either elements or compounds in pure or mixed form. This means that the more than 10 million compounds currently known and the thousands of new compounds made every day are a result of chemical reactions of the elements shown in the periodic table (Fig. 3–7). Fortunately, the chemistry of the elements can be organized and classified in a way that has helped both chemists and nonchemists. The periodic table is the single most important classification system of the elements because it summarizes, correlates, and predicts a wealth of chemical information.

Development of the Periodic Table

On the evening of February 17, 1869, at the University of St. Petersburg in Russia, a 35-year-old professor of general chemistry, Dimitri Ivanovich Mendeleev (1834–1907) was writing a chapter of his soon-to-be-famous textbook on chemistry. He had the properties of each element written on cards, with a separate card for each element. While he was shuffling the cards trying to gather his thoughts before writing his manuscript, Mendeleev realized that if the elements were arranged in the order of their atomic weights, there was a trend in properties which repeated itself several times! Thus the periodic law and table were born, although only 63 elements had been discovered by 1869 (for example, the noble gases were not discovered until after 1893), and the clarifying concept of the atomic number was not known until 1913. Mendeleev's idea and textbook achieved great success, and he rose to a position of prestige and fame as he continued to teach at St. Petersburg.

By 1871, Mendeleev published a more elaborate periodic table (Fig. 3–6). This version was the forerunner of the modern table currently seen in classrooms and textbooks. Two features of the 1871 version were especially interesting. Empty spaces were left in the table, and there was a problem with the positions of tellurium (Te) and iodine (I). The empty spaces showed the genius and daring of Mendeleev. He left the empty spaces to retain the rationale of ordered arrangement based on periodic recurrence of the properties. For example, in atomic weight order are copper (Cu), zinc (Zn), and then arsenic (As). If arsenic had been placed next to zinc, arsenic would have fallen under aluminum (Al). But arsenic forms compounds similar to those formed by phosphorus (P) and antimony (Sb), not aluminum. Mendeleev reasoned that two as yet undiscovered elements existed and moved arsenic over two spaces to the position below phosphorus. The two missing elements were soon discovered: gallium (Ga) in 1875 and germanium (Ge) in 1886. In later years the gaps in this 1871 periodic table were filled as the predicted elements were discovered.

Dmitri Mendeleev (1834–1907). Born in Siberia, Mendeleev rose to Professor of Chemistry at St. Petersburg and then to director of the Russian Bureau of Weights and Measures. Although a prolific writer, a versatile chemist and inventor, and a popular teacher, the fame of this brilliant scientist rests on his discovery of the periodic law. (*The World of Chemistry,* Program 7)

Mendeleev aided the discovery of the new elements by predicting their properties with remarkable accuracy, and he even suggested the geographical regions in which minerals containing the elements could be found. The properties of a missing element were predicted by consideration of the properties of its neighboring elements in the table. An example of Mendeleev's prediction of the properties of an undiscovered element is shown in Table 3–3. The term *eka* comes from Sanskrit and means "one"; thus, *ekasilicon* means "one place away from silicon." He also predicted the properties of ekaboron (scandium) and ekaaluminum (gallium).

The empty spaces in the table and Mendeleev's predictions of the properties of missing elements stimulated a flurry of prospecting for elements in

	Group I R_2O RCl	Group II RO RCl_2	Group III R_2O_3 RCl_3	Group IV RO_2 RCl_4	Group V R_2O_5 RH_3	Group VI RO_3 RH_2	Group VII R_2O_7 RH	Group VIII RO_4
1	H = 1							
2	Li = 7	Be = 9.4	B = 11	C = 12	N = 14	O = 16	F = 19	
3	Na = 23	Mg = 24	Al = 27.3	Si = 28	P = 31	S = 32	Cl = 35.5	
4	K = 39	Ca = 40	— = 44	Ti = 48	V = 51	Cr = 52	Mn = 55	Fe = 56, Co = 59 Ni = 59, Cu = 63
5	(Cu = 63)	Zn = 65	— = 68	— = 72	As = 75	Se = 78	Br = 80	
6	Rb = 85	Sr = 87	?Yt = 88	Zr = 90	Nb = 94	Mo = 96	— = 100	Ru = 104, Rh = 104 Pd = 106, Ag = 108
7	(Ag = 108)	Cd = 112	In = 113	Sn = 118	Sb = 122	Te = 125	I = 127	
8	Cs = 133	Ba = 137	?Di = 138	?Ce = 140	—	—	—	— — — —
9	(—)	—	—	—	—	—	—	
10	—	—	?Er = 178	?La = 180	Ta = 182	W = 184	—	Os = 195, Ir = 197 Pt = 198, Au = 199
11	(Au = 199)	Hg = 200	Tl = 204	Pb = 207	Bi = 208	—	—	
12	—	—	—	Th = 231	—	U = 240	—	— — — —

Figure 3–6 An 1871 version of Mendeleev's periodic table. The formulas for simple oxides, chlorides, and hydrides are shown under each group heading. R represents the element in each group.

the 1870s and 1880s. As a result, gallium (Ga) was discovered in 1875; scandium (Sc), samarium (Sm), holmium (Ho), and thulium (Tm) in 1879; gadolinium (Gd) in 1880; neodymium (Nd) and praseodymium (Pr) in 1885; and germanium (Ge) and dysprosium (Dy) in 1886. Many of these elements are not common even today, yet they are important as ingredients in catalysts and color television screens.

TABLE 3–3 Some of Mendeleev's Predicted Properties of Ekasilicon and the Corresponding Observed Properties of Germanium

	Ekasilicon (Es)	Germanium (Ge)
Atomic weight	72	72.6
Color of element	Gray	Gray
Density of element (g/mL)	5.5	5.36
Formula of oxide	EsO_2	GeO_2
Density of oxide (g/mL)	4.7	4.228
Formula of chloride	$EsCl_4$	$GeCl_4$
Density of chloride (g/mL)	1.9	1.844
Boiling point of chloride (°C)	Under 100	84

If Mendeleev had followed the atomic weight order precisely, some elements with similar properties would not have been in the same column. In the 1869 table, tellurium (Te) with an atomic weight of 128 was placed one position ahead of iodine (I), which has a lower atomic weight of 127. On the basis of its chemical properties, Te belonged with Sb, S, and O, and I belonged with F, Cl, and Br.

Mendeleev believed the atomic weight of tellurium was in error, but this was later shown not to be the case. In the 1871 table, the weight of Te had been changed from 128 to 125—an example of the unwise practice of changing data to fit a theory. The record is not clear as to why he changed the value. Other reversed pairs in the modern periodic table are uranium (U) before neptunium (Np), argon (Ar) before potassium (K), cobalt (Co) before nickel (Ni), and thorium (Th) before protactinium (Pa). When the atomic number concept was set forth in 1913, the question was resolved.

Building on the work of Mendeleev and others and using the clarifying concept of the atomic number, we are now able to state the modern periodic law:

When elements are arranged in the order of their atomic numbers, their chemical and physical properties show repeatable, or periodic, trends.

THE WORLD OF CHEMISTRY

The Periodic Chart

Among the most significant contributions to the modern periodic chart is that made by Nobel Laureate Glenn Seaborg. Among other things he demonstrated the importance of maintaining the courage of one's convictions.

Thanks to his insights, it is now very well established that the transuranium elements (atomic numbers greater than 92), a number of which he either discovered or helped to discover during the Manhattan Project, are members of the actinide series. Actinides are the elements following actinium and belonging in a grouping off the main periodic chart.

Until Seaborg offered his version of the periodic table, chemists were convinced that Th, Pa, and U belonged in the main body of the table, Th under Hf, Pa under Ta, and U under W. When Seaborg proposed that Th was the beginning of the actinides and that the transuranium elements belonged as a group under the rare earths, some prominent and famous inorganic chemists, many of them Seaborg's friends, tried to discourage his publication of this finding in the open literature. One very prominent inorganic chemist felt that he would ruin his scientific reputation. Nevertheless Seaborg, strongly convinced, persisted and as a result, properly placed this most important class of elements where they are today. Based on Seaborg's expansion of the periodic table, it was possible to predict accurately the properties of many of the as yet undiscovered transuranium elements. Subsequent preparation in atomic accelerators of these elements proved him right, and it was fitting that he was awarded the Nobel Prize in 1951 for his outstanding work.

The World of Chemistry (Program 7) "The Periodic Table."

Glenn Theodore Seaborg (1912–) began his college education as a literature major but changed to science in his junior year at the University of California. For his preparation and discovery of several transuranium elements, he shared the 1951 Nobel Prize in chemistry with E. M. McMillan (1907–), who started Seaborg in this area of research. (*The World of Chemistry,* Program 7)

The pattern in the properties of the elements, then, is *periodic;* hence the name *periodic law* or *table.* Other familiar periodic phenomena include the average daily temperature, which is periodic with time in a temperate climate. A shingle roof has the same pattern over and over and is, therefore, periodic.

So, to build up a periodic table according to the periodic law, line up the elements in a horizontal row in the order of their atomic numbers. Every time you come to an element with similar properties to one already in the row, start a new row. The columns then contain elements with similar properties.

A shingled roof with a repeatable pattern is one of many periodic phenomena.

Features of the Modern Periodic Table

Note the following features in the periodic table in Figure 3–7. The vertical columns are called **groups.** The A groups are the **representative** or **main group** elements. The B groups are the **transition** elements that link the two areas of representative elements. The **inner transition** elements are the lanthanide series and the actinide series. The inner transition elements are placed at the bottom of the periodic chart because the similarity of properties within the two series would require their placement between lanthanum and hafnium (lanthanide series) and between actinium and element 104 (actinide series).

Although this text uses the A and B group labels as described above (the American version), other group labels are used. For example, Group IIIA in European usage is Group IIIB in American practice. Since 1959, the International Union of Pure and Applied Chemistry (IUPAC) has been considering the differences between the European and American practices of labelling A and B groups. They have recommended that groups be labelled 1 through 18 consecutively from left to right. The periodic table on the inside back cover includes both 1 through 18 and A and B group labels (American version) for comparison. The preference for the American version in introductory chemistry texts is to allow the use of "A" for all "representative" elements and "B" for all "transition" elements. As a result, the group number of the "A" column is the number of electrons in the highest energy level for atoms of representative elements. This rule does not apply to the atoms of transition elements.

IUPAC resolves subjective issues related to chemistry.

The horizontal rows are called **periods.** The periods are not equal in size because the maximum number of electrons per energy level increases as the distance of the energy level from the nucleus increases. Periods one through seven have 2, 8, 8, 18, 18, 32, and 23 (incomplete) elements, respectively.

Eighty-four of the elements are **metals** (blue in Fig. 3–7) and are found in Groups IA, IIA, and IIIA, and the B groups. Characteristic physical properties of metals include malleability, ductility, and good conduction of heat and electricity. As we will see from the discussion of chemical bonding, chemical properties of metals are based on their tendency to give up electrons to form positive ions when they react with nonmetals.

Malleability **of metals means that they can be hammered into thin sheets.**

Ductile**—can be drawn into wires.**

Seventeen elements are **nonmetals** (in yellow), and, except for hydrogen, they are found in the upper right-hand corner of the periodic table. Hydrogen is shown in Group IA because its atoms have one electron. However, hydrogen is a nonmetal. For this reason, some versions of the periodic table

Hydrogen—the element without a home on the periodic table.

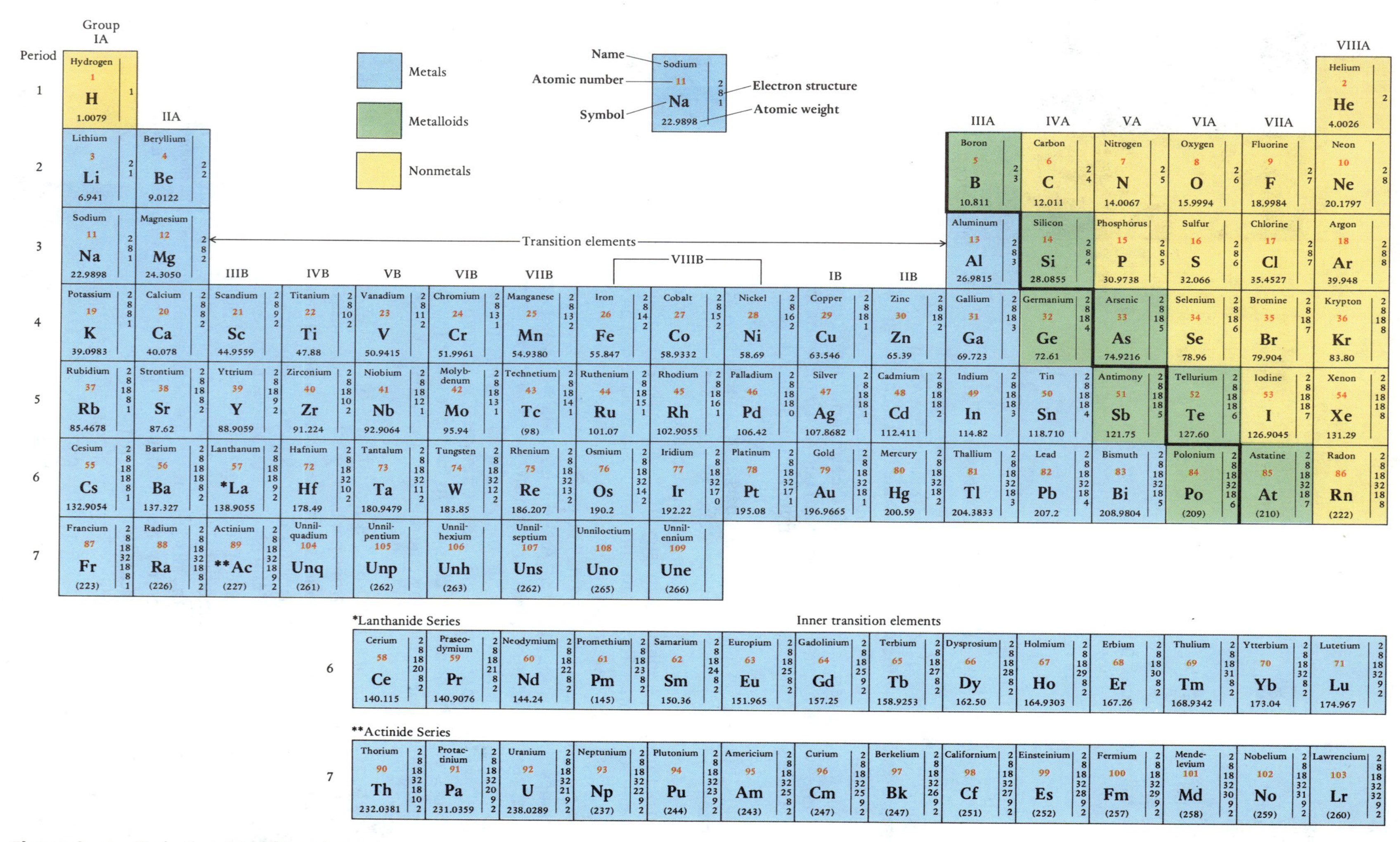

Figure 3–7 Periodic table of the elements.

place hydrogen in a box by itself centered above the other groups. The physical and chemical properties of nonmetals are opposite those of metals. For example, nonmetals are insulators; that is, they are extremely poor conductors of heat and electricity. Chemically, nonmetals have a tendency to gain electrons to form negative ions when they react with metals. Nonmetals also react with other nonmetals to form molecules.

Elements that border the staircase line in Figure 3–7 between metals and nonmetals are eight **metalloids** (in green). Their properties are intermediate between those of metals and nonmetals. For example, Si, Ge, and As are **semiconductors.** Semiconductors conduct electricity less than metals such as silver and copper but more than insulators such as sulfur.

Semiconductors are components of transistors.

The six **noble gases** in Group VIIIA have little tendency to undergo chemical reactions.

Elements in a Group Have Similar Properties

Some properties of elements in a group differ by degree in a regular pattern. For example, the melting points of Group IA elements beginning with Li and going down the column through Cs are (in °C) 179, 98, 64, 39, and 28, respectively. Lithium reacts slowly with water, sodium reacts faster, potassium still faster, and for the elements at the bottom of the group, mere exposure to moist air produces a vigorous explosion (Fig. 3–8).

Some properties differ by degree but not in a regular pattern. For example, the densities (in g/mL) of the solids Li through Cs are 0.53, 0.97, 0.86, 1.53, and 1.87, respectively.

$$\text{Density} = \frac{\text{mass}}{\text{volume}}$$

Some properties are the same for every member of a group. Elements in a group generally react with other elements to form similar compounds. *This is the most useful and powerful inference that can be made from the periodic table.* For example, if the formula for the compound composed of Li and Cl is LiCl, then probably there is a compound of Rb and Cl with the formula RbCl; the compound for Rb and Br would probably have the formula of RbBr. Likewise, if the formula Na_2O is known, then a compound with the formula K_2S predictably exists.

Figure 3–8 (a) Sodium reacts with water. (b) Potassium is a more active metal, reacting more vigorously with water. (*The World of Chemistry,* Program 7)

(a)

(b)

The Periodic Table and Chemical Behavior

Why do elements in the same group in the periodic table have similar chemical behavior? Why do metals and nonmetals have different properties? These relationships and properties exist because all of the elements in a group (particularly the representative Group A elements and the noble gases) have atoms with similar structural features. Note that in Group IA each element has atoms with one and only one electron in the outermost (highest) occupied energy level (electrons in the highest occupied energy level are called **valence electrons**). Atoms of Group IIA elements all have two valence electrons. Group IIIA atoms have three valence electrons, and the pattern continues through Group VIIIA.

The group number for A groups is the number of valence electrons.

Reactions of elements with one another to form compounds involve either transferring valence electrons to form **ionic compounds** or sharing valence electrons to form **covalent compounds.** In other words, *the chemical view of matter is primarily concerned with what valence electrons are doing because they are the ones involved in forming chemical bonds.*

CHEMICAL BONDS

How can we use the periodic table to predict what happens to valence electrons when atoms react? Probably the best place to start is to examine the structural feature of the noble gases which results in their having little or no chemical reactivity. The two valence electrons of helium and the eight valence electrons of the other noble gases seem to provide a balanced, stable electron arrangement that minimizes the tendency of a noble gas atom to react with other atoms. Perhaps if other elements achieved the same electron arrangement of a noble gas by losing, gaining, or sharing electrons, they would also be more stable. This idea occurred to chemists in the early 1900s and is known as the **octet rule.** When elements react, they tend to lose, gain, or share electrons to achieve the same electron arrangement as the noble gas nearest to them in the periodic table. Metals can achieve a noble gas electron arrangement by giving up electrons, and nonmetals can achieve a noble gas electron arrangement by adding or sharing electrons.

Ionic Bonds

When one or more electrons are transferred from a metal atom to a nonmetal atom, a positively charged metal ion and a negatively charged nonmetal ion are formed. The strong electrostatic attraction between the positive and negative ions is known as the **ionic bond.** If valence electrons are represented by dots, the formation of sodium chloride, common table salt, from sodium and chlorine atoms can be depicted as

Ionic bonds are the attractions between ions of opposite charge.

$$\mathrm{Na}\cdot + \cdot\ddot{\underset{\cdot\cdot}{\mathrm{Cl}}}\!: \longrightarrow \mathrm{Na}^+ + :\ddot{\underset{\cdot\cdot}{\mathrm{Cl}}}\!:^-$$

A chemical formula of a compound is electrically neutral, so the formula of sodium chloride is NaCl. (Note that the ionic charges are not indicated in the

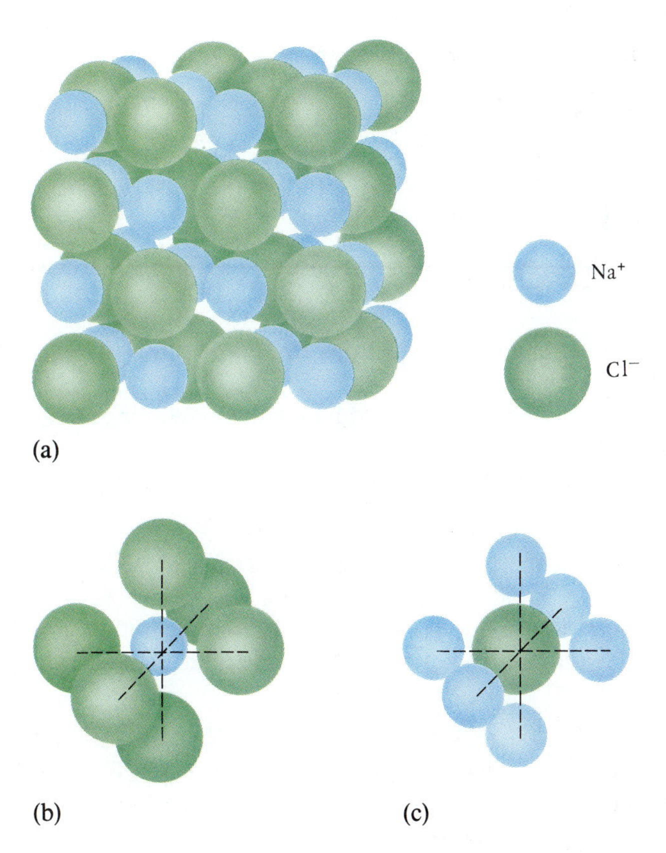

Figure 3-9 (a) Model of sodium chloride crystalline lattice. (b) Each Na^+ is surrounded by 6 Cl^-. (c) Each Cl^- is surrounded by 6 Na^+.

formula.) In ionic compounds the simplest ratio of oppositely charged ions that gives an electrically neutral unit is called a **formula unit.** The formula unit for sodium chloride is NaCl. Ionic crystals are made up of large numbers of formula units to form a regular three-dimensional crystalline lattice. A model of the sodium chloride crystalline lattice is shown in Figure 3-9. Note that each Na^+ ion has six Cl^- ions around it. Similarly, each Cl^- ion has six Na^+ ions around it. In this way, the one-to-one ratio of the singly charged ions is preserved. There is no *unique molecule* in ionic structures; no particular ion is attached exclusively to another ion, but each ion is attracted to all the oppositely charged ions surrounding it.

Ionic compounds have no molecules.

An ionic structure is a regular geometrical array of ions.

When atoms become ions, properties are drastically altered. For example, a collection of Br_2 molecules is red, but bromide ions (Br^-) contribute no color to a crystal of a compound such as NaBr. A chunk of sodium atoms (Fig. 3-8a) is soft, metallic, and violently reactive with water, but Na^+ ions are stable in water. A large collection of Cl_2 molecules constitutes a greenish-yellow poisonous gas, but chloride ions (Cl^-) produce no color in compounds and are not poisonous. For example, we use table salt (NaCl) for seasoning food. When atoms become ions, atoms obviously change their nature.

Formulas of thousands of ionic compounds can be predicted by using the periodic table and the following procedure.

1. Form a positive ion from a metal in an A group by removing the number of electrons equal to the group number.
2. Form a negative ion from a nonmetal in an A group by adding the number of electrons to the group number to give a total of eight.
3. Write the formula unit for the ionic compound that gives the simplest ratio needed to produce an electrically neutral unit.

For example, the formula of a compound of magnesium and nitrogen would be determined as follows:

1. Mg is in Group IIA, so remove two electrons from a Mg atom to give Mg^{2+}.
2. N is in Group VA, so add three electrons to give eight; the resulting ion is N^{3-}.
3. The lowest common denominator for 2 and 3 is 6; the formula is Mg_3N_2.

Let's check to make sure this formula is electrically neutral. Mg ion is $+2$ and three ions would give a total of $+6$ charge; N ion is -3 and two ions would give a total of -6 charge; $+6$ and $-6 = 0$.

Some simple, stable ions with a noble-gas electron configuration are listed in Table 3-4, and a memory aid tied to the periodic table is given in Figure 3-10 for ions formed by the elements.

Predictions for elements in A groups can be summarized as follows:

Positive ions are formed when metal atoms lose one electron (Group IA), two electrons (Group IIA), or three electrons (Group IIIA) to nonmetal atoms. The resulting ions have the same electron arrangement as a noble gas.

Nonmetals in the presence of metals tend to gain one, two, or three electrons from metal atoms to form negative ions that have all valence electrons paired and have the stable eight-electron arrangement of noble gases.

The group number is not reliable for predicting the positive ions formed by transition metals and inner-transition metals. The elements in Group IVA that form covalent compounds are discussed later in this chapter.

Note how Table 3-4 relates to the periodic table (Fig. 3-7).

TABLE 3-4 Electron Configurations of the Noble Gases and Ions with Identical Configurations

Species	Configuration
He, Li^+, Be^{2+}, H^-	2
Ne, Na^+, Mg^{2+}, F^-, O^{2-}	2-8
Ar, K^+, Ca^{2+}, Cl^-, S^{2-}	2-8-8
Kr, Rb^+, Sr^{2+}, Br^-, Se^{2-}	2-8-18-8
Xe, Cs^+, Ba^{2+}, I^-, Te^{2-}	2-8-18-18-8

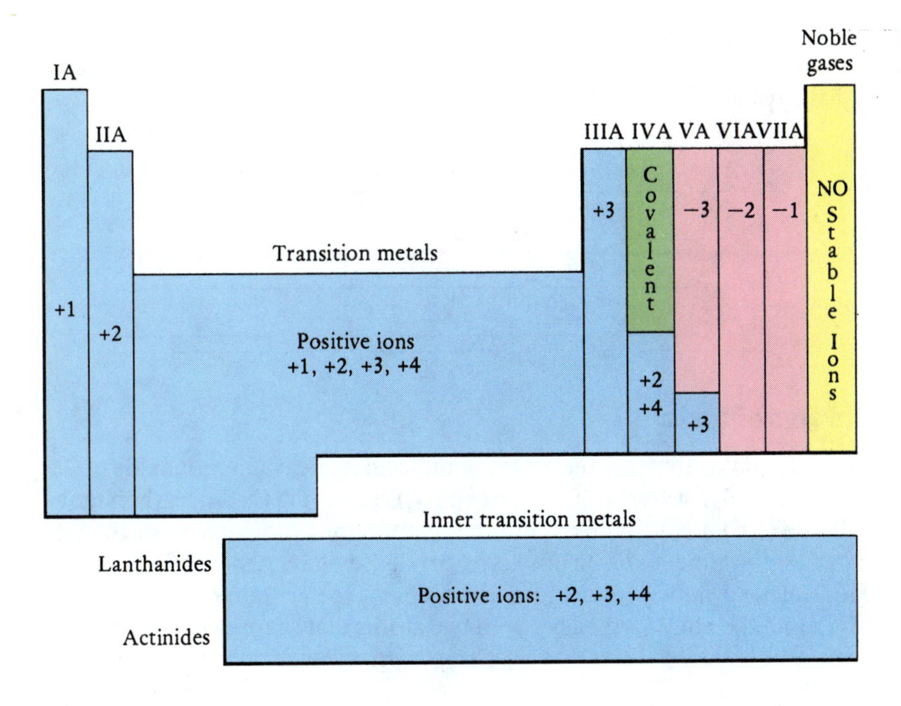

Figure 3–10 The periodic table and the formation of ions.

SELF-TEST 3B

1. Mark the following as transition elements (T), representative elements (R), noble gases (N), or inner transition elements (I):
   Be ____, P ____, Cr ____, Kr ____, Am ____.
2. Who was primarily responsible for formulating the periodic table?
   ____
3. According to the periodic law, when the elements are arranged in the order of their ________________, their properties show periodicity.
4. When a phenomenon shows the same pattern over and over, we say the pattern is ________________.
5. Charged atoms are called ________.
6. The attraction between positive and negative ions produces a(an) ________ bond.
7. A potassium atom loses ________ electron(s) in achieving a noble gas configuration.
8. What is the correct formula for calcium iodide (Ca^{2+} and I^{-})?
   ____
9. Which ion gained an electron in its formation: Na^{+} or Cl^{-}?
   ____
10. Electrons in the outer orbit may be called ________ electrons.
11. Positive ions are formed from neutral atoms by ( ) losing or ( ) gaining electrons.
12. Negative ions are formed from neutral atoms by ( ) losing or ( ) gaining electrons.

G. N. Lewis (1875–1946) proposed in 1916 that a *covalent bond* results from sharing an electron pair. He was a professor of chemistry at University of California, Berkeley. Many of his ideas about bonding are still applicable today.

13. Use the periodic table to predict what ions will form when the following metal-nonmetal pairs react. List the ions and the formula for the resulting compound in the appropriate column.

	Ion		Ion	Formula of Compound
Rb	______	S	______	______
Ca	______	O	______	______
Mg	______	P	______	______

Covalent Bonds

What holds together the atoms in molecules of carbon monoxide (CO), methane (CH_4), water (H_2O), quartz (SiO_2), ammonia (NH_3), carbon tetrachloride (CCl_4), and millions of other compounds in which all of the elements are nonmetals? The answer is **covalent bonds.** Atoms of nonmetals can achieve a noble gas electron arrangement (octet rule) by sharing electrons, and the bond formed between two atoms that share electrons is called a **covalent bond.**

Single Covalent Bonds

A single covalent bond is formed when two atoms share a single pair of electrons. The simplest examples are diatomic (two-atom) molecules such as H_2 (hydrogen), F_2 (fluorine), and Cl_2 (chlorine).

A hydrogen atom has one electron. If a hydrogen atom can share its electron with another atom that has an unpaired valence electron, a stable pairing of the two electrons can be achieved and the hydrogen atom can then have the electronic structure of helium, a noble gas. This arrangement can be achieved by two hydrogen atoms sharing their single electrons. The electron dot formula for the H_2 molecule is

To break a bond requires energy; when bonds are formed, energy is released.

$$2\,H\cdot \longrightarrow H{:}H + \text{energy}$$

Atoms Molecule

Since each fluorine atom has one unpaired electron ($:\ddot{\underset{..}{F}}\cdot$) two fluorine atoms also can share an electron each to form a single covalent bond and an F_2 molecule.

$$2\,:\ddot{\underset{..}{F}}\cdot \longrightarrow :\ddot{\underset{..}{F}}:\ddot{\underset{..}{F}}: + \text{energy}$$

Atoms Molecule

Electrons are all the same. The x's and dots distinguish the sources of the identical electrons.

Only the pair of electrons represented between the two symbols (the two Fs) are bonding electrons. The other pairs of electrons are called **nonbonding** valence electrons.

In a water molecule, two O—H single covalent bonds are formed. An oxygen atom has six valence electrons, of which two are unpaired ($\cdot\ddot{\underset{\cdot}{O}}:$). It needs two more electrons to pair up its electrons and produce the stable octet of a noble gas. Two hydrogen atoms supply the two electrons.

$$:\ddot{O}\cdot + 2H\times \longrightarrow :\ddot{O}\overset{\times}{\cdot}H + \text{energy}$$
$$\text{(with H below O)}$$

Water

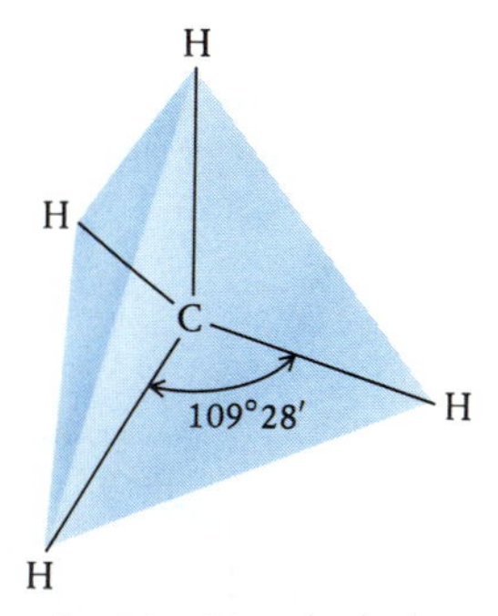

Figure 3–11 Tetrahedral geometry for methane.

An ammonia (NH_3) molecule has three N—H single covalent bonds. A nitrogen atom has five valence electrons, of which three are unpaired ($\cdot\ddot{N}\cdot$). The atom needs three more electrons to pair up its electrons and give it the stable eight. Three hydrogen atoms supply the three electrons to form the NH_3 molecule.

$$\cdot\ddot{N}\cdot + 3H\times \longrightarrow H\overset{\times}{\cdot}\ddot{N}\overset{\times}{\cdot}H + \text{energy}$$
$$\text{(with H below N)}$$

Ammonia

As we will see in Chapters 6 and 7, compounds of carbon and hydrogen (the **hydrocarbons**) are an important class of compounds. A carbon atom has four valence electrons. It can satisfy the octet rule by sharing its four electrons with electrons from four hydrogen atoms. This gives the simplest hydrocarbon molecule, methane. For convenience, an electron pair bond is often indicated by a line.

$$\begin{array}{ccc} & H & \\ & \overset{\times}{\cdot} & \\ H \times\cdot & C & \cdot\times H \\ & \overset{\cdot}{\times} & \\ & H & \end{array} \quad \text{or} \quad \begin{array}{ccc} & H & \\ & | & \\ H - & C & - H \\ & | & \\ & H & \end{array}$$

The methane molecule has a **tetrahedral** shape, with the carbon atom at the center of the **tetrahedron** and the four hydrogen atoms at the corners (Fig. 3–11).

The four faces of a tetrahedron are equilateral triangles, and the angle between two lines drawn from the center to two corners is 109.5 degrees.

Multiple Bonding

An atom with fewer than seven electrons in its valence shell can form covalent bonds in two ways. The atom may share a single electron with each of several other atoms that can each contribute a single electron. This leads to **single** covalent bonds. But the atom can also share two (or three) pairs of electrons with a single other atom. In this case there will be two (or three) bonds between these two atoms. When two shared pairs of electrons join together the same two atoms, we speak of a **double bond,** and when three shared pairs are involved, the bond is called a **triple bond.** Examples of these bonds are found in many compounds, such as those shown in Figure 3–12.

One pair of electrons is a *single* covalent bond.

A *double bond* consists of two electron pairs shared between two atoms.

As we can see from these structures, molecules may contain several types of bonds. Thus, ethylene (Fig. 3–12) contains a double bond between the carbon atoms and single bonds between the hydrogen atoms and the carbon atoms.

$$\begin{array}{ccccc} H & & & & H \\ & \diagdown & & \diagup & \\ & & C{=}C & & \\ & \diagup & & \diagdown & \\ H & & & & H \end{array}$$

One line: single bond
Two lines: double bond

Figure 3–12 Electron dot structures of some molecules containing multiple bonds. Line structures are shown for comparison.

Formula	Name	Electron dot structure	Line structure
Double bonds:			
CO_2	Carbon dioxide	:Ö::C::Ö:	O=C=O
C_2H_4	Ethylene	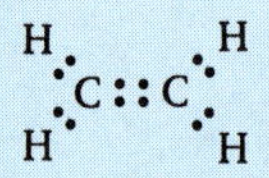 H:C::C:H (H above and below each C)	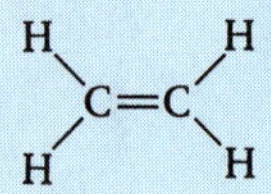 $H_2C{=}CH_2$
SO_3	Sulfur trioxide	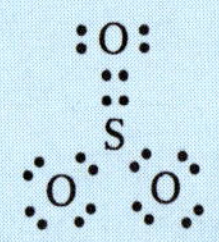 :O: :: S with two :O: bonded by single pairs	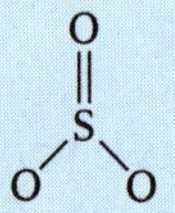 O=S with two O single-bonded
Triple bonds:			
N_2	Nitrogen	:N:::N:	N≡N
CO	Carbon monoxide	:C:::O:	C≡O
C_2H_2	Acetylene	H:C:::C:H	H—C≡C—H

Polar Bonds

In a *polar* bond, there is unequal sharing of bonding electrons.

In a molecule like H_2 or F_2, in which both atoms are alike, there is equal sharing of the electron pair, and the bond is a **nonpolar** covalent bond. Where two unlike atoms are bonded, however, the sharing of the electron pair is unequal and a **polar** covalent bond is formed. The more nonmetallic an element is, the more that element attracts electrons. Nonmetallic character increases across and up the periodic table toward fluorine, which is the most nonmetallic element. The bonds in HF, NO, SO_2, and H_2O are polar. The unequal sharing of the electron pair in a polar covalent bond causes the resulting molecule to have one end more positive than the other end. The water molecule is a good example. Because oxygen is more nonmetallic than hydrogen, oxygen has a greater attraction for the electron pair in the O—H

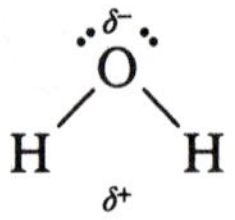

δ^- O (bonded to two H) δ^+

(The δ^- and δ^+ represent partial charges.)

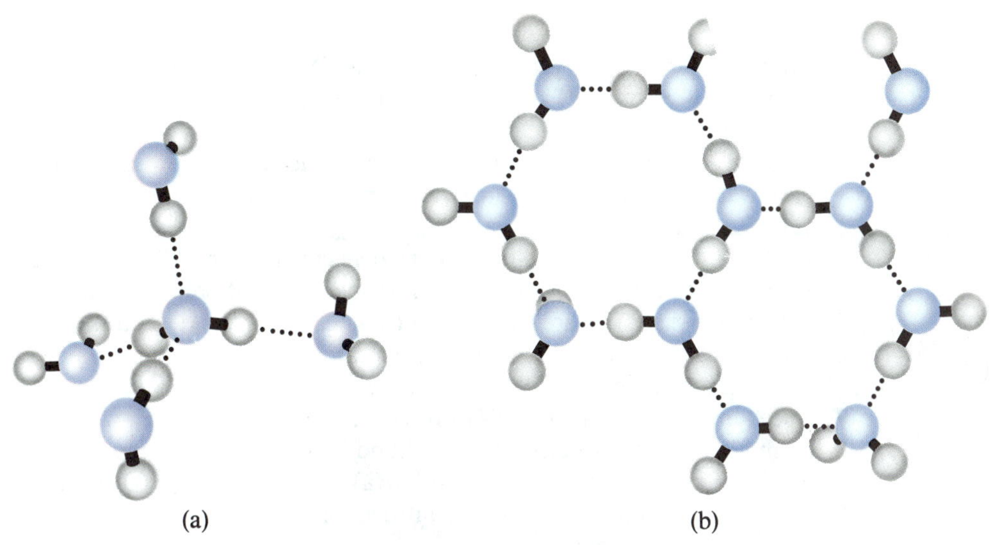

Figure 3–13 (a) Tetrahedral cluster of four water molecules around a fully hydrogen-bonded water molecule in the center. (b) Hydrogen bonding in the structure of ice. The hydrogen bonds are indicated by the dashed lines. In liquid water the hydrogen bonding is not as extensive as it is in ice. (*The World of Chemistry,* Program 12, "Water")

bond than does hydrogen. As a result, the oxygen end of the water molecule is more negative than the hydrogen end, and the water molecule is polar. Why is this important? Weak bonds between water molecules are formed by attraction of the positive end of one water molecule (hydrogen) to the negative end of another water molecule (oxygen). This is one example of **hydrogen bonding.** If it weren't for hydrogen bonding, water would be a gas at room temperature! Figure 3–13 illustrates how water molecules cluster together when hydrogen bonds are formed. Another important example is the hydrogen bonding found between base pairs in DNA (see Chapter 9).

Hydrogen bonding plays a key role in the chemistry of life.

Ionic Compounds with Polyatomic Ions

Two or more elements can also combine to form a polyatomic ion—a chemically distinct species with an electrical charge. Examples of polyatomic ions include hydroxide (OH^-), sulfate (SO_4^{2-}), phosphate (PO_4^{3-}), nitrate (NO_3^-), and ammonium (NH_4^+). The bonding between the atoms within these ions is covalent. However, compounds that contain these polyatomic ions are ionic, and formulas are written by the same procedure that was described above for Mg_3N_2. The only difference is that the polyatomic ion is enclosed in parentheses when the subscript is larger than one. Examples and some of their uses include magnesium hydroxide, $Mg(OH)_2$ (Milk of Magnesia); sodium phosphate, Na_3PO_4 (a food additive); calcium sulfate, $CaSO_4$ (gypsum, drywall); ammonium nitrate, NH_4NO_3 (fertilizers and explosives); and aluminum sulfate, $Al_2(SO_4)_3$ (water purification).

SELF-TEST 3C

1. Both hydrogen and oxygen are nonmetals. One compound they form is hydrogen peroxide (H_2O_2). Would you expect the HO bonds to be (a) ionic or (b) covalent?
2. A nitrogen atom has five electrons in its valence level. How many single covalent bonds would you expect a nitrogen atom to form?

   ____
3. How many electrons are shared in a single covalent bond? ________
4. How many electrons are shared in a double covalent bond? ________
5. How many electrons are shared in a triple covalent bond? ________
6. When a carbon atom is bonded to four chlorine atoms, what shape does the molecule have? ______________________
7. The element nitrogen consists of molecules of N_2. Since each nitrogen atom has five valence electrons, the bond between the nitrogen atoms in the nitrogen molecule must involve (a) 2, (b) 3, (c) 4, (d) 6 electrons.
8. The kind of covalent bond in the nitrogen molecule is called a ______________ bond.
9. The boiling point of water is about 200°C higher than would be predicted if ______________ ______________ were not present.

QUESTIONS

1. Explain how the discovery of natural radioactivity was at odds with Dalton's atomic theory.
2. Explain how the discovery of isotopes was at odds with Dalton's atomic theory.
3. Mendeleev's periodic table was arranged by what atomic property? What atomic property is used to arrange the modern periodic table?
4. Discuss the importance of Mendeleev's periodic table in the discovery of elements.
5. What is the difference between a group and a period in the periodic table?
6. Lithium reacts with water to form hydrogen gas and lithium ions in a basic solution. Write "word reactions" for the reactions between other Group IA elements and water.
7. Calcium (Ca) forms an oxide with the formula CaO. Sulfur is in Group VIA with oxygen. What is the formula for calcium sulfide?
8. What is the basis for giving importance to an octet of electrons in the bonding of atoms?
9. A sodium atom (atomic number 11) has one valence electron. Name two ways a sodium atom can achieve an octet of electrons in its outer energy level. Which one of these actually takes place?
10. In table salt, each sodium ion is surrounded by six chloride ions and each chloride ion is, in turn, surrounded by six sodium ions. What is the formula unit of table salt (sodium chloride)?
11. Write the nine possible formulas for the following positive ions combined with the negative ions. Positive ions: Na^+, Ca^{2+}, Al^{3+}. Negative ions: Cl^-, O^{2-}, N^{3-}.
12. Use the positive ions given in Question 11 and write the formula for their compounds with the polyatomic negative ions, OH^-; SO_4^{2-}; and PO_4^{3-}.
13. a. A carbon atom has four valence electrons. If a carbon atom forms a triple bond with another atom, how many single bonds can it form?
    b. If a carbon atom forms a double bond with another atom, how many single bonds can it form?
    c. If a carbon atom forms only single bonds with other atoms, how many single bonds can it form?
14. Use the information on the periodic chart to supply the following:
    a. The nuclear charge on cadmium (Cd)
    b. The atomic number of As
    c. The atomic mass (or mass number) of an isotope of Br having 46 neutrons
    d. The number of electrons in an atom of Ba
    e. The number of protons in an isotope of Zn
    f. The number of protons and neutrons in an isotope of Sr, atomic mass (or mass number) of 88
    g. An element forming compounds similar to those of Ga
15. There are more nonmetallic elements than metallic elements: True ( ) False ( )
16. Below are some selected properties of Li and K. Before

looking up the number, estimate values for the corresponding properties of Na.

	Lithium	Sodium	Potassium
Atomic weight	6.9	—	39.1
Density (g/mL)	0.53	—	0.86
Melting point (°C)	180	—	63.4
Boiling point (°C)	1330	—	757

17. Give the names and symbols for two elements most like selenium (Se).
18. What kind of bond (ionic, pure covalent, polar covalent) is likely to be formed by the following pairs of atoms?
    a. A Group IA element with a Group VIIA element
    b. A Group VIA element with a Group VIIA element
    c. Two Cl atoms
19. How is an ionic bond formed? How is a covalent bond formed?
20. How are ionic solids held together?
21. Why is water a liquid at room temperature?
22. Draw electron dot and line structures for
    a. nitrogen
    b. ethylene
    c. acetylene
    d. carbon dioxide
    e. carbon monoxide
23. Complete the following by writing the predicted formulas.

Element	F	O	Cl	S	Br	Se
Na						
K						
B						
Al						
Ga						
C						
Si						

James Abbott McNeill Whistler's *Nocturne in Black and Gold, the Falling Rocket* illustrates some of the more spectacular oxidation–reduction reactions. (The Detroit Institute of the Arts: Gift of Dexter M. Ferry Jr.)

CHAPTER

4

Chemical Reactions: Acids and Bases, Oxidation–Reduction

Many vitally important chemical reactions take place in liquid solutions. Two classes of these important reactions are acid/base reactions and oxidation/reduction reactions.

1. What are liquid solutions?
2. What is an electrolyte, and how is it different from a nonelectrolyte?
3. How is solution concentration measured?
4. What are acids and bases, and how important are they?
5. What is neutralization?
6. What does the term "pH" mean?
7. What causes indigestion, and how do antacids cure it?
8. What is oxidation?
9. What is reduction?
10. How do oxidation and reduction play a role in the chemistry of metals?
11. What is the chemistry of batteries and fuel cells?
12. What causes corrosion, and how can it be prevented?

The most important aspect of chemistry is chemical change. There are many types of chemical reactions, but two of the most important are those involving the exchange of protons (called acid–base reactions) and those involving the exchange of electrons (called oxidation–reduction reactions). These two classes of reactions are important in chemical reactions used in industry and in those going on in living organisms. Many acid–base and oxidation–reduction reactions take place in solutions.

LIQUID SOLUTIONS

Solution: homogeneous (uniform) mixture of atoms, ions, or molecules.

Recall from Chapter 2 that a solution is a homogeneous mixture. A liquid solution, then, is a uniform distribution of one substance in another, with the mixture having the properties of a liquid.

How many liquid solutions are familiar to you? How about sugar or salt dissolved in water, or oil paints dissolved in turpentine, or grease dissolved in gasoline? In each of these solutions, the substance present in the greater amount, the liquid, is the **solvent,** and the substance dissolved in the liquid, the one present in a smaller amount, is the **solute**(s). For example, in a glass of tea, water is the solvent, and sugar, lemon juice, and the tea itself are solutes. A theoretical concept of a solution of sugar in water pictures a collection of sugar molecules evenly dispersed among the water molecules (Fig. 4–1).

Aqueous solutions are water solutions.

In this discussion of acid–base and oxidation–reduction reactions, most of the chemistry studied will be in water or **aqueous** solutions, where water is the solvent.

Ionic Solutions (Electrolytes) and Molecular Solutions (Nonelectrolytes)

Solute particles may be ions or molecules. Solute particles bonded to solvent molecules are *solvated,* and such a solution process is termed *solvation.*

Solutes in aqueous solutions can be classified by their ability or inability to render the solution electrically conductive. When aqueous solutions are examined to see whether they conduct electricity, we find that solutions fall into one of two categories: **electrolytic** solutions, which conduct electricity, and **nonelectrolytic** solutions, which do not (Fig. 4–2).

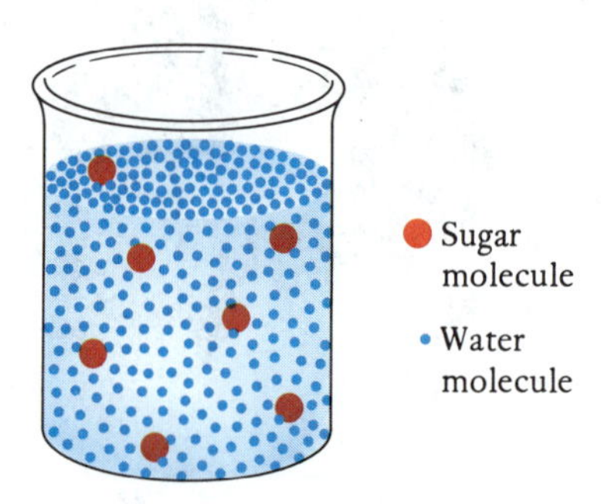

Figure 4–1 A schematic illustration at the molecular level of sugar solution in water. Large circles represent the sugar molecules and the small circles water. The sizes of the particles are not to scale.

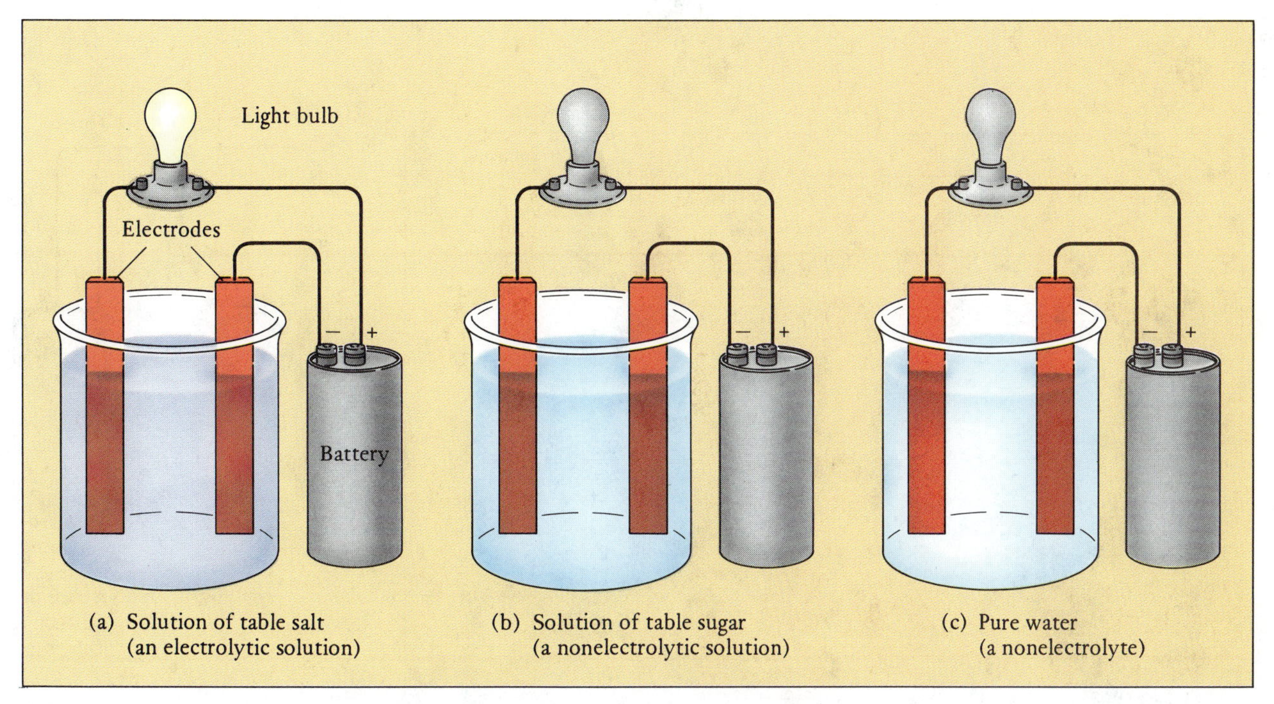

Figure 4–2 A simple test for an electrolytic solution. In order for the light bulb to burn (a) electricity must flow from one pole of the battery and return to the battery via the other pole. To complete the circuit, the solution must conduct electricity. A solution of table salt, sodium chloride, results in a glowing light bulb. Hence, sodium chloride is an electrolyte. In (b) the light bulb does not glow. Hence, table sugar is a nonelectrolyte. In (c) it is evident that the solvent, water, does not qualify as an electrolyte because it does not conduct electricity in this test.

The conductance of electrolytic solutions is caused by the solute particles in such solutions being ions rather than molecules. Recall from Chapter 3 that sodium chloride crystals are composed of sodium ions (Na^+), which are positively charged, and chloride ions (Cl^-), which are negatively charged. When sodium chloride dissolves in water, **ionic dissociation** occurs. The resulting solution (Fig. 4–3A) contains positive sodium ions and negative chloride ions dispersed in water. Each ion is surrounded and insulated by water molecules (Fig. 4–3A). This arrangement is represented by Na^+ (*aq*), where *aq* stands for aqueous. Of course, the solution as a whole is neutral, because the total numbers of positive and negative charges are equal, thereby canceling each other.

Ionic dissociation is the separation of ions of a solute when the substance is dissolved.

$$\underset{\text{(Solid)}}{Na^+Cl^-} \xrightarrow{\text{Water}} \underset{\substack{\text{(Aqueous)}\\ \text{Sodium}\\ \text{ion}}}{Na^+_{(aq)}} + \underset{\substack{\text{(Aqueous)}\\ \text{Chloride}\\ \text{ion}}}{Cl^-_{(aq)}}$$

The random motions of the sodium and chloride ions are not completely independent. The charges on the particles prevent all of the sodium

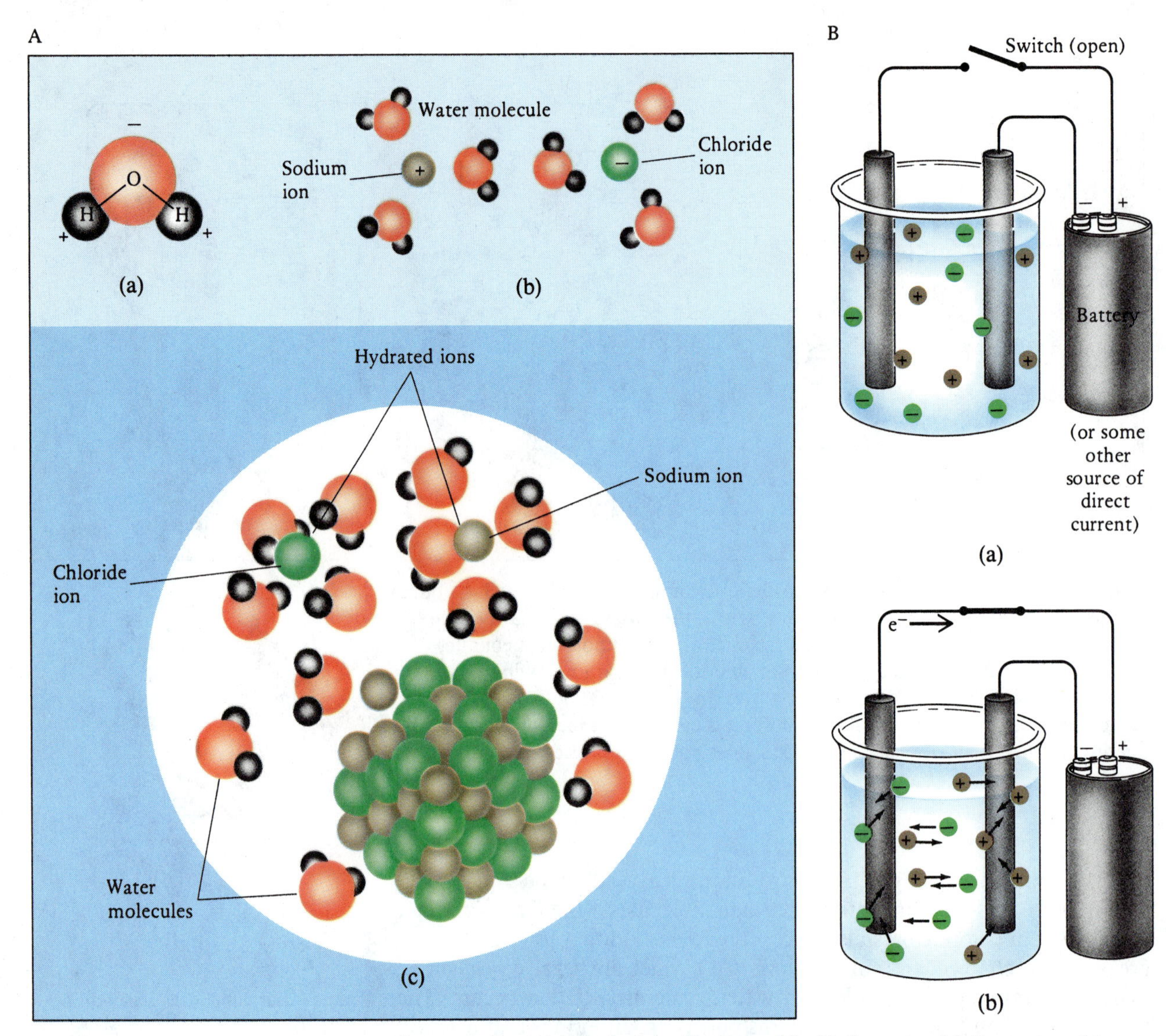

Figure 4–3 (A) Dissolution of sodium chloride in water. (a) Geometry of the polar water molecule. (b) Solvation of sodium and chloride ions due to interaction (bonding) between these ions and water molecules. (c) Dissolution occurs as collisions between water molecules and crystal ions result in the removal of ions from the crystal ions. In the process the ion becomes completely solvated.

(B) The hydrated ions are randomly distributed throughout the salt solution in (a); however, the net charge on the solution is zero. (b) The negative electrode attracts positive ions, and the positive electrode attracts negative ions. If electrons are transferred from the negative electrode to the positive electrode, the circuit is complete, and electricity flows through it.

ions from going spontaneously to one side of the container while all of the chloride ions are going to the other side. However, a net motion of ions occurs when charged electrodes are placed in an aqueous solution of an electrolyte (Fig. 4–3B). The electric current is carried through the solution by the movement of ions to oppositely charged electrodes. The positive ions move toward the negative electrode; the negative ions move toward the positive electrode. At the electrodes, electrons are interchanged between the ions and the electrodes to complete the circuit. We will say more about this later.

Ions migrate toward oppositely charged electrodes in an electric field.

Nonelectrolytic solutions are composed of solute molecules dispersed throughout solvent molecules, both of which are insensitive to negatively and positively charged electrodes unless the voltage is so great that molecules are changed into ions.

Sometimes ionic solutions arise when a molecular substance dissolves in water. For example, hydrogen chloride, HCl, is a gas composed of covalent, diatomic molecules. When hydrogen chloride dissolves in water an **ionization** reaction occurs, producing ions from molecules. The resulting solution is composed of hydrogen ions (protons) and chloride ions dispersed among the water molecules; consequently the solution conducts electricity and hydrogen chloride in water is properly termed an electrolyte.

When a molecular solute dissolves in water to produce ions, the process is called ionization.

$$\underset{\text{Molecule}}{HCl} \xrightarrow{\text{Water}} \underbrace{H^+_{(aq)} + Cl^-_{(aq)}}_{\text{Ions}}$$

The proton (H^+) in aqueous solution does not exist independently but becomes attached to the negative end of a water molecule. The proton in water is said to be hydrated and is often referred to as the **hydronium ion,** H_3O^+.

The H^+ ion is called a proton because it is a bare hydrogen nucleus consisting of a proton.

H_3O^+ is the hydronium ion.

$$H^+ + H:\ddot{\underset{\cdot\cdot}{O}}: \longrightarrow \underset{\text{Hydronium ion}}{\left[H:\ddot{\underset{\cdot\cdot}{O}}:H\right]^+}$$
(with an H bonded below each O)

Aqueous solutions of many acids, bases, and salts conduct electricity readily and are, therefore, electrolytes.

Concentrations of Solutions

When sugar, sodium chloride, alcohol, or any other readily soluble material dissolves in water, we can have either a concentrated or a dilute solution. Such a description of concentration is much less satisfactory and useful than one that tells us just how much of a given substance is dissolved in a specified volume.

Concentrations of solutions are often expressed in the number of *moles* of solute *per liter* of solution. **Molar** and **molarity** are used to denote this concentration unit. For example, 1 liter of a 1-molar solution contains 1 mole of solute. If a solution has a molarity of six, the solution has 6 moles of solute dissolved in 1 liter of solution. If a solution has a molarity of 0.1, the solution has 1/10 mole of solute dissolved in 1 liter of solution.

Molar concentration: number of moles of a substance per liter of solution.

Because of the heat generated when some acids are mixed with water, *acids should be added to water* to distribute the heat better. This is particularly important when mixing sulfuric acid with water.

Solutions of known molarity may be diluted to yield solutions whose molarity can be inferred by straightforward calculation.

ACIDS AND BASES

Taste is not a recommended practice for distinguishing acids and bases.

Litmus is but one acid–base indicator. Another, phenolphthalein, is colorless in acid and pink in base.

Acids and **bases** are classifications of chemicals dating back to antiquity. The word *acid* comes from the Latin *acidus,* meaning "sour" or "tart," since water solutions of acids have a sour or tart taste. Acids in water react with metals such as zinc and magnesium to liberate hydrogen; neutralize bases to produce salt and water; and change the color of litmus, a vegetable dye, from blue to red (Fig. 4–4). Citrus fruits offer a quick experience with natural acids because of their citric acid content, and any fruit with a high sugar content, such as apples, readily ferments to produce a vinegar containing acetic acid. Bases in water, or alkaline solutions, have a bitter taste, feel slippery or soapy to the touch, change litmus from red to blue, and neutralize acids to form salt and water.

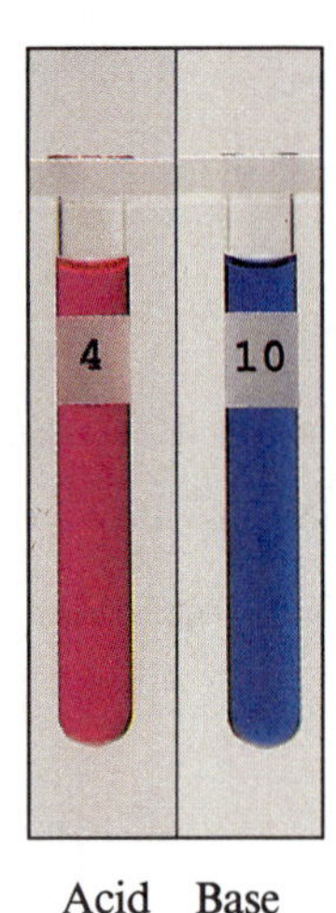

Acid Base

Litmus

Figure 4–4 Acid–base litmus test.

The variety of acids and bases in your life is illustrated by the following: Baking powder and baking soda are weak bases vital to cooking. Lye is a strong base often used as drain and toilet bowl cleaners. Lime, not as soluble in water as lye, is used to decrease the acidity of the soil. Antacids contain bases to neutralize excess acidity in the stomach. Acid skin tends to produce pimples. Your car battery depends on battery acid. Radiators corrode when antifreeze solutions acidify. Acid rain (see Chapter 12) and acid mine drainage are major threats to our environment. Your digestion and body metabolism are critically dependent on narrow controls of acidity and alkalinity in body fluids and tissues. The list could go on and on.

Acids and bases are defined in terms of a particular type of chemical reaction. **Any species that dissolves in water to produce hydronium ions, H_3O^+, is an acid.** Acids may be characterized as *strong* or *weak.* **Strong acids,** like hydrochloric acid, are completely ionized in water.

$$HCl + H_2O \longrightarrow H_3O^+ + Cl^-$$

In addition to noting the hydronium ion produced, observe that double arrows are not used because there is no evidence for a measurable back reaction; the ionization is 100%! Two other strong acids are sulfuric acid (H_2SO_4) and nitric acid (HNO_3). Solutions of strong acids can be harmful to human tissue (see Chapter 10). Most metals dissolve in solutions of strong acids. Hydrogen gas is one of the reaction products.

$$\text{Metal} + \text{Strong acid} \longrightarrow \text{Hydrogen gas} + \text{Metal salt in solution}$$

The organic acids, so important to the metabolism of living things, are weak acids (see Chapter 7).

Acetic acid, $HC_2H_3O_2$, is a **weak acid.** Weak acids ionize only to a slight degree in water.

$$HC_2H_3O_2 + H_2O \rightleftharpoons H_3O^+ + C_2H_3O_2^-$$

Some eyewashes contain boric acid (H_3BO_3), a weak acid.

Notice that the reverse reaction arrow is much longer than the forward reaction arrow, in keeping with the fact that only about 1% of the acetic acid molecules are ionized. One hydrogen atom is set apart in the acetic acid formula because only one of the four can be donated to the water to produce hydronium ions.

One can test the strength of an acid by testing its conductivity. Moderate concentrations of strong acid solutions are good conductors of electricity because of the high concentration of ions. Similar concentrations of weak acid solutions, which have relatively few ions, are poor conductors.

Weak acids dissolve many metals, but more slowly than strong acids do. Many weak acids are not generally harmful to human tissue like strong acids are. Weak acids like acetic acid, citric acid, and lactic acid are found naturally in some foods and they are added to others (see Chapter 14). Another weak acid, phosphoric acid—made by the burning of the element phosphorus and reacting the phosphorus oxide with water, is used in many cola soft drinks. The phosphate ion (PO_4^{3-}) is important in many of the reactions taking place in living cells (see Chapter 9).

Some metals like gold, silver, and platinum resist acids. They can be made to dissolve in certain acid mixtures, however.

The properties of an aqueous base are due to the OH^- ions in solution. **Any species that dissolves in water to produce OH^- ions is a base.** Bases may be characterized as *strong* or *weak.* The hydroxides of the alkali metals are **strong bases.** For example, NaOH and KOH are already ionic in the solid state and dissociate readily in water solutions.

$$Na^+OH^-_{(s)} \xrightarrow{H_2O} Na^+_{(aq)} + OH^-_{(aq)}$$

The subscripts are used to indicate solid (*s*) and solution (*aq*) species.

The hydroxides of the alkaline earth metals, such as $Ca(OH)_2$ and $Ba(OH)_2$, are also strong bases, but they are not as caustic to the skin as the alkali metal hydroxides because the alkaline earth hydroxides are not as soluble in water. Concentrated (greater than 1 molar) solutions of strong bases can be harmful to human tissue (see Chapter 10). In addition, many metals, including aluminum, dissolve in solutions of strong bases.

Ammonia dissolved in water is a **weak base.** The reaction between ammonia and water produces relatively few ions, so ammonia is a weak electrolyte. Ammonia, like acetic acid, remains mostly in the molecular form when dissolved in water. In the reaction below, this is indicated by the unequal length of the double arrow.

Ammonia is the number six commercial chemical in quantity produced, sodium hydroxide is number eight and nitric acid is number thirteen. (See inside front cover.)

$$\underset{\text{Ammonia}}{NH_3} + \underset{\text{Water}}{H_2O} \rightleftharpoons \underset{\text{Ammonium ion}}{NH_4^+} + \underset{\text{Hydroxide ion}}{OH^-}$$

The Happy Medium Between Acids and Bases—Neutralization

When acids react with bases, the properties of both species disappear. The process involved is called **neutralization.** The products of the reaction are water and a salt. The hydrochloric acid solution contains H_3O^+ and Cl^- ions; the sodium hydroxide solution contains Na^+ and OH^- ions. When these two solutions are mixed, a reaction occurs between H_3O^+ and OH^-.

$$Na^+ + Cl^- + \underset{\text{Acid}}{H_3O^+} + \underset{\text{Base}}{OH^-} \longrightarrow H_2O + H_2O + \underbrace{Na^+ + Cl^-}_{\text{Salt}}$$

The chemical compounds known as salts play a vital role in nature, in plant and animal growth and life, and in the manufacture of various chemicals for human use. Most salts contain ions held together by ionic bonding (see Chapter 3).

THE pH SCALE

Chemists have devised a scale, the pH scale, which is useful in describing the acidity of aqueous solutions. Pure water is neither acidic nor basic and yet it contains small and equal quantities of hydronium ions and hydroxide ions, formed by the reaction

$$H_2O + H_2O \rightleftharpoons H_3O^+ + OH^-$$

One water molecule acts as an acid while another acts as a base. The double arrow of unequal length in this reaction indicates that, by far, most water molecules are unreacted at equilibrium and that few H_3O^+ and OH^- ions are found in the solution. In pure water at room temperature, the concentration of the H_3O^+ ion is 0.0000001 (10^{-7}) mole/liter. The concentration of the OH^- ion is the same as the concentration of the H_3O^+ ion because the two are produced in equal amounts as the water ionizes. When the concentrations of H_3O^+ and OH^- are equal, the solution is *neutral.*

A mole of water is 18 g.

pH is defined as the negative log (exponent of ten) of the proton concentration expressed as $[H^+]$ or $[H_3O^+]$. Brackets, [ ], are used as a shorthand notation for mole/L.

$$pH = -\log[H^+]$$

At pH = 7, the number of H_3O^+ ions equals the number of OH^- ions.

Pure water has a pH of 7 and is a neutral solution. If the pH is below 7, the solution is acidic, and each drop of one pH unit represents a tenfold increase in acidity, or hydronium ion concentration. A pH above 7 is alkaline, with

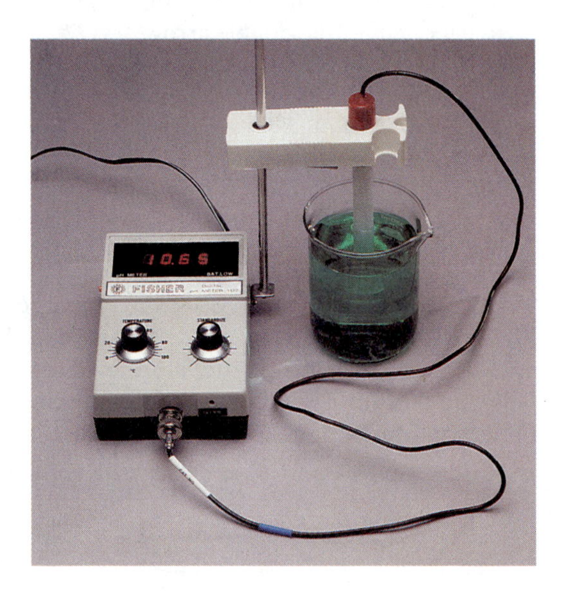

Figure 4–5 An electric pH meter detects the hydrogen ion concentration and expresses it as the negative exponent of 10. The positive exponent is the pH. The solution shown has a pH of 10.86, which means the solution is basic. (Marna G. Clarke)

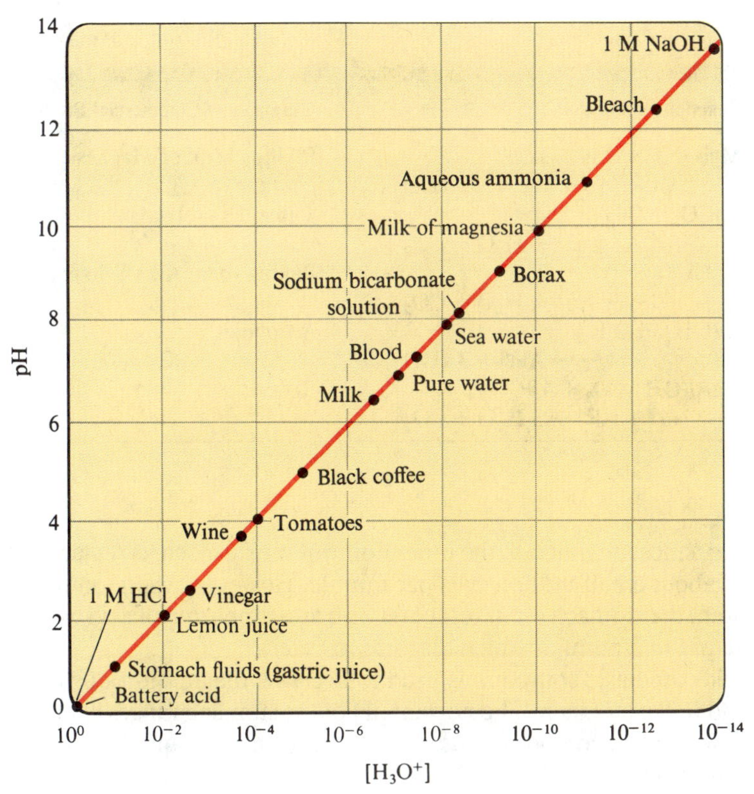

Figure 4–6 A plot of pH versus hydrogen ion concentration (H_3O^+). Note that the pH increases as the (H_3O^+) decreases. The pH values of some common solutions are given for reference. (A solution in which $(H_3O^+) = 1$ M has a pH of 0, since $1 = 10^{\circ}$.)

each unit of increase on the exponent scale representing a decrease in the hydronium ion concentration by a factor of one-tenth. In summary:

Pure water is neutral. Acids form H_3O^+ ions in water; bases form OH^- ions in water.

if $pH < 7.0$, solution is acidic
if $pH = 7.0$, solution is neutral
if $pH > 7.0$, solution is basic

The pH can be approximated by acid–base indicators or measured more exactly by electronic pH meters (Fig. 4–5).

Consumers are frequently asked to deal with pH. Figure 4–6 gives the pH of some common materials.

INDIGESTION: WHY REACH FOR AN ANTACID?

The contents of the stomach are highly acidic.

The walls of a human stomach contain thousands of cells that secrete hydrochloric acid, the main purposes of which are to suppress the growth of bacteria and to aid in the hydrolysis (digestion) of certain foodstuffs. Normally the stomach's inner lining is not harmed by the presence of hydro-

TABLE 4–1 The Chemistry of Some Antacids

Compound	Reaction in Stomach	Examples of Commercial Products
Milk of magnesia: $Mg(OH)_2$ in water	$Mg(OH)_2 + 2\ H^+ \longrightarrow Mg^{2+} + 2\ H_2O$	Phillips Milk of Magnesia
Calcium carbonate: $CaCO_3$	$CaCO_3 + 2\ H^+ \longrightarrow Ca^{2+} + H_2O + CO_2$	Tums, Di-Gel
Sodium bicarbonate: $NaHCO_3$	$NaHCO_3 + H^+ \longrightarrow Na^+ + H_2O + CO_2$	Baking soda, Alka-Seltzer
Aluminum hydroxide: $Al(OH)_3$	$Al(OH)_3 + 3\ H^+ \longrightarrow Al^{3+} + 3\ H_2O$	Amphojel
Dihydroxyaluminum sodium carbonate: $NaAl(OH)_2CO_3$	$NaAl(OH)_2CO_3 + 4\ H^+ \longrightarrow Na^+ + Al^{3+} + 3\ H_2O + CO_2$	Rolaids

chloric acid, since the mucosa, the inner lining of the stomach, is replaced at the rate of about a half million cells per minute. However, when too much food is eaten, the stomach often responds with an outpouring of acid, which lowers the pH to a point at which discomfort is felt.

If the reduction of acidity is too great, the stomach responds by secreting an excess of acid. This is "acid rebound."

Antacids are basic compounds used to decrease the amount of hydrochloric acid in the stomach. The normal pH of the stomach ranges from 0.9 to 1.5. Some alkaline compounds used for antacid purposes and their modes of action are given in Table 4–1.

SELF-TEST 4A

1. In a solution, the substance present in the greater amount (generally the liquid) is called the __________.
2. In a solution, the substance present in the lesser amount is the __________.
3. When a substance dissolves and produces a solution that conducts electricity, that substance is a(n) __________.
4. Another name for the hydrogen ion (H^+) is __________.
5. One liter of a 0.5 molar solution contains ____ mole(s) of dissolved solute.
6. An acid (donates/accepts) protons. __________ A base (donates/accepts) protons. __________
7. When an acid reacts with a base so that all the acid and base properties are gone, this reaction is called a __________ reaction.
8. Which is more acidic, a solution with a pH of 2 or a solution with a pH of 11? ____
9. Approximately what is the pH of stomach fluid? __________
10. When a solution has a pH of 7, the solution is said to be __________.
11. All antacids are (acids, bases) __________.

Stomach Acidity

The pH of stomach fluids, even in the normal stomach, is 1. The acid is hydrochloric acid. The stomach produces a small amount of acid all the time, but then that amount of acid may be stimulated by food. Even the sight and smell of appetizing food is enough to make the stomach produce more acid.

What actually happens when you eat a meal? Both hydrogen ions and chloride ions, maintaining an electrochemical balance, move through the stomach lining from the surrounding blood plasma. In the stomach the result is a highly acidic medium. That's what it takes to activate certain enzymes for the process of digestion. If the acidity of the stomach becomes excessive, problems can occur, problems that often need antacid solutions.

An expert on stomach acid is Dr. Paul Maton of the National Institutes of Health; he says that:

There is a variety of different compounds, basically bases, that can function as antacids. For example, sodium bicarbonate could be used as an antacid, or magnesium hydroxide or calcium carbonate.

All antacids neutralize acid. And if they're given in sufficient amounts, any antacid is as good as another at neutralizing acid.

One hears a lot of talk about hyperacidity—too much acid. In fact, there's very little or no evidence that people with ulcers actually produce more acid than many of the rest of us.

For most of those acid stomach discomforts, then, the tried and tested over-the-counter remedies do their job of neutralization well enough.

The World of Chemistry (Program 16) "The Proton in Chemistry."

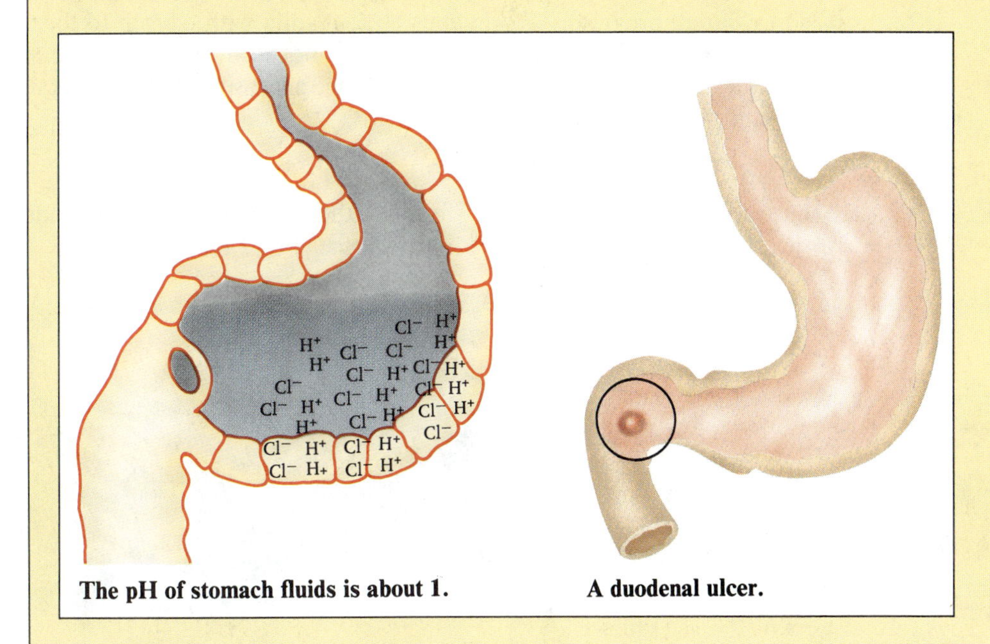

The pH of stomach fluids is about 1.

A duodenal ulcer.

OXIDATION-REDUCTION REACTIONS

Equal in importance to the proton transfer in acid-base chemistry are the processes called oxidation and reduction. As we shall find out later, oxidation and reduction reactions always occur together, so they are often named together as **oxidation-reduction.** Oxidation got its name from the chemical changes associated with the element oxygen combining with other elements. In fact, oxygen combines with every element except helium, neon, and argon. Prior to the discovery of the electron, oxidation was considered a simple combination of two elements that produced a compound called an **oxide.**

Oxidation

When oxygen combines with another element, heat is almost always produced. If this energy (as heat) is given off rapidly enough, the oxidation is called *combustion,* or *burning.* A tragic example of rapid combustion is shown in Figure 4-7.

Combustion is always accompanied by heat and light.

Neither oxidation nor combustion is limited to oxygen combining with just elements. Compounds may be oxidized as well. Automobile engines burn hydrocarbon fuels (Chapter 7) and produce the oxides of hydrogen (water) and carbon (carbon monoxide and carbon dioxide). Oxides of nitrogen are produced as well. These nitrogen oxides come from the oxidation of some of the nitrogen in the air that is mixed with the fuel and ignited in the combustion chamber.

If the atmosphere were composed of a greater concentration of oxygen, then fires could more readily get out of control; rates of chemical reactions are related to the concentrations of the reactants.

Whenever oxygen combines with another element or compound, the chemical reaction is one form of **oxidation.** The products of the reaction are called **oxidation products.** Most metals react readily with oxygen to form oxides. A few metals like gold and platinum do not readily oxidize but can form oxides using indirect means. When iron, an easily oxidized metal, reacts with oxygen, a red-brown oxide forms.

$$4\ Fe + 3\ O_2 \longrightarrow 2\ Fe_2O_3$$

In the presence of moisture, usually found in the air, a *hydrate* of iron oxide forms. This iron oxide hydrate is known as **rust** (Fig. 4-8).

A hydrate is a stable molecular or ionic substance associated with water.

Figure 4-7 The hydrogen-filled dirigible *Hindenburg,* May 1939 at Lakehurst, New Jersey. (United Press International)

$$4\ Fe + 3\ O_2 + xH_2O \longrightarrow 2\ Fe_2O_3 \cdot xH_2O$$

In the formula for rust, the x represents a varying number of water molecules.

Oxygen also combines with nonmetals to form oxides. Carbon burns to form carbon monoxide and carbon dioxide.

$$2\ C + O_2 \longrightarrow 2\ CO$$

$$C + O_2 \longrightarrow CO_2$$

The formation of carbon monoxide when carbon dioxide could be formed is called **incomplete combustion.** In a limited supply of oxygen, carbon monoxide is the likely product. The carbon monoxide can be further oxidized to carbon dioxide.

Carbon monoxide is highly toxic; see Chapter 10.

$$2\ CO + O_2 \longrightarrow 2\ CO_2$$

Carbon monoxide formed by the incomplete combustion of hydrocarbon fuels is a major component of urban air pollution.

While providing about 90% of all the energy needs for our society through the combustion of fuels, oxygen combines with other elements, either in the air or in the fuels themselves, to produce air pollutants (see Chapter 12).

Figure 4–8 Rusting of iron and steel objects costs billions of dollars each year. (*The World of Chemistry,* Program 15, "The Busy Electron")

The carbon in carbon dioxide is more oxidized than the carbon in carbon monoxide (CO). In general, when elements form several different compounds with oxygen, there is a **degree of oxidation.** Elements that are highly oxidized are often themselves capable of causing oxidation to occur. One name used for the compounds of these highly oxidized elements is **oxidizing agent.** Table 4–2 shows several of these oxidizing agents. Note the oxygen in their formulas.

The second and more general definition of oxidation involves **electron loss.** An element is said to be oxidized when it loses electrons. When a neutral atom becomes a positive ion, it has lost electrons and has been oxidized. Sodium is oxidized by bromine to produce sodium ions and bromide ions (Fig. 4–9).

Oxidation is the loss of electrons.

$$2\ Na + Br_2 \longrightarrow 2\ Na^+ + 2\ Br^-$$

These ions, in equal numbers, form sodium bromide (NaBr), a white solid.

Figure 4–9 The burning of metallic sodium in bromine vapors. The product is white sodium bromide. (*The World of Chemistry,* Program 8, "Chemical Bonds")

TABLE 4–2 Some Oxidizing Agents and Their Uses

Name	Formula	Uses as Oxidizing Agent
Potassium dichromate	$K_2Cr_2O_7$	Tests for alcohol in breath
Potassium nitrate	KNO_3	Gunpowder
Calcium hypochlorite	$Ca(OCl)_2$	Bleach, swimming pool disinfectant
Lead dioxide	PbO_2	Lead storage batteries
Manganese dioxide	MnO_2	Alkaline and lithium batteries
Hydrogen peroxide	H_2O_2	Disinfectant, antiseptic
Potassium peroxydisulfate	$K_2S_2O_8$	Denture cleansers

In addition to oxygen and bromine, the elements chlorine (Cl) and fluorine (F) combine with elements and compounds in ways that can be called oxidation. Fluorine is rather exotic in its applications, but Cl_2 is commonly used in oxidation applications such as disinfecting water supplies and in bleaches and cleaning compounds.

Reduction—The Opposite of Oxidation

Oxidation is always accompanied by reduction and vice versa. One cannot occur without the other.

Reduction always accompanies oxidation. When something is oxidized, something else is reduced. When something gains oxygen and gets oxidized, something else loses oxygen and gets reduced. When something loses electrons and gets oxidized, something else gains electrons and gets reduced. As an oxidizing agent causes oxidation, a **reducing agent** *causes* reduction.

Substances called **fuels** are reducing agents. Common fuels like gasoline and diesel fuel contain compounds of carbon and hydrogen and are reducing agents. Table 4–3 lists some common and some uncommon fuels.

Often oxidizers and fuel react out of control. Forest and building fires are examples. Grain dust explosions are another. Grain dust, a carbohydrate similar in composition to the major component found in wood, is the fuel, and the oxygen in the air mixed with the dust is the oxidant.

The tragic explosion of the space shuttle *Challenger* in January, 1986, in which seven astronauts perished was caused by hot exhaust gases rupturing the hydrogen fuel tanks. The released hydrogen burned in the atmosphere with explosive force. The pure oxygen in the adjacent oxygen tank accelerated the reaction even more when the tank ruptured.

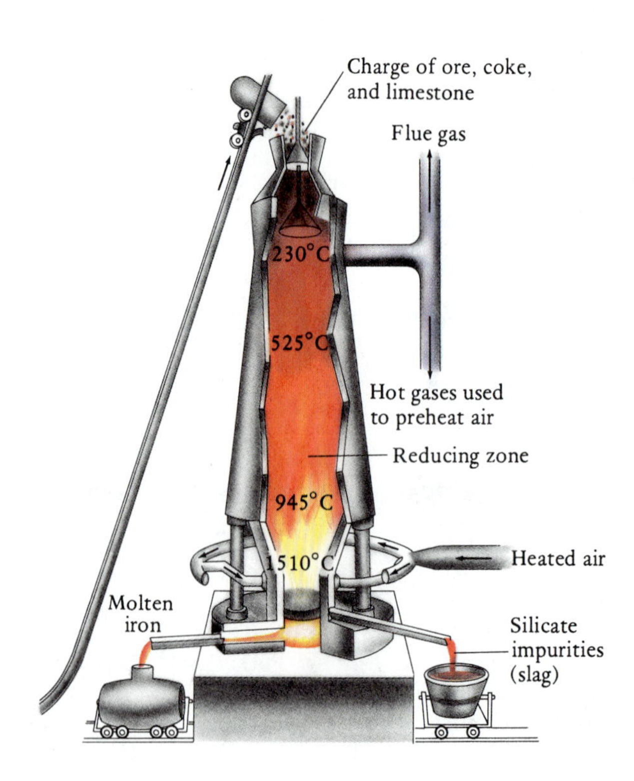

Figure 4–10 Diagram of a blast furnace used for the reduction of iron from iron ore.

TABLE 4-3 Some Fuels and Their Reactions

Fuel	Application	Oxidizer	Products
Gasoline	Passenger cars	Oxygen in air	$CO_2 + H_2O$ + some CO
Ethanol	Race cars	Oxygen in air	$CO_2 + H_2O$
Hydrogen	Space shuttle main engine	Pure Oxygen	H_2O
Propane	Home heating	Oxygen in air	$CO_2 + H_2O$
Coal	Home heating	Oxygen in air	$CO_2 + H_2O$ + some CO

Reduction of Iron from Its Ore

Iron ores are reduced to metal by using carbon, in the form of coke, as the reducing agent. Iron ore is reduced in a blast furnace (Fig. 4-10). The solid material fed into the top of the blast furnace consists of a mixture of an oxide of iron (Fe_2O_3), coke (C), and limestone ($CaCO_3$). A blast of heated air is forced into the furnace near the bottom. Much heat is liberated as the coke burns, and the heat speeds up the reaction, the speed being important in making the process economical. The reactions that occur within the blast furnace are

Iron ores are mixtures containing iron compounds. To get iron from the ores, the iron in the compounds must be reduced.

$$\underset{\text{Carbon}}{2\,C} + \underset{\text{Oxygen}}{O_2} \longrightarrow \underset{\text{Carbon monoxide}}{2\,CO} + \text{heat}$$

$$\underset{\text{Iron oxide}}{Fe_2O_3} + \underset{\text{Carbon monoxide}}{3\,CO} \longrightarrow \underset{\text{Iron}}{2\,Fe} + \underset{\text{Carbon dioxide}}{3\,CO_2} + \text{heat}$$

Limestone (calcium carbonate) is added to remove the silica (SiO_2) impurity. The overall reaction is

$$2\,Fe_2O_3 \cdot xH_2O + 3\,C \longrightarrow 4\,Fe + 3\,CO_2 + 2x\,H_2O$$

Looking at the equation for iron oxide reduction in the blast furnace, we see that the iron in iron oxide loses oxygen (gets reduced). The reducing agent is carbon, which has no oxygen associated with it at the start of the reaction. At the end of the reaction, carbon is combined with oxygen as carbon dioxide. The carbon gets oxidized.

As it comes from the blast furnace, the iron contains too much carbon for most uses. If some of the carbon is removed, the mixture becomes structurally stronger and is known as **steel.** Steel is an alloy of iron with a relatively small amount of carbon (less than 1.5%); it may also contain small percentages of other metals. In order to convert iron into steel, the excess carbon is burned out with oxygen.

An alloy is a metal mixture consisting of two or more elements.

ELECTROLYSIS

Several metals either are separated from their ores or are purified afterward by electrolysis. **Electrolysis** is a type of chemical reaction caused by the application of electrical energy.

The suffix *-lysis* means "splitting" or "decomposition"; electrolysis is decomposition by electricity.

Aluminum from the Electrolytic Reduction of Al^{3+} Ions

Aluminum, in its oxidized form of Al^{3+} ions, constitutes 7.4% of Earth's crust. Aluminum metal (the reduced form of aluminum) is soft and has a low density; many of its alloys, however, are very strong. Hence, aluminum is an excellent choice when a lightweight, strong metal is required. Aluminum is also used as a decorative metal and as a conductor in high-voltage transmission lines. Larger diameter wires must be used because of the lower electrical conductivity of aluminum, which is about 60% of the conductivity of copper. Aluminum competes with copper as an electrical conductor because of the lower cost of aluminum. Aluminum metal is easily oxidized by oxygen in the air. A transparent, hard film of aluminum oxide, Al_2O_3, forms over the surface, which protects the aluminum from further oxidation:

When aluminum was first made, it was very expensive and rare. A bar of aluminum was displayed next to the Crown Jewels at the Paris Exposition in 1855.

$$4\ Al + 3\ O_2 \longrightarrow 2\ Al_2O_3$$

The principal ore of aluminum contains the mineral bauxite, a hydrated aluminum oxide, $Al_2O_3 \cdot xH_2O$.

The top of the Washington Monument is a casting of aluminum made in 1884.

Metallic aluminum is obtained from purified oxide by the Hall-Heroult process, an electrolytic process that uses molten cryolite, Na_3AlF_6 (melting point 1000°C). Cryolite dissolves considerable amounts of aluminum oxide, which in turn lowers the melting point of the cryolite solution. This mixture

THE WORLD OF CHEMISTRY

A Better Aluminum Foil

The forte of the materials scientist is to try to control, using changes in chemistry and processing, the production of a material that possesses the unique properties required for a particular application. It is relatively easy to have ideas and dreams for new materials for new technological applications, but these applications must be facilitated or limited by our ability to transform the raw materials into the new material required.

One example is the story of a new kind of aluminum foil developed by researchers at Allied Signal Corporation. Chemically and physically, it's very different from household aluminum foil. The new kind of aluminum foil is an alloy containing iron, silicon, and other elements. In ordinary aluminum casting, the metal cools slowly and any added elements can separate partially and form different kinds of crystals within the foil. The interfaces between the different crystalline forms are weak points along which cracks could easily develop. Researchers at Allied Signal Corporation found that if they cooled the molten aluminum alloy quickly, up to a hundred degrees in a fraction of a second, the aluminum solidified before the different crystals could form. This quick cooling of the metal casting created a new aluminum alloy. Its enhanced strength comes from the iron and its heat resistance comes from the silicon.

The World of Chemistry (Program 19) "Metals."

of cryolite and aluminum oxide is electrolyzed in a cell with carbon electrodes and a carbon cell lining that serves as the surface on which aluminum is deposited. As the operation of the cell proceeds, molten aluminum sinks to the bottom of the cell. From time to time the cell is tapped and molten aluminum is allowed to run off into molds.

$$2\,Al_2O_3 \xrightarrow{\text{Electricity}} 2\,Al + 3\,O_2$$

About ten times more energy is needed to produce a ton of aluminum than to produce a ton of steel. Recycled aluminum stock from aluminum cans requires approximately one-half the energy needed to produce the same amount of the metal from the ore.

As a college chemistry student at Oberlin College, Charles Martin Hall was intrigued by the potential uses of aluminum and the difficulties involved in reducing this chemically active metal from its oxide. In electrolysis experiments in his family woodshed, Hall used batteries and a blacksmith's fire in 1886 to reduce the Al_2O_3 dissolved in a high-melting salt, cryolite, to metallic aluminum. Hall was 22 years old when he made this great discovery. Later he founded the Aluminum Corporation of America and died a multimillionaire in 1914. Independently, Paul Héroult, a Frenchman, made the same discovery at approximately the same time.

RELATIVE STRENGTHS OF OXIDIZING AND REDUCING AGENTS

When a piece of metallic zinc is placed in a solution containing hydrated copper ions (Cu^{2+}), an oxidation–reduction reaction occurs:

$$Zn + Cu^{2+} \longrightarrow Zn^{2+} + Cu$$

Evidence for this reaction is the deposit of copper on the zinc (Fig. 4–11). The gradual decrease in the intensity of the blue color of the solution indicates removal of the Cu^{2+} ions.

The oxidation of zinc by copper ions can be thought of as a competition between zinc ions (Zn^{2+}) and copper ions (Cu^{2+}) for electrons. Since the reaction proceeds almost to completion, the Cu^{2+} ions obviously win out in the competition. Other metals can compete similarly for electrons.

The **activity** of a metal is a measure of its tendency to lose electrons. Zinc is a more active metal than copper on the basis of the experiment just described.

Experiments of this type with various pairs of metals and other reducing agents yield an **activity series** of the elements, which ranks each oxidizing and reducing agent according to its *strength* or *tendency* for the electron

Figure 4–11 Metallic copper plates onto a zinc rod from a blue copper solution. After some time the blue color of the solution disappears, indicating that no more copper ions are in solution. (Marna G. Clarke)

transfer to take place. An iron nail will be partly dissolved in a solution of a copper salt containing Cu^{2+} ions, with copper being deposited on the nail that remains. From this, it is determined that iron, like zinc, is more active than copper. The reaction that occurs is

$$Fe + Cu^{2+} \longrightarrow Fe^{2+} + Cu$$

Now, which is more active, zinc or iron? This question can be answered by placing an iron nail in a solution containing Zn^{2+} ions and, in a separate container, a strip of zinc in a solution containing Fe^{2+} ions. The zinc strip is found to be eaten away in the solution containing Fe^{2+} ions. The reaction, then, is

$$Zn + Fe^{2+} \longrightarrow Fe + Zn^{2+}$$

Nothing happens to the iron nail in the solution of Zn^{2+} ions. We deduce that Zn loses electrons more readily than Fe.

Knowledge of the relative strengths of oxidizing and reducing agents can be used to make batteries.

BATTERIES

One of the most useful applications of oxidation–reduction reactions is the production of electrical energy in a device called an electrochemical cell. The term *battery* is used for a single electrochemical cell or a group of cells linked together.

The most common battery in use is the "dry cell," invented by Georges Leclanché in 1866. The container of the cell is made of zinc, which serves as the reducing agent. Being a metal, the zinc also conducts electrons and is the negative terminal of the battery (Fig. 4–12). The positive terminal is a metal

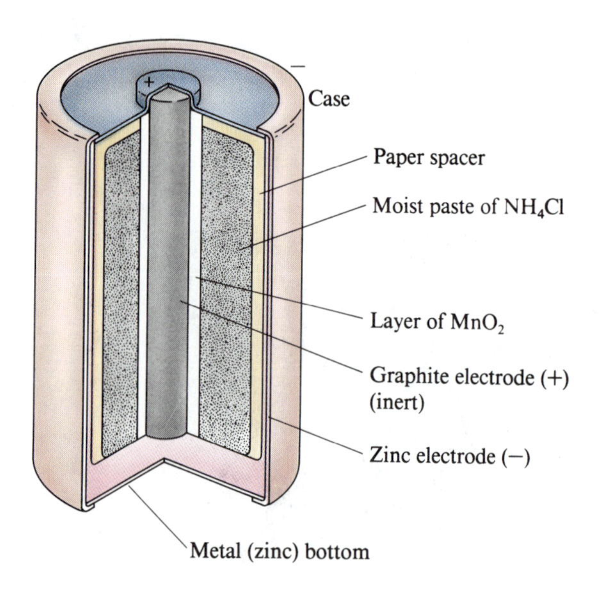

Figure 4–12 A common dry cell battery found in such items as flashlights.

button attached to a piece of graphite, a good electrical conductor, which is surrounded by a paste containing ammonium chloride and manganese dioxide (MnO_2), the oxidizer. As electrons flow from the cell through a device like a radio, the zinc is oxidized and the ammonium ions of the ammonium chloride are reduced. The products of the oxidation–reduction reaction mix in the paste inside the battery, making it impossible to recharge the battery by reversing the reaction.

Batteries in which the stored chemical energy is simply used up are called **primary** batteries. In such batteries, the oxidation products are allowed to

THE WORLD OF CHEMISTRY

The Pacemaker Story

Sometimes an advance in science can come from an unlikely source. Several years ago, the inventor Wilson Greatbatch had an outrageous dream to prolong human life. His story is fascinating:

I quit all my jobs and with two thousand dollars, I went out in the barn in the back of my house and built 50 Pacemakers in two years.

I started making the rounds of all the doctors in Buffalo who were working in this field, and I got consistently negative results. The answer I got was, well, these people all die in a year, you can't do much for them, why don't you work on my project, you know.

When I first approached Dr. Shardack with the idea of the Pacemaker, he, alone, thought that it really had a future. He looked at me sort of funny, and he walked up and down the room a couple of times. He said, "you know, . . . if you can do that, . . . you can save a thousand lives a year."

In 1958, a medical team implanted the first heart pacemaker, but for the next few years there was one major problem.

After the first ten years, we were still only getting one or two years out of pacemakers, two years on average, and the failure mechanism was always the battery. It didn't just run down, it failed. The human body is a very hostile environment, it's worse than space, it's worse than the bottom of the sea. You're trying to run things in a warm salt water environment. The first pacemakers could not be hermetically sealed, and the battery just didn't do the job. Well, after ten years, the battery emerged as the primary mode of failure, and so we started looking around for new power sources. We looked at nuclear sources, we looked at biological sources, of letting the body make its own electricity, we looked at rechargeable batteries, and we looked at improved mercury batteries. And we finally wound up with this lithium battery. It really revolutionized the pacemaker business. The doctors have told me that the introduction of the lithium battery was more significant than the invention of the Pacemaker in the first place.

The World of Chemistry (Program 15) "The Busy Electron."

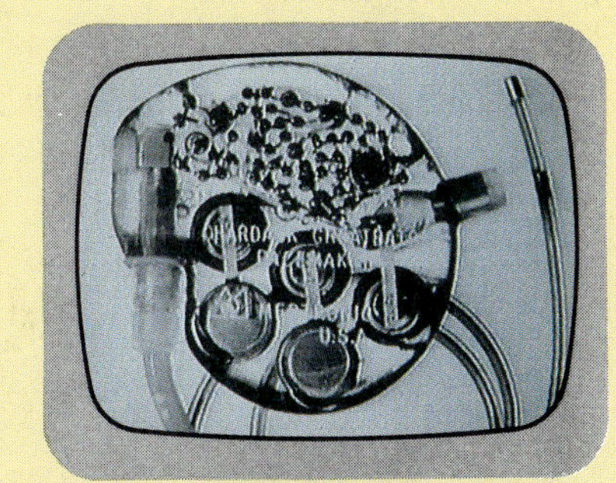

Pacemaker. (*The World of Chemistry,* Program 15, "The Busy Electron")

mingle with the reduction products. Because of this mixing, the battery cannot be recharged. Many of the less expensive batteries used to power flashlights, toys, radios, watches, cameras, and hand-held calculators are primary batteries.

Alkali is another name for base.

Another primary battery is the alkaline battery. Alkaline dry cells (alkaline batteries) are similar to Leclanché dry cells, except that the electrolyte mixture contains potassium hydroxide (KOH), a strong base, and the surface area of the zinc electrode is increased. The mercury battery (Fig. 4–13) is similar to the alkaline battery in that a zinc electrode is oxidized. The oxidizing agent is mercuric oxide. Mercury batteries should never be disposed of in fire because the mercury will vaporize and rupture the sealed container. Mercury's toxicity means that these batteries represent an environmental hazard if disposed of improperly (see Chapter 12).

Lithium metal has the lowest density of any nongaseous element: 0.534 g/mL.

Lithium batteries, which are so popular because of their light weight and high energy content per pound, contain a lithium electrode that is oxidized. Some lithium batteries use MnO_2 as the oxidizer, and others, like some pacemaker batteries, use exotic chemicals like sulfuryl chloride (SO_2Cl_2).

Some batteries are constructed in such a way that the oxidation–reduction reaction products remain separated during the discharge reaction. These batteries can be recharged (the electrode reactions are reversed) and are called **secondary batteries.** Under favorable conditions, secondary batteries can be discharged and recharged many times. Because of this fact, secondary batteries are very important in devices where large electrical flows are needed, such as in cranking an automobile engine, or where replacement is inconvenient, such as in orbiting satellites.

One of the most widely used secondary batteries is the lead storage battery (Fig. 4–14). As this battery is discharged, metallic lead is oxidized to lead sulfate and lead dioxide is reduced to lead sulfate. The reaction is

The double arrows in the reactions shown here indicate discharge (→) and recharge (←).

$$Pb + PbO_2 + 2H_2SO_4 \rightleftharpoons 2\ PbSO_4 + 2\ H_2O + \text{electrical energy}$$

The lead sulfate formed at both electrodes is an insoluble compound, so it stays on the electrode surface rather than dissolving in the battery acid. Because the reaction is reversible, the battery can be recharged by putting electricity back into it. Normal charging of an automobile lead storage battery occurs during driving. A voltage regulator senses the output from the alternator, and when the alternator voltage exceeds that of the battery, the battery is charged. During the charging cycle in most lead storage batteries, some water is electrolyzed to hydrogen and oxygen.

The lead–acid battery was first presented to the French Academy of Science in 1860 by Gaston Pianté.

$$2\ H_2O \xrightarrow[\text{energy}]{\text{Electrical}} 2\ H_2 + O_2$$

These reactions produce a mixture of hydrogen and oxygen in the atmosphere in the top of the battery. If this mixture is accidentally sparked, an explosion results. For this reason it is a good idea to be careful of sparks when working near a lead storage battery.

Figure 4–13 The mercury battery. (Photo courtesy of Eveready Battery Company)

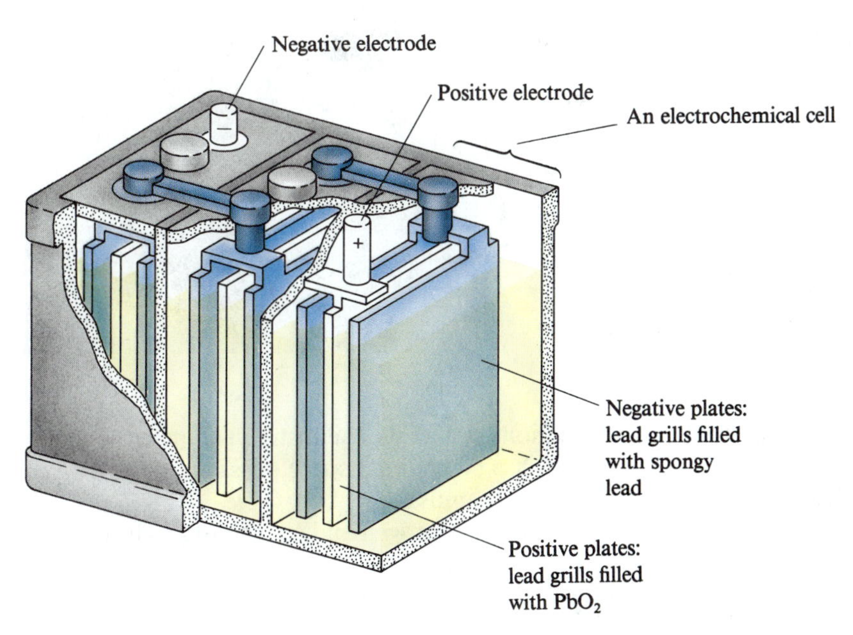

Figure 4–14 The lead storage battery. A typical 12-volt automobile battery contains six electrochemical cells. Three are shown here.

Nickel-cadmium batteries ("Ni-Cad") are another popular secondary battery. Being lightweight and producing a constant voltage until discharge, these batteries are popular in cordless appliances, video camcorders, portable radios, and other applications. NiCd batteries can be recharged because the reaction products are insoluble hydroxides that remain at the electrode surface.

Edison invented the NiCd battery in 1900.

FUEL CELLS

Unlike batteries, which are energy storage devices, fuel cells are energy conversion devices. Most fuel cells convert the energy of oxidation–reduction reactions of gaseous reactants directly into electricity. They are a

Fuel cells are energy conversion devices. Batteries are energy storage devices.

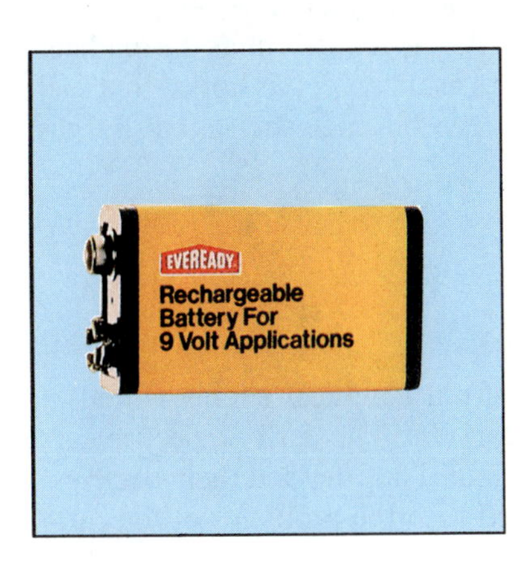

Nickel-cadmium battery. (Courtesy of Eveready Battery Company)

Figure 4–15 A hydrogen-oxygen fuel cell like the kind used on spacecraft. (United Technologies)

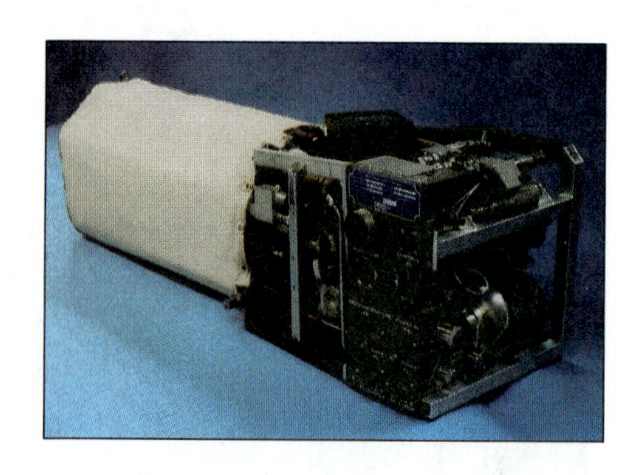

special application of oxidation–reduction chemistry. The most popularized application of fuel cells has been in the space program on board the Gemini, Apollo, and Space Shuttle missions.

Consider the reaction between hydrogen and oxygen to produce water and energy.

$$2\ H_2 + O_2 \longrightarrow 2\ H_2O + \text{energy}$$

As mentioned earlier in this chapter, if a mixture of hydrogen and oxygen is sparked, the energy is released suddenly in the form of a violent explosion. In the presence of a platinum gauze, these gases react at room temperature, heating the catalytic surface to incandescence. In a fuel cell the oxidation of hydrogen by oxygen takes place in a controlled manner, with the electrons lost by the hydrogen molecules flowing out of the fuel cell and back in again at the electrode where oxygen is reduced. This electron flow powers the electrical needs of a spacecraft, or whatever else is connected to the fuel cell. The water produced in the fuel cell can be purified for drinking purposes.

Fuel cells are about 60% efficient in converting chemical energy to electricity.

Potable means drinkable.

Because of their light weight and their high efficiencies compared with batteries, fuel cells like the one shown in Figure 4–15 have proved valuable in the space program. Beginning with Gemini 5, fuel cells have logged over 10,000 hours of operation in space. The fuel cells used aboard the Space Shuttle deliver the same power that batteries weighing ten times as much would provide. On a typical seven-day mission, the Shuttle fuel cells consume 1500 lb of hydrogen and generate 190 gal of potable water.

Other types of fuel cells that have been developed use air as the oxidizer and less-pure hydrogen or carbon monoxide as the fuel. It is hoped that fuel cells capable of direct air oxidation of cheap gaseous fuels such as natural gas will eventually be developed.

ELECTRIC AUTOMOBILES

Automobiles that derive their power from chemical energy stored in batteries have been around since the late 1800s. The earliest electric cars were very elegant and expensive for their time. They derived their electrical power from lead–acid batteries. Compared with the problems of the early gasoline-powered cars, such as hard starting and temperamental engines that

The prototype Impact electric automobile from General Motors. Introduced in 1990, this lightweight car accelerates from 0 to 60 in 8 seconds, has a top speed around 100 miles per hour, and has a range of 125 miles. (Courtesy of General Motors, Inc.)

sometimes didn't run very well on the gasoline of the day, electric cars seemed like the better choice—just get in, push the accelerator, and go. By about 1920, however, electric cars had all but disappeared from the roads as gasoline engines became more powerful and offered far greater speed and range. By then the electric starter had become standard equipment.

Electric cars can be manufactured now. In fact large fleets of electric trams, postal vehicles, delivery vans, and recreational vehicles are currently in use. The problem with these vehicles is the same as when the electric car lost out to the gasoline car—ease of use, speed, range, and now, three generations of drivers who have grown up with these features and who don't seem to want to give them up.

CORROSION—UNWANTED OXIDATION-REDUCTION

In the United States alone, more than $10 billion is lost each year to corrosion. Much of this corrosion is the rusting of iron and steel, although other metals may oxidize as well. The problem with iron is that its oxide, rust, does not adhere strongly to the metal's surface once the rust is formed. Because the rust flakes off or is rubbed off easily, the metal surface becomes pitted. The continuing loss of surface iron by rust formation eventually causes structural weakness.

The corrosion of metals involves oxidation and reduction. The driving forces behind corrosion are the activity of the metal as a reducing agent and the strength of the oxidizing agent. Whenever a strong reducing agent (the metal) and a strong oxidizing agent (like oxygen) are together, a reaction between the two substances is likely. Factors governing the rates of chemical reaction such as temperature and concentration affect the rate of corrosion as well. Consider the corrosion of an iron spike (Fig. 4-16). The surface of the iron is far from perfect. Microcrystals of loosely bound iron atoms occupy the surface of the metal. The iron can readily ionize into any water present on the surface of the metal (Fig. 4-17). The ionization of iron atoms into Fe^{2+} ions is an oxidation process. Iron is a fairly active metal; that is, it

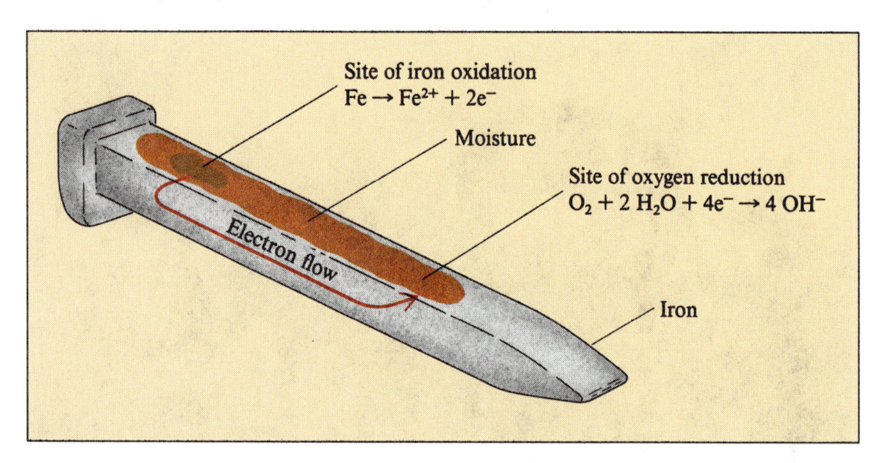

Figure 4–16 The site of iron oxidation may be different from the point of oxygen reduction owing to the ability of the electrons to flow through the iron. The point of oxygen reduction can be located with an acid–base indicator because of the OH^- ions produced.

tends to give up its electrons rather easily. Because iron is a good conductor of electricity, the electrons produced at this site can migrate to a point where they can reduce something. If these electrons did not migrate, the corrosion of iron would come to an abrupt halt as a result of a build-up of excessive negative charge. One location on the surface of the iron where electrons can be used would be any tiny drop of water containing dissolved oxygen (Fig. 4–17.) This reduction of oxygen occurs so readily that when Fe^{2+} ions are

Figure 4–17 (a) A schematic of how cathodic protection works for an underground storage tank. The more active magnesium metal protects the steel (iron). (b) A ship's hull is cathodically protected by strips of titanium (shown as four strips along the propeller axis).

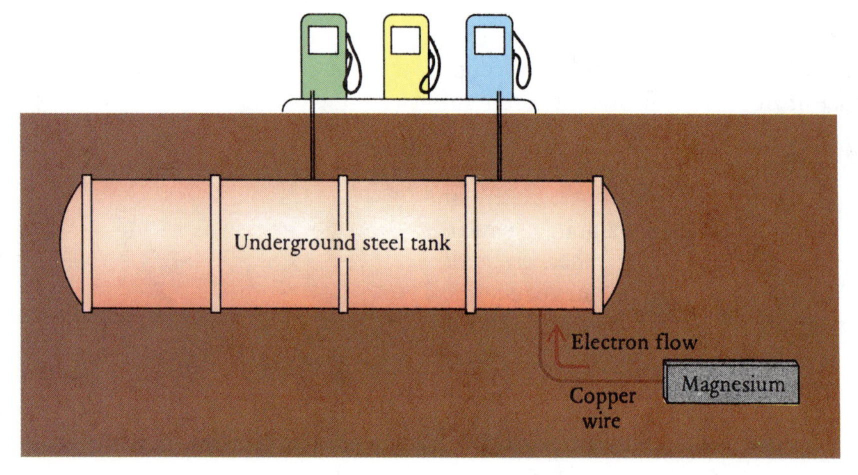

(a)

(b)

encountered, they are further oxidized to Fe^{3+} ions. Finally, the Fe^{3+} ions combine with hydroxide ions (OH^-) to form the iron oxide we call rust.

$$2\ Fe^{3+} + 6\ OH^- \longrightarrow \underset{\text{Rust}}{Fe_2O_3 \cdot 3\ H_2O}$$

The rate of rusting is enhanced by salts, which dissolve in the water on the surface of the iron and conduct electrical currents. The hydroxide ions and Fe^{2+} and Fe^{3+} ions migrate more easily in the ionic solutions produced by the presence of the dissolved salts. Automobiles rust out more quickly when exposed to road salts in wintery climates. If road salts are used in your driving area, it's a good idea after snowy seasons to wash the undersides of automobiles to remove the accumulated salts. For similar reasons, it is a good idea to wash off sea water from the surfaces of automobiles.

Rusting can be prevented by protective coatings such as paint, grease, oil, enamel, or some corrosion-resistant metal like chromium. Some metals are more active than iron, but when these metals corrode, they form adherent oxide coatings. Coatings with these metals provide corrosion protection. One of these metals is zinc. Zinc coating of iron and steel is called **galvanizing** and may be done by dipping the object into a molten bath of zinc metal or by electroplating zinc onto the surface of an iron or steel object.

In galvanized objects in which the zinc coating is exposed to air and water, a thin film of zinc oxide forms that protects the zinc from further oxidation. Galvanizing is a type of **cathodic protection.** As the name implies, a **cathode** is protected by using a more active metal in good electrical contact with the metal to be protected. The electrons for the reduction of oxygen are supplied by the more active metal. Thus, a more active metal, electrically connected to a piece of iron, would be oxidized before the iron is oxidized.

Cathode is the name given to the electrode in an electrochemical cell where reduction takes place.

Some cathodic protection relies on the cathode being sacrificed. An important application is the cathodic protection of underground steel storage tanks (Fig. 4-17) that hold gasoline and other hazardous liquids. These tanks must be protected because leakage would contaminate groundwater supplies. Since 1986, these tanks have been cathodically protected under new federal regulations designed to protect groundwater.

The importance of groundwater purity is discussed in Chapter 11.

Chrome plating for beauty and corrosion protection. (Tom Stack and Associates)

SELF-TEST 4B

1. When oxygen combines with a fuel rapidly enough that heat and light are given off, we call that reaction a ____________ reaction.
2. Oxygen can combine with some elements in varying ratios. Which oxide is more highly oxidized, CO (carbon monoxide) or CO_2 (carbon dioxide)? ____________
3. In terms of electrons, oxidation is electron (loss/gain). ____________
4. Which is the product of incomplete combustion, CO or CO_2? ____________
5. The opposite of oxidation is ____________.
6. A fuel is (a) an oxidizing agent, (b) a reducing agent.
7. In the blast furnace, the reducing agent is ____________.
8. Electrolysis is a kind of chemical reaction caused by the application of ____________ energy.
9. When water is electrolyzed, the products are ____________ and ____________.
10. When aluminum oxide is reduced in molten cryolite, the metal produced is ____________.
11. All metals have the same tendency to lose or gain electrons: True ( ) False ( )
12. Which battery is rechargeable: (a) dry cell, (b) mercury battery, (c) lead storage battery?
13. What is the name associated with unwanted oxidation? ____________
14. When zinc is used to coat a metal, the process is called ____________.

QUESTIONS

1. Give five examples of solutions with which you are familiar. Name the solvent in each case. Name the solutes if you can.
2. What is an electrolyte? Give an example.
3. What is meant by the term 1 *molar*?
4. Many acid–base indicators change color in the presence of acids and bases. Explain what is meant by the term *litmus test*. Give an example of how this term is used in a nonchemistry setting.
5. Define acid–base reactions in terms of protons.
6. Describe the test to determine if a solution is acidic or basic.
7. What is neutralization, as the term is used with acids and bases?
8. What is the difference between a strong electrolyte and a weak electrolyte?
9. Name the acid in stomach acid.
10. What is the purpose of antacids?
11. Describe how the pH of pure water changes as a small amount of acid is added.
12. Describe how the pH of pure water changes as a small amount of base is added.
13. Name three weak acids commonly found in foods.
14. Name three commonly used fuels. What is the oxidizing agent for these fuels?
15. How is combustion a special case of oxidation?
16. Explain how reduction is the opposite of oxidation.
17. Look at the reaction for the corrosion of iron. What are the three chemical reactants?
18. Using the answer to Question 17, explain how grease can help prevent corrosion of iron.
19. When an active metal is connected to an underground storage tank, what does this accomplish? What happens to the piece of active metal?
20. What is the main difference between a primary battery and a secondary battery?

21. Name two commonly used primary batteries and two commonly used secondary batteries.
22. Which type of battery makes more sense environmentally, the primary battery or the secondary battery? Explain your answer.
23. How is a fuel cell different from a battery? How is it similar?
24. Which common battery produces hydrogen and oxygen during recharging? How is this hazardous?
25. Cite two dangers associated with mercury batteries.

Vincent van Gogh's *The Huth Factories at Asnieres, 1887* captures the technological appetite for energy. (The Saint Louis Art Museum: Gift of Mrs. Mark C. Steinberg)

CHAPTER

5 What Every Consumer Should Know About Energy

Energy is stored in matter, both in the reality that matter can be converted into energy, and in energetic relationships between units of matter. Though during most of human history, available energy resources appeared limitless (The problem was finding ways to harness the energy to do useful work.), we now find serious limitations on available energy for human activities.

1. What is energy, and what is its primary source?
2. What are the fossil fuels, and how long will they last?
3. What is the difference between energy and power?
4. How can energy be stored in chemical bonds between atoms?
5. What are the first and second laws of thermodynamics?
6. Can the quality of life be preserved with less energy consumption?
7. Why is acid rain the result of burning coal and oil?
8. What are the advantages of coal liquification and gasification?
9. Are electrical systems the most efficient way to transport energy?
10. Why is nuclear energy attractive relative to chemical energy?
11. What are the problems associated with nuclear energy?
12. Is progress being made in harnessing nuclear fusion energy?
13. What are the best ways to harness solar energy?

In our industrialized, high-tech, appliance-oriented society, the average use of energy per individual is near the highest point in the history of the world. In the United States alone, with only 5% of the world's population, we consume 30% of the daily supply of energy. Only Canadians use more energy per capita. Such nations are highly dependent on a huge supply of energy. Since 1958, the United States has consumed more energy than it has produced. However, our voracious appetite has been curbed somewhat in recent years as a result of increased energy costs and the prospects of actual shortages. Emphasis has been placed on more efficient use of energy rather than on curtailment of our standard of living. Since the oil shortage of 1973, the energy required to produce $1 of the U.S. gross national product has fallen by 28% and is now essentially flat as we move into the middle 1990s.

Who is using energy?

Figure 5–1 When water falls, potential energy is changed to kinetic energy. *(The World of Chemistry,* Program 5, "A Matter of State")

What is energy?

What is it we use? **Energy**—defined as the ability to do work, which is accomplished only by moving things—is involved every time anything moves or changes. Types of energy involved with matter in motion include heat (molecules in motion), electricity (electrons in motion), sound (compression and expansion of the space between molecules), and mechanical energy (macroscopic objects in motion). All matter in motion involves **kinetic energy,** the energy of motion (Fig. 5–1).

Energy can be stored. Examples include energy stored in chemical bonds (Figs. 5–2 and 5–3), (as in wood and food), in the nuclei of atoms (atomic energy), and in gravitational systems (rocks on the top of a hill). Stored energy is **potential energy.**

Figure 5–2 Chemical potential energy is released when zinc reacts with solid ammonium nitrate.

What are the practical sources of energy? At first it was **biomass,** the mass of material produced by living organisms and containing significant potential chemical energy. The biomass, mostly wood, still provides one third of the energy requirements for some developing countries. After animal power, the wind was employed for transportation. It was the fossil fuels—petroleum, coal, and natural gas—that made possible the Industrial Revolution and are to this day providing most of the energy utilized in the developed nations. For 200 years coal was the principal fuel for industrializing nations, only to be replaced by petroleum and natural gas around the middle of this century. Petroleum and natural gas are easier to handle and cleaner in use. The world appetite for energy has greatly inflated the prices of petroleum and natural gas because the supply of these fluids is quite limited. With the greater amount of coal available, attention is shifting back toward coal for energy and as a new source for many of the petrochemicals formerly obtained from petroleum. Relatively smaller but still massive amounts of energy are now supplied by hydroelectric sources and nuclear energy, and still smaller amounts are provided in the form of direct solar energy, geo-

Figure 5–3 A rocket upon take-off requires much energy in a short time, that is, much power. (NASA)

thermal energy, wind currents, and ocean currents. These **primary sources** of energy may be converted into electricity (a **secondary** source), the form of energy we find more useful.

Our ultimate energy source is the Sun. Energy from the Sun is available to us through the biomass as a result of photosynthesis, the fossil fuels that are believed to have been formed from the biomass through transformations that required millions of years, the wind and water that store kinetic and potential energy as a result of solar energy, direct absorption of solar energy by our bodies and the materials around us, and, of late, the photovoltaic cells capable of transforming solar energy into electrical energy. A **renewable energy source** such as biomass or hydroelectric energy can figure in long-range human energy planning, in contrast to the fossil fuels which, when used, are gone.

Our dependence on the Sun is immense and multifaceted.

Per capita energy consumption: (a) Primal human without fire—2000 kcal/day; (b) Primitive agriculture—12,000 kcal/day; (c) Early industrial revolution—70,000 kcal/day; (d) Modern peak (1979)—243,000 kcal/day.

The purposes of this chapter are to explain some consumer interest principles about energy, to describe how energy is extracted from the environment, and to summarize our energy situation.

FUNDAMENTAL PRINCIPLES OF ENERGY

The Distinction Between Energy and Power

Energy is the ability to move matter and may be expressed in units of calories, or joules; in contrast, **power** is the rate at which energy is used. Power is expressed in units of energy used per time, such as calories per second or joules per second (watts). Some consumers confuse energy and power when looking at their electrical bills; consumers pay for the amount of energy they use (kilowatt-hours), not for how fast they use it (kilowatts, or

Power is the rate at which energy is used.

kilojoules per second). Some industries that consume huge amounts of electricity are given restrictions on the rate at which they can use energy.

When Shopping for Energy, Choices Are Limited

Oxidation was discussed in Chapter 4.

By far most substances around us are low-energy compounds. Most mixtures, such as rocks, dirt, earth, and water, are chemically oxygen-saturated. Only a few products of photosynthesis (food, fossil fuels, wood) are in the chemical position of providing energy through oxidation, the principal way in which we obtain energy from chemicals. Our choices of sources of energy from chemical reactions are limited and dwindling.

Energy in Chemical Bonds

A house on fire is normally an unwanted form of reaction of wood with oxygen at temperatures to give off heat and light—combustion. (*The World of Chemistry*, Program 15, "The Busy Electron")

Energy is absorbed (is endothermic) when a chemical bond is broken to yield isolated atoms. The energy required to break 1 mole of a particular kind of bond is the **bond energy** (Table 5-1). The same amount of energy is released (is exothermic) when 1 mole of these bonds is formed from isolated atoms.

A given chemical reaction is exothermic if the formation of new bonds liberates more energy than is required to break the bonds in the reaction; a reaction is endothermic if the bonds in the reactants are stronger than those in the products. In summary, we receive energy from a chemical reaction only when more energy is produced by bond-making than is required for bond-breaking.

For example, consider the oxidation of methane (CH_4), the principal component of natural gas:

$$\mathrm{H{-}\overset{\overset{\displaystyle H}{|}}{\underset{\underset{\displaystyle H}{|}}{C}}{-}H} + 2\,\dot{\mathrm{O}}{-}\dot{\mathrm{O}} \longrightarrow \mathrm{O{=}C{=}O} + 2\,\mathrm{H{\diagup}\overset{O}{}{\diagdown}H} + 211\ \text{kcal/mole}\ \mathrm{CH_4}$$

The O_2 molecule has two unpaired electrons, as indicated by the dots in $\dot{\mathrm{O}}{-}\dot{\mathrm{O}}$.

The energy released in this exothermic reaction can be used to heat houses, drive gas turbines, and generate electricity. The fact that energy is released means that it takes less energy to break the C—H and $\dot{\mathrm{O}}{-}\dot{\mathrm{O}}$ bonds than is produced when the C=O and O—H bonds are formed. The bond energies in Table 5-1 bear this out. It takes only 632 kcal to break 2 moles of

TABLE 5-1 Approximate Bond Energies

Bond	Energy (kcal/mole)
H—C	99
$\dot{\mathrm{O}}{-}\dot{\mathrm{O}}$ in O_2	118
C=O in CO_2	192
H—O	111

TABLE 5-2 Heat Produced by the Combustion of Some Organic Materials

Substance	Heat (kcal/g)
Methane (principal component of natural gas)	13.2
Gasoline, kerosene, crude petroleum, tallow	9.5-11.5
Lipids (fats)	9.0-9.5
Carbon (coal)	7.8
Ethyl alcohol	7.1
Proteins	4.4-5.6
Carbohydrates (sugars and starches)	3.6-4.2

These figures correspond to laboratory combustion to yield CO_2, H_2O, and oxides of nitrogen. In the body, proteins are oxidized to CO_2, H_2O, and urea. For the latter process, the heat yield is less than the value indicated above. Thus, proteins and carbohydrates yield (per gram) about the same energy in the body.

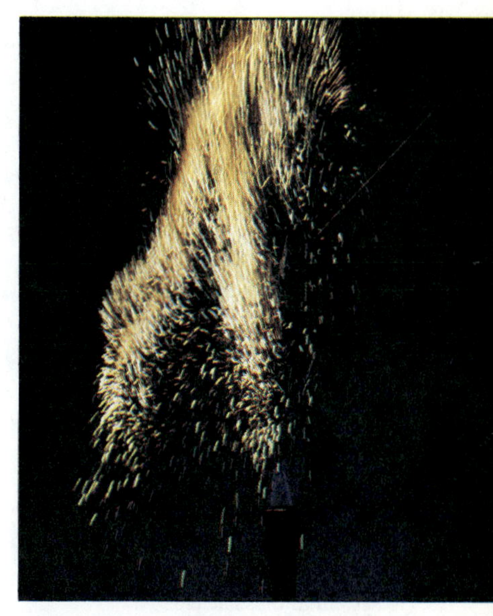

Powdered iron sprinkled in a flame combines with oxygen to give off heat and light—combustion.

$\dot{O}$—$\dot{O}$ bonds and 4 moles of C—H bonds [2(118) + 4(99) = 632]. The making of bonds produces 828 kcal generated by the formation of 2 moles of C=O bonds and 4 moles of O—H bonds [2(192) + 4(111) = 828]. This gives a net release of 211 kcal (828 − 632 = 196), which closely agrees with the experimental value of 192 kcal/mole of CH_4 burned.

If the supply of oxygen is sufficient, **combustion** of fossil fuels and wood produces principally carbon dioxide (CO_2) and water (H_2O). The net energy received from the burning of the fuel is the difference between the energy given off in making bonds in CO_2 and H_2O and the energy required to break bonds in O_2 and the fuel. Water and carbon dioxide have their capacities for oxygen satisfied and therefore cannot be burned to extract more energy. The amounts of energy derived from the combustion of some of nature's storehouses of energy are given in Table 5-2.

The Law of Conservation of Energy—The First Law of Thermodynamics

Also known as the first law of thermodynamics, the law of conservation of energy asserts that energy is neither lost nor gained in all energy processes. When a beaker of water is heated on a burner, all of the energy given off by the flame can be accounted for in the increased energy of the water and its surroundings; no energy is lost or gained in the transformation of chemical energy to heat energy. Furthermore, the transformation is quantitative, in that a certain amount of gas burned produces a certain amount of energy (see Table 5-2); when one kind of energy is changed into another, the exchange rate is definite, reliable, and reproducible.

Thermodynamics is the movement of energy.

The law of conservation of energy also implies that the total amount of energy in the universe is constant. Energy is transformed regularly from one kind to another, but the total remains the same. This means that the Sun and the energy stored in chemicals on Earth are what we have to use—that is all! There is no creation of new energy.

First statement of the first law of thermodynamics: "A force [translated: energy] once in existence cannot be annihilated"—Julius Robert Mayer, a ship's doctor, 1840.

The law of conservation of energy was extended by Albert Einstein, who showed the interrelationship between mass and energy in the equation $E = mc^2$. This required the inclusion of matter in a more general law:

The total amount of matter and energy in the universe is constant.

Energy Is Conserved in Quantity but Not in Quality— The Second Law of Thermodynamics

Usable energy is not conserved.

What does this second law of thermodynamics mean: that energy is conserved in quantity but not in quality? Perhaps two examples will clarify the concept. Consider first the release of some energy by the burning of coal, petroleum, or wood. Recall from our previous discussion that the main products of these combustion reactions, CO_2 and H_2O, do not burn and release more energy. In the burning process, both matter and energy are conserved, as required by the laws of conservation of matter and energy, respectively. However, the reactants and their stored energy are more useful in energetic terms than the products and their spent energy.

No matter how we try, we can never convert all of the stored energy in a system into usable energy.

As a second example, consider an electric motor. The electricity that runs the motor is more useful than the heat that comes from the warm motor. Again, energy is conserved in the process of running an electric motor, but the usable energy is not conserved.

In all processes, then, some energy is wasted—not lost—by conversion into energy that is not usable in doing work. The wasted (or unusable) energy is represented by **entropy,** a measure of the disorder in a physical system (Fig. 5-4). Entropy is not energy per se, but it is a function of energy with units of energy per degree, such as calories/degree.

Another statement of the second law of thermodynamics is based on entropy:

***Entropy* means disorder, and measures nonuseful energy.**

In all natural processes, entropy is increased.

Figure 5-4 A familiar sight is deteriorating automobiles, which in their assembled splendor are prized possessions. After going through a metal shredder (note the pile on the right), the once-valued automobile is now a pile of mixed-up metal junk. The metal shredder created more disorder for the automobile, and, hence, created more entropy. Junk yards epitomize entropy . . . a more disordered state.

Taken to its extreme, this means that the entropy (disorder) of the whole universe is increasing. As the material in the universe is scattered, there is less usable energy. This is not a reason for worry, because the universe is so vast that enough usable energy exists for all conceivable purposes for many billions of years. However, sources of usable energy that are not limitless are the so-called fossil fuels (coal, petroleum, and natural gas), which when gone are not easily restored. And if they were restored, more energy would be required in the restoration than would be available from the restored fuels.

Consider a burning match (Fig. 5–5). Some of the energy can be used to ignite another object, or to heat an object, or to provide light; this is the organized energy. However, while the usable energy is being used, some of the total energy emanating simply heats molecules in the vicinity, or at some distant point, and thereby increases the entropy of the universe. Eventually, all of the heat and light coming from the match becomes increased random motion of the molecules.

Figure 5–5 Where do the heat and light of a flame go? Is all of the emitted energy useful? See the text. (*The World of Chemistry,* Program 13, "The Driving Forces")

What does the second law of thermodynamics mean to the informed citizen? Simply stated, when usable energy-rich chemicals such as coal and petroleum are consumed, the usable energy is permanently reduced.

When fossil fuels are gone, then what?

The Efficiency of Energy Use Is Low

In every energy process, the efficiency of energy use is less than 100%, usually far less. Automobiles are about 20% to 25% efficient; that is, about 80% of the useful energy available to do work is lost and not applied to the turning of the wheels. Some fuel cells are about 70% efficient. The human body is about 45% efficient in converting the energy of glucose metabolism to muscle movement. Photosynthesis is about 30% efficient in converting absorbed sunlight into glucose, steam turbines for producing electricity are about 38% efficient, heating homes with electricity is about 38% efficient, and heating homes with natural gas is about 70% efficient. The efficiency is usually greater when a **primary source** of energy is used (gas) than when a **secondary source** is used (electricity). For example, it takes about 2500 kcal to produce 1 kilowatt-hour of electricity. If this kilowatt-hour is then used for heating, only 960 kcal of heat are produced. Natural gas burned on site would be more efficient than natural gas burned in a steam generator plant to produce electricity.

Efficiency = used energy ÷ available energy.

Primary source of energy: one transformation on site (e.g., chemical → heat via combustion). *Secondary source:* usually more than one transformation, plus long-distance transport (e.g., chemical → heat via combustion → steam → mechanical → electricity).

Conservation — Energy Not Lost Is "Energy Gained"

It is possible in industrial processes, home lighting, space heating and cooling, and the operation of home and commercial appliances to get the desired result and save large fractions of the energy expended in these energy uses. The reason why these energy efficiencies have not been achieved is that it has been cheaper to buy the extra energy than to redesign the equipment involved. Lighting uses 25% of U.S. electricity, and a 55% immediate savings would be gained if we switched to fluorescent lights with improved starters. Electric motors use over one half of the electricity generated in the United States, and 20% of this could be saved by using variable-speed motors with improved windings; the motor would require energy depending on work load rather than consuming energy at a fixed rate even if the energy had to be dumped when no work was required. Available savings for refrigerators, television sets, photocopiers, and computers are between 75% and 95% of 1990 use rates. Commercial buildings and homes use 30% of U.S. energy (total energy, not just electrical energy) consumption. Estimates range up to a 50% savings based on existing technology by the year 2010. For example, in the United States the energy drain from windows alone is equal to the energy flow through the Alaska oil pipeline. This drain could be cut by one fourth by using a double-pane coated glass window filled with argon gas. The large argon atoms have relatively poor thermal conductivity, and the surface coating reflects the heat radiation in the desired direction. At least for a period of time it is reasonable to predict that the expected increase in the efficiency of energy use will offset the increased price of the energy unit.

A better window has an increased initial energy cost which must be recovered before real savings are realized. For example, energy is required to obtain argon from air. See Chapter 12.

FOSSIL FUELS

The oxidation of coal, petroleum, and natural gas provides 88% of all of the energy purchased in the world. For each material, the products of complete combustion are principally CO_2 and H_2O. Compounds of other elements are also produced depending on the composition of the particular fossil fuel mixture. For example, the sulfur in coal or petroleum burns to form sulfur dioxide (SO_2), a major cause of acid rain. Urban smog and global warming are also of concern as a consequence of burning such massive amounts of fossil fuels.

Acid rain is discussed in Chapter 12.

No significant ongoing coal or petroleum production underground has been detected. It is true that small droplets of new oil have been discovered in the Guaymas Basin along the San Andreas Fault as a result of 5000 years of plankton deposits in the heated ocean fault zone. Such reports offer little hope of replenishment of the fossil fuels in the Earth, which are thought to have required millions of years to form. The fossil fuels that have already been found may be all there are in any reasonable human time frame. At current usage rates, world stores of coal, natural gas, and oil are estimated to last 1500, 120, and 60 years, respectively.

As the prices of the fossil fuels are increased because of shrinking reserves, a greater percentage of the actual fuel in the ground can be economically removed. For example, less than one half of the oil in an oil field can currently be pumped at a financial profit. If a higher percentage of the oil could be removed, the 60-year supply estimate for oil would be extended.

Figure 5–6 A petroleum refinery tower. (Courtesy of Standard Oil)

Petroleum

Petroleum was first discovered in the United States (in Pennsylvania) in 1859 and in the Middle East (in Iran) in 1908. Today petroleum is pumped from the ground in many parts of the world. As an energy source, petroleum was first used as kerosene for lighting, then as gasoline, aircraft fuel and lubricants for transportation, and most recently as fuel oil for heating and the production of electricity. However, the increased price of oil is curtailing fuel oil uses.

Petroleum is a mixture of many hydrocarbons, a class of organic compounds composed of hydrogen and carbon described in Chapter 6. By refining (separation of the components of the mixture by distillation) and subsequent conversions, much crude oil is turned into gasoline (43%, or 6.7 million barrels per day in the United States), and lesser amounts are turned into fuel oil (27%), jet fuel (7.4%), and other miscellaneous fuels (Fig. 5-6). From the small fraction of petroleum that is not burned come thousands of chemicals known as the petrochemicals; they are discussed in Chapter 7.

There are 42 gallons of oil per barrel.

Only in the Middle East and Latin America can current oil production rates be reasonably projected beyond 30 years. The United States, with only 4% of global reserves in 1990, produced 12% of the world output of oil. The United States and the Soviet Union, still the world's leading producers of oil, are now declining in production rates as most of the cheap oil is gone. The average U.S. well produces 15 barrels of oil per day, in contrast to the average of 9000 barrels for a well in Saudi Arabia. Current oil use is approximately 4.5 barrels of oil a year for each person, with the figure ranging from 24 barrels in the United States to less than 1 in sub-Saharan Africa. The use rate is expected to fall to 1.5 barrels per person per year by the year 2030, which will necessitate extensive changes in the global energy economy.

One half of the world's oil consumption is used by 500 million vehicles in order to transport people and freight. Based on present growth rates, there should be 1 billion vehicles by the year 2030! Fuel consumption can be cut by building more efficient engines, but it is evident that other fuels such as the alcohols, liquefied and compressed natural gas, hydrogen, and electricity will have to be employed. The technology is available in each of these areas to meet transportation needs, but the economic factors have not thus far been favorable.

Alcohol fuels for transportation are discussed in Chapter 7.

Coal

Coal is an even more complex mixture of hydrocarbons than petroleum. Molecular structures in coal contain rings of carbon atoms (see Chapter 6) as well as the long-chain molecules characteristic of petroleum. Complex linkages between the rings and between the rings and chains make coal an extremely complex mixture. Coal is much more likely than petroleum to have relatively high sulfur content, but sulfur can range up to 4% in both coal and petroleum. Burning coal or oil containing sulfur is a major source of air pollution (see Chapter 12). Industrial quantities of coal represent vast amounts of stored chemical potential energy (Fig. 5-7).

Minable coal is defined as 50% of all coal in a seam at least 12 inches thick and within 4000 feet of the Earth's surface. In the United States the minable coal reserves are divided among anthracite (2%), lignite (8%), subbituminous coal (38%), and bituminous coal (52%). Some properties of the different kinds of coal are listed in Table 5-3.

The largest portion of mined coal (about 75%) is burned to produce electricity. Only about 1% is used for residential and commercial heating. Although the use of coal is on the rise, coal's decline as a heating fuel was caused by its being a relatively dirty fuel, bulky to handle, and a major cause of air pollution (because of its sulfur content).

Most coal is burned to make electricity.

Problems with coal.

Figure 5-7 Coal is carried by many forms of transportation to where it is often stockpiled at electrical power plants and various kinds of industries. (Visuals Unlimited/Albert Copley)

World coal reserves are vast relative to the other fossil fuels. Figure 5-8 shows that to date we have used only a small fraction of the world coal that is available.

Heating coal at high temperatures in the absence of air produces **coke, coal tar,** and **coal gas.** This process is an example of **pyrolysis,** which means decomposition by heat. One ton of bituminous (soft) coal yields about 1500 lb of coke, 8 gal of coal tar, and 10,000 cubic feet of coal gas. Coke is relatively pure carbon and is an excellent nonpolluting fuel. However, because of the energy required in its production, it cannot compete economically except in industrial operations requiring pure carbon. Coal tar is a valuable source of organic compounds (see Chapter 7). The coal gas is a mixture of H_2, CH_4, CO, C_2H_6, NH_3, CO_2, H_2S, and other gases. Coal gas, because of its nitrogen and sulfur content, does not burn cleanly and has been abandoned as a gaseous fuel.

Coal can also be converted completely into a combustible gas (coal gasification) or a liquid fuel (coal liquefaction). In each case environmental problems can be averted, but at additional costs per energy unit obtained from these fuels.

TABLE 5-3 Some Properties and Characteristics of Types of Coal

Characteristics	Anthracite	Bituminous Coal	Subbituminous Coal	Lignite
Heat content	High	High	Medium	Low
Sulfur content	Low	High	Low	Low
Hydrogen/carbon mole ratio	0.5	0.6	0.9	1.0
Major deposits	New York, Pennsylvania	Appalachian Mts., Midwest, Utah	Rocky Mts.	Montana

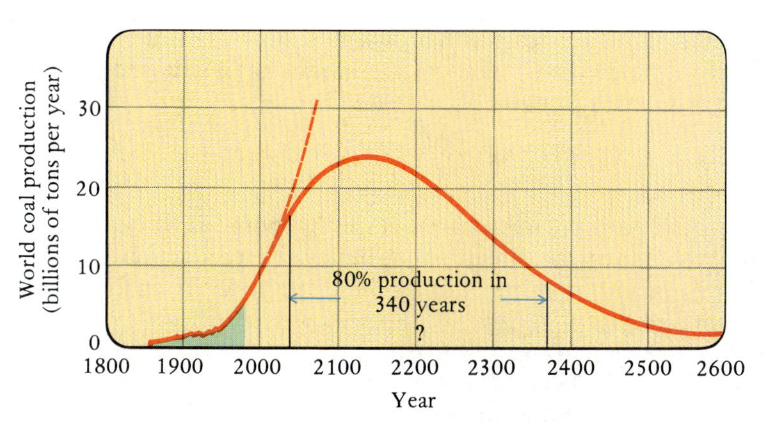

Figure 5–8 The coal mined to date (shaded area) represents only a small fraction of minable coal. The rate of increase in coal consumption (dashed line) is 4% per year. It is obvious that such an exponential rise cannot continue long after the year 2000. Known reserves = 1020 billion short tons (global estimate).

Coal Gasification

Coal can be converted into a relatively clean-burning fuel by a process known as **gasification** (Fig. 5–9). In this process, coal is made to react with a limited supply of either hot air or steam. In the reaction of coal with air, the product is a gaseous mixture known as **"power gas,"** and the reaction is exothermic.

Gasification can make coal cleaner to burn and easier to handle from supplier to user.

$$\text{Coal} + \text{Air} \longrightarrow \underbrace{CO(g) + H_2(g) + N_2(g)}_{\text{Power gas}} + 26.39 \text{ kcal/mole C}$$

Power gas contains up to 50% nitrogen (N_2) by volume and is consequently a relatively poor fuel. In fact, power gas of this composition has only one-sixth the heat content of methane.

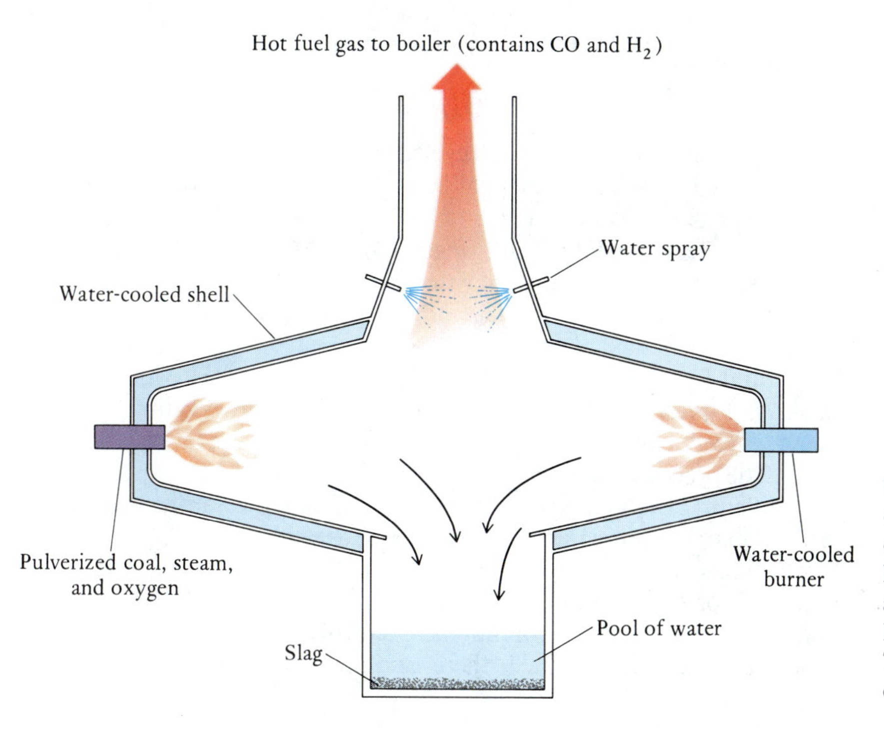

Figure 5–9 Schematic drawing of coal gasifier. A relatively cool combustion of powdered coal in a limited supply of oxygen produces a mixture of carbon monoxide and hydrogen along with other gases. The mineral content in the coal collects in the slag.

If the coal is allowed to react with high-temperature steam, a mixture of carbon monoxide and hydrogen known as **synthesis gas** is obtained. Unlike power gas, this mixture contains no nitrogen.

This reaction is endothermic.

$$\underset{\text{Coal}}{C} + \underset{\text{Steam}}{H_2O(g)} \xrightarrow{\text{Heat}} \underset{\text{Synthesis gas}}{CO(g) + H_2(g)}$$

When air and steam are mixed in the correct proportions, the reaction of the mixture with coal can be self-sustaining because the production of power gas is exothermic and produces enough energy to drive the endothermic production of coal gas.

In both power gas and coal gas mixtures, the CO and H_2 are burned by oxygen (O_2) in the air to produce heat. The heat produced is about one-third that of an equal volume of methane (natural gas).

These reactions are exothermic.

$$2\,CO + O_2 \longrightarrow 2\,CO_2 + 135.3 \text{ kcal } (67.6 \text{ kcal/mole CO})$$

$$2\,H_2 + O_2 \longrightarrow 2\,H_2O + 115.6 \text{ kcal } (57.8 \text{ kcal/mole } H_2)$$

In a newer coal gasification process, high-energy methane is the end product. The process uses a catalyst (usually potassium hydroxide or potassium carbonate) and is thermally neutral (neither exothermic nor endothermic) at 700°C, the temperature at which the process is usually run.

In the process, crushed coal is mixed with an aqueous catalyst; the mixture is then dried and sent to a gasifier chamber where CO and H_2 are added. The mixture is then heated to 700°C.

Reactions that occur in the gasifier are (numbers are values at 25°C):

The three-reaction sequence is at the expense of 2 kcal/mole of CH_4 produced ($-64 + 8 + 54 = -2$).

$$2\,C + 2\,H_2O \longrightarrow 2\,CO + 2\,H_2 - 64 \text{ kcal/2 mole C}$$

$$CO + H_2O \longrightarrow CO_2 + H_2 + 8 \text{ kcal}$$

$$CO + 3\,H_2 \longrightarrow CH_4 + H_2O + 54 \text{ kcal}$$

The overall (or net) reaction is:

$$2\,C + 2\,H_2O \longrightarrow CH_4 + CO_2 - 2 \text{ kcal/mole } CH_4$$

Any unreacted CO and H_2 are cycled back through the gasifier. Recycled steam is used to help dry the coal before the coal enters the gasifier. The catalyst is recovered and reused.

This process converts solid, messy coal into easily transported, efficiently burned methane, the chief component of natural gas. The energy consumed by the process is small, and, best of all, the combustion of methane is environmentally clean.

Liquid fuels are made from coal by reacting the coal with hydrogen under high pressure in the presence of catalysts (hydrogenating the coal). Some sources of hydrogen are power gas, synthesis gas, and the electrolysis of water. The process produces more straight chains of carbon atoms, like those in petroleum. The resulting crude-oil type of material can be fractionally distilled like petroleum into diesel fuel, gasoline, and chemical raw materials for plastics, medicine, and other commodities. About 5.5 barrels of liquid are produced for each ton of coal fed to the liquefication plants. Present economic conditions do not favor the conversion of coal into gaseous and liquid fuels. However, the potential is there, and the reserves of coal are vast.

Natural Gas

Natural gas burns with a high heat output (see Table 5–2), produces little or no residue or pollution from burning, and is transported easily. Practically the only pollution produced by the combustion of natural gas is CO_2, which contributes to the greenhouse effect discussed in Chapter 12. However, the amount of CO_2 produced per energy unit is less for natural gas than for other fossil fuels.

Natural gas is mostly methane, CH_4.

Typical equations for the burning of natural gas are illustrated by those for the complete oxidation of methane and ethane in air:

$$CH_4 + 2\ O_2 \longrightarrow CO_2 + 2\ H_2O + 211\ \text{kcal/mole}\ CH_4$$

$$2\ C_2H_6 + 7\ O_2 \longrightarrow 4\ CO_2 + 6\ H_2O + 373\ \text{kcal/mole}\ C_2H_6$$

Natural gas is a mixture of gases trapped with petroleum in Earth's crust and is recoverable from oil wells or gas wells where the gases have migrated through the rock. Relative to petroleum and coal, natural gas is a simple mixture of hydrocarbons. In North America the natural gas mixture is composed of 60% to 90% methane (CH_4), 5% to 9% ethane (C_2H_6), 3% to 18% propane (C_3H_8), 1% to 2% butane (C_4H_{10}), along with a number of other gases, such as CO_2, N_2, H_2S, and noble gases present in varying amounts. In Europe and Japan the natural gas is essentially all methane. Natural gas is easily stored and transported across continental distances through pipelines (Fig. 5–10).

About half of the homes in the United States are heated by natural gas, followed by electricity (18.5%), fuel oil (14.9%), wood (4.8%), and liquefied gas such as butane and propane (4.6%). Coal and kerosene come in at a low 0.5%, and solar heating of homes is even lower.

It is estimated that about 60% of the North American natural gas deposits are now depleted. Larger reserves are known to exist in the Middle East, Eastern Europe, and the former Soviet Union. It remains to be seen if methane will be produced in commercial quantities from coal and then used as a fuel as the reserves of natural gas diminish.

Figure 5–10 Liquefied natural gas storage tanks. (*The World of Chemistry*, Program 5, "A Matter of State")

SELF-TEST 5A

1. Energy associated with moving matter is ____________ energy.
2. There is ____________ energy in a mixture of natural gas and air that is sufficient to cause an explosion.
3. Examples of fossil fuels are ____________, ____________, and ____________.
4. Natural gas and petroleum react with ____________ to produce carbon dioxide and ____________.
5. The ultimate fate of all types of energy release is an increase in ____________.
6. Energy is (a) required or (b) released when chemical bonds between atoms are broken.
7. Which fossil fuel produces the most energy per gram during combustion? ____________
8. Which is conserved in any energy exchange? (a) quality or (b) quantity of the energy involved
9. What is the efficiency of energy use in a typical automobile? ____________
10. Petroleum is composed mostly of what two elements? ____________ and ____________
11. What is the greatest use (about 75%) in burning mined coal? ____________
12. Synthesis gas is composed of what two combustible substances? ____________ and ____________
13. Which fossil fuel remains in greatest natural supply? ____________
14. What fuel is most used in heating homes in the United States? ____________

ELECTRICITY: THE MAJOR SECONDARY SOURCE OF ENERGY

A secondary source of energy is made from a primary source on the way to the end user. Coal is used to produce more electricity than all of the other primary sources combined. Following in order are nuclear energy, natural gas, hydroelectric along with geothermal energy, and oil.

The use of electricity continues its rapid growth. Electric motors, the force behind enormous gains in technological innovation and industrial productivity over the last half century, consume two thirds of America's electricity. Electronics, communication devices, elevators, air conditioners, electric irons, microwaves, toothbrushes, flashlights, and typewriters are some of the thousands of electric gadgets that we "cannot do without" in our modern society.

The quad is a very large unit of energy. 1 quad = 1 quadrillion British thermal units of energy (Btus) = 2.52×10^{14} kcal

$$\frac{\text{8.8 quads of electricity}}{\text{28.6 quads required}} = 31\% \text{ efficiency}$$

1 quadrillion = 10^{15}

About 36% of all the energy consumed in the United States is used in the production of electricity. In 1988 the 28.6 quads of energy put into the production of electricity yielded about 8.8 quads of electricity in the home or factory using the electricity. The 19.8-quad difference was lost in the production and the transmission of electricity. At least part of this loss is ex-

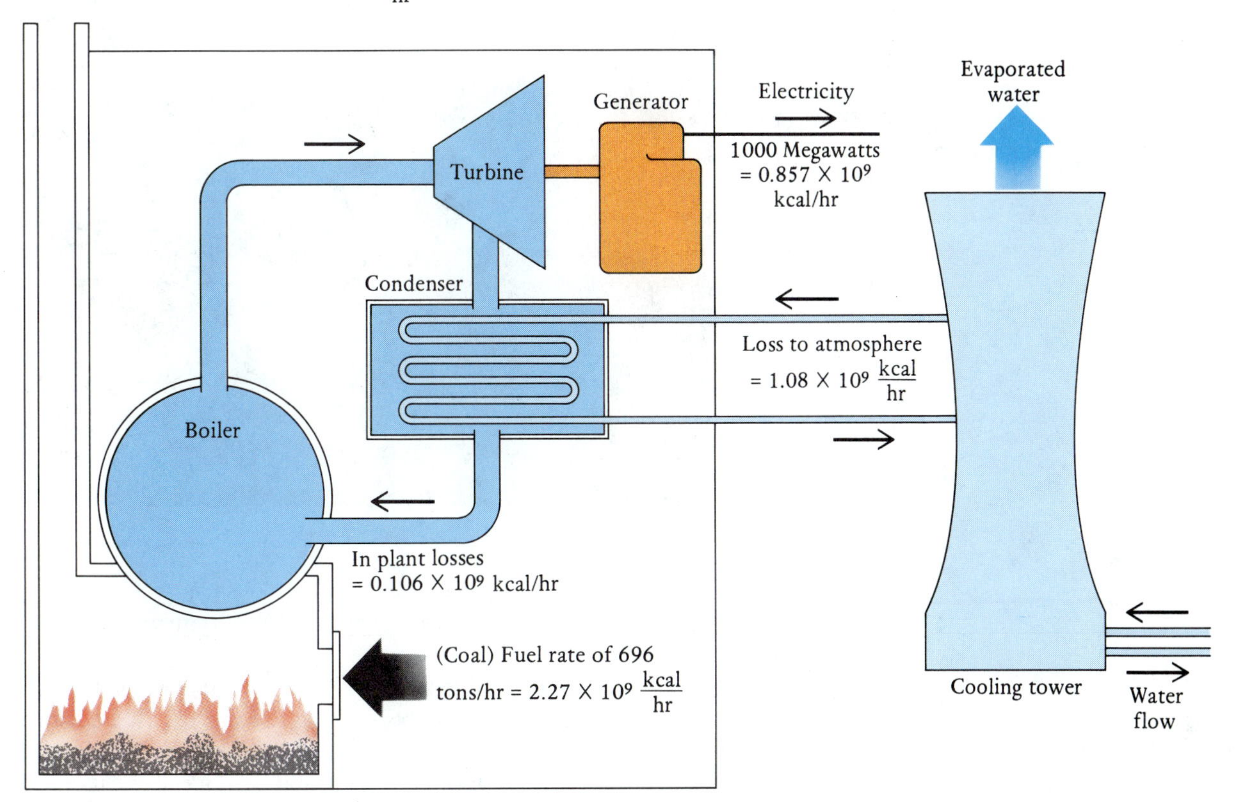

Figure 5-11 The heat balance of a 1000-megawatt coal-burning electric generating plant. Note that the 696 tons of coal burned per hour furnish 2.27×10^9 kcal of heat energy, but only 0.857×10^9 kcal of energy, or 38%, is converted to electricity. Note also the large amounts of heat energy lost to the cooling water and atmosphere.

pected because of the second law of thermodynamics (discussed earlier in this chapter), which states that a natural process loses some nonuseful energy as entropy (disorder) is increased.

A specific example of energy loss in the production of electricity in a coal-burning power plant is shown in Figure 5-11 and summarized below. For a 1000-megawatt coal-burning plant, one hour of operation might look like this:

Coal consumed	696 tons producing 2.270 billion kcal
Smokestack heat loss	0.227 billion kcal
Heat loss in plant	0.106 billion kcal
Heat loss in evaporator to cool condenser	1.080 billion kcal
Electrical energy delivered to power lines	0.857 billion kcal
Percentage of energy delivered as electricity before transmission losses	$\frac{0.857}{2.27} \times 100\% = 37.8\%$

There is a further energy loss in the power lines and the transformers (Fig. 5-12), lowering the useful output of the plant to 30% of the energy consumed. This is the **efficiency** figure for the overall operation. It is important to note that we pay for 300 kcal of heat energy in the form of coal or fuel oil

When we pay the electricity bill, we pay for energy (kilowatt-hours), not for power (how fast we use the energy, or kilowatts).

Figure 5–12 Electrical transmission lines are necessary to carry the electricity, but they incur energy loss to the consumer. You have noted radiant energy from such lines as static on your automobile radio. Currently there is considerable debate relative to the possible harmful effects of the radiant energy from electrical transmission lines.

but receive less than 100 kcal of energy in the form of electricity. Obviously, it requires much less fuel to heat homes with the fuel itself than with electricity made from the fuel.

Electricity, having found considerable use in public transportation as an energy source for trains and trollies, may yet be significant in private transportation, where the energy has to be stored rather than a contact made with an on-line power supply. Because of battery limitations, electric cars have low driving ranges and long recharging times. However, the major automobile makers such as Fiat, Peugeot, Ford, and General Motors are actively developing, and in some cases producing, experimental cars that could someday become the principal means of private urban transportation.

NOVEL, NONMAINSTREAM SOURCES OF ENERGY

Some energy sources offer more potential than they now supply, although most of these sources are available only in certain areas. For example, in some locales (e.g., Boise, Idaho), it is possible to heat homes and make electricity from geothermal sources, such as hot springs and geysers. California produces 5% of its electricity from geothermal energy. Political debate in Hawaii is intense, where over 60% of its electricity is generated by burning foreign oil, relative to possible environmental damage in tapping the endless supply of geothermal energy readily available on the Big Island. Other areas near oceans derive energy from ocean currents and temperature differences between warm surface water and colder, deeper water. For many years wind currents and windmills have provided energy for pumping water and generating electricity. Wind turbines can now produce electricity at favorable sites for 7 to 9 cents per kilowatt-hour, and the prediction is that this cost will be reduced to 4 to 6 cents as the rate of production of the equipment is in-

Today almost all U.S. wind power is harnessed in California where more than 15,000 wind turbines produce over 1% of California's electricity.

creased. Although the potential energy from wind is vast, it is not reliable as a *constant* source on call at any time.

One novel source of energy is available in every populated area. Garbage can be used to produce energy by burning or fermentation. Plants for burning garbage to extract energy are in operation in several countries, including France and the United States; there are about 130 operating plants in the United States. Bridgeport, Connecticut, generates about 10% of its electricity by burning trash.

Everybody has it; nobody wants it. Why not use it to provide electricity? Great idea!

The Nashville Thermal Transfer Corporation in downtown Nashville, Tennessee, began operation in 1974 (Fig. 5-13). The plant supplies steam and/or cold water through underground pipes to buildings in the central city. The ash from the burned garbage goes to the landfill, where it takes up only 10% of the volume that would have been occupied by the raw garbage. Oil and gas are available as back-up energy sources if needed in the Nashville plant, but they are rarely needed.

The advantage of burning garbage for energy production is offset if careful control is not maintained in the proper disposal of the ash, which is likely to contain toxic heavy metals (see Chapter 10) and harmful flue gases (see Chapter 12). Plans to increase the Nashville plant from a consumption of 400 tons of garbage a day to 800 tons have prompted years of political debate. Such environmental concern is typical even though in theory, if not always in practice, the operation can be carried out with only the emission of carbon dioxide as an environmental negative.

A product of the fermentation of garbage is extracted from the world's largest garbage dump at Fresh Kills on Staten Island, New York. Underneath the huge mounds of garbage, bacteria turn some of the old buried garbage into methane (natural gas). The Brooklyn Union Gas Company has tapped this gas, which provides enough methane to fuel 16,000 homes on Staten Island.

The U.S. Department of Energy is supporting a research program aimed at obtaining liquid fuels, mostly alcohols, directly from recently grown bio-

Maine generates about 20% of its electricity by burning wood chips, and Hawaii obtains as much as one third of its electricity by burning sugarcane residues. However, biomass accounts for less than 1% of U.S. electricity.

Figure 5-13 The Nashville, Tennessee, thermal energy plant.

mass. Research has centered on the black cottonwood tree, the fastest growing hardwood in the Washington State area. Field tests for over six years show that through selective breeding and cloning, the dry weight of wood can be more than doubled relative to normal growth of this species. The ability to obtain alcohol fuels through the degradation and fermentation of wood grown for this purpose will add new meaning to the efforts to reverse the deforestation of the world. Indeed, the key to having combustible fuels for the indefinite future may be through the capture of solar energy in the biomass, which can then be processed for fuel production.

NUCLEAR ENERGY

Few issues have captured the awe, imagination, and scrutiny of mankind to quite the extent that nuclear energy has in the past five decades. Nuclear energy has been acclaimed, on the one hand, as the source of all our energy needs and accused, on the other hand, of being our eventual destroyer.

Part of the interest in nuclear power is the tremendous amount of energy generated by a relatively small amount of fuel. Nuclear change, unlike chemical change, involves rearrangement within atomic nuclei. As a consequence, an atom can be changed from one element to another. An example is the uranium-238 isotope; each atom contains 92 protons and 146 neutrons in the nucleus and 92 electrons in energy levels outside the nucleus. In nuclear change an atom of an isotope is represented with a mass number as a superscript to the left of the elemental symbol and the nuclear charge (atomic number for the element involved) as a subscript, also to the left of the symbol. The representation for the uranium-238 isotope is:

Isotopes were introduced in Chapter 3.

$$^{238}_{92}U$$

The atomic number subscript is redundant to the elemental symbol but is commonly used. Atoms of uranium-238 are not completely stable, as some of them spontaneously give off an alpha particle ($^{4}_{2}He$). After the alpha particle has been lost, the resulting nucleus has a mass of 234 and a nuclear charge of 90. Atoms containing 90 protons in the nucleus are atoms of thorium (Th), not uranium. This spontaneous nuclear reaction then has changed an atom of one element into an atom of another element.

Review the discussion of radioactivity and alpha particles in Chapter 3.

The decomposition of the $^{238}_{92}U$ nucleus is stated briefly by the following **nuclear equation:**

$$^{238}_{92}U \longrightarrow {}^{4}_{2}He + {}^{234}_{90}Th$$

In nuclear reactions the sum of the atomic numbers on the left side of the equation equals the sum of the atomic numbers on the right side of the equation. Likewise for the atomic masses.

Vast amounts of energy are released when heavy atomic nuclei split, called the **fission process,** and when small atomic nuclei combine to make heavier nuclei, called the **fusion process.** Consider the energy contrast between combustion of a fossil fuel and a nuclear fusion reaction. When 1 mole (6.022×10^{23} molecules, or 16 g) of methane is burned, over 200 kcal of heat are liberated:

$$CH_4 + 2\ O_2 \longrightarrow CO_2 + 2\ H_2O + 211\ kcal\ (kcal/mole\ CH_4)$$

In contrast, a lithium (Li) nucleus can be made to react with a H nucleus to form two helium (He) nuclei in a nuclear reaction. The energy released per mole of lithium in this reaction is 23,000,000 kcal. This means that 7 g of Li and 1 g of H produce 100,000 times more energy through fusion of nuclei than 16 g of CH_4 and 64 g of O_2 produce by electron exchange.

Energy changes associated with nuclear events may be many thousands of times larger than those associated with chemical events.

$$^{7}_{3}Li + ^{1}_{1}H \longrightarrow 2\ ^{4}_{2}He + 23{,}000{,}000\ \text{kcal/mole of}\ ^{7}_{3}Li$$

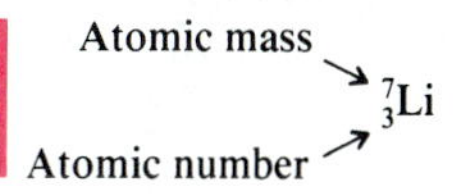

Realizing that nuclear changes could involve giant amounts of energy relative to chemical changes for a given amount of matter, Otto Hahn, Fritz Strassman, Lise Meitner, and Otto Frisch discovered in 1938 that $^{235}_{92}U$ is fissionable. Subsequently the dream of controlled nuclear energy became a reality, followed by the bomb and nuclear power plants. In the 1950s it was hoped that nuclear energy would soon relieve the shortage of fossil fuels. To date this has not been accomplished, although the production of nuclear energy has grown to the point of providing 20% of the electricity generated in the United States at the present time. In France 80% of its total energy production in 1988 was nuclear, and in Japan it was 67%.

There are 113 nuclear power plants in the United States, but no new ones have been ordered since 1978. Reasons: (a) financial concerns, (b) safety problems, and (c) radioactive waste disposal.

Fission Reactions

Fission is the break-up of heavy nuclei.

Fission can occur when a thermal neutron (with a kinetic energy about the same as that of a gaseous molecule at ordinary temperatures) enters certain heavy nuclei with an odd number of neutrons ($^{235}_{92}U$, $^{233}_{92}U$, $^{239}_{94}Pu$). The splitting of the heavy nucleus produces two smaller nuclei, two or more neutrons (an average of 2.5 neutrons for $^{235}_{92}U$), and much energy. Typical nuclear fission reactions may be written:

$$^{235}_{92}U + ^{1}_{0}n \longrightarrow ^{141}_{56}Ba + ^{92}_{36}Kr + 3\ ^{1}_{0}n + \text{energy}$$

$$^{235}_{92}U + ^{1}_{0}n \longrightarrow ^{103}_{42}Mo + ^{131}_{50}Sn + 2\ ^{1}_{0}n + \text{energy}$$

Note that the same nucleus may split in more than one way. The fission products, such as $^{141}_{56}Ba$ and $^{92}_{36}Kr$, emit beta particles ($_{-1}^{0}e$) and gamma rays ($^{0}_{0}\gamma$) until stable isotopes are reached.

$^{1}_{0}n$ represents a neutron.

$_{-1}^{0}e$ or $_{-1}^{0}\beta$ represents a beta particle.

$$^{141}_{56}Ba \longrightarrow _{-1}^{0}e + ^{0}_{0}\gamma + ^{141}_{57}La$$

$$^{92}_{36}Kr \longrightarrow _{-1}^{0}e + ^{0}_{0}\gamma + ^{92}_{37}Rb$$

A low-energy neutron will disrupt some large nuclei.

The products of these reactions emit beta particles, as do their products. After several such steps, stable isotopes are reached: $^{141}_{59}Pr$ and $^{90}_{40}Zr$, respectively.

The neutrons emitted can cause the fission of other heavy atoms if they are slowed down by a moderator, such as graphite. For example, the three neutrons emitted in the preceding uranium reaction could produce fission in three more uranium atoms, the nine neutrons emitted by those nuclei could produce nine more fissions, the 27 neutrons from these fissions could produce 81 neutrons, the 81 neutrons could produce 243, the 243 neutrons could produce 729, and so on. This process is called a **chain reaction** (Fig. 5-14), and it occurs at a maximum rate when the uranium sample is large enough for most of the neutrons emitted to be captured by other nuclei

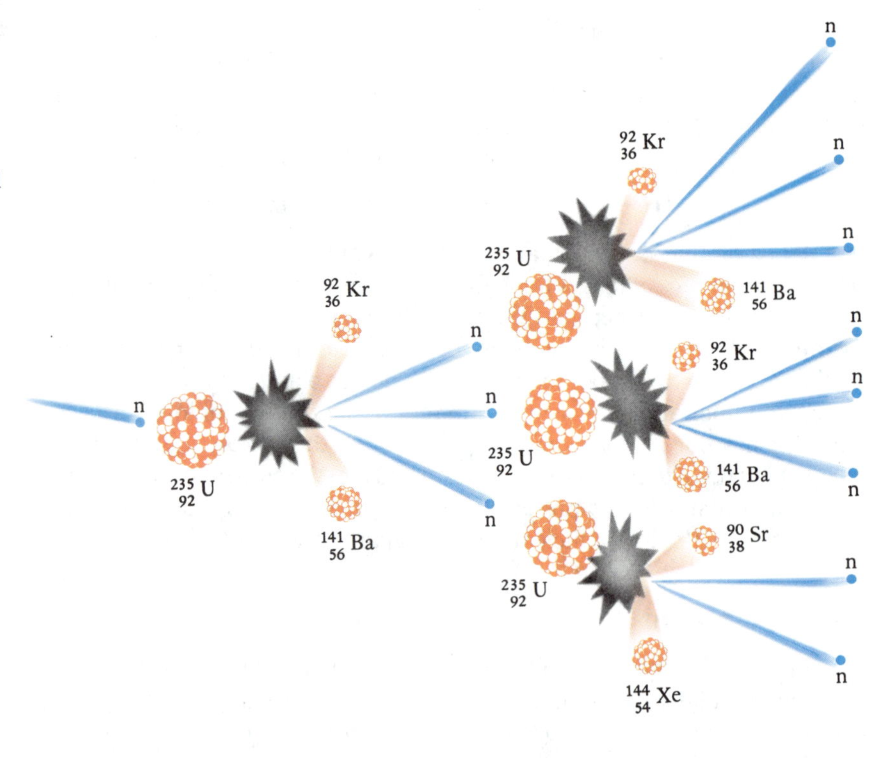

Figure 5–14 A chain reaction. A thermal neutron collides with a fissionable nucleus, and the resulting reaction produces three additional neutrons. If enough fissionable nuclei are present, a chain reaction will be sustained.

before passing out of the sample. Sufficient sample in a certain volume to sustain a chain reaction is termed the **critical mass.**

An atomic bomb explosion. (*The World of Chemistry,* Program 6, "The Atom")

In the atomic bomb the critical mass is kept separated into several smaller subcritical masses until detonation, at which time the masses are driven together by an implosive device. It is then that the tremendous energy is liberated and everything in the immediate vicinity is heated to temperatures of 5 to 10 million degrees. The sudden expansion of hot gases literally explodes everything nearby and scatters the radioactive fission fragments over a wide area. In addition to the movement of gases, there is the tremendous vaporizing heat that makes the atomic bomb so devastating.

There is no danger of an atomic explosion in the uranium mineral deposits in the Earth for two reasons. First, uranium is not found pure in nature—it is found only in compounds, which in turn are mixed with other compounds. Second, less than 1% of the uranium found in nature is fissionable $^{235}_{92}U$. The other 99% is $^{238}_{92}U$, which is not fissionable by thermal neutrons. In order to make nuclear bombs or nuclear fuel for electricity generation, a purification enrichment process must be performed on the uranium isotopes, thus increasing the relative proportion of $^{235}_{92}U$ atoms in a sample. Ordinary uranium such as that found in ores is only 0.7% $^{235}_{92}U$.

Separation of uranium isotopes had to precede the control of atomic energy.

Mass Defect—The Ultimate Nuclear Energy Source

What is the source of the tremendous energy of the fission process? It ultimately comes from the conversion of mass into energy, according to Einstein's famous equation, $E = mc^2$, where E is energy that results from the

The mass that is lost leaves in the form of energy: $E = mc^2$.

loss of an amount of mass m, and c^2 is the speed of light squared. If separate neutrons, electrons, and protons are combined to form any particular atom, there is a loss of mass called the **mass defect.** For example, the calculated mass of one ${}^{4}_{2}He$ atom from the masses of the constituent particles is 4.032982 amu:

amu = atomic mass unit

$2 \times 1.007826 = 2.015652$ amu, mass of two protons and two electrons
$2 \times 1.008665 = \underline{2.017330}$ amu, mass of two neutrons
total = 4.032982 amu, calculated mass of one ${}^{4}_{2}He$ atom

6.02×10^{23} amu/g
or
1 mole amu/g

Since the measured mass of a ${}^{4}_{2}He$ atom is 4.002604 amu, the mass defect is 0.030378 amu:

$$4.032982 \text{ amu} - 4.002604 \text{ amu} = 0.030378 \text{ amu, mass defect}$$

Because the atom is more stable than the separated neutrons, protons, and electrons, the atom is in a lower energy state. Hence, the 0.030378 amu lost per atom would be released in the form of energy if the ${}^{4}_{2}He$ atom were made from separate protons, electrons, and neutrons.

Separated nuclear particles have more mass than when combined in a nucleus.

It takes only about 1 kg of ${}^{235}_{92}U$ or ${}^{239}_{94}Pu$ undergoing fission to be equivalent to the energy released by 20,000 tons (20 kilotons) of ordinary explosives like TNT. The energy content in matter is further dramatized when it is realized that the atomic fragments from the 1 kg of nuclear fuel weigh 999 g, so only one-tenth of 1% of the mass is actually converted to energy. The fission bombs dropped on Japan during World War II contained approximately this much fissionable material.

Controlled Nuclear Fission

The fission of a ${}^{235}_{92}U$ nucleus by a slow-moving neutron to produce smaller nuclei, extra neutrons, and large amounts of energy suggested to Enrico Fermi and others that the reaction could proceed at a moderate rate if the number of neutrons could be controlled. If a neutron control could be found, the concentration of neutrons could be maintained at a level sufficient to keep the fission process going but not high enough to allow an uncontrolled explosion. It would then be possible to drain the heat away from such a reactor on a continuing basis to do useful work. In 1942, Fermi, working at the University of Chicago, was successful in building the first atomic reactor, called an **atomic pile.**

Atomic pile:
1. Carefully diluted fissionable material
2. Moderator to control fission reaction
3. Coolant to control heat
4. Shielding to limit radiation

The first one was piled together at the University of Chicago in 1942.

An atomic reactor has several essential components. The charge material (fuel) must be fissionable or contain significant concentrations of a fissionable isotope such as ${}^{235}_{92}U$, ${}^{239}_{94}Pu$, or ${}^{233}_{92}U$. Ordinary uranium, which is mostly the nonfissionable ${}^{238}_{92}U$, cannot be used because it has a small concentration of the ${}^{235}_{92}U$ isotope. A moderator is required to slow the speed of the neutrons produced in the reactions without absorbing them. Graphite, water, and other substances have been used successfully as moderators. A substance that absorbs neutrons, such as cadmium or boron steel, is present in order to have a fine control over the neutron concentration. Shielding, to protect the workers from dangerous radiation, is an absolute necessity. Shielding tends to make reactors heavy and bulky installations. A heat-

A few nuclear reactors, such as some at Oak Ridge, Tennessee, are used to produce radioactive isotopes for medicine and industry, not to produce electricity.

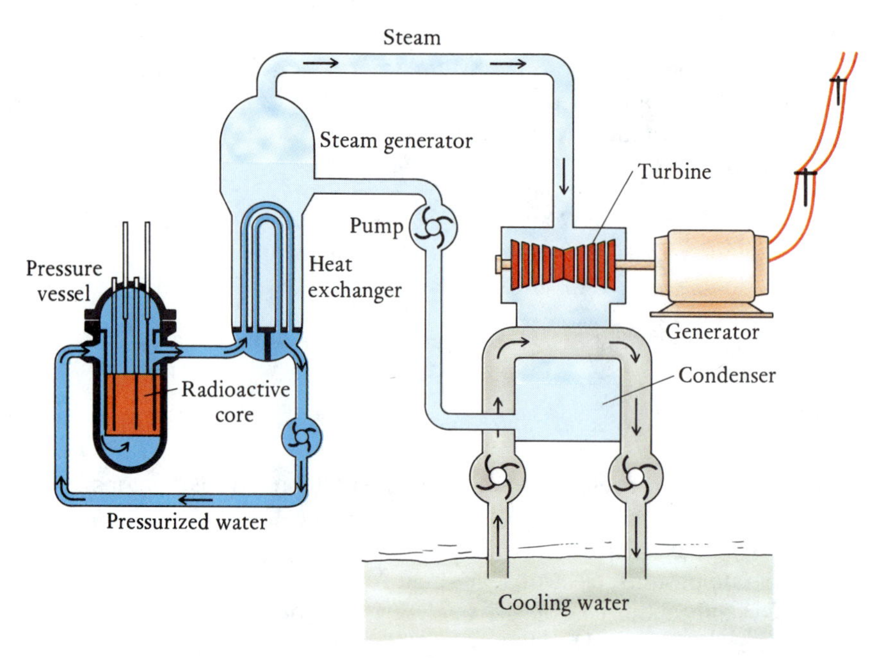

Figure 5-15 Schematic illustration of a nuclear power plant.

transfer fluid provides a large and even flow of heat away from the reaction center.

Once the heat is produced in a nuclear reactor and safety measures are employed to protect against radiation, conventional technology allows this energy to be used to generate electricity, to power ships, or to operate any device that uses heat energy. A system for the nuclear production of electricity is illustrated in Figure 5-15.

What are the fuel requirements in nuclear fission energy production? A typical fission event and the energy released is

$$^{1}_{0}n + ^{235}_{92}U \longrightarrow ^{93}_{37}Rb + ^{141}_{55}Cs + 2\,^{1}_{0}n + 7.7 \times 10^{-12} \text{ cal/atom of U}$$

Since 1 g of pure $^{235}_{92}U$ contains 2.56×10^{21} atoms, the total energy release for 1 g of uranium-235 undergoing fission would be

$$1.0\,\cancel{g} \times 2.56 \times 10^{21} \frac{\cancel{\text{atoms}}}{1\,\cancel{g}} \times 7.7 \times 10^{-12} \frac{\text{cal}}{\cancel{\text{atom}}} = 2.0 \times 10^{10} \text{ cal}$$

This is the amount of energy that would be released if 5.95 tons of coal were burned, or if 13.7 barrels of oil were burned to produce heat to power a boiler. This means that about 3 kg of $^{235}_{92}U$ fuel per day would be required for a 1000-megawatt electric generator. The fuel used, however, is not pure $^{235}_{92}U$, but **enriched** uranium containing up to 3% $^{235}_{92}U$.

A megawatt is equal to 1 million watts.

It is possible to convert the nonfissionable $^{238}_{92}U$ and $^{232}_{90}Th$ into fissionable fuels by using a **breeder reactor.** In such a reactor, a blanket of nonfissionable material is placed outside the fissioning $^{235}_{92}U$ fuel, which serves as the source of neutrons in the breeder reactions. The two breeder reaction sequences are

$^{235}_{92}U$ is in very short supply.

$$^{238}_{92}U + ^{1}_{0}n \longrightarrow ^{239}_{92}U \xrightarrow{\beta} ^{239}_{93}Np \xrightarrow{\beta} ^{239}_{94}Pu$$

$$^{232}_{90}Th + ^{1}_{0}n \longrightarrow ^{233}_{90}Th \xrightarrow{\beta} ^{233}_{91}Pa \xrightarrow{\beta} ^{233}_{92}U$$

The products of the breeder reactions, $^{233}_{92}U$ and $^{239}_{94}Pu$, are both fissionable with slow neutrons, and neither is found in Earth's crust.

$^{239}_{94}Pu$ is toxic from a radiation as well as a chemical point of view.

Another approach to the breeder reactor is the emerging technology of electrical breeding. In this technique high-speed protons are electrically accelerated into a target of nonfissionable isotopes such as uranium-238. Again, fissionable products, which can be used as fuel, are the result.

At one time it was thought that our appetite for energy might use up the fissionable isotopes available in nature, thus limiting the future of atomic energy. It is now apparent that breeder reactions can be used to tap vast supplies of energy from the fission of selected isotopes of the heavy elements.

The primary problem with nuclear energy is the radiation danger to all life forms. No amount of high-energy radiation, such as X and gamma radiation, is known to be associated with zero risk to living cells. Moderate and major exposures result in sickness and death, and even minor exposures are statistically associated with cancer and mutations. Only very small exposures such as in the case of medical X rays are generally believed to be "safe" or at least "worth the risk." The emission of gamma radiation is associated with the nuclear changes in nuclear reactors and nuclear explosions. Although all humans are exposed to a low level of natural background radiation coming from natural events in the universe, human-provoked radiation can be intense enough to cause major health problems, severe illness, and death. Our experience thus far includes:

Tens of thousands of Japanese killed by radiation from planned atomic explosions at the end of World War II.

An as yet unmeasured disaster at Chernobyl, with a reactor meltdown and steam explosion. Thirty-one men died from radiation on site and hundreds became sick from radiation damage. Long-range estimates are that thousands of individuals will have long-term health problems as a result of this incident.

Low-level radiation contamination around numerous processing and weapons plants around the world.

Individual accidental radiation exposures in laboratories.

To balance the picture, one should take note that 16% of the commercially produced electricity in the world is safely produced by 429 nuclear power plants, and numerous naval vessels are powered by nuclear reactors. Even though these reactors produce relatively close to lethal levels of radiation, the operators are "adequately shielded" from them and their harmful effects.

Radiation safety is complicated by the long lives of some of the radioactive materials produced in nuclear reactors. Some have **half-lives** measured in thousands of years. The half-life is the time required for one half of the material to decompose through radiation processes. At present, there is no generally accepted plan for permanently storing nuclear wastes. The U.S. Department of Energy has spent $2 billion in generating plans for a radioactive waste disposal site at Yucca Mountain 100 miles northwest of Las Vegas. This site would store 77,000 tons of wastes "safely for 10,000 years." Nevada has opposed the plan in court on three grounds: a possibility of (a) volcanic activity, (b) earthquakes, and (c) a rising ground water level. Managed temporary storage is currently accomplished with radiation exposure at "acceptable" low levels. Temporary storage of radioactive wastes in the United States will reach 44,000 tons by the year 2000.

It should be remembered, when measuring the dangers of nuclear radiation, that no commercial energy production is risk free and that risks associated with controlled processes have been reducible if values and efforts are associated with that risk reduction.

Fusion Reactions

Fusion is the combination of very light nuclei.

When very light nuclei, such as H, He, and Li, are combined, or **fused,** to form an element of higher atomic number, energy must be given off consistent with the greater stability of the elements in this intermediate atomic number range. This energy, which comes from a decrease in mass, is the source of the energy released by the Sun and by hydrogen bombs. Typical examples of fusion reactions are:

$^{2}_{1}H$ = Deuterium
$^{3}_{1}H$ = Tritium
$^{\ 0}_{+1}e$ = Positron

$$4\,^{1}_{1}H \longrightarrow \,^{4}_{2}He + 2\,^{\ 0}_{+1}e + 6.22 \times 10^{8} \text{ kcal for 4 moles of } ^{1}_{1}H \text{ fused}$$

$$^{2}_{1}H + \,^{2}_{1}H \longrightarrow \,^{3}_{2}He + \,^{1}_{0}n + 0.75 \times 10^{8} \text{ kcal}$$

$$^{2}_{1}H + \,^{2}_{1}H \longrightarrow \,^{3}_{1}H + \,^{1}_{1}H + 0.93 \times 10^{8} \text{ kcal}$$

$$^{3}_{1}H + \,^{2}_{1}H \longrightarrow \,^{4}_{2}He + \,^{1}_{0}n + 4.1 \times 10^{8} \text{ kcal}$$

The net reaction for the last three reactions is:

$$5\,^{2}_{1}H \longrightarrow \,^{4}_{2}He + \,^{3}_{2}He + \,^{1}_{1}H + 2\,^{1}_{0}n + 5.078 \times 10^{8} \text{ kcal for 5 moles of } ^{2}_{1}H \text{ fused}$$

Materials for fusion reactions are available in enormous amounts.

The critical mass for ^{235}U tends to limit the size of a fission bomb, but a fusion bomb with more LiH can be made much more powerful.

Deuterium is a relatively abundant isotope—out of 6500 atoms of hydrogen in sea water, for example, one is a deuterium atom. What this means is that the oceans are a potential source of fantastic amounts of deuterium. There are 1.03×10^{22} atoms of deuterium in a single liter of sea water. In a single cubic kilometer of sea water, therefore, there would be enough deuterium atoms with enough potential energy to equal the burning of 1360 billion barrels of crude oil, and this is approximately the total amount of oil originally present on this planet.

A plasma is a gaseous state composed of ions.

Fusion reactions occur rapidly only when the temperature is of the order of 100 million degrees or more. At these high temperatures atoms do not exist as such; instead, there is a **plasma** consisting of unbound nuclei and electrons. In this plasma nuclei merge or combine. In order to achieve the high temperatures required for the fusion reaction of the hydrogen bomb, a fission bomb (atomic bomb) is first set off.

One type of hydrogen bomb depends on the production of tritium ($^{3}_{1}H$) in the bomb. In this type, lithium deuteride ($^{6}_{3}Li^{2}_{1}H$, a solid salt) is placed around an ordinary $^{235}_{92}U$ or $^{239}_{94}Pu$ fission bomb. The fission is set off in the usual way. A $^{6}_{3}Li$ nucleus absorbs one of the neutrons produced and splits into tritium, $^{3}_{1}H$, and helium, $^{4}_{2}He$.

$$^{6}_{3}Li + \,^{1}_{0}n \longrightarrow \,^{3}_{1}H + \,^{4}_{2}He$$

The temperature reached by the fission of $^{235}_{92}U$ or $^{239}_{94}Pu$ is sufficiently high to bring about the fusion of tritium and deuterium:

$$^{3}_{1}H + ^{2}_{1}H \longrightarrow ^{4}_{2}He + ^{1}_{0}n + 4.1 \times 10^{8} \text{ kcal}$$

A 20-megaton bomb usually contains about 300 lb of lithium deuteride, as well as a considerable amount of plutonium and uranium.

Attempts at Controlled Nuclear Fusion

Three critical requirements must be met for controlled fusion. First, the temperature must be high enough for fusion to occur. The fusion of deuterium ($^{2}_{1}H$) and tritium ($^{3}_{1}H$) requires a temperature of 100 million degrees or more.

$$^{3}_{1}H + ^{2}_{1}H \longrightarrow ^{4}_{2}He + ^{1}_{0}n + 4.1 \times 10^{8} \text{ kcal/mole } ^{2}_{1}H$$

Second, the plasma must be confined long enough to release a net output of energy. Third, the energy must be recoverable in some usable form.

Attractive features that encourage research in controlled nuclear fusion are the rather limited production of dangerous radioactivity and the great abundance of hydrogen fuel (in water), a most abundant resource. Most radioisotopes produced by fusion have short half-lives and therefore are a serious hazard for only a short time.

Fusion reactions have not been "controlled." No physical container can contain the plasma without cooling it below the critical fusion temperature. Magnetic "bottles," enclosures in space bounded by a magnetic field, have confined the plasma, but not for long enough periods.

Containment is one of the biggest problems in developing controlled fusion.

A newer confinement method is based on a laser system that simultaneously strikes tiny hollow glass spheres called **microballoons,** which enclose the fuel, consisting of equal parts of deuterium and tritium gas at high pressures (Fig. 5-16).

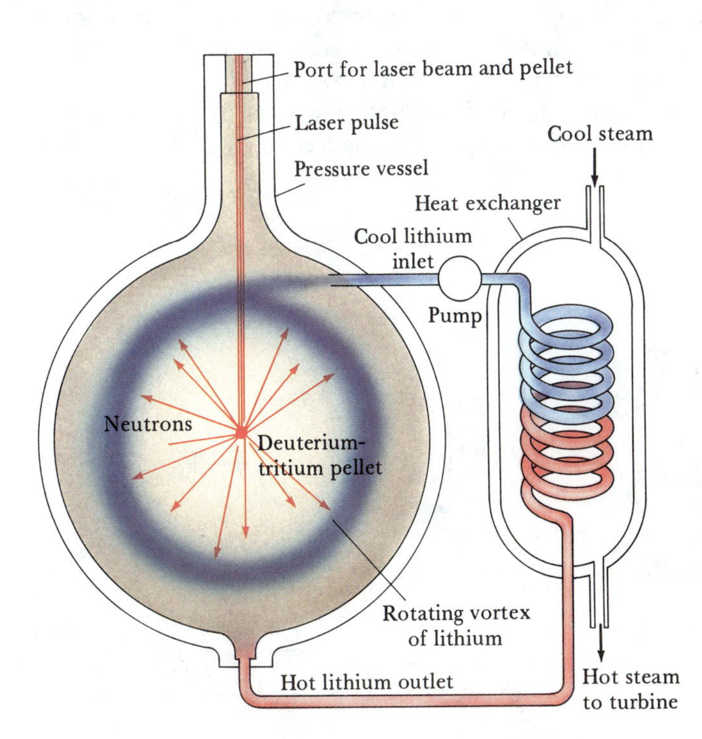

Figure 5-16 Schematic diagram of an apparatus for laser-induced fusion. Tiny glass pellets (microballoons about 0.1 mm in diameter) filled with frozen deuterium and tritium are subjected to a powerful laser beam, and the contents undergo nuclear fusion.

Two other attempts to achieve fusion are aneutronic fusion and the particle-beam fusion accelerator (PBFA). Neither approach has delivered enough concentrated energy to sustain fusion.

Aneutronic fusion, also called migma (a Greek word for "mixture"), was presented theoretically by the American scientist, Bogdan Maglich, in 1973. Since this process does not involve neutrons either as products or reactants, a penetrating, hard-to-capture, potentially damaging particle is eliminated. Fusion is achieved by accelerating nuclei (such as deuterons, ^2_1H) in linear accelerators to an energy equivalent to a temperature of 7×10^9°C. The high-energy ions are directed on a lithium target. The fusion of a deuteron and a lithium nucleus produces two helium nuclei and energy.

In November 1991 the Joint European Torus fusion reactor in Culham, England provided almost 2 million watts during a two-second controlled fusion reaction. The fuel in the fusion reactor was a mixture of 14% tritium and 86% deuterium.

In the PBFA process, electrical charge is stored in capacitors and discharged in 40-nsec (nanosecond) pulses. The energy accelerates lithium ions to a kinetic energy of between 250 and 500 kcal. The lithium ions impinge on a target of deuterium (^2_1H) and tritium (^3_1H). Lithium nuclei fuse with one or the other of the hydrogen isotopes and produce energy.

After more than 40 years of intense research, it does not appear that controlled hot fusion is even on the verge of making any contribution to our energy needs in the foreseeable future.

The present projection for commercial energy from fusion reactors is the year 2040.

SOLAR ENERGY

The amount of energy that enters Earth's atmosphere from the Sun each day is enormous, about 15,000 times the world's present daily energy demand, although it is only about three ten-millionths of the total energy released by the Sun. About 30% of the incoming energy is reflected, 50% is radiated as heat, and 20% is consumed by the hydrologic cycles that produce the weather on our planet. Only 0.06% of the energy from the Sun goes into photosynthesis. Although the actual amount of solar energy that reaches the surface of the earth depends on location, season, and weather conditions, the roof of an average-sized house receives enough energy per day to equal

Solar energy is transmitted nuclear energy.

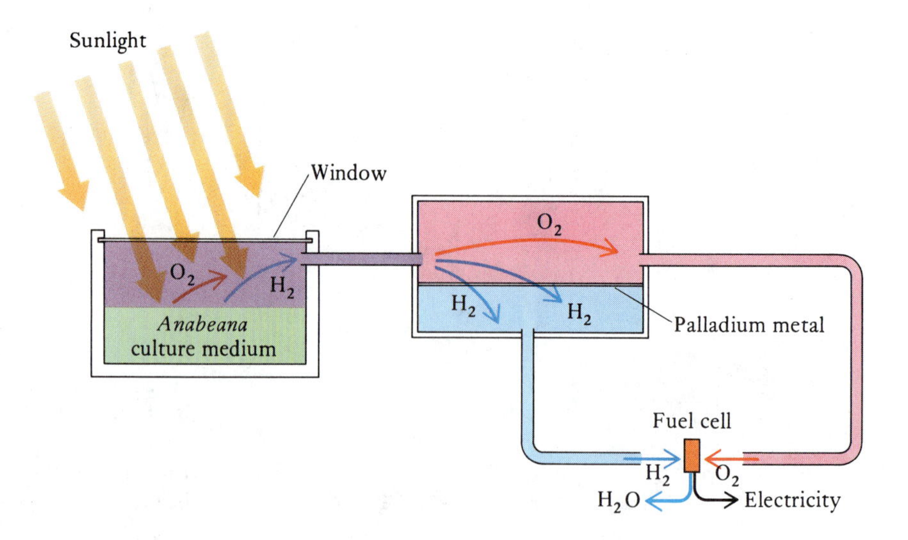

Figure 5–17 Schematic diagram of an electricity-producing photosynthesis process. H_2 and O_2 produced by the blue-green algae, *Anabeana cylindrica,* are separated by palladium metal, which is permeable to H_2 but not to O_2. The H_2 and O_2 are then combined in the fuel cell to produce electricity.

the burning of 32 lb of coal, or 120 kilowatt-hours of electrical energy—more than enough energy to heat an average American home in the winter.

Efforts are now being made in several directions to use more of the Sun's gift to us. One technique is to use algae to produce hydrogen (Fig. 5–17), which then can be used in fuel cells to produce electricity. Another device is the solar collector, which uses the warmth of sunlight to heat water and air to heat homes. A third device uses photosensitive materials to make a solar electric cell, such as the type that is commonly used to power hand calculators.

95% of U.S. solar-power electricity comes from Luz International in California's Mojave Desert, where 1.5 million reflector mirrors are used to focus the solar energy.

Solar Energy and the Solar Cell

Another approach to the direct utilization of solar energy is the *solar cell,* known as a photovoltaic device (Fig. 5–18). A solar cell converts energy from the sun into electron flow. During the 1980s the efficiency of solar cells doubled to the value of 23% routinely and as high as 40% in the laboratory. Routinely, then, solar cells convert sunlight into electric power at the rate of at least 100 watts per square yard of illuminated surface. Solar cells are now used in calculators, watches, space-flight applications, communication satellites, power for remote water pumps, signals for automobiles and trains, light-weight power supplies for boats and golf carts, and as the source of electricity in utility power plants throughout the world.

A 100-megawatt utility plant would cover about 600 acres.

Silicon is the key element in the structure of solar cells. One outstanding property of silicon in a high state of purity is its electrical conductivity. Unlike a metal, which easily conducts electricity, and unlike a nonmetal, which does not, silicon is a **semiconductor.** That is, it does not conduct until a certain electrical voltage is applied, but beyond that it conducts moderately. By placing other atoms in a crystal of pure silicon, a process known as **doping,** experimenters have found that its conductivity properties can be changed. One type of solar cell consists of two layers of almost pure silicon. The lower, thicker layer contains a trace of boron, and the upper, thinner layer a trace of arsenic. The As-enriched layer is an *n*-type semiconductor. The term *n-type* (negative-type) is used to describe this layer because some electrons are present that readily move under the influence of an electrical voltage. These

Figure 5–18 A bank of photovoltaic, or solar, cells. (*The World of Chemistry,* Program 15, "The Busy Electron")

mobile electrons are in addition to the regular bonding array of electrons in the silicon structure. Just the opposite occurs in the B-enriched layer. This layer is a *p*-type semiconductor (positive-type) in that there are positive spots ("holes") in the bonding electron array (Fig. 5–19).

Hyperpure silicon is manufactured by zone refining in only three plants in the world—two in Japan and one in Germany.

Silicon has four valence electrons and is covalently bonded to four other silicon atoms in a crystal of pure silicon. Arsenic has five valence electrons. When arsenic is included in the silicon structure, only four of the five valence electrons of arsenic are used for bonding with four silicon atoms; one electron is relatively free to move. Boron has three valence electrons. When boron atoms are included in the silicon structure, there is a deficiency of one electron around the boron atom; this creates the "holes" in the boron-enriched layer.

There is a strong tendency for the "free" electrons in the arsenic layer to pair with the unpaired silicon electrons in the "holes" in the boron layer. If the two layers are connected by an external circuit and light of sufficient energy strikes the surface, excited electrons can leave the arsenic layer and flow through the external circuit to the boron layer. As the boron layer becomes more negative because of added electrons, electrons are repelled *internally* back to the arsenic layer, which is now positive and attracts the electrons from the boron layer. The process can continue indefinitely as long as the cell is exposed to a light source.

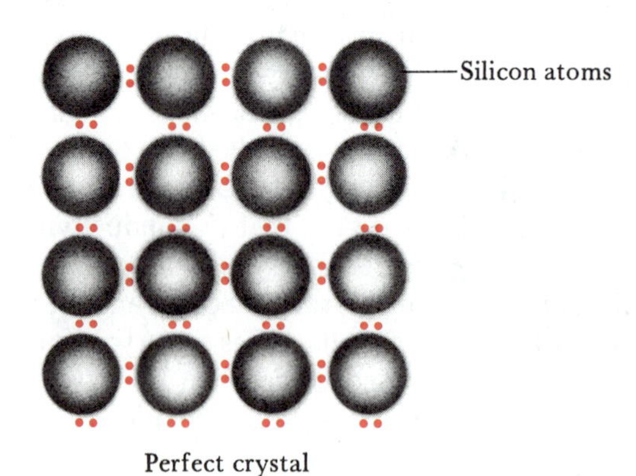

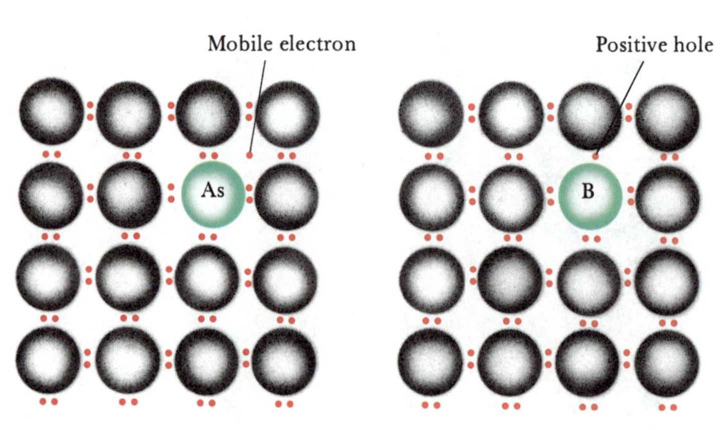

Figure 5–19 Schematic drawing of semiconductor crystal layers derived from silicon.

A typical solar cell is constructed on a sheet of plastic or glass (Fig. 5-20). Next to the plastic or glass is a thin sheet of metal that is the electrode giving electrons to the *p*-type semiconductor layer. The topmost *n*-type semiconductor layer, which receives the sun's rays, is nearly transparent. The solar cell is covered with a thin film of indium tin oxide ($InSnO_2$), which acts as an antireflection coating. A metallic grid structure on top of the cell allows as much light as possible to strike the *n*-type layer while functioning as an electrode. The efficiency of the solar cell depends on its ability to absorb photons of light and convert the light to electrical energy. Photons that are reflected, pass through, or produce only heat, decrease the efficiency of the cell.

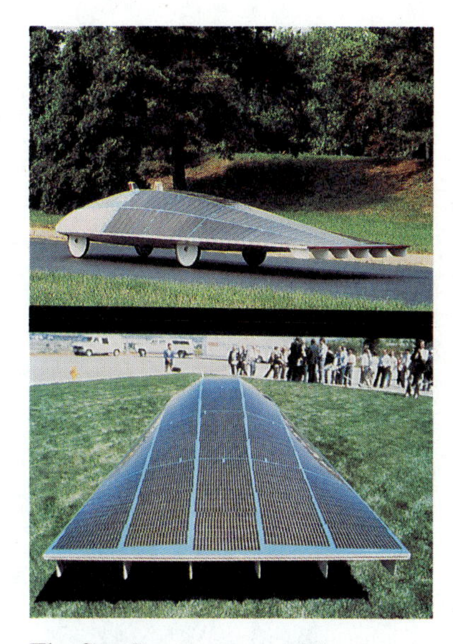

The Sun Raycer automobile was built by General Motors. Powered by solar cells, it also has silver-zinc batteries to help on a hill or on a cloudy day. (Courtesy of GM Hughes Electronics)

Older solar cells were made from single-crystal silicon, but the photovoltaic cells developed in the 1980s use amorphous silicon (a-Si). The amorphous silicon has an irregular array of Si atoms instead of the regular pattern of atoms in a single crystal of Si. The a-Si is as efficient in absorbing light in very thin layers (0.5 μm thick) as single 300-μm-thick crystals were. Other advantages of a-Si include its relatively low production cost (5% of the cost of single-crystal Si) and its production in larger sheets.

Efficiencies are improved, but the expense of producing the cell is increased, by using concentrator cells, which have a lens to concentrate the sunlight onto the solar cell. An added computer turns concentrator cells to track the sun through the sky. The cost of electricity from a new photovoltaic plant would be 22¢ per kilowatt-hour compared with 7¢ from an *existing* coal-fueled power plant and 9¢ from an *existing* petroleum-fueled power plant.

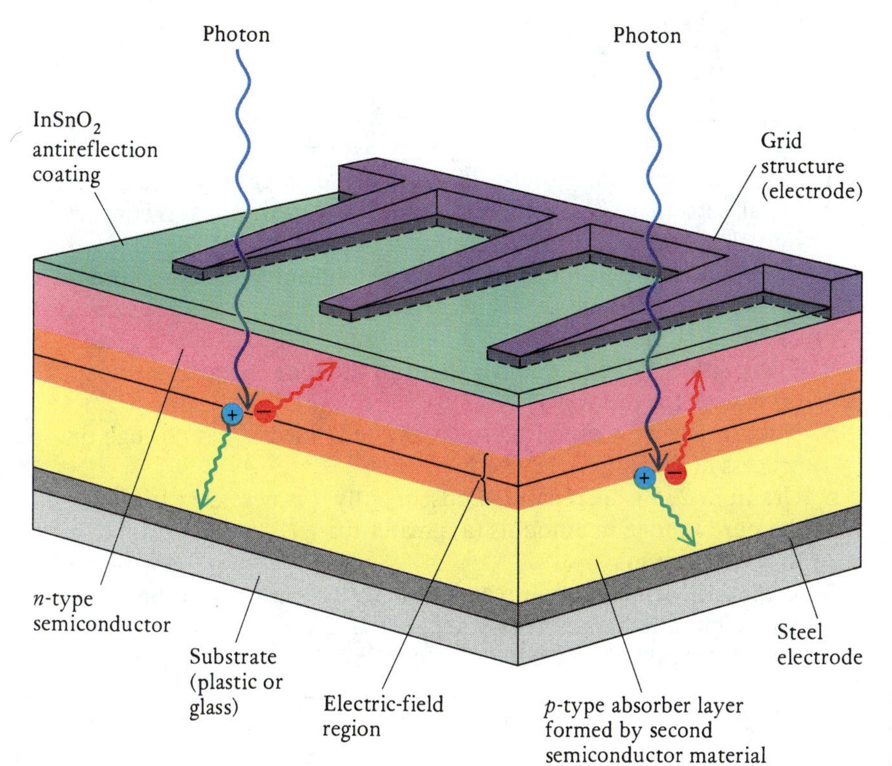

Figure 5-20 Typical photovoltaic cell using crystals of silicon. (Adapted from *Scientific American*)

A Hydrogen Economy

Hydrogen can be burned in most devices that now burn natural gas.

Some hydrogen-powered buses and cars are now operating on an experimental basis.

When solar cells become cheap enough, the electricity they supply could be used to electrolyze water to yield hydrogen and oxygen. Hydrogen can be transported through pipes far more efficiently than electricity is transported through transmission lines. The hydrogen could be piped to where the energy is needed and burned to heat water to steam, which in turn could generate electricity. Such an arrangement could give rise to a hydrogen economy and further release us from our dependence on fossil fuels and nuclear energy. Liquid hydrogen packs more energy per pound than any other nonnuclear fuel. It is the only fuel that ignites fast enough to boost an aircraft to orbital velocity (about 25,000 feet per second). The principal product is the raw material water.

ENERGY PLANNING FOR THE FUTURE

Prototype car running on hydrogen gas stored in a metal hydride. (Courtesy of Mercedes-Benz)

More than perhaps any other single physical factor, the availability of energy determines the quality of our lives. Because of the lack of planning, persons have been observed making firewood from the siding of their homes. It is possible for the human race to degrade the human environment and unwisely use up resources in immediate grabs for quick energy supplies. Political debate seeks to evaluate the relative merits of using currently available energy sources at environmental costs, finding new energy sources, and using less energy through conservation efforts to achieve the desired living standards. You will likely have an opportunity to participate in the energy decisions that lie ahead.

SELF-TEST 5B

1. The major secondary source of commercial energy is ________.
2. What percentage of the energy released by burning coal is converted into electrical energy in a coal-burning electric plant? ________.
3. Cite an advantage and a disadvantage in burning garbage to produce heat energy. ________________________________
____
4. The splitting of an unstable nucleus to produce energy is termed (a) fission, (b) fusion.
5. Nuclear energy is produced today in the United States through the process of (a) fission, (b) fusion.
6. Uranium-235 is fissionable. Consequently it is necessary to store the pure isotope in amounts (a) greater than, (b) less than, (c) equal to the critical mass.
7. In an exothermic nuclear reaction there is a mass defect that is equivalent to an amount of ________ ________.
8. An atomic pile or an atomic reactor allows us to control the rate of a ________ ________ ________.
9. A breeder reactor is able to change a ________ isotope of

uranium or thorium into a fissionable isotope that can be fuel for an atomic bomb or an atomic reactor.

10. Controlled nuclear fusion is now the principal means by which nuclear energy is produced in the United States: True ( ) False ( )
11. Solar energy on the roof of an average house in the United States is sufficient to provide about what percent of the energy needed to heat the house in the winter months? (a) 10%, (b) 50%, (c) 75%, (d) 100%
12. The solar cell contains a small amount of ____________ in the *n*-type silicon layer and a small amount of ____________ in the *p*-type silicon layer.

QUESTIONS

1. What two countries are first and second on the list of energy used per capita?
2. What is the ultimate physical source of energy in the universe?
3. Which theoretically yields the greatest energy per mole?
   a. The burning of gasoline
   b. The fission of uranium-235
   c. The burning of methane (natural gas)
4. Which is the more efficient use of energy: burning coal in a house to heat it, or heating the house electrically with energy produced in a coal-burning power plant?
5. Is the electrical energy where you live produced by burning fossil fuels? If not, what is the energy source? Are there pollution problems associated with the generation of the electrical energy?
6. What was the original source of the energy that is tied up in fossil fuels?
7. Define power. Give a unit in which power is measured.
8. What is the difference between power and energy?
9. Which is a low-energy system: (a) wood and air or (b) water and air? Explain.
10. Entropy is a measure of the disorder of a physical system. Is the entropy of the universe as we know it increasing or decreasing? Explain.
11. What major problem is associated with harnessing the energy from a fusion reaction?
12. Suggest several ways in which solar energy might be harnessed.
13. Define efficiency in the conversion of one form of energy to another. Give an example.
14. Explain how useful energy might be obtained from garbage.
15. Which is more fundamental: a supply of energy or a supply of food? Explain.
16. Which fuel has the greatest energy content per gram of fuel burned: coal, natural gas, or petroleum? Is this factor more important to you, the consumer, than the economics of energy use or pollution caused by fuel use?
17. Which fuel—coal, petroleum, or natural gas—burns naturally with the least amount of pollution?
18. What chemical elements are found in coal?
19. What are two dangerous properties of plutonium-239?
20. What is the persistent problem relative to radioactive wastes in the production of nuclear energy?
21. What is the difference between nuclear fusion and nuclear fission?
22. Which is the more efficient transport of energy: gas through pipes or electricity through wires?
23. If solar energy is so clean, why are we so slow in moving to its use?
24. What has been the limiting factor in the failure thus far to produce a controlled nuclear fusion reactor?
25. Is electricity a primary or secondary source of energy? Explain your answer.
26. What is the difference between coal gas and power gas?
27. What problems should be controlled for environmental reasons if we are to burn garbage for energy production?
28. Explain why coal gasification could be a route to a cleaner environment.
29. How does a photovoltaic cell produce electricity?
30. As a project, update the energy situation in the United States and the world by consulting the most recent edition of the *Annual Energy Review* (published by the Department of Energy, Energy Information Administration), a copy of which is probably in your library.

The Alchemist's Experiment Takes Fire by Hendrick Heerschop is a representation of perhaps one of humankind's less successful ventures into organic chemistry. (Courtesy of Fisher Scientific)

CHAPTER

6

An Introduction to Organic Chemistry

Why are there so many organic compounds? Reasons include the unique ability of carbon atoms to form strong bonds to other carbon atoms, the occurrence of isomers, and the variety of functional groups that can bond to carbon atoms. The hydrocarbon classes include alkanes, alkenes, alkynes, and aromatics. Replacement of one or more hydrogen atoms in any of these millions of possible hydrocarbons with functional groups gives rise to additional millions of organic compounds.

1. What are the elemental forms of carbon?
2. Why are there so many organic compounds?
3. What are the structural differences among alkanes, alkenes, and alkynes?
4. How do aromatic compounds differ from alkanes, alkenes, and alkynes?
5. What are straight-chain and branched-chain isomers?
6. What are some common functional groups?

Carbon compounds hold the key to life on Earth. Consider what the world would be like if all carbon compounds were removed; the result would be much like the barren surface of the moon. If carbon compounds were removed from the human body, there would be nothing left except water and a residue of minerals. The same would be true for all forms of living matter. Carbon compounds are also an integral part of our lifestyle. Fossil fuels, foods, and most drugs are made of carbon compounds. Because we live in an age of plastics and synthetic fibers, our clothes, appliances, and most other consumer goods contain a significant portion of carbon compounds.

Organic chemistry is the study of the nonmineral compounds of carbon.

About 9 million of the over 10 million known compounds are carbon compounds, and thousands of new ones are reported every year. The very large and important branch of chemistry devoted to the study of carbon compounds is **organic chemistry.** The name *organic* is actually a relic of the past, when chemical compounds produced from once-living matter were called "organic" and all other compounds were called "inorganic."

WHY ARE THERE SO MANY ORGANIC COMPOUNDS?

An important reason for the millions of organic compounds is the unique ability of carbon to form stable covalent bonds with other carbon atoms. We will examine how the different forms of elemental carbon illustrate this property before discussing organic compounds.

Forms of Elemental Carbon

Sprinkle some graphite powder on a key before inserting it in a lock to lubricate the lock mechanism.

Graphite has layers of fused six-membered rings of carbon atoms (Fig. 6–1a). The carbon atoms are bonded to three other carbon atoms in the rings. The covalent bonds between carbon atoms within the rings are strong, but the bonds between layers are weak. Because the layers slide past each other easily, graphite is soft (the "lead" in pencils), a good lubricant, and a good conductor of electricity.

The carbon atoms in **diamond** (Fig. 6–1b) are linked to four other carbon atoms with strong covalent bonds to give an interconnected tetrahedral network of carbon atoms. As a result, diamond is hard and a good electrical insulator.

In 1985 Richard Smalley and co-workers at Rice University and Harry Kroto of the University of Sussex, England, detected another form of carbon in the soot formed from laser vaporization of graphite. They proposed that the new form was a 60-atom cluster in the shape of an icosahedron (Fig. 6–1c). In 1990 researchers verified their prediction for the shape of C_{60}. This form of carbon was named "buckminsterfullerene" (or simply "buckyball")

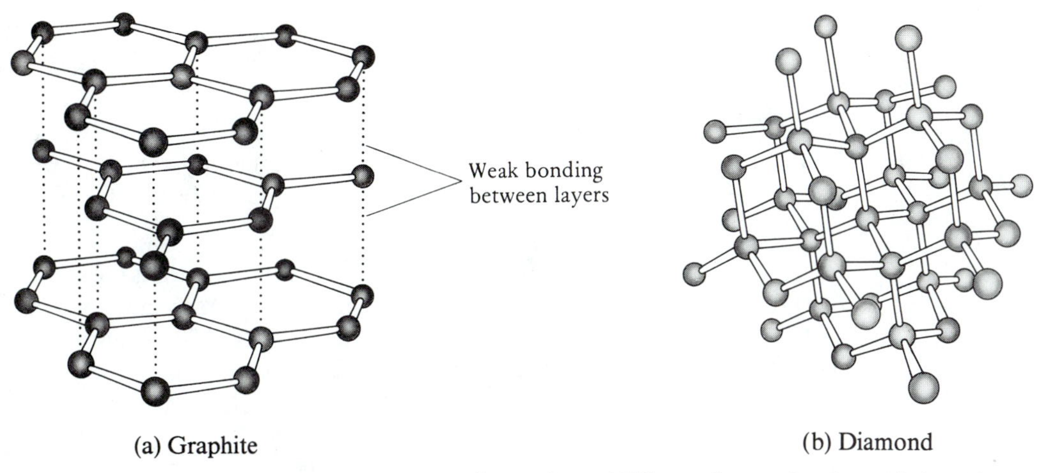

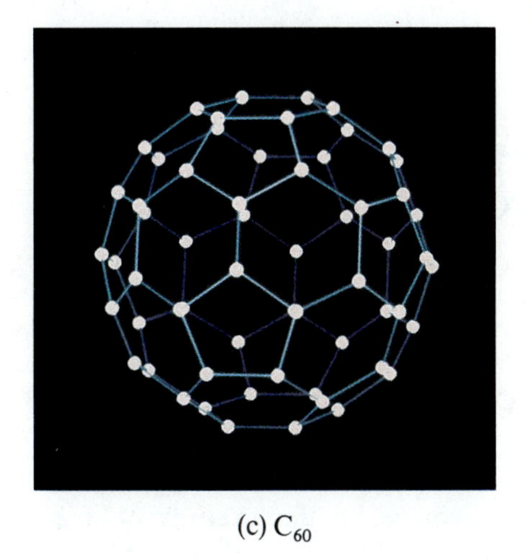

Figure 6–1 Different forms of carbon. (a) Graphite. (b) Diamond. (c) C_{60}. (Richard Smalley, Rice University)

after Buckminster Fuller, who popularized the icosahedral shape by using it in his patented geodesic dome. The icosahedral structure of C_{60} can be pictured by imagining a soccer ball (Fig. 6–2), with the seams representing covalent bonds between carbon atoms at the end of each seam.

C_{60} is isolated from soot, and recent research indicates that C_{60} is the first of a series of "fullerenes" that contain an even number of carbon atoms arranged in closed, hollow cages. Others that have been discovered include C_{70}, C_{76}, C_{90}, and C_{94}. The next time you see some soot, think about the buckyballs it contains!

Figure 6–2 Soccer ball as a model of C_{60}. Seams represent covalent bonds with carbon atoms at the intersection of each seam with other seams.

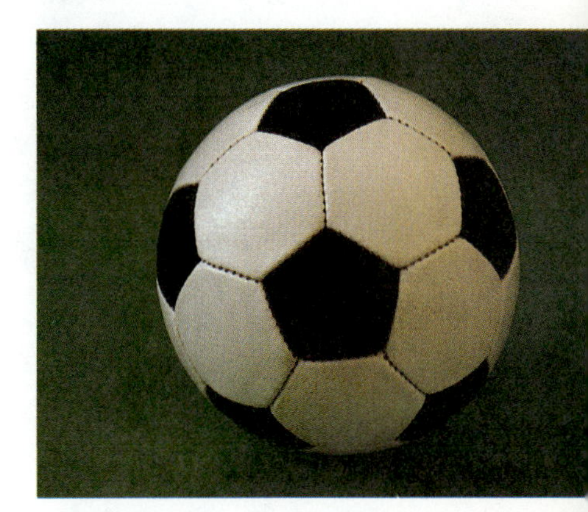

Carbon–Carbon Bonds in Organic Compounds

The tendency of carbon atoms to link together in elemental carbon carries over to compounds of carbon. An organic molecule may contain from two

```
  |  |
—C—C—
  |  |
```

to thousands of carbon-carbon bonds.

```
  |  |  |  |  |  |  |  |  |  |  |  |  |  |  |  |  |  |  |  |
—C—C—C—C—C—C—C—C—C—C—C—C—C—C—C—C—C—C—C—C—
  |  |  |  |  |  |  |  |  |  |  |  |  |  |  |  |  |  |  |  |
```

Another reason for the large number of organic compounds is the ability of a given number of atoms to combine in more than one molecular pattern and hence produce more than one compound. For example, a compound with a formula of C_4H_{10} could have a straight-chain pattern

```
   H  H  H  H
   |  |  |  |
H—C—C—C—C—H
   |  |  |  |
   H  H  H  H
```

n-butane

n stands for ***normal*** and is used for "straight-chain" hydrocarbons.

or a branched-chain pattern.

```
          H
          |
   H  H—C—H   H
   |      |     |
H—C——————C——————C—H
   |      |     |
   H      H     H
```

Isobutane

These molecules have different properties even though they have the same number of atoms per molecule. Such compounds, each of which has molecules containing the same number and kinds of atoms, but arranged differently relative to each other, are called **isomers.** Another example of how a difference in molecular pattern can cause a large difference in properties can be seen by considering two molecules that have the formula C_2H_6O. Possible arrangements are

Isomers are two or more different compounds with the same number of each kind of atom per molecule.

```
   H  H                H      H
   |  |                |      |
H—C—C—O—H          H—C—O—C—H
   |  |                |      |
   H  H                H      H
```

The molecular structure on the left represents a molecule of ethyl alcohol whereas the molecular structure on the right represents a molecule of dimethyl ether—two completely different compounds.

A final factor explaining the large number of organic compounds is the ability of the carbon atom to form strong covalent bonds with atoms of numerous other elements, such as nitrogen, oxygen, sulfur, chlorine, fluorine, bromine, iodine, silicon, boron, and even many metals. As a result, there are large classes of organic compounds. A distinguishing **functional group,** a particular combination of atoms, appears in each member of a class. For example, ethyl alcohol and dimethyl ether are members of the alcohol and ether classes, respectively.

In summary, the large number of carbon compounds is due to

1. The ability of carbon atoms to form stable bonds to other carbon atoms and to other nonmetals.
2. The occurrence of isomers.
3. The variety of functional groups that bond to carbon atoms.

HYDROCARBONS

The largest class of organic compounds is the **hydrocarbons,** compounds of hydrogen and carbon. Complex mixtures of hydrocarbons occur in enormous quantities as natural gas, petroleum, and coal **(fossil fuels).**

See Chapter 5 for a discussion of fossil fuels.

Classes of hydrocarbons include the **alkanes,** which contain C—C bonds; the **alkenes,** which contain one or more C=C bonds; the **alkynes,** which contain one or more C≡C bonds; and the **aromatics,** which consist of benzene (Fig. 6–11), benzene derivatives, and fused benzene rings.

Alkanes

Alkanes are hydrocarbons that contain only single bonds. They are often referred to as **saturated hydrocarbons** because they contain the highest ratio of hydrogen to carbon possible. The simplest alkane is **methane,** CH_4, the principal component of natural gas. The tetrahedral structure of methane was discussed in Chapter 3 (Fig. 3–11). Models for the methane molecule are shown in Figure 6–3.

The next member of the alkane family is **ethane,** C_2H_6, a hydrocarbon gas with two carbon atoms (Fig. 6–4). The third member is **propane,** C_3H_8, one of the principal components of bottled gas. In Figure 6–5 note that the three carbon atoms in propane do not lie in a straight line because of the tetrahedral bonding about each carbon atom.

These bonding concepts can be extended to a four-carbon molecule and to a "limitless" number of larger hydrocarbon molecules. Actually, many such compounds are known; some, such as natural rubber, are known to contain over a thousand carbon atoms in a chain.

The first ten straight-chain alkanes are listed in Table 6–1. Remember that straight-chain means that the carbon atoms are bonded together in succession, not in a straight line. Notice that each succeeding formula in

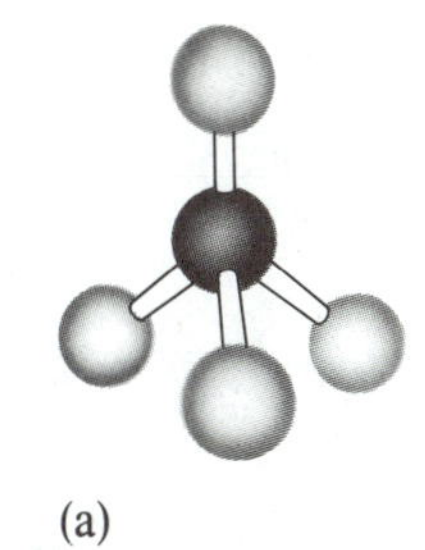

(a)

(b)

Figure 6–3 Methane. (a) Ball-and-stick model showing tetrahedral structure. (b) Space-filling model showing relative size of atoms in relationship to interatomic distances. (Charles Steele)

Figure 6–4 Ball-and-stick model of ethane. (Charles Steele)

Figure 6–5 Ball-and-stick and space-filling models of propane, C_3H_8. (Charles Steele)

Table 6–1 is obtained by adding CH_2 to the previous formula. Alkanes are an example of a **homologous series**—a series of compounds of the same chemical type that differ only by a fixed increment. In this case the fixed increment is CH_2. The alkane homologous series can be represented by the general formula C_nH_{2n+2}, where n is the number of carbon atoms for a member of the series.

The names of the first ten straight-chain alkanes are given in Table 6–1. Much attention has been given to naming organic compounds in a consistent way, and several international conventions have been held to work out a satisfactory system that can be used throughout the world. The International Union of Pure and Applied Chemistry has given its approval to a systematic nomenclature system (IUPAC system), which is now in general use. The name of each of the members of the hydrocarbon classes has two parts. The first part—*meth-*, *eth-*, *prop-*, *but-*, and so on—reflects the number of carbon atoms. When more than four carbons are present, the Greek number prefixes are used: *pent-*, *hex-*, *hept-*, *oct-*, *non-*, and *dec-*. The second part of the name, or the suffix, tells the class of hydrocarbon. Alkanes have carbon-carbon single bonds, alkenes have carbon-carbon double bonds, and alkynes have carbon-carbon triple bonds.

Chemists from all over the world belong to IUPAC. Nomenclature recommendations from this organization are accepted worldwide.

Structural Isomers: Straight- or Branched-Chain Variations

The two possible structural arrangements for **butane,** C_4H_{10}, were given earlier in this chapter. Four carbon atoms can be bonded in either a straight chain or a branched chain. These are different compounds with different properties, but because they have the same formula, they are examples of structural isomers. Ball-and-stick models of these two isomers are shown in Figure 6–6. Structural formulas for the two isomers can be written in expanded form or in condensed form.

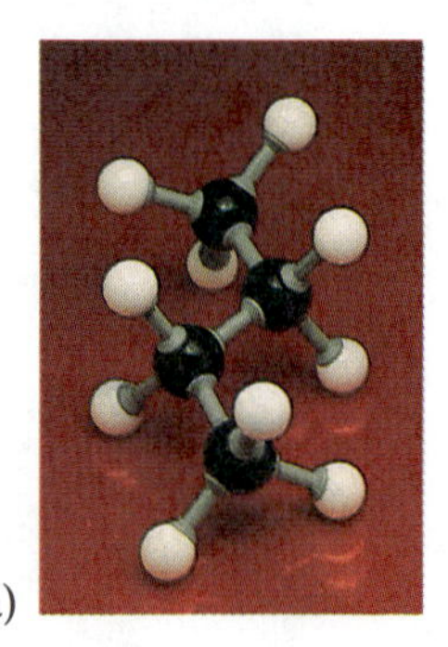
(a)

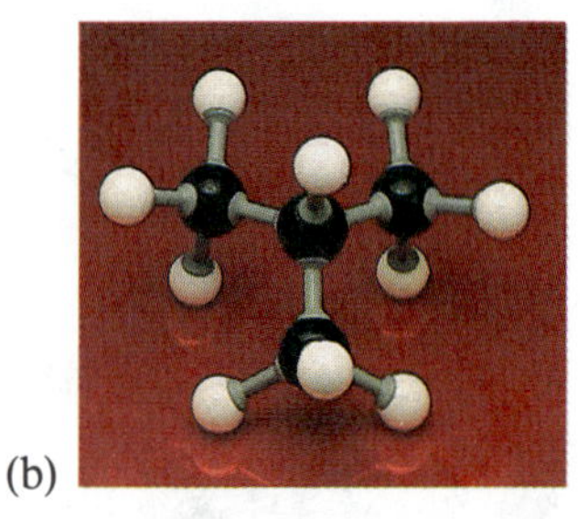
(b)

Figure 6–6 Ball-and-stick models of the two isomeric butanes C_4H_{10}. (a) Normal butane, usually written *n*-butane. (b) Methylpropane (isobutane). (Charles D. Winters)

TABLE 6–1 The First Ten Straight-Chain Saturated Hydrocarbons

Name	Formula	Boiling Point, °C	Structural Formula	Use
Methane	CH_4	−162	H H—C—H H	Principal component in natural gas
Ethane	C_2H_6	−88.5	H H H—C—C—H H H	Minor component in natural gas
Propane	C_3H_8	−42	H H H H—C—C—C—H H H H	Bottled gas for fuel
n-Butane	C_4H_{10}	0	H H H H H—C—C—C—C—H H H H H	
n-Pentane	C_5H_{12}	36	H H H H H H—C—C—C—C—C—H H H H H H	Some of the components of gasoline
n-Hexane	C_6H_{14}	69	H H H H H H H—C—C—C—C—C—C—H H H H H H H	
n-Heptane	C_7H_{16}	98	H H H H H H H H—C—C—C—C—C—C—C—H H H H H H H H	
n-Octane	C_8H_{18}	126	H H H H H H H H H—C—C—C—C—C—C—C—C—H H H H H H H H H	
n-Nonane	C_9H_{20}	151	H H H H H H H H H H—C—C—C—C—C—C—C—C—C—H H H H H H H H H H	
n-Decane	$C_{10}H_{22}$	174	H H H H H H H H H H H—C—C—C—C—C—C—C—C—C—C—H H H H H H H H H H H	Found in kerosene

EXPANDED STRUCTURAL FORMULAS:

CONDENSED STRUCTURAL FORMULAS:

	$CH_3CH_2CH_2CH_3$ *n*-butane	$CH_3CH(CH_3)CH_3$ 2-Methylpropane (isobutane)
Melting point	−138.3°C	−160°C
Boiling point	−0.5°C	−12°C
Density (at 20°C)	0.579 g/mL	0.557 g/mL

Structural isomerism can be compared to the results you might expect from a child building many different structures with the same collection of building blocks, and using all of the blocks in each structure.

Consider the isomeric pentanes, C_5H_{12}. There are three of these (Fig. 6-7).

EXPANDED STRUCTURAL FORMULAS:

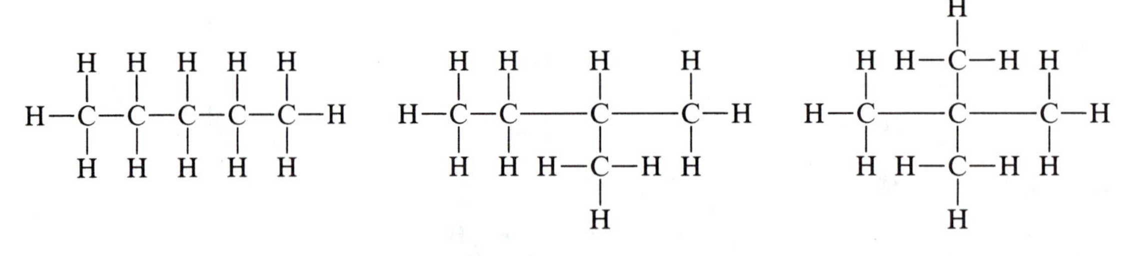

CONDENSED STRUCTURAL FORMULAS:

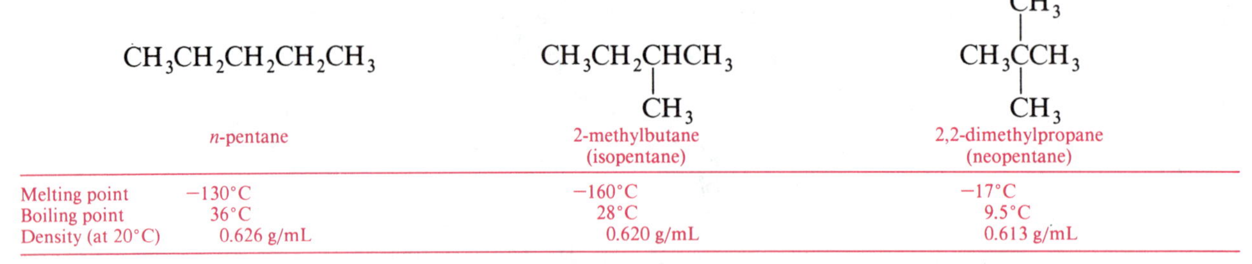

	$CH_3CH_2CH_2CH_2CH_3$ *n*-pentane	$CH_3CH_2CH(CH_3)CH_3$ 2-methylbutane (isopentane)	$CH_3C(CH_3)_2CH_3$ 2,2-dimethylpropane (neopentane)
Melting point	−130°C	−160°C	−17°C
Boiling point	36°C	28°C	9.5°C
Density (at 20°C)	0.626 g/mL	0.620 g/mL	0.613 g/mL

The IUPAC names are given under the structures with the common names in parentheses. There are no other ways to unite the 17 atoms in the isomeric pentanes and still follow bonding rules for carbon (eight valence shell electrons) and hydrogen (two valence shell electrons). These three isomers of pentane are well known, and no others have been found.

The IUPAC rules for naming structural isomers are illustrated with the names for the pentane isomers. The italic prefix *n*- stands for *normal* and identifies the straight-chain isomer. For branched-chain isomers, it becomes

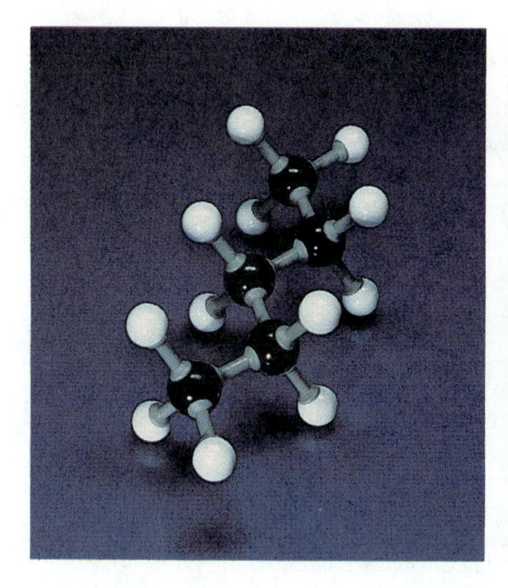

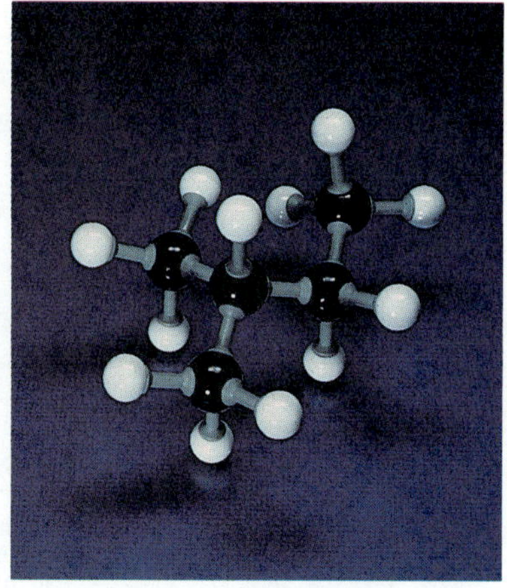

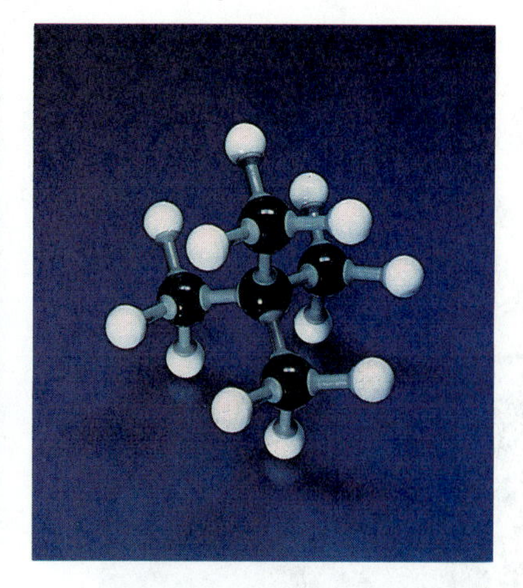

Figure 6-7 Ball-and-stick models of the three isomeric pentanes C_5H_{12}. (Charles Steele)

necessary to name submolecular groups. The $—CH_3$ group is called the methyl group; this name is derived from methane by the deletion of the *-ane* and the addition of *-yl*. Any of the alkanes can give rise to a similar subgroup, referred to generally as an alkyl group, by using the first part of the alkane name and the substitution of *-yl* for *-ane*. Some examples are given in Table 6-2.

The rules for naming branched chain isomers are (1) find the longest continuous chain of carbon atoms and use this chain as the parent name for the compound, (2) use numbers to designate the location of attached groups and number the longest chain beginning with the end of the chain nearest the branching, and (3) use prefixes di-, tri-, tetra-, and so on for two or more identical substituents. The two branched-chain isomers of pentane have a *butane* parent chain and a *propane* parent chain. Hence, the names of the branched isomers are 2-methylbutane and 2,2-dimethylpropane.

Table 6-3 gives the number of structural isomers predicted for some larger alkane molecules, starting with C_6H_{14}. Every predicted isomer, *and*

TABLE 6-2 Some Alkyl Groups

Name	Condensed Structural Representation*
Methyl	$CH_3—$
Ethyl	$CH_3CH_2—$ or $C_2H_5—$
n-propyl	$CH_3CH_2CH_2—$ or $C_3H_7—$
Isopropyl	$CH_3CH(CH_3)—$ (CH₃ drawn below the CH)

* $CH_3—$ represents $H—\underset{H}{\overset{H}{|}}\!\!\!\!C—$ (H—C— with H above and below C). Although a more accurate representation would be $—CH_3$ to show that the available bond is to carbon, not hydrogen, the conventional representation is $CH_3—$.

TABLE 6-3 Structural Isomers of Some Hydrocarbons*

Formula	Isomers Predicted	Found
C_6H_{14}	5	5
C_7H_{16}	9	9
C_8H_{18}	18	18
$C_{15}H_{32}$	4,347	—
$C_{20}H_{42}$	366,319	—
$C_{30}H_{62}$	4,111,846,763	—
$C_{40}H_{82}$	62,491,178,805,831	—

* R. E. Davies and P. J. Freyd: "$C_{167}H_{336}$ Is the Smallest Alkane with More Realizable Isomers than the Observed Universe Has 'Particles'," *Journal of Chemical Education,* Vol. 66, pp. 278–281, 1989.

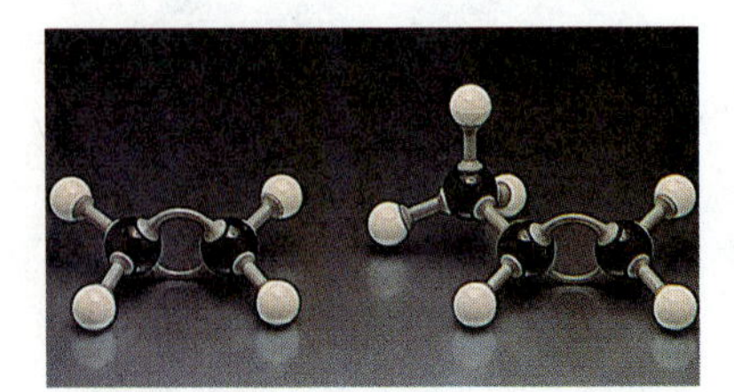

Figure 6–8 Ball-and-stick models of ethene *(left)* and propene *(right).* (Charles D. Winters)

no more, has been isolated and identified for the C_6, C_7, and C_8 groups. Although not all of the C_{15} molecules have been isolated, there is reason to believe that with enough time and effort they could be, so structural isomerism certainly helps to explain the vast number of carbon compounds. However, as the number of carbon atoms gets larger than 17, the number of isomers that could be expected to be stable enough for isolation is much smaller than the predicted number. The reason for this difference is that the calculation of predicted isomers does not include a consideration of space requirements of atoms within the molecules for the various isomers. As a result, many of the predicted isomers for these larger molecules would require such overcrowding of atoms within the molecules that they would not be stable. (See the reference listed in Table 6–3 for further discussion of this point.)

Alkenes

Molecules of alkenes have one or more carbon-carbon double bonds (C=C). The general formula for alkenes with one double bond is C_nH_{2n}. The first two members of the homologous alkene series are ethene (C_2H_4) and propene (C_3H_6) and their structural formulas are:

```
H       H        H        H
 \     /          \      /
  C=C              C=C
 /     \          /      \
H       H     H3C         H
 Ethene          Propene
(ethylene)      (propylene)
```

Ball-and-stick models for these formulas are shown in Figure 6–8. The common names, ethylene and propylene, are often used, particularly when referring to polyethylene and polypropylene, plastics prepared from ethylene and propylene, respectively. (See Fig. 7–1.)

▮ Plastics are discussed in Chapter 8.

The structural formulas illustrate why alkenes are said to be **unsaturated hydrocarbons.** They contain fewer hydrogen atoms than the corresponding alkanes and can be made to react with hydrogen to form alkanes. Platinum is used as a catalyst.

See Chapter 7 for a discussion of platinum as a surface catalyst.

$$\underset{\text{(unsaturated)}}{\underset{\text{Ethene}}{H_2C{=}CH_2}} + H_2 \xrightarrow{\text{Platinum}} \underset{\text{(saturated)}}{\underset{\text{Ethane}}{H_3C{-}CH_3}}$$

Alkenes are named by using the prefix to indicate the number of carbons and the suffix, *-ene* to indicate one or more double bonds. The first member, ethene, is the most important raw material used in the organic chemical industry. It ranks fourth in the top 50 chemicals (see inside front cover) and is the number-one organic chemical. Over 37 billion pounds were produced in 1990 for use in making polyethylene, antifreeze (ethylene glycol), ethyl alcohol, and other chemicals.

The third alkene in the series, butene, has two possible locations for the double bond.

$$\underset{\text{1-butene}}{H_2C{=}CH{-}CH_2{-}CH_3} \qquad \underset{\text{2-butene}}{H_3C{-}CH{=}CH{-}CH_3}$$

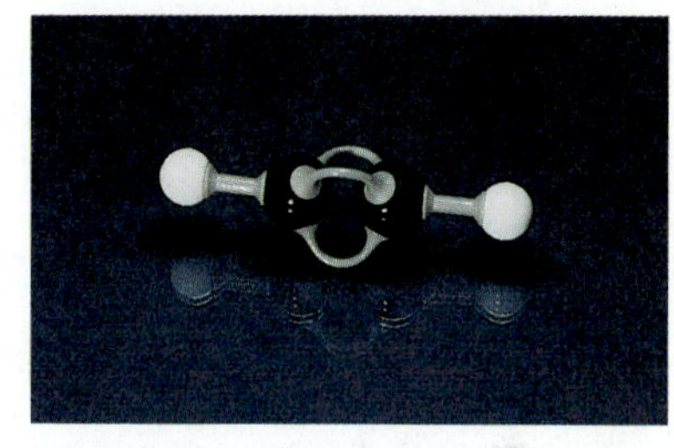

Figure 6–9 Ball-and-stick model of acetylene. (Charles Steele)

Alkynes

The alkynes have one or more triple bonds (—C≡C—) per molecule and have the general formula C_nH_{2n-2}. The simplest one is ethyne, commonly called acetylene (C_2H_2). Acetylene is a linear molecule (Fig. 6–9).

$$H{-}C{\equiv}C{-}H$$

A mixture of acetylene and oxygen burns with a flame hot enough to cut steel (3000°C).

Cutting steel with an oxyacetylene torch. (© Joseph Nettis, Photo Researchers, Inc.)

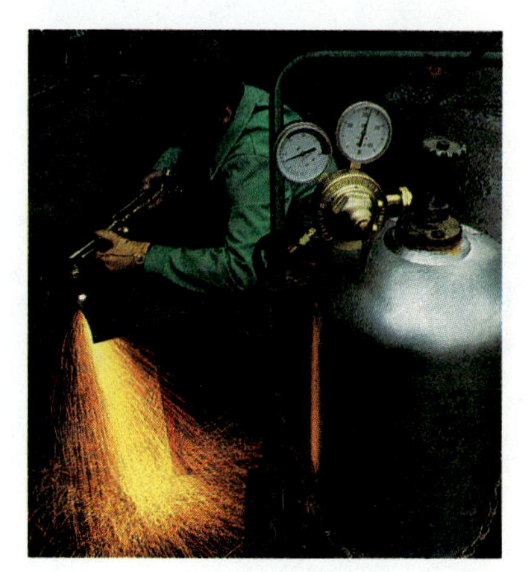

SELF-TEST 6A

1. The form of carbon that is soft and a good lubricant is ______________.
2. The form of carbon that is the hardest naturally occurring substance is ______________.
3. Each carbon in a saturated hydrocarbon has ______________ geometry.
4. Butane and isobutane are examples of ______________ isomers.
5. The number-one organic chemical produced in United States is ______________.
6. An alkene always has at least one ______________ bond.
7. The first member of the alkyne series of hydrocarbons is ______________.
8. The formula for the ethyl group is ______________.

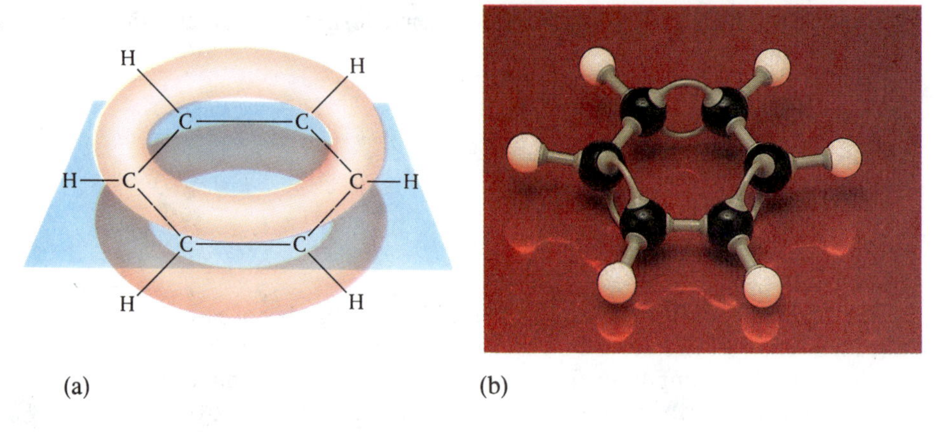

Figure 6–10 (a) Bonding in benzene. (b) Ball-and-stick model of benzene. (Charles D. Winters)

THE CYCLIC HYDROCARBONS

Hydrocarbons can form rings as well as straight chains and branched chains. Two important classes of cyclic hydrocarbons are the **cycloalkanes** (all single bonds) and the **aromatics** (a unique merging of single bonds and double bonds to give electron delocalization around the ring).

Cycloalkanes

Cycloalkanes have the same general formula as alkenes: C_nH_{2n}.

Symbols like △ are just more chemical shorthand.

The simplest cycloalkane is cyclopropane, a highly strained ring compound:

```
      H   H
       \ /
   H    C    H
    \  / \  /
     C — C
    /     \
   H       H
```

or △ C_3H_6

The ring is strained because of the 60-degree angles in the ring; angles above 90-degrees show a much greater stability. Cyclopropane, a volatile, flammable gas (b.p. is −32.7°C), is a rapidly acting anesthetic. A cyclopropane-oxygen mixture is useful in surgery on babies, small children, and "bad risk" patients because of its rapid action and the rapid recovery of the patient. Helium gas is added to the cyclopropane-oxygen mixture to reduce the danger of explosion in the operating room.

The cycloalkanes are commonly represented by a polygon. Each corner represents a carbon atom and two hydrogen atoms, and the lines represent C—C bonds. The C—H bonds are not shown, but are understood. Other common homologous cycloalkanes include cyclobutane, cyclopentane, and cyclohexane. These are represented as:

Aromatic Compounds

Hydrocarbons containing one or more benzene rings are called **aromatic compounds.** The word *aromatic* was derived from *aroma,* which describes the rather strong and often pleasant odor of these compounds. However, benzene and some other aromatic compounds, such as benzo(α)pyrene, are both toxic and carcinogenic. The main structural feature, which is responsible for the distinctive chemical properties of the aromatic compounds, is the benzene ring (Fig. 6 – 10). Figure 6 – 10(a) illustrates the delocalization of six electrons above and below the plane of the ring. In other words, all six carbon-carbon bonds are equivalent, and benzene is a planar molecule. Benzene can be represented as

Carcinogenic means cancer-causing. The type of cancer caused may vary from one carcinogen to another. Benzene causes a form of leukemia. See Chapter 10.

where the circle represents the evenly distributed, delocalized electrons.

When hydrogen and carbon atoms are not shown, benzene is represented by a circle in a hexagon. Each corner in the hexagon represents one carbon atom and one hydrogen atom. Remember that this diagram of benzene stands for C_6H_6.

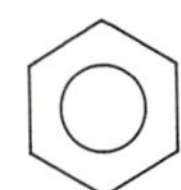

Examples of some aromatic hydrocarbons are shown in Figure 6 – 11. Many aromatic compounds, such as benzene, toluene, and the xylenes, are on the list of top 50 chemicals (see inside front cover) because of their use in the manufacture of plastics, gasoline, detergents, pesticides, drugs, and other organic chemicals.

The three xylenes shown in Figure 6 – 11 are structural isomers. Each of these isomers has two methyl groups substituted for hydrogen atoms on the benzene ring. For two groups, either the prefixes *ortho-*, *meta-*, *para-* or numbers are used.

Xylenes rank numbers 26 and 28 among commercial chemicals.

If more than two groups are attached to the benzene ring, numbers must be used. Consider the following compounds:

1,2,3-trichlorobenzene 1,2,4-trichlorobenzene 1,3,5-trichlorobenzene

These are the only possibilities for isomers of trichlorobenzene.

Note that 1,2,3- is the same as 4,5,6- if the molecule is flipped over and turned around. Molecules don't know external directions.

Figure 6–11 Examples of aromatic hydrocarbons.

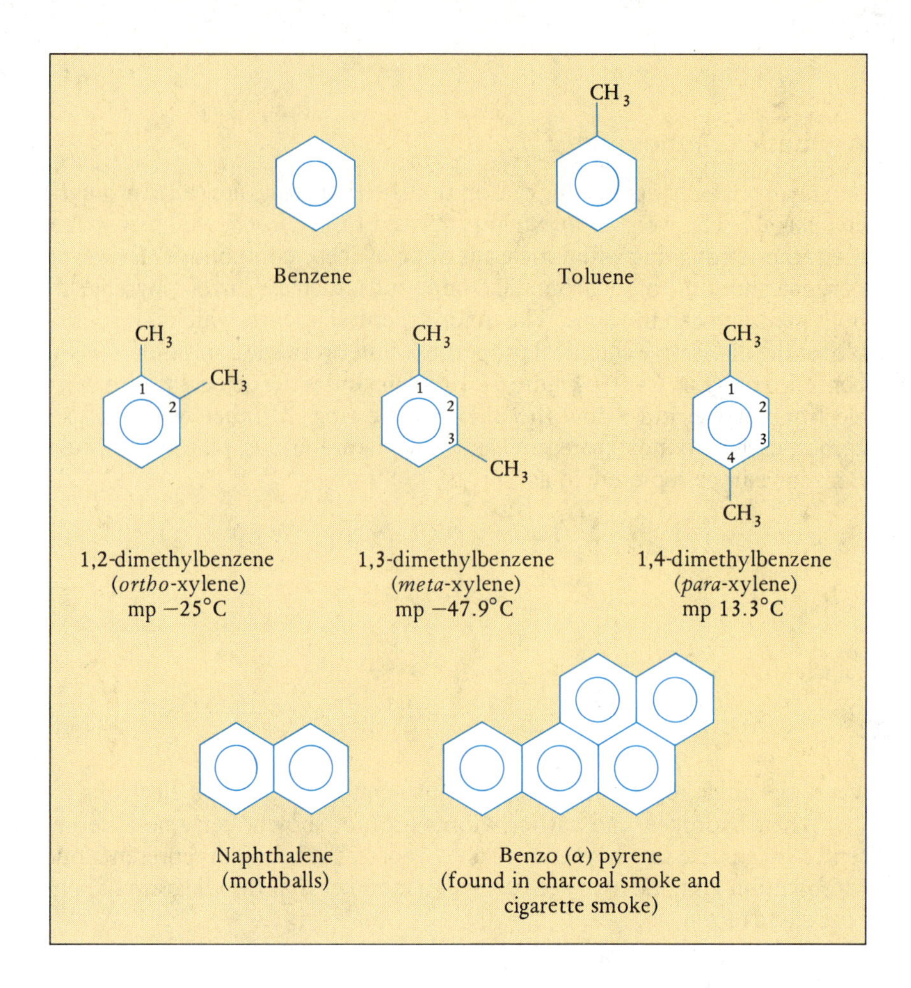

FUNCTIONAL GROUPS

The millions of organic compounds include classes of compounds that are obtained by replacing hydrogen atoms of hydrocarbons with atoms or groups of atoms known as **functional groups.** The important classes of compounds that result from attaching functional groups to a hydrocarbon framework are shown in Table 6–4. The "R" attached to the functional group represents the nonfunctional, hydrocarbon framework with one hydrogen atom removed for each functional group added. Refer to Table 6–2 for some examples of alkyl groups that are represented by R.

Both common names and the IUPAC name for each class of functional group are illustrated in Table 6–4. The consumer needs to know both the common names and the IUPAC names because both are widely used. The IUPAC system provides a systematic method for naming all members of a given class. For example, alcohols end in *-ol* (methan*ol*); aldehydes end in *-al* (methan*al*); carboxylic acids end in *-oic* (ethan*oic* acid); and ketones end in *-one* (propan*one*). Note that in each of these cases the prefix represents the total number of carbon atoms in the molecule.

Isomers are also possible for molecules containing functional groups. For example, a single hydrocarbon molecule can give rise to several alcohols

TABLE 6–4 Classes of Organic Compounds Based on Functional Groups*

General Formulas of Class Members	Class Name	Typical Compound	Compound Name	Common Use of Sample Compound
R—X	Halide	CH_2Cl_2 (H—C(H)(Cl)—Cl)	Methylene chloride (dichloromethane)	Solvent
R—OH	Alcohol	CH_3—OH	Wood alcohol (methanol)	Solvent
R—C(=O)—H	Aldehyde	H—C(=O)—H	Formaldehyde (methanal)	Preservative
R—C(=O)—OH	Carboxylic acid	CH_3—C(=O)—OH	Acetic acid (ethanoic acid)	Vinegar
R—C(=O)—R′	Ketone	CH_3—C(=O)—CH_3	Acetone (propanone)	Solvent
R—O—R′	Ether	C_2H_5—O—C_2H_5	Ethyl ether (diethyl ether)	Anesthetic
R—O—C(=O)—R′	Ester	CH_3—CH_2—O—C(=O)—CH_3	Ethyl acetate (ethyl ethanoate)	Solvent in fingernail polish
R—N(H)—H	Amine	CH_3—NH_2	Methylamine	Tanning (foul odor)
R—C(=O)—N(H)—R′	Amide	CH_3—C(=O)—NH_2	Acetamide	Plasticizer

* R stands for an H or a hydrocarbon group such as $—CH_3$ or $—C_2H_5$. R′ could be a different group from R. IUPAC names are given in parentheses.

if there are different isomeric positions for the —OH group. Three different alcohols result when a hydrogen atom is replaced by an —OH group in *n*-pentane, depending on which hydrogen atom is replaced (Table 6–5).

Alcohols are classified as primary, secondary, and tertiary based on the number of other carbons bonded to the —C—OH carbon.

TABLE 6–5 Alcohols Derived from Pentane (C_5H_{12})

```
   H  H  H  H  H                                 H  H  H  H  H
   |  |  |  |  |                                 |  |  |  |  |
H—C—C—C—C—C—H←                        gives  H—C—C—C—C—C—OH
   |  |  |  |  |                                 |  |  |  |  |
   H  H  H  H  H                                 H  H  H  H  H
```

Substitution of an —OH for an end hydrogen — 1-pentanol

```
   H  H  H  H  H                                 H  H  H  H  H
   |  |  |  |  |                                 |  |  |  |  |
H—C—C—C—C—C—H                         gives  H—C—C—C—C—C—H
   |  |  |  |  |                                 |  |  |  |  |
   H  H  H  H  H                                 H  H  H  O  H
            ↑                                             H
```

Substitution of an —OH for a 2-carbon hydrogen — 2-pentanol

```
   H  H  H  H  H                                 H  H  H  H  H
   |  |  |  |  |                                 |  |  |  |  |
H—C—C—C—C—C—H                         gives  H—C—C—C—C—C—H
   |  |  |  |  |                                 |  |  |  |  |
   H  H  H  H  H                                 H  H  O  H  H
         ↑                                             H
```

Substitution of an —OH for a 3-carbon hydrogen — 3-pentanol

The use of R, R′, and R″ indicates all R groups can be different.

The common name of 2-propanol is isopropyl alcohol.

The common name of 2-methyl-2-propanol is tertiary-butyl alcohol.

```
   PRIMARY          SECONDARY          TERTIARY
      H                 R                  R
      |                 |                  |
  R—C—OH           R′—C—OH            R′—C—OH
      |                 |                  |
      H                 H                  R″

                      CH3                CH3
                      |                  |
  CH3CH2OH        CH3—C—OH           CH3—C—OH
                      |                  |
                      H                  CH3

   Ethanol         2-propanol       2-methyl-2-propanol
(grain alcohol) (rubbing alcohol)  (gasoline additive)
```

Characteristic chemistry for important members of each functional group class is given in Chapter 7.

SELF-TEST 6B

1. The simplest aromatic compound is ______________.
2. Identify the functional groups present in each of the following molecules:

   a. R—OH ______________

   b. R—C(=O)—OH ______________

   c. R—C(=O)—H ______________

   d. R—C(=O)—R′ ______________

3. **a.** Write the structural formulas for ethane, ethanol, ethanal, ethanoic acid, diethyl ether, and ethyl amine.

b. Give common names where possible.
c. What R group is present in these compounds?

4. Give examples of the following:
   a. An ether ______________________
   b. An alcohol ______________________
   c. An organic acid ______________________
   d. A ketone ______________________

5. How many atoms does the symbol [benzene ring] represent?
   ____

QUESTIONS

1. Saturated hydrocarbons are so named because they have the maximum amount of hydrogen present for a given amount of carbon. The saturated hydrocarbons have the general formula C_nH_{2n+2}, where n is a whole number. What are the names and formulas of the first four members of this series of compounds?
2. Give three reasons why there are over 9 million known organic compounds.
3. Draw the three trichlorobenzene structures and name them.
4. Are more organic or inorganic compounds known?
5. Why is carbon the "central" element in organic compounds?
6. What is buckminsterfullerene?
7. Why are the properties of graphite, diamond, and buckminsterfullerene different?
8. Draw the tetrahedral structure of the methane molecule and label the bond angles.
9. Defend or refute the statement: "The tetrahedral angle is the most common angle found in naturally occurring substances."
10. The discovery of buckminsterfullerene is regarded as a major breakthrough in chemistry that will likely open up a whole new area of chemistry. Why?
11. Draw the expanded and condensed structural formulas for 2,3-dimethylpentane.
12. What is a functional group? Why are they important?
13. Write the structural formulas for:
    a. 2-methylpentane
    b. 4,4-dimethyl-5-ethyloctane
    c. methylbutane
    d. 2-methyl-2-hexene
14. Which propanol is used as rubbing alcohol?
15. What unique bond is present in an alkyne hydrocarbon?
16. How do primary, secondary, and tertiary alcohols differ?
17. Draw the structure of 1,1,1-trichlorethane.
18. What is the structural formula for 1-pentene?
19. Give an example of
    a. an alkane
    b. an amine
    c. a carboxylic acid
    d. an ether
    e. an ester
    f. an alkene
    g. an alkyne
    h. an alcohol
    i. a ketone
    j. an aldehyde
20. Indicate the functional groups present in the following molecules.
    a. $CH_3CH_2CH_2COOH$
    b. $CH_3CH_2NH_2$
    c. $CH_3CHCH_2CH_2COOH$ (with NH_2 on the second C)
    d. CH_3CHCH_2COOH (with OH on the second C)
    e. $CH_3CCH_2CH_2COOH$ (with =O on the second C)

    f. CH_3CHCH_2OH (with NH_2 on the second C)
21. Draw condensed structural formulas for the five isomers of hexane, C_6H_{14}.

Gasoline from oil—how long will it last? (*Standard Station* by Edward Ruscha. Collection, The Museum of Modern Art, New York: John B. Turner Fund)

CHAPTER

7

Organic Chemicals: Materials for Society

Organic chemicals are important in the production of thousands of consumer products ranging from plastics to medicines. Current topics include petroleum refining, octane rating of gasolines, chemistry of alcohols, carboxylic acids, and esters. Petroleum is currently the principal source of organic chemicals used as starting materials in the synthesis of consumer products. The first complete chemicals-from-coal plant is an example of the possible use of coal as a major source of organic chemicals after petroleum reserves are depleted.

1. Why is the organic chemical industry referred to as the petrochemical industry?
2. How is petroleum refined?
3. What is the octane rating of gasoline?
4. How is the 55 octane rating of straight-run gasoline raised to 90 octane?
5. What is the status of using methanol as a replacement for gasoline?
6. What are some commercial uses of ethanol and methanol?
7. What is coal gasification?
8. How can coal be used to produce acetic anhydride?
9. What synthesis steps are involved in the commercial synthesis of aspirin?

The millions of organic compounds are either hydrocarbons or derivatives of hydrocarbons. In Chapter 6 the hydrocarbon derivatives were organized into classes based on functional groups (see Table 6-4) that replace one or more hydrogen atoms in the parent hydrocarbon. The importance of organic chemicals in the production of plastics, synthetic rubber, synthetic fibers, fertilizers, drugs, and thousands of other consumer products can be seen by looking at the list of the top 50 chemicals produced in the United States. Of the top 50 listed on the inside front cover of this text, 29 are organic chemicals.

Fossil fuels are the major source of hydrocarbons, and among these, about 10% of the petroleum refined today is the source of most of the organic chemicals used to make consumer products. For this reason the organic chemical industry is often referred to as the **petrochemical industry.** However, natural gas and coal are important sources of organic chemicals, and coal will likely become more important as petroleum reserves are depleted (see Chapter 5). Synthesis gas, a mixture of CO and H_2 produced by treating coal or natural gas with steam, is already widely used in the preparation of organic chemicals. Coal tar, obtained by the pyrolysis of coal, is an important source of aromatic compounds. Figure 7-1 summarizes the organic chemicals obtained from fossil fuels and their uses in the synthesis of a wide range of commercial products.

Review the discussion of synthesis gas and coal tar in Chapter 5.

REFINING PETROLEUM

Petroleum is a complex mixture of alkanes, cycloalkanes, alkenes, and aromatic hydrocarbons. Thousands of compounds are present in crude petroleum, and the composition of petroleum can be classified according to boiling point ranges (Table 7-1). Each of these ranges has important uses. Because 90% of the petroleum fractions are used for fuels and the gasoline fraction accounts for about half of this use, petroleum refining is discussed first. The **refining of petroleum** is the separation of fractions with a certain boiling point range by a process called **fractional distillation.**

Figure 7-2 is a schematic drawing of a fractional distillation tower used in the petroleum refining process; a picture of a fractionating tower is shown in Figure 7-3(a). The crude oil is heated to about 400°C to produce a hot vapor and liquid mixture that enters the fractionating tower. The vapor rises and condenses at various points along the tower. The lower boiling fractions (those that are more volatile) remain in the vapor stage longer than the higher boiling fractions. These differences in boiling point ranges allow the separation of fractions in the same way that simple distillation allows the partial separation of water and ethanol. Some of the gases do not condense and are

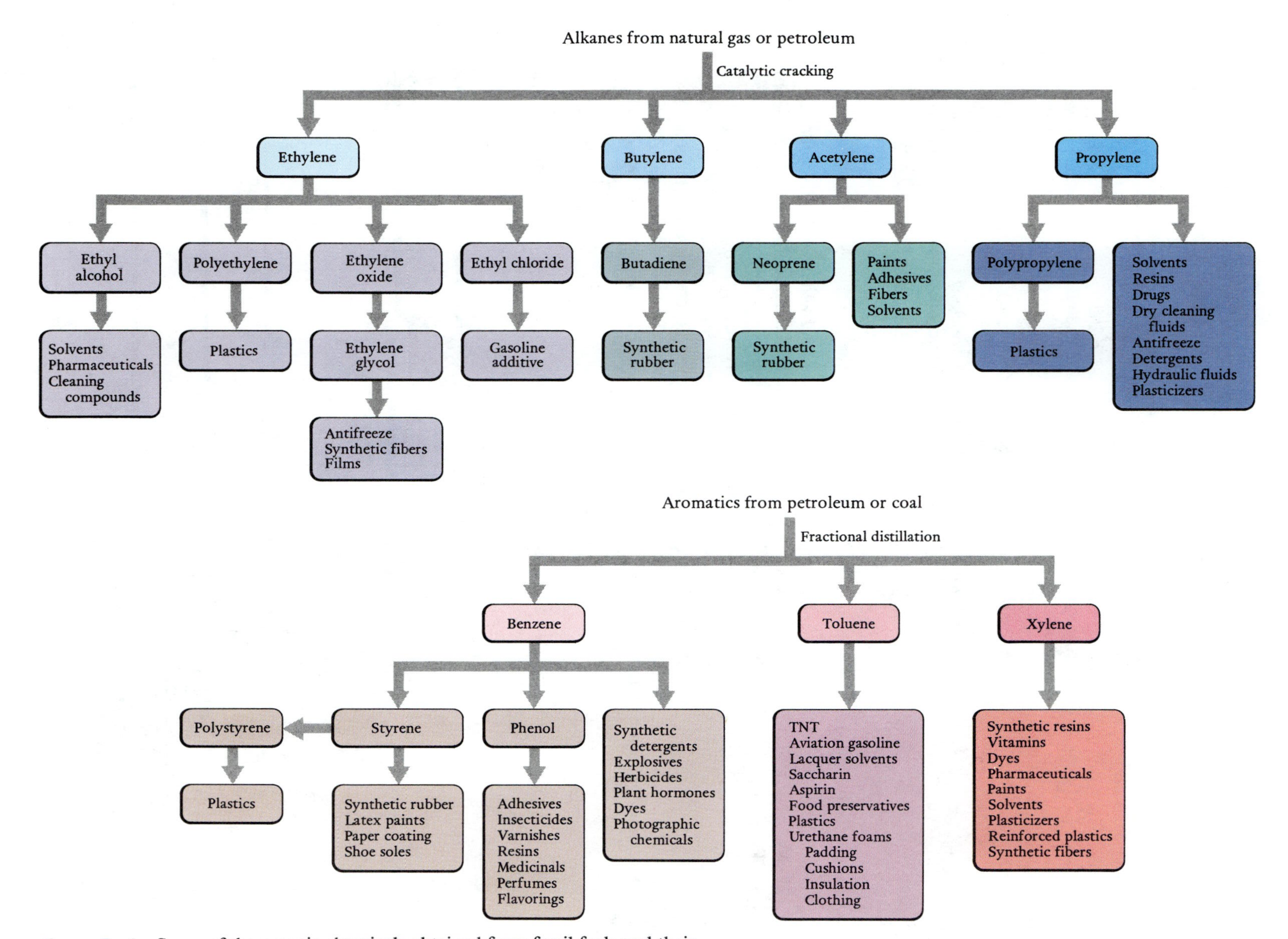

Figure 7–1 Some of the organic chemicals obtained from fossil fuels and their uses as raw materials.

TABLE 7-1 Hydrocarbon Fractions from Petroleum

Fraction	Size Range of Molecules	Boiling Point Range, °C	Uses
Gas	C_1-C_4	0 to 30	Gas fuels
Straight-run gasoline	C_5-C_{12}	30 to 200	Motor fuel
Kerosene	$C_{12}-C_{16}$	180 to 300	Jet fuel, diesel oil
Gas-oil	$C_{16}-C_{18}$	Over 300	Diesel fuel, cracking stock
Lubricants	$C_{18}-C_{20}$	Over 350	Lubricating oil, cracking stock
Paraffin wax	$C_{20}-C_{40}$	Low-melting solids	Candles, wax paper
Asphalt	above C_{40}	Gummy residues	Road asphalt, roofing tar

drawn off the top of the tower. The unvaporized residual oil is collected at the bottom of the tower. Typical products of the fractionation of petroleum are listed in Table 7-1.

Part of the refinement process involves adjusting the percentage of each fraction to match the demand. For example, the demand for gasoline is higher than that for kerosene. As a result, chemical reactions are used to convert the larger kerosene fraction molecules into molecules in the gasoline range ("cracking" process). The **catalytic cracking** process involves heating

Cracking breaks larger molecules into smaller ones.

Figure 7-2 Schematic diagram of a fractionating column for distilling petroleum.

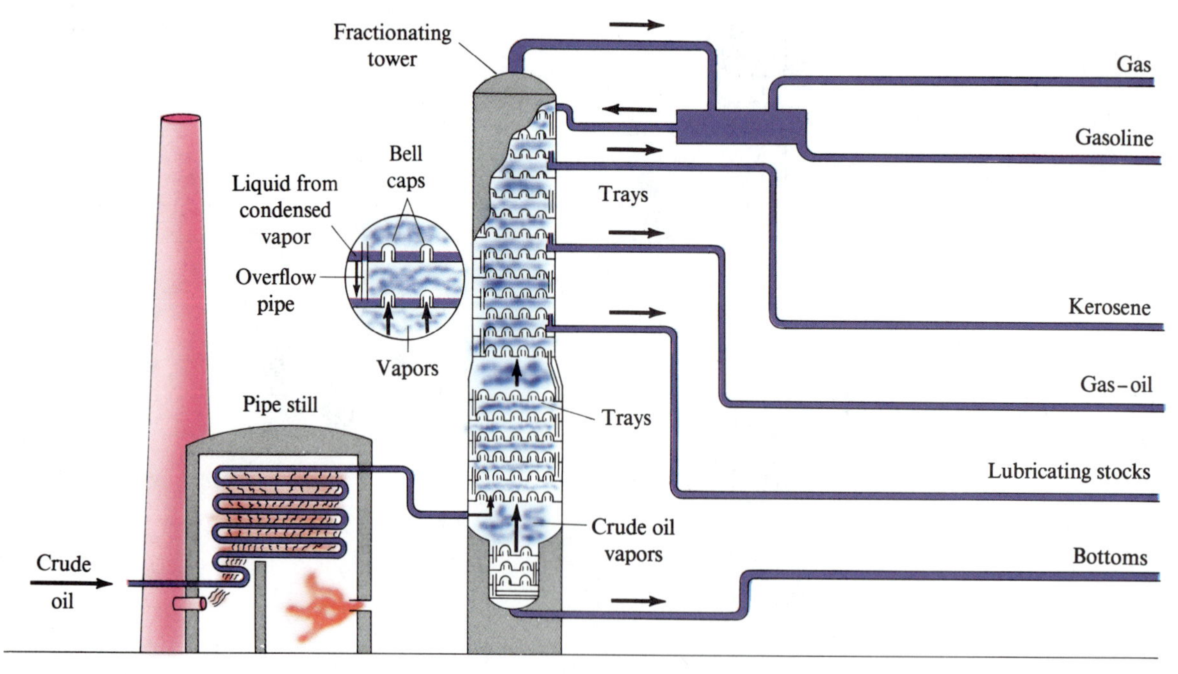

(a)

(b)

Figure 7–3 (a) A view of the towers at an oil refinery used for the fractional distillation of petroleum. (© Four by Five.) (b) A view of the catalytic cracking unit. (*The World of Chemistry,* Program 22, "The Age of Polymers")

saturated hydrocarbons under pressure in the absence of air (Fig. 7–3b). The hydrocarbons break into shorter-chain hydrocarbons—both alkanes

$$\underset{\text{An alkane}}{C_{16}H_{34}} \xrightarrow[\text{Heat}]{\text{Catalyst, Pressure}} \underset{\substack{\text{An alkane}\\\text{in the gasoline range}}}{C_8H_{18}} + \underset{\text{An alkene}}{C_8H_{16}}$$

and alkenes, some of which are in the gasoline range. The catalysts include specially processed clays, known as **zeolites.** The "straight-run" gasoline fraction obtained from the fractional distillation of petroleum contains primarily straight-chain hydrocarbons that burn too rapidly to be suitable for use as a fuel in internal combustion engines. Rapid ignition causes a "knocking" or "pinging" sound in the engine that reduces engine power and may damage the engine. Cracking of petroleum fractions brought about not only an increase in the quantity of gasoline available from a barrel of crude oil, but also an increase in *quality.* That is, gasoline from a cracking process can be used at higher efficiency (in a high-compression engine) than can straight-run gasoline, because the molecular structures of the hydrocarbons in the cracked gasoline allow them to oxidize more smoothly at high pressure without knocking.

A catalyst increases the speed of a reaction without being consumed in the reaction.

Review the discussion of catalysis in Chapter 2.

Knocking in gasoline engines is a sign of improper combustion.

Octane Rating

An arbitrary scale for rating the relative knocking properties of gasolines has been developed based on the operation of a standard test engine. Normal heptane knocks considerably and is assigned an octane rating of 0:

$$\underset{n\text{-heptane}}{CH_3CH_2CH_2CH_2CH_2CH_2CH_3} \qquad \text{(Octane rating = 0)}$$

The octane scale measures the ability of a mixture to burn without knocking in a gasoline engine.

whereas 2,2,4-trimethylpentane (isooctane) is far superior in this respect and is assigned an octane rating of 100:

$$CH_3-\underset{CH_3}{\overset{CH_3}{\underset{|}{\overset{|}{C}}}}-CH_2-\underset{H}{\overset{CH_3}{\underset{|}{\overset{|}{C}}}}-CH_3 \qquad \text{(Octane rating = 100)}$$

2,2,4-trimethylpentane

The octane rating of a gasoline is determined by first using the gasoline in a standard engine and recording its knocking properties. This is compared to the behavior of mixtures of *n*-heptane and isooctane, and the percentage of isooctane in the mixture with identical knocking properties is called the octane rating of the gasoline. Thus, if a gasoline has the same knocking characteristics as a mixture of 9% *n*-heptane and 91% isooctane, it is assigned an octane rating of 91. This corresponds to a regular grade of gasoline. Since the octane rating scale was established, fuels superior to isooctane have been developed, so the scale has been extended well above 100.

Table 7–2 lists octane ratings for some hydrocarbons and octane enhancers. The straight-run gasoline fraction obtained from the fractional distillation of petroleum has an octane rating of 50 to 55, which is too low for use as a fuel in vehicles. From Table 7–2 we can see that the octane rating of a gasoline can be increased either by increasing the percentage of branched-chain and aromatic hydrocarbon fractions or by adding octane enhancers (or a combination of both).

The **catalytic re-forming** process is used to produce branched-chain and aromatic hydrocarbons. Under the influence of certain catalysts, such as finely divided platinum, straight-chain hydrocarbons with low octane numbers can be re-formed into their branched-chain isomers, which have higher octane numbers.

TABLE 7–2 Octane Numbers of Some Hydrocarbons and Gasoline Additives

Name	Octane Number
n-Heptane	0
n-Hexane	25
n-Pentane	62
1-Pentene	91
2,2,4-Trimethylpentane (isooctane)	100
Benzene	106
o-Xylene	107
Methanol	107
Ethanol	108
t-Butyl alcohol	113
Methyl *t*-butyl ether	116
p-Xylene	116
Toluene	118
Ethyl *t*-butyl ether	118

Catalytic re-forming is also used to produce aromatic hydrocarbons such as benzene, toluene, and xylenes by using different catalysts and petroleum mixtures. For example, when the vapors of straight-run gasoline, kerosene, and light oil fractions are passed over a catalyst at 650°C, a high percentage of the original material is converted into a mixture of aromatic hydrocarbons, from which benzene, toluene, xylenes, and similar compounds may be separated by fractional distillation. For example, *n*-hexane is converted into benzene

$$CH_3CH_2CH_2CH_2CH_2CH_3 \xrightarrow[650^\circ C]{\text{Catalyst}} C_6H_6 + 4\,H_2$$

n-hexane Benzene

The hydrogen produced here can be used in the synthesis of ammonia by the Haber process. (See Chapter 13.)

and *n*-heptane is changed into toluene.

$$CH_3CH_2CH_2CH_2CH_2CH_2CH_3 \xrightarrow[650^\circ C]{\text{Catalyst}} C_6H_5CH_3 + 4\,H_2$$

n-heptane Toluene

This process is one example of the use of expensive noble metals (platinum, palladium, rhodium, iridium, gold, and silver) as catalysts in many industrial processes. The surfaces of these metals are often used to catalyze gas-phase reactions. For example, the platinum catalyst in the catalytic re-forming process adsorbs low-octane alkanes such as *n*-hexane and converts them to compounds with higher octane numbers, such as benzene and branched or cyclic alkanes. The catalytic converter found in American automobiles produced since 1975 is also a good example of a catalytic process that uses platinum as a surface catalyst.

The octane number of a given blend of gasoline can also be increased by adding "antiknock" agents or octane enhancers. Prior to 1975, the most widely used antiknock agent was tetraethyllead, $(C_2H_5)_4Pb$. The addition of 3 g of $(C_2H_5)_4Pb$ per gallon increases the octane rating by 10 to 15, and before the Environmental Protection Agency (EPA) required reductions in lead content, both regular and premium gasoline contained an average of 3 g of $(C_2H_5)_4Pb$ or $(CH_3)_4Pb$ per gallon.

Because tetraethyllead can no longer be used, other octane enhancers are being added to gasoline to increase the octane rating. These include toluene, 2-methyl-2-propanol (also called **tertiary,** or *t*-butyl alcohol), methyl-*t*-butyl ether (MTBE), methanol, and ethanol. In 1990, the most popular octane enhancer was MTBE, which joined the top-50 chemical list for the first time in 1984 and was number 24 in 1990.

$$CH_3-\underset{CH_3}{\overset{OH}{C}}-CH_3$$

t-butyl alcohol

$$CH_3-O-\underset{CH_3}{\overset{CH_3}{C}}-CH_3$$

MTBE

Gasoline blends that contain methanol and ethanol are also being used as fuels. The EPA and all U.S. car manufacturers have approved the use of ethanol–gasoline blends up to 10% ethanol (known as **gasohol** when introduced in the 1970s). However, methanol is receiving much attention because it offers several advantages as an octane enhancer. When properly blended, methanol is more economical, has a higher octane rating, and can reduce emission levels of particulates, hydrocarbons, carbon monoxide, and nitrogen oxides. However, the biggest disadvantage of methanol relates to moisture. Small amounts of moisture destabilize the methanol–gasoline mixture, and metal corrosion of the engine becomes a serious problem.

A cosolvent aids in dissolving a solute.

The methanol moisture problem is solved by using another alcohol (ethanol, propanols, butanols) as a cosolvent in methanol blends. The EPA has approved several methanol blends that meet the vehicle emission standards and provide a high-octane gasoline. Most methanol blends contain about 2.5% methanol, 2.5% *t*-butyl alcohol, 95% gasoline, and a corrosion inhibitor.

SELF-TEST 7A

1. The fractions of petroleum are separated by ________________.
2. The kerosene fraction has a ( ) higher ( ) lower boiling point range than the gasoline fraction.
3. The ________________ process is used to produce branched-chain and aromatic hydrocarbons from straight-chain hydrocarbons.
4. The major source of organic chemicals used as starting materials in the synthesis of a wide range of commercial products is __________.
5. The __________ process is used in refining petroleum to convert molecules in the higher boiling fractions to molecules in the gasoline fraction.
6. A widely used octane enhancer is ________________.
7. Gasohol contains 90% gasoline and 10% ________________.
8. Which of the following hydrocarbons would be expected to have the highest octane number?

a. $CH_3CH_2CH_2CH_2CH_2CH_2CH_3$

b. $CH_3CH_2\underset{}{\overset{CH_3}{\overset{|}{C}}}HCH_2CH_2CH_3$

c. $CH_3{-}\underset{CH_3}{\underset{|}{\overset{CH_3}{\overset{|}{C}}}}{-}\underset{H}{\underset{|}{\overset{CH_3}{\overset{|}{C}}}}{-}CH_3$

ORGANIC SYNTHESIS

The preparation of new and different organic compounds through chemical reactions is called **organic synthesis.** Millions of organic compounds have been synthesized in the laboratories of the world during the past 150 years. Organic compounds were obtained originally from plants, animals, and fossil fuels, and these are still direct sources for many important chemicals such as sucrose from sugar cane or ethanol from fermented grain mash. However, the development of organic chemistry led to cheaper methods for the synthesis of both naturally occurring substances and new substances. Prior to 1828, it was widely believed that chemical compounds synthesized by living matter could not be made without living matter—a

Figure 7–4 Friedrich Wöhler (1800–1882) was professor of chemistry at the University of Berlin and later at Göttingen. His preparation of the organic compound urea from the inorganic compound ammonium cyanate did much to overturn the theory that organic compounds must be prepared in living organisms. He was also one of the first to study the properties of aluminum and the first to isolate the element beryllium, among many other outstanding contributions to chemistry.

"vital force" was necessary for the synthesis. In 1828, a young German chemist, Friedrich Wöhler (Fig. 7–4), destroyed the vital force myth and opened the door to modern organic syntheses. Wöhler heated a solution of silver cyanate and ammonium chloride, neither of which had been derived from any living substance. From these he prepared urea, a major animal waste product found in urine.

$$\underset{\text{Silver cyanate}}{AgOCN} + \underset{\text{Ammonium chloride}}{NH_4Cl} \longrightarrow \underset{\text{Silver chloride (precipitate)}}{AgCl} + \underset{\text{Ammonium cyanate}}{NH_4OCN}$$

$$\underset{\text{Ammonium cyanate}}{NH_4OCN} \xrightarrow{\text{Heat}} \underset{\text{Urea}}{H_2N\overset{\overset{\displaystyle O}{\|}}{C}NH_2}$$

The notion of a mysterious vital force declined as other chemists began to synthesize more and more organic chemicals without the aid of a living system. Soon it was shown that chemists could do more than imitate the products of living tissue; they could form unique materials of their own design.

Organic synthesis is based on an understanding of the classification of organic compounds by functional groups, a concept introduced in Chapter 6, and the study of the chemical reactions of those groups. Advances in understanding the structure of organic compounds also gave organic synthesis a tremendous boost. The organic chemist who knows the structure of compounds can predict, by analogy with simpler molecules, what reactions might take place when organic reagents are used. Very elegant and reliable schemes of synthesis can then be constructed that lead to the synthesis of new and useful compounds or to the more economical synthesis of known compounds; both are central functions of modern organic chemistry.

Acquaintance with a few representative compounds of each functional group class will give you a feel for the subject of organic chemistry. Keep in mind that the functional group classes represented in Table 6–4 and discussed here are examples of the most important functional groups, but there are many other functional groups.

ALKYL HALIDES

Alkyl halides are molecules in which one or more hydrogen atoms are replaced by halogen atoms. In the IUPAC system for naming these compounds, the halogen is specified as fluoro, chloro, bromo, or iodo. Structural

The halogens are group VIIA elements (F, Cl, Br, I).

formulas and names, including familiar common names, of some alkyl halides are:

```
     H                Cl               Cl  H
     |                |                |   |
 Cl—C—Cl          Cl—C—Cl          Cl—C—C—H
     |                |                |   |
     Cl               Cl               Cl  H
```

Trichloromethane (chloroform) | Tetrachloromethane (carbon tetrachloride) | 1,1,1-trichloroethane (methylchloroform)

```
     H              H   H
     |              |   |
 Cl—C—H         H—C—C—H
     |              |   |
     Cl             Br  Br
```

Dichloromethane (methylene chloride) | 1,2-dibromoethane (ethylene dibromide)

See Chapter 10 for a discussion of the toxicity and carcinogenicity of these compounds.

At one time chloroform and carbon tetrachloride were widely used as solvents in the laboratory and in industrial cleaning, but their toxicity and carcinogenicity have led them to be removed from use. Many chlorinated hydrocarbons are on either the carcinogen or the suspected carcinogen list of the EPA. Only 1,1,1-trichloroethane appears to be safe, and it has replaced the others in many solvent applications.

CFCs are perhaps better known as Freons® (the Du Pont trade name for a variety of CFCs used as refrigerants in refrigerators and air conditioning units and propellants in aerosol cans).

Substituted alkanes that contain both fluorine and chlorine are called chlorofluorocarbons (CFCs). Chlorofluorocarbons are relatively nontoxic, nonflammable, noncorrosive, odorless gases or liquids. These nonreactive properties led to the use of CFC-11 and CFC-12 as propellants in aerosol spray cans and as refrigerants in refrigerators

```
     F               F               H
     |               |               |
 Cl—C—Cl         F—C—Cl         F—C—Cl
     |               |               |
     Cl              Cl              F
```

CFC-11 | CFC-12 | CFC-22

The numbers in CFC-11, CFC-12, and CFC-22 are industrial code numbers.

and air conditioning units for buildings and vehicles. However, recent studies have shown that CFCs do react in the upper atmosphere, decreasing stratospheric ozone that is essential for filtering out harmful ultraviolet radiation. The reactions of ozone with CFCs and proposed reductions in CFC production are discussed in Chapter 12.

ALCOHOLS

Alcohols among the top 50 chemicals include methanol (21), ethylene glycol (29), and isopropyl alcohol (49).

Methanol

The production of synthesis gas, a mixture of carbon monoxide and hydrogen, from coal was discussed in Chapter 5.

Methanol, CH_3OH, is the simplest of all alcohols. Over 7 billion pounds of methanol are produced each year from synthesis gas. High pressure, high temperature, and a mixture of catalysts are used to increase the yield.

$$\underset{\text{Coal}}{C} + \underset{\text{Steam}}{H_2O} \longrightarrow \underset{\text{Synthesis gas}}{CO + H_2}$$

$$CO + 2\,H_2 \xrightarrow[300^\circ C]{ZnO,\ Cr_2O_3} \underset{\text{Methanol}}{CH_3OH}$$

An old method of producing methanol involved heating a hardwood such as beech, hickory, maple, or birch in a retort in the absence of air. For this reason methanol is sometimes called *wood alcohol.*

About 50% of the methanol produced in the United States is used in the production of formaldehyde (used in plastics, embalming fluid, germicides, and fungicides); 30% is used in the production of other chemicals; and the remaining 20% is used for jet fuels, antifreeze mixtures, solvents, as a gasoline additive, and as a denaturant (a poison added to make ethanol unfit for beverages). Methanol is a deadly poison that causes blindness in less than lethal doses. Many deaths and injuries have resulted from the accidental substitution of methanol for ethanol in beverages.

Methanol is the main ingredient in many windshield washer fluids.

Methanol will likely move upward in the ranking of top 50 chemicals when petroleum and natural gas become too expensive as sources of both energy and chemicals. Although most of the world's methanol currently comes from synthesis gas made from natural gas, coal gasification will become a more important source of methanol as the natural gas reserves are used up. Since methanol is relatively cheap, its potential as a fuel and as a starting material for the synthesis of other chemicals is receiving more attention.

Methanol as a Fuel

Methanol is being considered as a replacement for gasoline, especially in urban areas with extremely high levels of air pollution caused by motor vehicles. For example, methanol-powered cars and buses have been tested in southern California since 1980. The positive results with these vehicles and the high levels of air pollutants in southern California have led to a proposal for the gradual elimination of gasoline-powered vehicles in southern California over the next 20 years. The first five-year step of this proposal is to require owners of fleet vehicles to replace existing vehicles with electric- or methanol-powered vehicles. This would involve an estimated 1 million vehicles.

Methanol model. (Charles D. Winters)

What are the advantages and disadvantages of switching to methanol-powered vehicles? Since methanol burns more completely, levels of troublesome pollutants such as carbon monoxide, unreacted hydrocarbons, nitrogen oxides, and ozone would be reduced. Estimates are that these reductions would be to levels below those recommended by the EPA (see Chapter 12). Levels of methanol and formaldehyde would be higher. Present projections indicate acceptable levels of methanol and formaldehyde, but careful monitoring of these would be necessary.

The technology for methanol-powered vehicles has existed for many years, particularly for racing cars, which burn methanol because of its high-octane rating. Since methanol has about one-half the energy content of gasoline, it takes almost 2 gallons of methanol to go as far as 1 gallon of

Cars at the Indianapolis 500 are powered by methanol.

gasoline. Methanol costs about half as much as gasoline so the price per mile would be competitive. However, the size of fuel tanks will need to be doubled. In addition, methanol corrodes regular steel so the fuel system will need to be made out of stainless steel or have a methanol-resistant coating. Until sufficient numbers of methanol-powered vehicles are on the road, cars equipped to run on either methanol or gasoline will be necessary because of the lack of service stations selling methanol. Ford Motor Company is producing such a vehicle, and California plans to purchase 5000 of them. ARCO has also announced plans to build 20 service stations in southern California for marketing methanol. As the problems of distribution and storage are solved, better engineered methanol-fueled engines will be designed and produced, which will lead to higher efficiency utilization of methanol as a fuel.

Methanol to Gasoline

Another option is to use methanol to make gasoline. Mobil Oil Company has developed a methanol-to-gasoline process, which is currently not competitive with refined gasoline prices in the United States, but is competitive in those regions of the world, such as New Zealand, where the price of gasoline is much higher. In fact, the production of 92-octane gasoline from methanol is now taking place in New Zealand.

New Zealand Synthetic Fuels Company is operating a plant based on a methanol-to-gasoline process developed by Mobil Oil Company. The process starts with the production of synthesis gas from natural gas and then uses the synthesis gas to make methanol. The key reaction for the production of gasoline from methanol is the dehydration of methanol with a clay catalyst developed by Mobil; this catalyst is known as the ZSM-5 zeolite catalyst. The

Wine *(right)* is produced from the glucose in grape juice *(left)* by fermentation. The fermentation jug *(center)* has a bubble chamber to allow CO_2 to escape but prevent oxygen from entering and oxidizing ethanol to acetic acid (vinegar). (Charles D. Winters)

catalyst aids the dehydration to yield short-chain alkenes, which then cyclize and polymerize to give a mixture of C_5 to C_{12} hydrocarbons made up of branched chains, straight chains, and aromatics, similar to the 92-octane gasoline currently obtained by the refinement of straight-run gasoline from petroleum refining. The dehydration and subsequent polymerization can be represented by the following reactions:

Polymerization is discussed in Chapter 8.

$$2\ CH_3OH \xrightarrow[\text{Catalyst}]{\text{ZSM-5}} \underset{\text{Dimethyl ether}}{(CH_3)_2O} + H_2O$$

$$2\ (CH_3)_2O \xrightarrow[\text{Catalyst}]{\text{ZSM-5}} \underset{\text{Ethylene}}{2\ C_2H_4} + 2\ H_2O$$

$$C_2H_4 \xrightarrow[\text{Catalyst}]{\text{ZSM-5}} \underset{\text{Gasoline}}{\text{Hydrocarbon mixture in the } C_5\text{–}C_{12} \text{ range}}$$

The New Zealand plant is currently producing 14,000 barrels per day of gasoline with an octane rating of 92 to 94. This is about one-third the amount of gasoline used in New Zealand.

Ethanol

Ethanol, also called ethyl alcohol or grain alcohol, can be obtained by the fermentation of carbohydrates (starch, sugars). For example, glucose is converted into ethanol and carbon dioxide by the action of yeast in the absence of oxygen.

Ethanol can be prepared by the fermentation of grains.

$$\underset{\text{Glucose}}{C_6H_{12}O_6} \xrightarrow{\text{Yeast}} \underset{\text{Ethanol}}{2\ C_2H_5OH} + 2\ CO_2$$

A mixture of 95% ethanol and 5% water can be recovered from the fermentation products by distillation. Ethanol is the active ingredient of alcoholic beverages. Some of the most commonly encountered alcoholic beverages and their characteristics are presented in Table 7–3. The "proof" of an

95% ethanol (190 proof) is a strong dehydrating agent. Never drink it straight.

TABLE 7–3 Common Alcoholic Beverages

Name	Source of Fermented Carbohydrate	Amount of Ethyl Alcohol	Proof
Beer	Barley, wheat	5%	10
Wine	Grapes or other fruit	12% maximum, unless fortified*	20–24
Brandy	Distilled wine	40–45%	80–90
Whiskey	Barley, rye, corn, etc.	45–55%	90–110
Rum	Molasses	~45%	90
Vodka	Potatoes	40–50%	80–100

* The growth of yeast is inhibited at alcohol concentrations over 12%, and fermentation comes to a stop. Beverages with a higher concentration are prepared either by distillation or by fortification with alcohol that has been obtained by the distillation of another fermentation product.

TABLE 7–4 Alcohol Blood Level and Effect

Blood Alcohol Level (Percentage by Volume)	Effect
0.05–0.15	Lack of coordination
0.15–0.20	Intoxication
0.30–0.40	Unconsciousness
0.50	Possible death

alcoholic beverage is twice the volume percent of ethanol; 80 proof vodka, for example, contains 40% ethanol.

Although ethanol is not as toxic as methanol, 1 pint of pure ethanol, rapidly ingested, would kill most people. Ethanol is a depressant. The effects of different blood levels of alcohol are shown in Table 7–4. Rapid consumption of two 1-oz "shots" of 90-proof whiskey or of two 12-oz beers can cause one's blood alcohol level to reach 0.05%.

The breathalyzer test used to detect drunken drivers is based on the color change that occurs when ethanol is oxidized to acetic acid by dichromate anion ($Cr_2O_7^{2-}$) in acidic solution.

$$16\ H^+ + \underset{\text{Yellow-orange}}{2\ Cr_2O_7^{2-}} + 3\ CH_3CH_2OH \longrightarrow 3\ CH_3COOH + \underset{\text{Green}}{4\ Cr^{3+}} + 11\ H_2O$$

Ethanol is quickly absorbed by the blood and metabolized by enzymes produced in the liver. The rate of detoxification is about 1 oz of pure alcohol per hour. The ethanol is oxidized to acetaldehyde, which is further oxidized to acetic acid; eventually CO_2 and H_2O are produced and eliminated through the lungs and kidneys.

$$\underset{\text{Ethanol}}{CH_3CH_2OH} \xrightarrow[\text{Enzymes}]{\text{Liver}} \underset{\text{Acetaldehyde}}{CH_3CHO}$$

Ethanol model. (Charles D. Winters)

Industrial Use of Ethanol

The Federal tax on alcoholic beverages is about $20 per gallon. Since the cost of producing ethanol is only about $1 per gallon, ethanol intended for industrial use must be *denatured* to avoid the beverage tax. **Denatured alcohol** contains small amounts of a toxic substance, such as methanol or gasoline, that cannot be removed easily by chemical or physical means.

Apart from being used in the alcoholic beverage industry, ethanol is used widely in gasohol as a gasoline additive (described earlier in this chapter), as a solvent, and in the preparation of many other organic compounds. Although fermentation of grains is an important source of ethanol, over 1 billion pounds of ethanol are synthesized each year from ethylene and water vapor.

Where does the ethylene come from?

$$\begin{array}{c} H \quad\quad H \\ \diagdown\;\;\diagup \\ C{=}C \\ \diagup\;\;\diagdown \\ H \quad\quad H \end{array} + HOH \xrightarrow[300^\circ C]{70\ atm.} \begin{array}{c} H\ \ H \\ |\ \ \ | \\ H{-}C{-}C{-}OH \\ |\ \ \ | \\ H\ \ H \end{array}$$

Ethylene — Ethanol

Propanols

When one considers the possible structures for propyl alcohol (propanol), it is apparent that two isomers are possible. The most common one is 2-propanol (rubbing alcohol).

$$\begin{array}{c} H\ \ H\ \ H \\ |\ \ \ |\ \ \ | \\ H{-}C{-}C{-}C{-}OH \\ |\ \ \ |\ \ \ | \\ H\ \ H\ \ H \end{array} \qquad \begin{array}{c} H\ \ H\ \ H \\ |\ \ \ |\ \ \ | \\ H{-}C{-}C{-}C{-}H \\ |\ \ \ |\ \ \ | \\ H\ \ O\ \ H \\ H \end{array}$$

1-propanol
n-propyl alcohol

2-propanol
Isopropyl alcohol
(rubbing alcohol)

Ethylene Glycol and Glycerol

More than one alcohol group (—OH) can be present in a single molecule. Glycerol and ethylene glycol, the principal component of permanent antifreeze, are examples of such compounds.

$$\begin{array}{c} H \\ | \\ H{-}C{-}OH \\ | \\ H{-}C{-}OH \\ | \\ H \end{array} \qquad \begin{array}{c} H \\ | \\ H{-}C{-}OH \\ | \\ H{-}C{-}OH \\ | \\ H{-}C{-}OH \\ | \\ H \end{array}$$

Ethylene glycol — Glycerol (glycerin)

Products that contain ethylene glycol or glycerin. (Charles Steele)

Glycerol is a byproduct in the manufacture of soaps. Because of its moisture-holding properties, glycerol has many uses in foods and tobacco as a digestible and nontoxic humectant (gathers and holds moisture), and in the manufacture of drugs and cosmetics. It is also used in the production of nitroglycerin and numerous other chemicals. Perhaps the most important compounds of glycerol are its natural esters (fats and oils), which are discussed in Chapter 9.

ALDEHYDES AND KETONES

Aldehydes and ketones contain a **carbonyl group,** $>C{=}O$ and are generally obtained by the oxidation of alcohols. In aldehydes the carbonyl group is on an end carbon, whereas in ketones the carbonyl group is bonded to two carbon atoms.

$$\underset{\text{Aldehyde}}{R{-}\overset{\overset{\large O}{\|}}{C}{-}H} \qquad \underset{\text{Ketone}}{R{-}\overset{\overset{\large O}{\|}}{C}{-}R}$$

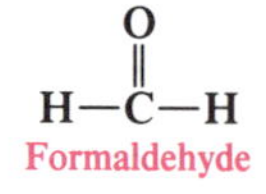

Formaldehyde

Formaldehyde is a suspected carcinogen.

Formaldehyde, the simplest aldehyde, has a foul odor. It is the starting material in the production of several plastics and is used in the laboratory as a preservative for dead animals. Aldehydes with an aromatic ring have pleasant odors, and some are used in food flavors and perfumes.

Benzaldehyde (bitter almonds) — Vanillin (vanilla bean) — Cinnamaldehyde (cinnamon)

The simplest ketone is acetone, an important commercial solvent. Methyl ethyl ketone is a solvent in model-airplane glue.

Formaldehyde model. (Charles D. Winters)

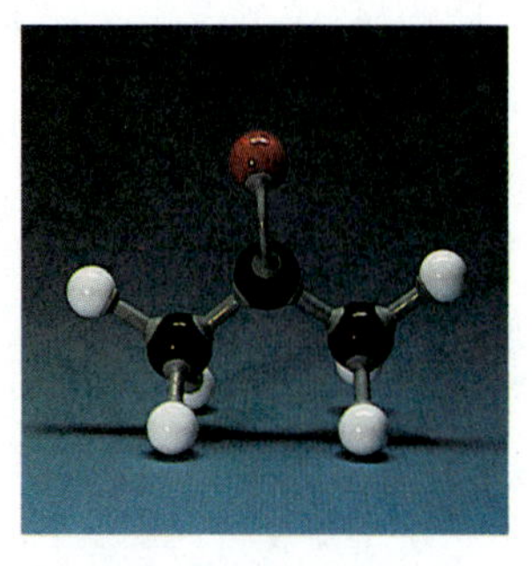

Acetone model. (Charles D. Winters)

$$\underset{\text{Acetone}}{H_3C-\overset{\overset{\displaystyle O}{\|}}{C}-CH_3} \qquad \underset{\text{Methyl ethyl ketone}}{H_3C-\overset{\overset{\displaystyle O}{\|}}{C}-CH_2CH_3}$$

CARBOXYLIC ACIDS

Organic, or carboxylic, acids contain the **carboxyl group,** —COOH, and can be prepared by the oxidation of alcohols or aldehydes. These reactions occur quite readily, as evidenced by the souring of wine, which is the oxidation of ethanol to acetic acid in the presence of oxygen from the air.

Carboxylic acids are found in both the plant and animal kingdoms. The first six carboxylic acids, with their sources, common names, and odors, are given in Table 7–5. Longer chain carboxylic acids do not smell as bad, in part because they are less volatile. Some of the other common carboxylic acids found in nature are given in Table 7–6. As can be seen in the table, some organic acids have more than one carboxyl group as well as other groups, usually hydroxyl groups.

Oxidation of ethanol to acetic acid by oxygen in air is responsible for the souring of wine.

Formic acid, found in ants and other insects, is part of the irritant that produces itching and swelling after an insect bite.

Organic acids are generally weak acids and have the following structure:

$$R-C\begin{matrix}\nearrow O \\ \searrow OH\end{matrix}$$

Acetic Acid

Acetic (ethanoic) acid is the most widely used of the organic acids. It is found in vinegar, an aqueous solution containing 4% to 5% acetic acid. Flavor and colors are imparted to vinegars by the constituents of the alcoholic solutions from which they are made. Ethanol in the presence of certain bacteria and air is oxidized to acetic acid:

$$\underset{\text{Ethanol}}{CH_3CH_2OH} + \underset{\text{Oxygen}}{O_2} \xrightarrow{\text{Bacteria}} \underset{\substack{\text{Ethanoic acid}\\ \text{(acetic acid)}}}{CH_3COOH} + \underset{\text{Water}}{H_2O}$$

The bacteria, called mother of vinegar, form a slimy growth in a vinegar solution. The growth of bacteria can sometimes be observed in a bottle of commercially prepared vinegar after it has been opened to the air.

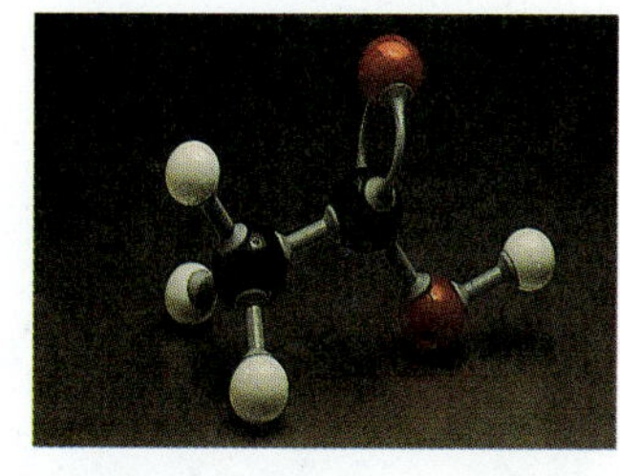

Acetic acid model. (Charles D. Winters)

TABLE 7–5 First Six Carboxylic Acids

Formula	Source	Common Name	IUPAC Name	Odor
$HCOOH$	Ants (Latin, *formica*)	Formic acid	Methanoic acid	Sharp
CH_3COOH	Vinegar (Latin, *acetum*)	Acetic acid	Ethanoic acid	Sharp
CH_3CH_2COOH	Milk (Greek, *protos pion,* "first fat")	Propionic acid	Propanoic acid	Swiss cheese
$CH_3(CH_2)_2COOH$	Butter (Latin, *butyrum*)	Butyric acid	Butanoic acid	Rancid butter
$CH_3(CH_2)_3COOH$	Valerian root (Latin, *valere,* "to be strong")	Valeric acid	Pentanoic acid	Manure
$CH_3(CH_2)_4COOH$	Goats (Latin, *caper*)	Caproic acid	Hexanoic acid	Goat

TABLE 7–6 Some Other Naturally Occurring Carboxylic Acids

Name	Structure	Natural Source
Citric acid	$HOOC-CH_2-C(OH)(COOH)-CH_2-COOH$	Citrus fruits
Lactic acid	$CH_3-CH(OH)-COOH$	Sour milk
Malic acid	$HOOC-CH_2-CH(OH)-COOH$	Apples
Oleic acid	$CH_3(CH_2)_7-CH{=}CH-(CH_2)_7-COOH$	Vegetable oils
Oxalic acid	$HOOC-COOH$	Rhubarb, spinach, cabbage, tomatoes
Stearic acid	$CH_3(CH_2)_{16}-COOH$	Animal fats
Tartaric acid	$HOOC-CH(OH)-CH(OH)-COOH$	Grape juice, wine

Sources of some naturally occurring carboxylic acids. (Charles D. Winters)

Acetic acid is an important starting substance for making textile fibers, vinyl plastics, and other chemicals and is a convenient choice when a cheap organic acid is needed.

ESTERS

Organic esters are compounds of the type

$$R-O-\underset{\underset{O}{\|}}{C}-R'$$

formed by the reaction of organic acids and alcohols.

In the presence of strong mineral acids, organic acids react with alcohols to form compounds called **esters** (Table 7–7). For example, when ethyl alcohol is mixed with acetic acid in the presence of sulfuric acid, ethyl acetate is formed. This reaction is a dehydration in which sulfuric acid acts as a catalyst and dehydrator.

$$CH_3CH_2O\boxed{H + HO}\underset{\underset{O}{\|}}{C}CH_3 \xrightleftharpoons{H_2SO_4} CH_3CH_2O\underset{\underset{O}{\|}}{C}CH_3 + H_2O$$

Ethyl acetate

Ethyl acetate is a common solvent for lacquers and plastics and is often used as fingernail polish remover.

Some odors of common fruits are due to the presence of mixtures of volatile esters (Table 7–7). In contrast, esters of higher molecular weight often have a distinctly unpleasant odor.

TABLE 7-7 Some Alcohols, Acids, and Their Esters

Alcohol	Acid	Ester	Odor of the Ester
$CH_3CH(CH_3)CH_2CH_2OH$ Isopentyl alcohol	CH_3COOH Acetic acid	$CH_3CH(CH_3)CH_2CH_2{-}O{-}C({=}O){-}CH_3$ Isopentyl acetate	Banana
$CH_3CH(CH_3)CH_2CH_2OH$ Isopentyl alcohol	$CH_3CH_2CH_2CH_2COOH$ Pentanoic acid	$CH_3CH(CH_3)CH_2CH_2{-}O{-}C({=}O){-}CH_2CH_2CH_2CH_3$ Isopentyl pentanoate	Apple
$CH_3CH_2CH_2CH_2OH$ *n*-Butyl alcohol	$CH_3CH_2CH_2COOH$ Butanoic acid	$CH_3CH_2CH_2CH_2{-}O{-}C({=}O){-}CH_2CH_2CH_3$ Butyl butanoate	Pineapple
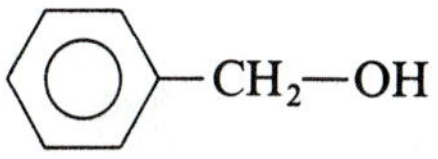 $C_6H_5{-}CH_2{-}OH$ Benzyl alcohol	$CH_3CH_2CH_2COOH$ Butanoic acid	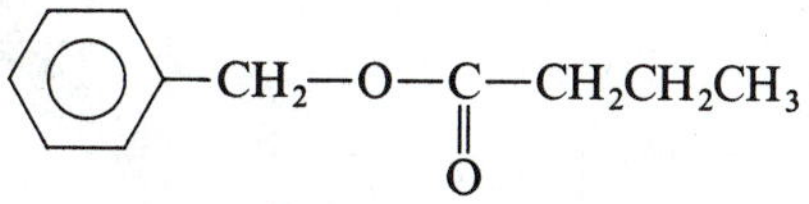 $C_6H_5{-}CH_2{-}O{-}C({=}O){-}CH_2CH_2CH_3$ Benzyl butanoate	Rose

AMINES AND AMIDES

Organic amines can be considered derivatives of ammonia (NH_3). The nitrogen in the amine may be attached to R groups or may be part of a ring.

$CH_3{-}NH_2$

Methyl amine (fish odor)

Pyridine (foul odor)

Alkaloids (alkali-like) are amines derived from plants. The amine nitrogen is usually part of a ring. Caffeine, nicotine, morphine, and coniine are examples of alkaloids (Fig. 7-5).

The functional group in amides is $-C(=O)-NH_2$. The functional groups of amines and amides are found in many important biological compounds (discussed in Chapter 9).

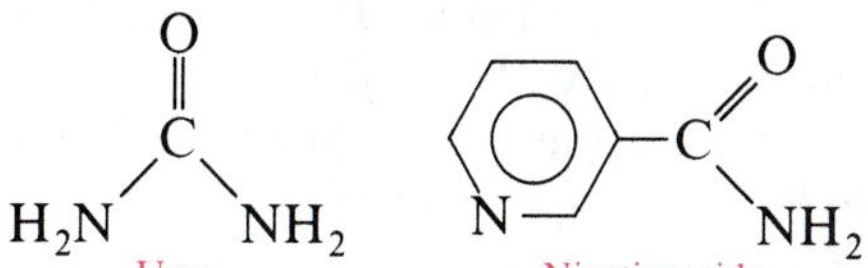

$H_2N{-}C({=}O){-}NH_2$

Urea

Nicotinamide

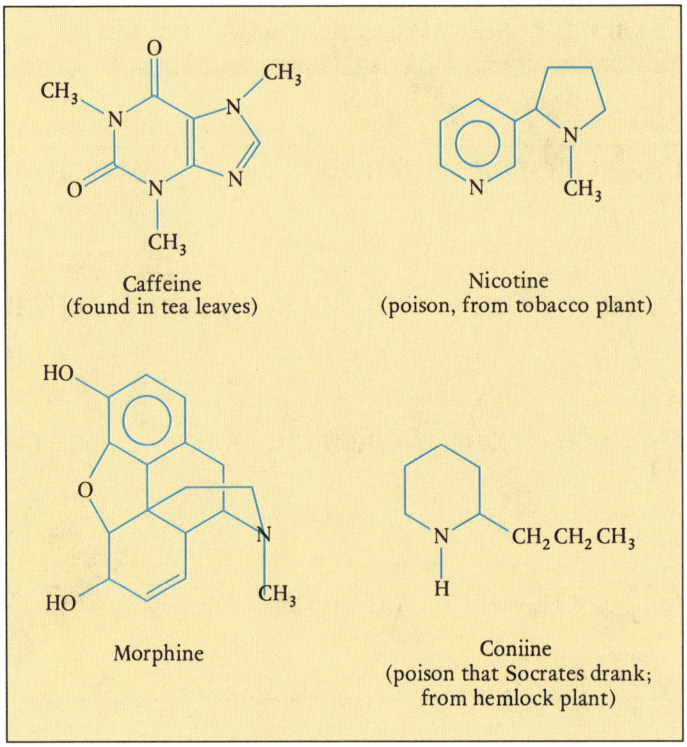

Caffeine
(found in tea leaves)

Nicotine
(poison, from tobacco plant)

Morphine

Coniine
(poison that Socrates drank;
from hemlock plant)

Figure 7–5 Some common alkaloids.

WILL COAL BECOME THE MAJOR SOURCE OF CHEMICALS?

Although petroleum is now the source of over 90% of the organic chemicals used to synthesize consumer products, the projected depletion of petroleum reserves mentioned in Chapter 5 has increased the interest in coal as a source of organic chemicals.

Synthesis gas from coal is already receiving increased attention as a starting material for the production of organic chemicals that are among the top 50 produced in the United States (inside front cover). For example, the first complete chemicals-from-coal plant, built by Eastman Kodak in Kingsport, Tennessee, started production in 1983. Figure 7–6(a) is a schematic drawing of the various components of the plant, which is pictured in Figure 7–6(b). The basic reactions are to produce synthesis gas from coal, to use the synthesis gas to make methanol, and to use the methanol in the synthesis of acetic anhydride. Acetic anhydride is used by Eastman Kodak to make cellulose acetate, a polymer used in the manufacture of photographic film base, synthetic fibers, plastics, and other products.

Within the complex pictured in Figure 7–6(b) are nine separate plants, four related to the gasification of coal, two for synthesis-gas preparation, and

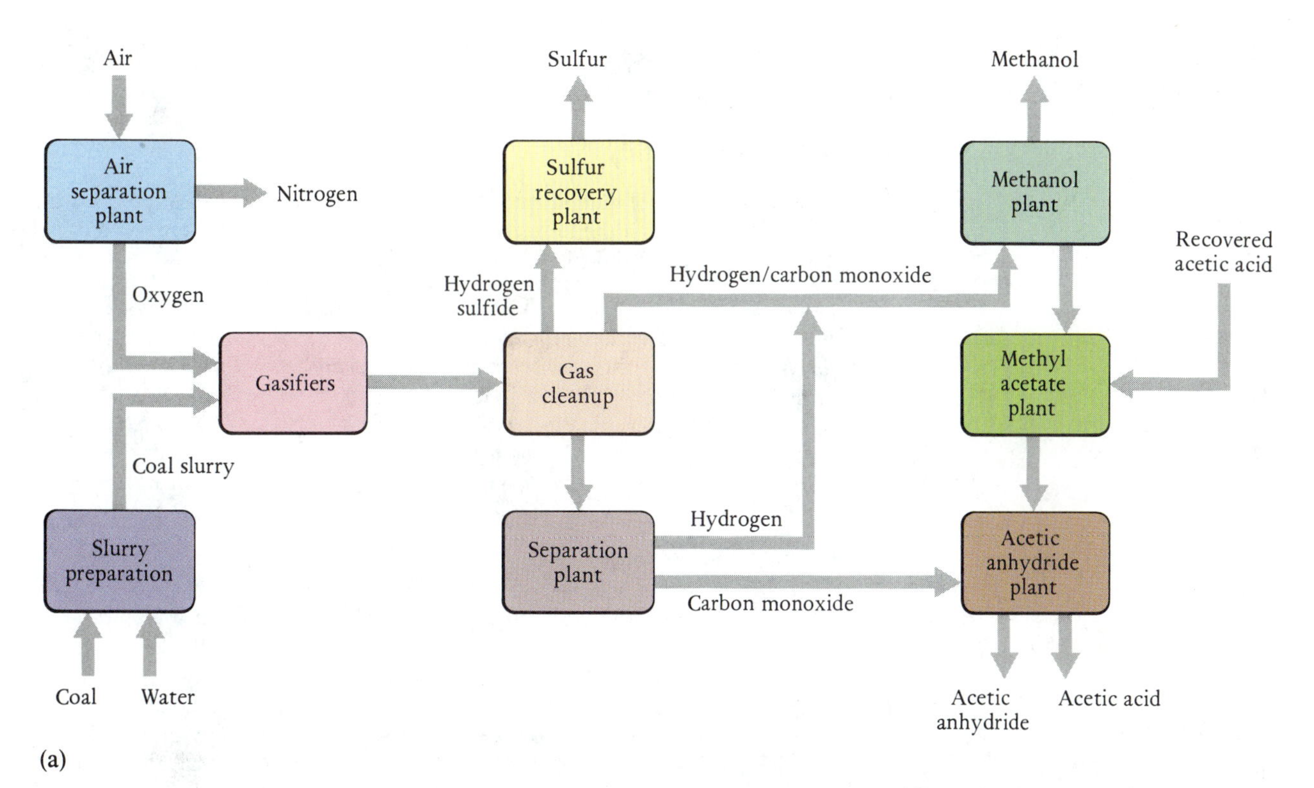

(a)

(b)

Figure 7–6 (a) Schematic drawing showing the production of acetic anhydride from coal. (b) Eastman Kodak's chemicals-from-coal facility in Kingsport, Tennessee. Numbers on the photograph represent different parts of the plant: 1, coal unloading; 2, coal silos; 3, steam plant; 4, slurry preparation; 5, coal gasification plant; 6, gas cleanup and separation; 7, sulfur recovery plant; 8, gas flare stack; 9, chemical storage; 10, methanol plant; 11, methyl acetate plant; 12, acetic anhydride plant. (Courtesy of Tennessee Eastman)

THE WORLD OF CHEMISTRY

A Catalyst in Action

Because of their remarkable chemical properties, catalysts are used throughout the industry. Dr. Norman Hochgraf, vice-president of Exxon Chemical, shares with us his viewpoint:

Within the petroleum industry, where most of the feedstocks for chemicals come from, catalysts are used to produce motor gasoline, heating oil, and feedstocks for chemicals. And within the chemical industry, catalysts are used not only to purify those feedstocks, but they're also used to polymerize the feedstocks, to make plastics, to make rubbers, and to make synthetic fibers. This industry not only wouldn't be profitable without catalysts, but in a sense, it wouldn't exist. Almost everything we do is the result of catalytic activity, which has been carefully designed, carefully selected, and built into our commercial operations.

Let's look at Eastman Kodak's group of plants in Kingsport, Tennessee. Synthetic products—fabrics, plastics, films, and aspirin—are all made there. These products are all made using acetic anhydride, which is also manufactured at the same location. Yet without the rare South African metal, rhodium, there would be no plant at all. Eastman Kodak used to make acetic anhydride from oil. With the rise in oil prices in the early 1970s, the company began looking for alternative inexpensive materials. Coal was the obvious choice. First it is gasified, producing hydrogen and carbon monoxide. At a later stage, carbon monoxide is reacted with methyl acetate to produce acetic anhydride. This crucial step requires rhodium as a catalyst.

Solution of rhodium chloride catalyst. (*The World of Chemistry,* Program 14, "Molecules in Action")

If rhodium is so expensive, how can it be profitable to use the catalyst in such a large-scale reaction? The answer lies in the fact that catalysts are not used up in reactions. Each catalyst molecule may react with thousands and thousands of molecules of reactant. So this catalyst need only be present in tiny amounts. In addition, as the mixture is drawn off, the catalyst is carefully separated from the acetic anhydride and recycled back into the reactor. Very little rhodium is lost.

Without the development of a viable catalyst system for producing acetic anhydride from methyl acetate and carbon monoxide, the chemicals-from-coal complex at Kingsport, Tennessee, would not have been built.

The World of Chemistry (Program 14) "Molecules in Action."

Acetic anhydride reacts with water to give acetic acid (number 33 in the top 50 chemical list).

three for the synthesis of methanol, methyl acetate, and acetic anhydride. The main chemical reactions used in the process are shown below:

$$\text{1. } \underset{\text{Coal}}{C} + \underset{\text{Steam}}{H_2O} \longrightarrow \underset{\text{Synthesis gas}}{CO + H_2}$$

$$\text{2. } CO + 2\,H_2 \longrightarrow \underset{\text{Methanol}}{CH_3OH}$$

$$\text{3. } CH_3OH + \underset{\text{Acetic acid}}{CH_3\overset{\overset{\displaystyle O}{\|}}{C}OH} \longrightarrow \underset{\text{Methyl acetate}}{CH_3\overset{\overset{\displaystyle O}{\|}}{C}OCH_3} + H_2O$$

4. $CH_3\overset{\overset{O}{\|}}{C}OCH_3 + CO \longrightarrow CH_3-\overset{\overset{O}{\|}}{C}-O-\overset{\overset{O}{\|}}{C}-CH_3$

Acetic anhydride

About 900 tons per day of high-sulfur coal from nearby Appalachian coal mines are ground in water to form a slurry of 55% to 65% by weight of coal in water. The slurry is fed into two gasifiers to make synthesis gas. To produce the same amounts of these chemicals by conventional means would require the annual equivalent of 1 billion barrels of oil.

Plant design uses the latest environmental control technologies to protect the environment. For example, the sulfur recovery unit converts the hydrogen sulfide gas that was removed during the gasification of coal into free sulfur. This process removes over 99% of the sulfur from the coal, and this sulfur is sold to chemical companies.

Synthesis gas will become increasingly important as a raw material because it can be used to make a variety of hydrocarbons that can then be converted into other commercially important organic chemicals.

DESIGN OF ORGANIC SYNTHESIS REACTIONS

In this chapter we have given examples of useful compounds in some major functional group classes. Many chemists are engaged in the synthesis of organic compounds. In college, university, government, and industrial laboratories throughout the world, they prepare new and different compounds on a small scale. If the new compound has commercial value, the preparation is subsequently adapted for full-scale plant operations. Because multistep syntheses are often used to make important commercial products, the synthesis of aspirin is given here as an example.

Over half of the estimated 140,000 chemists in the United States are working in the organic chemical industry.

Aspirin (acetylsalicylic acid) was first synthesized for medical use in 1893 by Felix Hofmann, a German chemist working for the F. Bayer Company. Aspirin is still the leading pain killer and the standard treatment to reduce fever and swelling. Over 30 million pounds of aspirin, or 150 tablets per person, are consumed in the United States each year, and worldwide use exceeds 100,000 tons per year.

The starting point for the synthesis of aspirin is benzene, which is obtained from coal tar. The steps in the synthesis are shown in Figure 7–7. Only the principal organic substance is shown for each step. Other products such as sodium chloride (a coproduct with phenol) and water (a coproduct with sodium phenoxide) are sometimes important in the synthesis because they have to be removed to avoid interference with subsequent steps. However, in a broad outline of the synthetic process, the coproducts are generally omitted; only those products made in a previous step and required for subsequent steps are included. In Figure 7–7 the step-by-step structural changes can be followed by noting groups in color. Conditions and additional reactants for each step are written with the arrow. These conventions are generally used to summarize organic syntheses.

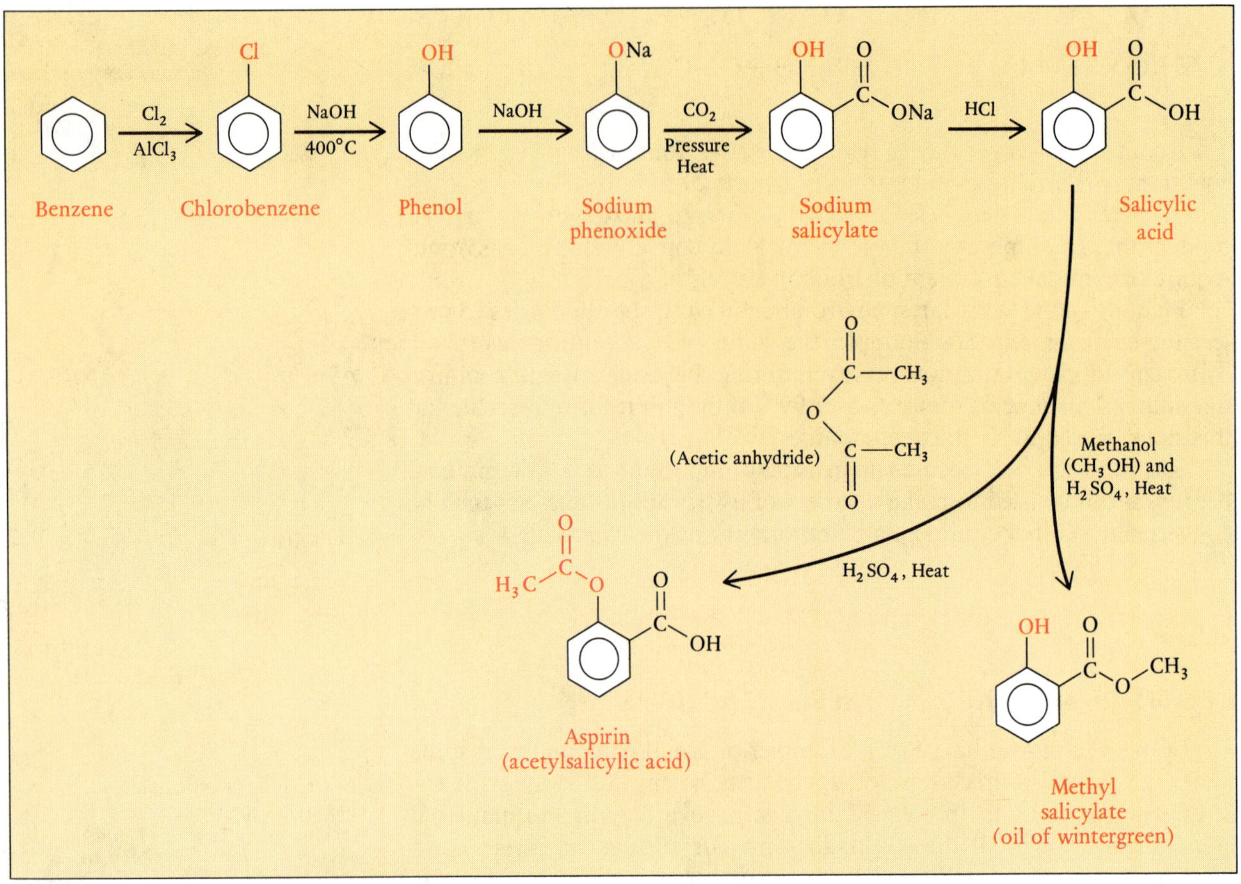

Benzene $\xrightarrow[AlCl_3]{Cl_2}$ Chlorobenzene $\xrightarrow[400°C]{NaOH}$ Phenol $\xrightarrow{NaOH}$ Sodium phenoxide $\xrightarrow[\text{Pressure, Heat}]{CO_2}$ Sodium salicylate $\xrightarrow{HCl}$ Salicylic acid

Salicylic acid + (Acetic anhydride), H_2SO_4, Heat → Aspirin (acetylsalicylic acid)

Salicylic acid + Methanol (CH_3OH) and H_2SO_4, Heat → Methyl salicylate (oil of wintergreen)

Figure 7–7 Preparation of aspirin and oil of wintergreen. A discussion of the synthesis is given in the text.

SELF-TEST 7B

1. Gin that is 84 proof contains what percentage of ethanol? ________
2. Ethanol is quickly absorbed by the blood and oxidized to ____________ in the liver.
3. Ethanol intended for industrial use is ____________ by the addition of small amounts of a toxic substance.
4. 2-Propanol is commonly known as ____________ alcohol.
5. Souring of wine is caused by the oxidation of ethanol to ________.
6. The organic acid found in vinegar is ____________.
7. The simplest aldehyde is ____________.
8. The alcohol being considered as a replacement for gasoline in heavily polluted urban areas is ____________.
9. One third of the gasoline used in ____________ is produced by a methanol-to-gasoline process developed by Mobil Oil Company.
10. Organic acids react with ____________ to form esters.
11. The class of organic compounds that include some compounds with fruity fragrances is ____________.
12. Amines derived from plants are known as ____________. An example of a compound in this class is ____________.

13. The first chemicals-from-coal plant in the United States uses a series of chemical reactions to obtain acetic anhydride. A key step is the synthesis of ____________ from synthesis gas.

QUESTIONS

1. Describe how petroleum is refined, starting with a barrel of crude oil.
2. Explain how fractional distillation is used in the refinement of petroleum.
3. What is "straight-run" gasoline?
4. List three gasoline additives that increase the octane rating of gasoline.
5. What is gasohol?
6. What is meant by the following terms?
   a. Proof rating of an alcohol
   b. Denatured alcohol
   c. Catalytic re-forming
   d. Catalytic cracking
   e. Octane rating
7. What volume percent of ethanol does 90 proof gin contain?
8. What is the difference in structure between ethanol, ethylene glycol, and glycerol?
9. What is vinegar? How is it made?
10. Many naturally occurring carboxylic acids have more than one acid group (Table 7–6). What other functional group is often present?
11. Give examples of
    a. two naturally occurring esters and where they are found.
    b. two naturally occurring carboxylic acids and where they are found.
12. Why is coal receiving increased attention as a source of organic compounds?
13. Outline the steps used by Eastman Kodak to make acetic anhydride from coal.
14. What is synthesis gas? How can it be used to produce chemicals?
15. Explain the common names of *wood alcohol* for methanol and *grain alcohol* for ethanol.
16. What alcohol is commonly referred to as rubbing alcohol?
17. Pure ethanol is what proof?
18. What are two advantages and two disadvantages of using methanol as a fuel for vehicles?
19. Explain how the liver detoxifies ethanol.
20. Explain how gasoline can be made from methanol.
21. In 1990 the price of rhodium metal ranged from $1625 per ounce to $5300 per ounce. If rhodium is so expensive, why is it used as a catalyst in the Eastman Kodak chemicals-from-coal process?
22. Methanol is now number 21 in the list of top 50 chemicals produced in the United States. What factors are likely to lead to an increased demand for methanol in the next decade?

The Formica counter and the vinyl-covered stools in Ralph Goings's *Pee Wee's Diner, Warnerville, NY* represent the extent to which plastics are part of our everyday life. (Courtesy of O.K. Harris works of art, New York, NY)

CHAPTER

Giant Molecules—The Synthetic Polymers

Polymers are giant molecules. A variety of synthetic polymers is used in consumer products including plastics, synthetic fibers, synthetic rubber, oils and lubricants, and building materials. Addition polymers are made by polymerizing ethylene or an ethylene derivative. Examples of addition polymers include polyethylene, polypropylene, polystyrene, and polyvinyl chloride. Condensation polymers can form when functional groups on the ends of one molecule react with functional groups on the ends of another molecule. An example is a polyester formed from the reaction of a dialcohol with a diacid. Plastic recycling is growing as a result of public awareness and increased efforts to enhance collection of plastic wastes through curbside recycling.

1. What are polymers?
2. What are some common addition polymers?
3. What are polyesters and polyamides?
4. What is the difference between natural and synthetic rubber?
5. What are silicones?
6. What types of plastics are being recycled now?
7. Why is the percentage of recycled plastics so low?

It is impossible for us to get through a day without using a dozen or more synthetic **polymers.** The word *polymer* means "many parts" (Greek, *poly* meaning "many," *meros* meaning "parts"). Polymers are *giant molecules* with molecular weights ranging from thousands to over a million. Our clothes are polymers; our food is packaged in polymers; our appliances and cars contain a number of polymer components. However, polymers were not invented by humans. Both *inorganic* and *organic* polymers exist in nature. The mineral silicates and silica sand are examples of inorganic polymers. The structural materials of all forms of life are organic polymers—proteins, nucleic acids (DNA and RNA), cellulose, lignin, and starch are examples.

Natural polymers are essential to life.

A plastic is a substance that flows under heat and pressure, and hence is capable of being molded into various shapes. All plastics are polymers, but not all polymers are plastic.

The average production of synthetic polymers in the United States exceeds 200 lb per person annually.

Many synthetic polymers are **plastics** of one sort or another. Examples include plastic dishes and cups; plastic containers; telephones; plastic bags for packaging foods and trash; plastic pipes and fittings; plastic water-dispersed paints; automobile steering wheels and seat covers; and cabinets for appliances, radios, and television sets. In fact, these plastics along with textile fibers and synthetic rubbers are so widely used that they are usually taken for granted. The prominence of synthetic polymers in consumer products is indicated by the fact that 24 of the top 50 chemicals (inside front cover) are used in the production of plastics, fibers, and rubbers.

This "flood of plastic objects" did not arise accidentally; it slowly became a reality during the last 50 years because (1) natural resources like wood were dwindling, (2) with rising labor costs, many items could be made less expensively and more uniformly by molding than by sawing, shaping, sanding, and gluing, and (3) for many applications, the properties of the new synthetic materials were superior to those of metals, wood, or other natural materials. We have moved from the Iron Age to the Plastics Age, as evidenced by the fact that since 1976, plastic has exceeded steel as the nation's most widely used material.

Some of our most useful polymer chemistry has resulted from copying giant molecules in nature. Rayon is remanufactured cellulose; synthetic rubber is copied from natural latex rubber. As useful as they may be, however, polymer chemistry is not restricted to nature's models. Polystyrene, nylon, and Dacron® are a few examples of synthetic molecules that do not have exact duplicates in nature. We have gone to school on nature and extended our knowledge to produce polymers that are more useful than natural ones.

The purpose of this chapter is to investigate the structural chemistry of polymers to see just why they have such useful properties. Are these properties the result of stronger bonds or of groups of molecules acting together, or is there some other explanation? In the next chapter we shall study some of nature's polymers.

WHAT ARE GIANT MOLECULES?

In the early 20th century many chemists were reluctant to accept the concept of giant molecules, but in the 1920s a persistent German chemist, Hermann Staudinger (1881–1965; he won the Nobel Prize in chemistry in 1953), championed the idea and introduced a new term, **macromolecule,** for these giant molecules. Staudinger devised experiments that yielded accurate molecular weights, and he synthesized "model compounds" to test his theory. One of his first model compounds was prepared from styrene, a chemical made from ethylene and benzene.

A *macromolecule* is a molecule with a very high molecular weight.

Styrene is 20th on the list of the top 50 commercial chemicals.

$$H_2C=CH$$

Styrene

Under the proper conditions, styrene molecules use the "extra" electrons of the double bond to undergo a **polymerization** reaction to yield polystyrene, a material composed of giant molecules (Fig. 8–1). The molecules of styrene are the **monomers** (Greek, *mono* meaning "one"); they provide the repeating units in the giant molecule analogous to identical railroad cars coupled together to make a long train.

The macromolecule polystyrene is represented as a long chain of monomer units bonded to each other. Each unit is bonded to the next by a covalent bond. The polymer chain is not an endless one; some polystyrenes made by Staudinger were found to have molecular weights of about 600,000, corresponding to a chain of about 5700 styrene units. The polymer chain can be indicated as

$$R-CH_2-CH-\left(CH_2-CH-\right)_n CH_2-CH-R$$

where R represents some terminal group, often an impurity, and n is a large number, in this case $n = 5700$.

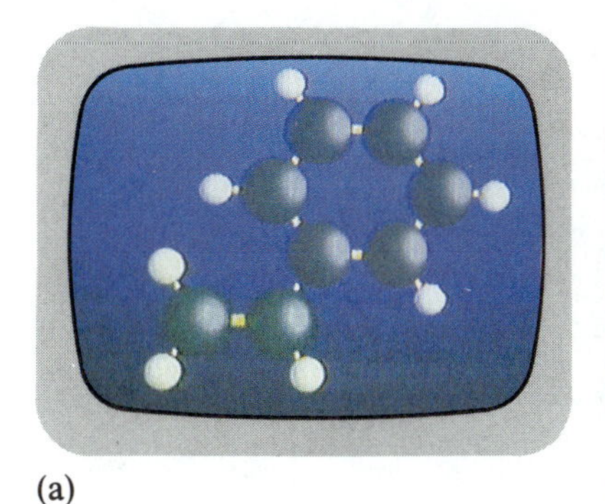

(a)

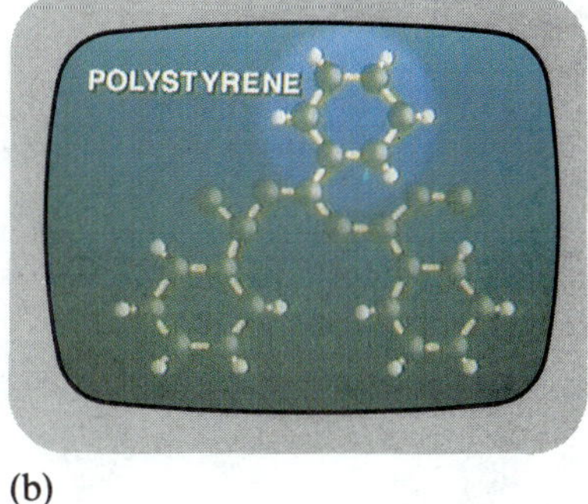

(b)

Figure 8–1 (a) Model of a styrene molecule. (b) Model of a polystyrene molecule. *(The World of Chemistry,* Program 22, "The Age of Polymers.")

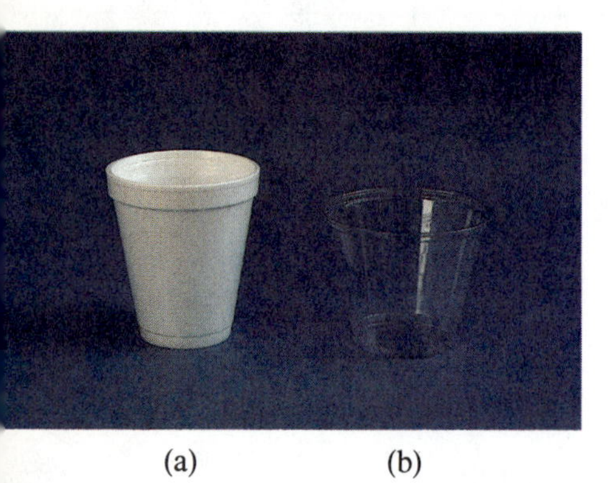

(a) (b)

Figure 8–2 (a) Polystyrene coffee cup. (b) Clear polystyrene "glass."

Polystyrene is a clear, hard, colorless solid at room temperature that can be molded easily at 250°C. Commercial production of polystyrene began in Germany in 1929, and today its U.S. production exceeds 5 billion pounds per year. Polystyrene is used to make food containers, toys, electrical parts, insulating panels, appliance and furniture components, and many other items. The variation in properties shown by polystyrene products is typical of synthetic polymers. For example, a clear polystyrene drinking glass that is brittle and breaks into sharp pieces somewhat like glass is much different from the polystyrene coffee cup that is soft and pliable (Fig. 8–2).

Styrofoam® is produced by "expansion molding." In this process, polystyrene beads are placed in a mold and heated with steam or hot air. The beads, 0.25 to 1.5 mm in diameter, contain 4% to 7% by weight of a low-boiling liquid such as pentane. The steam causes the low-boiling liquid to vaporize and this expands the beads and the foamed particles are then molded in the shape of the mold cavity (Fig. 8–3). Styron®, another form of expanded polystyrene, is used for egg cartons, meat trays, coffee cups, and packing material.

Thermoplastic polymers can be repeatedly softened by merely heating.

Thermosetting polymers form cross-linking bonds when heated, resulting in a rigid structure.

There are two broad categories of plastics. One, when heated repeatedly, softens and flows; when it is cooled, it hardens again. Materials that undergo such reversible changes when heated and cooled are called **thermoplastics;** polystyrene is a thermoplastic. The other type is plastic when first heated, but when heated further it forms a set of interlocking bonds. When reheated, it cannot be softened and reformed without extensive degradation. These materials are called **thermosetting plastics** and include such familiar names as Bakelite® and rigid-foamed polyurethane, a polymer that is finding many new uses as a construction material.

To gain a better understanding of polymers, we must look at representative examples of the different types of polymerization processes. We shall see in the sections that follow that synthetic polymers can be **addition polymers,** in which monomer units are joined directly, or **condensation polymers,** in which monomer units combine by splitting out a small molecule, usually water.

Figure 8–3 Large piece of Styrofoam® (*The World of Chemistry,* Program 22, "The Age of Polymers.")

ADDITION POLYMERS

In the previous section it was noted that some polymers, such as polystyrene, are made by adding monomer to monomer to form a polymer chain of great length. Perhaps the addition reactions that are easiest to understand chemically are those involving monomers containing double bonds. The simplest monomer of this group is ethene (C_2H_4). When ethene (ethylene) is heated under pressure in the presence of oxygen, polymers with molecular weights of approximately 1 million are formed (Fig. 8–4). To enter into reaction, the carbon-carbon double bond in the ethene molecule must be partially broken. This forms **reactive sites** composed of unpaired electrons at either end of the molecule.

Ethylene (ethene) is the number four commercial chemical.

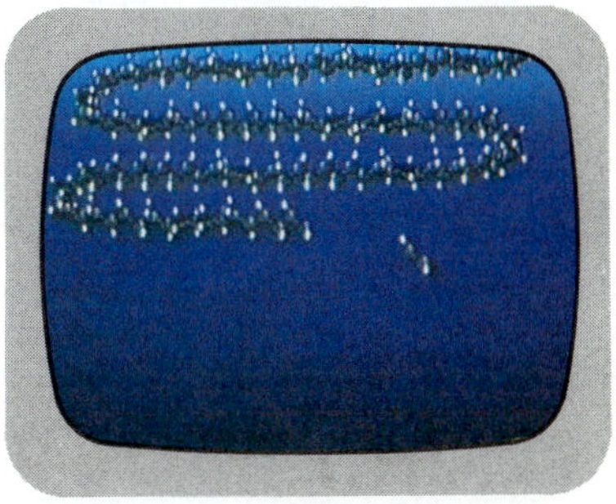

Figure 8–4 Model of polyethylene chain. (*The World of Chemistry*, Program 22, "The Age of Polymers.")

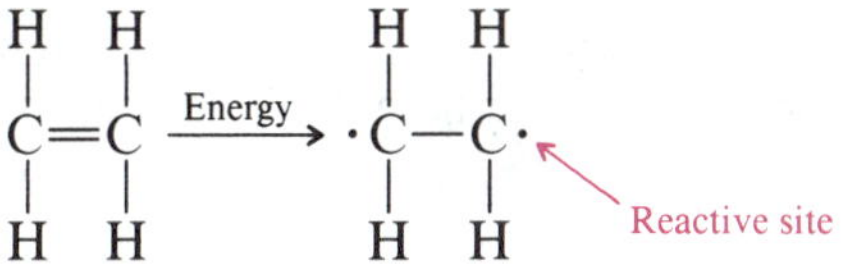

The reactive species join to form long chains:

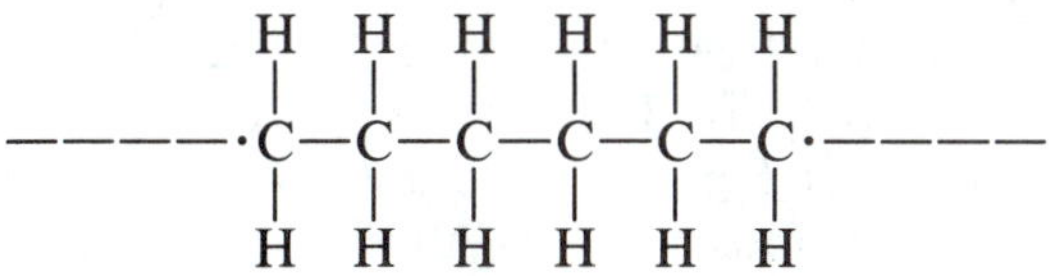

Polyethylenes formed under various pressures and catalytic conditions have different molecular structures and hence different physical properties. For example, chromium oxide as a catalyst yields almost exclusively the linear polyethylene shown below. A kinked structure represents more closely the tetrahedral carbon in the saturated polyethylene chain (Fig. 8–5).

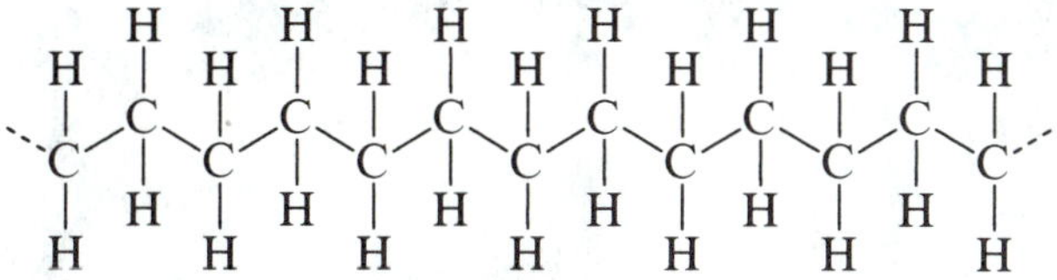

A portion of a polyethylene molecule

or

where each point is a CH_2 group.

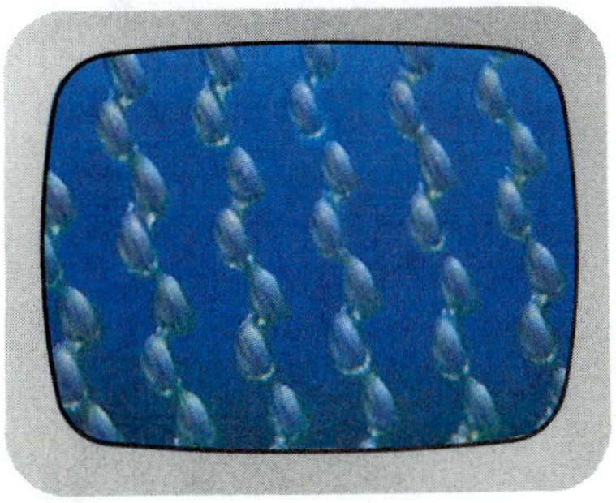

Figure 8–5 Model of linear chains of polyethylene. (*The World of Chemistry*, Program 22, "The Age of Polymers.")

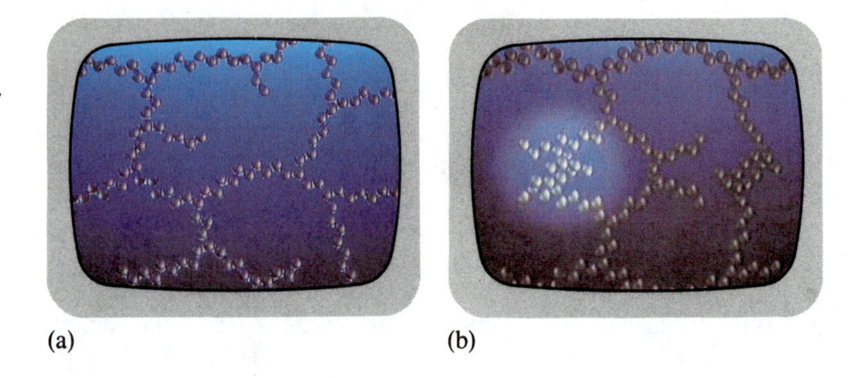

(a) (b)

Figure 8–6 (a) Model of branched-chain polyethylene. (b) Model of cross-linked polyethylene. (*The World of Chemistry,* Program 22, "The Age of Polymers.")

Branched-chain polyethylene and cross-linked polyethylene are obtained by heating ethylene under pressure (Fig. 8–6).

Uses of Polyethylene

Polyethylene is the world's most widely used polymer. Over 19 billion pounds were produced in 1990 in the United States alone. What are some of the reasons for this popularity? The wide range of properties of polyethylene leads to its variety of uses (Fig. 8–7). The key to the range of properties is the variety of structures and molecular weights that are possible. The formation of linear, branched, and cross-linked polyethylene gives polymer materials

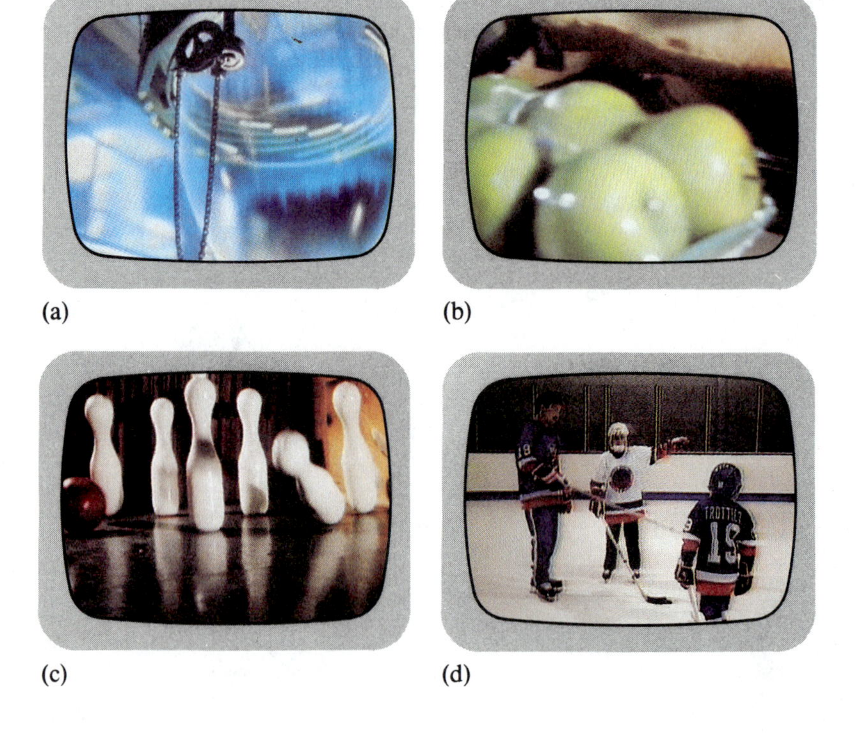

(a) (b) (c) (d)

Figure 8–7 (a) Production of polyethylene. (b), (c), (d) The wide range of properties of different structural types of polyethylene lead to a variety of applications. (*The World of Chemistry,* Program 22, "The Age of Polymers.")

THE WORLD OF CHEMISTRY

Discovery of a Catalyst for Polyacrylonitrile Production

Oil companies invest considerable sums in equipment and human effort to develop new catalysts to make fuels and chemicals from petroleum. Research and development in this area is a multimillion dollar gamble. There is no guarantee the money spent will produce anything useful. But if it does, the payoff can be enormous. Just one catalyst breakthrough made more than half a billion dollars for Standard Oil of Ohio, now part of BP America. In the late 1950s, SOHIO researchers came up with a new catalytic process that soon dominated all others in the production of polyacrylonitrile, the polymer used to make textiles, tires, and car bumpers. Oddly enough, SOHIO researchers weren't even trying to produce acrylonitrile at first. They simply wanted to make a metal oxide catalyst to convert waste propane gas from petroleum refining into something more valuable. As Dr. Jeanette Grasselli said:

The theory, the hypothesis at the time, was that we could take the oxygen from the catalyst and insert it into the propane, a relatively unreactive molecule. So this was a tough technical objective. And, in turn, we wanted to generate or take the catalyst back to its original oxidized form by using oxygen from the air.

But the theory didn't hold up. Propane was too stable to react, and the catalyst particles broke down in service. Management gave them three more months to show results, so they made some changes. They replaced propane with a more reactive refinery gas, propylene. They made their catalyst out of different metal oxides, and they added ammonia to promote, or speed up, the reaction.

To their surprise, ammonia reacted. Rather than just encouraging the reaction to go faster, as a promoter, oxygen reacted and became part of the reaction sequence, and acrylonitrile was made in one step. The researchers had struck pay dirt. Their new catalyst, combined with ammonia, had made a valuable product out of a cheap gas. Management quickly saw the value of the new process and wasted no time building a plant to use it. As Dr. Grasselli recalls,

In 1960, when our plant came on stream, acrylonitrile was selling for 28¢/lb. We were making it for 14¢/lb. And we shut down every other commercial process. Today 90% of the world's acrylonitrile is manufactured by the SOHIO process.

The World of Chemistry (Program 20) "On the Surface."

Jeanette Grasselli. (*The World of Chemistry,* Program 20, "On the Surface.")

with widely different properties. Long, linear chains of polyethylene can pack closely together (Fig. 8–5) and give a material with high density and high molecular weight, referred to as high-density polyethylene (HDPE). This material is hard, tough, and rigid. The plastic milk carton is a good example of an application of HDPE. Branched chains of polyethylene can-

The bottle on the left is made of flexible, low-density branched polyethylene. The one on the right is made of rigid, high-density, linear polyethylene. (Marna G. Clarke.)

not be packed closely together (Fig. 8–6); so the resulting material is low-density polyethylene (LDPE). This material is soft and flexible. Sandwich bags are made from LDPE. If the linear chains of polyethylene are treated in a way that causes cross-links between chains to form cross-linked polyethylene (CLPE), a very tough form of polyethylene is produced. The plastic caps on soft-drink bottles are made from CLPE. These examples of the influence of polyethylene structure on the properties of polyethylene illustrate how polymer chemists and engineers can obtain polymers with differing properties by varying the temperature, pressure, catalyst, time, and the order of mixing the various reactants.

A Variety of Addition Polymers

A large number of addition polymers can be made by using monomers that are derivatives of ethylene, in which one or more of the hydrogen atoms have been replaced with either atoms of halogens or organic functional groups. If the formation of polyethylene is represented as

$$n \begin{array}{ccc} \mathrm{H} & & \mathrm{H} \\ & \mathrm{C{=}C} & \\ \mathrm{H} & & \mathrm{H} \end{array} \longrightarrow \left(\begin{array}{ccc} \mathrm{H} & & \mathrm{H} \\ | & & | \\ -\mathrm{C} & - & \mathrm{C}- \\ | & & | \\ \mathrm{H} & & \mathrm{H} \end{array} \right)_n$$

then the general reaction is

$$n \begin{array}{ccc} \mathrm{H} & & \mathrm{H} \\ & \mathrm{C{=}C} & \\ \mathrm{H} & & \mathrm{X} \end{array} \longrightarrow \left(\begin{array}{ccc} \mathrm{H} & & \mathrm{H} \\ | & & | \\ -\mathrm{C} & - & \mathrm{C}- \\ | & & | \\ \mathrm{H} & & \mathrm{X} \end{array} \right)_n$$

where X is Cl, F, or an organic group that can be used to represent a number of other important addition polymers.

Polypropylene, used in making indoor-outdoor carpeting, bottles, and battery cases, is made from propylene.

$$n\ \mathrm{H_2C{=}CH(CH_3)} \longrightarrow \left(\begin{array}{cc} \mathrm{H} & \mathrm{H} \\ | & | \\ \mathrm{-C-} & \mathrm{C-} \\ | & | \\ \mathrm{H} & \mathrm{CH_3} \end{array}\right)_n$$

Propylene Polypropylene

Polyvinyl chloride (PVC), used for making floor tile, garden hoses, plumbing pipes, and trash bags, has a chlorine atom substituted for one of the hydrogen atoms in ethene.

$$n\ \mathrm{H_2C{=}CHCl} \longrightarrow \left(\begin{array}{cc} \mathrm{H} & \mathrm{H} \\ | & | \\ \mathrm{-C-} & \mathrm{C-} \\ | & | \\ \mathrm{H} & \mathrm{Cl} \end{array}\right)_n$$

Vinyl chloride Polyvinylchloride

Although the representation

$$\left(\begin{array}{cc} \mathrm{H} & \mathrm{H} \\ | & | \\ \mathrm{-C-} & \mathrm{C-} \\ | & | \\ \mathrm{H} & \mathrm{H} \end{array}\right)_n$$

saves space, keep in mind how large the polymer molecules are. Generally n is 500 to 10,000, and this gives molecules with molecular weights ranging from 10,000 to over 1 million. The molecules that make up a given polymer sample are of different lengths and thus are not all of the same molecular weight. As a result, only the average molecular weight can be determined.

Teflon-coated pans. (Beverly March.)

In summary, the variation in substituents, length, branching, and cross-linking gives a variety of properties for each addition polymer. These examples illustrate why the uses of polymers continue to increase. The chemists and chemical engineers can fine-tune the properties of the polymer to match desired properties by appropriate selection of monomer and reaction conditions for making the polymer. Table 8-1 summarizes information about common addition polymers.

Natural and Synthetic Rubbers Are Addition Polymers

Natural rubber, a product of the *Hevea brasiliensis* tree, is a hydrocarbon with the composition C_5H_8; when decomposed in the absence of oxygen, the monomer isoprene is obtained:

$$\mathrm{CH_2{=}C(CH_3){-}CH{=}CH_2} \quad \text{or} \quad \mathrm{H_2C{=}C(CH_3){-}CH{=}CH_2}$$

Isoprene

TABLE 8–1 Ethylene Derivatives That Undergo Addition Polymerization

Formula	Monomer Common Name (Top 50 Rank)	Polymer Name (Trade Names)	Uses	Polymer U.S. Production (Tons/Yr)
$H_2C{=}CH_2$	Ethylene (4)	Polyethylene (Polythene)	Squeeze bottles, bags, films, toys and molded objects, electrical insulation	10 million
$H_2C{=}CH{-}CH_3$	Propylene (9)	Polypropylene (Vectra, Herculon)	Bottles, films, indoor–outdoor carpets	4 million
$H_2C{=}CH{-}Cl$	Vinyl chloride (18)	Polyvinyl chloride (PVC)	Floor tile, raincoats, pipe, phonograph records	4.5 million
$H_2C{=}CH{-}CN$	Acrylonitrile (38)	Polyacrylonitrile (Orlon, Acrilan)	Rugs, fabrics	1 million
$H_2C{=}CH{-}C_6H_5$	Styrene (20)	Polystyrene (Styrene, Styrofoam®, Styron®)	Food and drink coolers, building material insulation	2.5 million
$H_2C{=}CH{-}O{-}C({=}O){-}CH_3$	Vinyl acetate (40)	Polyvinylacetate (PVA)	Latex paint, adhesives, textile coatings	0.5 million
$H_2C{=}C(CH_3){-}C({=}O){-}O{-}CH_3$	Methyl methacrylate	(Plexiglas, Lucite)	High-quality transparent objects, latex paints, contact lenses	0.5 million
$F_2C{=}CF_2$	Tetrafluoroethylene	(Teflon)	Gaskets, insulation, bearings, pan coatings	7000

Natural rubber occurs as latex (an emulsion of rubber particles in water) that oozes from rubber trees when they are cut. Precipitation of the rubber particles yields a gummy mass that is not only elastic and water-repellent but also very sticky, especially when warm. In 1839, after five years' work on this

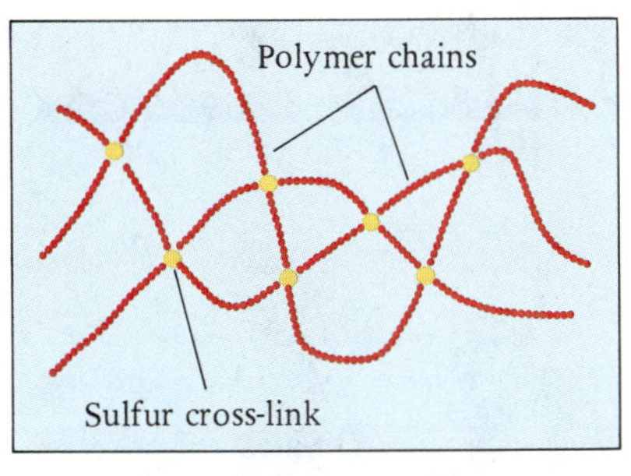

(a) Before stretching

(b) Stretched

Figure 8–8 Stretched vulcanized rubber springs back to its original structure, an elastomeric property.

material, Charles Goodyear (1800–1860) discovered that the heating of gum rubber with sulfur produced a material that was no longer sticky but was still elastic, water-repellent, and resilient.

Vulcanized rubber contains short chains of sulfur atoms that bond together the polymer chains of the natural rubber and reduce its unsaturation. The sulfur chains help align the polymer chains, so the material does not undergo a permanent change when stretched but springs back to its original shape and size when the stress is removed (Fig. 8–8). Substances that behave this way are called **elastomers.**

Vulcanization

$$>C{=}C< \; + \; S_8 \; + \; >C{=}C< \;\longrightarrow\; -\overset{|}{\underset{|}{C}}-\overset{|}{\underset{|}{C}}- \text{ cross-linked through } -S-(S)_6-S- \text{ to } -\overset{|}{\underset{|}{C}}-\overset{|}{\underset{|}{C}}-$$

Sulfur, or S_8 molecule

In later years chemists searched for ways to make a synthetic rubber so we would not be completely dependent on imported natural rubber during emergencies, such as during the first years of World War II. In the mid-1920s, German chemists polymerized butadiene (obtained from petroleum and structurally similar to isoprene, but without the methyl group side chain). The product was buna rubber, so named because it was made from butadiene (Bu—) and catalyzed by sodium (—Na).

The behavior of natural rubber (polyisoprene), it was learned later, is due to the specific arrangement within the polymer chain. We can write the formula for polyisoprene with the CH_2 groups on opposite sides of the double bond (the *trans* arrangement):

$$-CH_2(CH_3)C{=}CH-CH_2-CH_2-(CH_3)C{=}CH-CH_2-CH_2-(CH_3)C{=}CH-CH_2-$$

Poly-*trans*-isoprene (the $-CH_2-CH_2-$ groups are *trans*)

The formula can also be written with the CH_2 groups on the same side of the double bond (the *cis* arrangement, from Latin meaning "on this side").

$$-CH_2(CH_3)C{=}CH-CH_2-CH_2-(CH_3)C{=}CH-CH_2-CH_2-(CH_3)C{=}CH-CH_2-$$

Poly-*cis*-isoprene (the $-CH_2-CH_2-$ groups are *cis*)

Natural rubber is poly-*cis*-isoprene. However, the *trans* material also occurs in nature in the leaves and bark of the sapotacea tree and is known as *gutta-percha.* It is used as a thermoplastic for golf ball covers, electrical insulation, and other such applications. Without an appropriate catalyst,

TABLE 8–2 A Rubber Formulation

Rubber	Poly-*cis*-isoprene	62.0	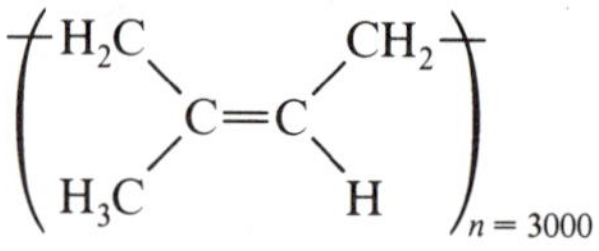 $\left(-H_2C(H_3C)C{=}CH-CH_2- \right)_{n = 3000}$	Elastomer
Activators	Zinc oxide stearic acid	2.7 0.6	ZnO $C_{17}H_{35}COOH$	Activates vulcanizing agents; stearic acid acts as a lubricant in processing
Vulcanizing agent	Sulfur	1.5	S_8	Cross-links polymer chains
Filler	Carbon black	30.5	C	Provides strength and abrasion resistance
Accelerator	Dibenzthiozole disulfide	1.1	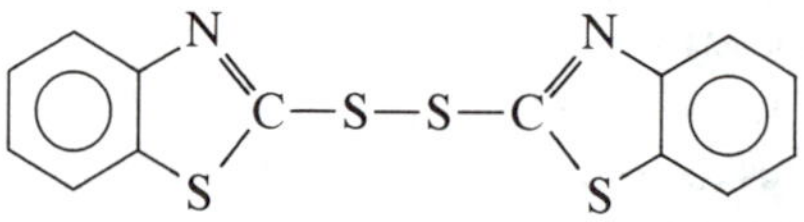 (benzothiazolyl)C—S—S—C(benzothiazolyl)	Catalyzes vulcanization
Antioxidant	Alkylated diphenylamine	1.1	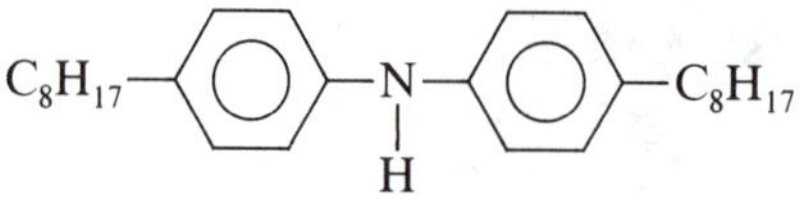 $C_8H_{17}-C_6H_4-NH-C_6H_4-C_8H_{17}$	Inhibits attack by oxygen or ozone in the air
Processing oil	Hydrocarbon oil	0.5	C_nH_{2n+2}	Plasticizer

polymerization of isoprene yields a solid that is like neither rubber nor gutta-percha. Neither the *trans* polymer nor the randomly arranged material is as good as natural rubber *(cis)* for making automobile tires.

In 1955, chemists at the Goodyear and Firestone companies almost simultaneously discovered how to use stereoregulation catalysts to prepare synthetic poly-*cis*-isoprene. This material is structurally identical to natural rubber. Today, synthetic poly-*cis*-isoprene can be manufactured cheaply and is used almost equally well (there is still an increased cost) when natural rubber is in short supply. More than 2.4 million tons of synthetic rubber are produced in the United States yearly. Table 8–2 gives a typical rubber formulation as it might be used in a tire.

Neoprene

One of the first synthetic rubbers produced in the United States was neoprene, an addition polymer of the monomer 2-chlorobutadiene:

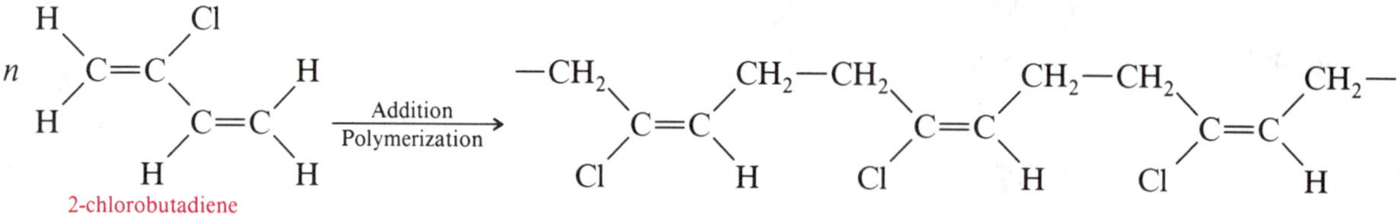

$$n\ H_2C{=}C(Cl)-CH{=}CH_2 \xrightarrow[\text{Polymerization}]{\text{Addition}} -CH_2-C(Cl){=}CH-CH_2-CH_2-C(Cl){=}CH-CH_2-CH_2-C(Cl){=}CH-CH_2-$$

2-chlorobutadiene

which has a chlorine atom substituted for the methyl group in isoprene. Neoprene is used in the production of gaskets, garden hoses, and adhesives.

Polybutadiene

Polybutadiene, a synthetic rubber used in the production of tires, hoses, and belts, is an addition polymer of the monomer 1,3-butadiene:

$$n\ \mathrm{H_2C{=}CH{-}CH{=}CH_2}\ \xrightarrow[\text{Polymerization}]{\text{Addition}}\ \mathrm{{-}CH_2{-}CH{=}CH{-}CH_2{-}CH_2{-}CH{=}CH{-}CH_2{-}CH_2{-}CH{=}CH{-}CH_2{-}}$$

1,3-butadiene

SELF-TEST 8A

1. The individual molecules from which polymers are made are called ____________.
2. Plastics that undergo reversible changes when heated and cooled are ( ) thermosetting ( ) thermoplastic.
3. Draw the formulas of the monomers used to prepare the following polymers.
   a. Polypropylene
   b. Polystyrene
   c. Teflon
   d. Polyvinyl chloride
4. Draw the repeating unit for the following polymers.
   a. Polyethylene
   b. Polyacrylonitrile
5. Natural rubber is a polymer of ____________________.

(a)

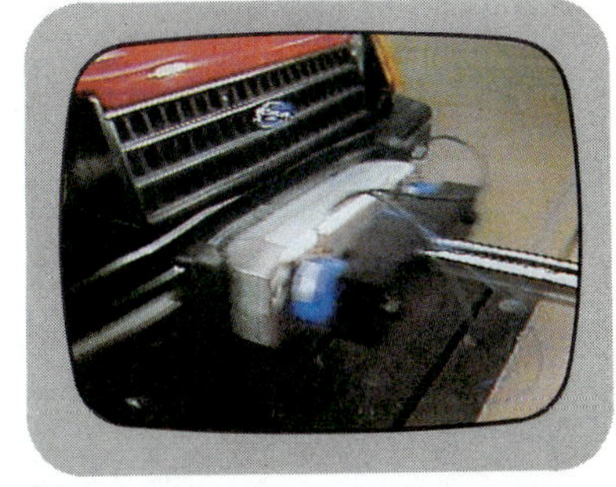

(b)

Tires and car bumpers are examples of products made from synthetic rubber polymers. (*The World of Chemistry*, Program 22, "The Age of Polymers.")

CONDENSATION POLYMERS

Polyesters

A chemical reaction in which two molecules react by splitting out or eliminating a small molecule is called a **condensation reaction.** For example, acetic acid and ethyl alcohol react, splitting out a water molecule, to form ethyl acetate, an **ester.**

$$\underset{\text{Acetic acid}}{CH_3\overset{\overset{\displaystyle O}{\|}}{C}{-}OH} + \underset{\text{Ethanol}}{HOCH_2CH_3} \xrightarrow[\text{Catalyst}]{H^+} \underset{\text{Ethyl acetate (an ester)}}{CH_3\overset{\overset{\displaystyle O}{\|}}{C}{-}OCH_2CH_3} + H_2O$$

See Chapter 7 for a discussion of esters.

This important type of chemical reaction does not depend on the presence of a double bond in the reacting molecules. Rather, it requires the presence of two kinds of functional groups on two different molecules. If

THE WORLD OF CHEMISTRY

Inventor of the Poly(Ethylene Terephthalate) Bottle

The inventor of the poly(ethylene terephthalate) soft-drink bottle is Nathaniel Wyeth, who comes from the internationally famous family of artists. His brother, Andrew Wyeth, expresses his creativity on canvas, but Nat Wyeth expresses his through chemical engineering. He has an intriguing story:

I got to thinking about the work that Wallace Carothers did for Du Pont way back in the days when nylon was born, where he found that, if you took a thread of nylon when it was cold, that is, below the melt point, and stretched it, it would orient itself. That is, the molecules of the polymer would align themselves. This is what you're doing to the molecules when you orient them, you're lining them up so they can give you the most strength. They're all pulling in the direction you want them to pull in.

But the bottles kept splitting. Wyeth estimates that he made 10,000 tries and 10,000 failures before he made a simple observation.

Well, then I realized what we've got to do now is to align these molecules in the sidewall of the bottle; not only in one direction, but in two directions. So I thought I'd play a trick on this mold, on this problem. I took two pieces of polyethylene and turned one of them ninety degrees with the other. So then I had one that would split in this direction, and one that would split in that direction. Well, one piece reinforced the other. As soon as I did that, I could blow bottles. That seems almost dirt simple. But as I've often said, quoting Einstein, the biggest part of a problem and the easiest way to solving a problem is to understand it, have the problem in a form you can understand what's going on. And what I was doing here was learning about what was going on. Once I knew, it was simple to solve.

The World of Chemistry (Program 22) "The Age of Polymers."

Nathaniel Wyeth. (*The World of Chemistry*, Program 22, "The Age of Polymers.")

each reacting molecule has *two* functional groups, both of which can react, it is then possible for condensation reactions to lead to a long-chain polymer. If we take a molecule with two carboxyl groups, such as terephthalic acid, and another molecule with two alcohol groups, such as ethylene glycol, each molecule can react at each end. The reaction of one acid group of terephthalic acid with one alcohol group of ethylene glycol initially produces an ester molecule with an acid group left over on one end and an alcohol group left over on the other:

Terephthalic acid is the number 22 commercial chemical.

$$\underset{\text{Terephthalic acid}}{HO{-}\overset{\displaystyle O}{\overset{\|}{C}}{-}C_6H_4{-}\overset{\displaystyle O}{\overset{\|}{C}}{-}OH} + \underset{\text{Ethylene glycol}}{HO{-}CH_2{-}CH_2{-}OH} \longrightarrow$$

$$\underset{\text{(an ester)}}{HO{-}\overset{\displaystyle O}{\overset{\|}{C}}{-}C_6H_4{-}\overset{\displaystyle O}{\overset{\|}{C}}{-}OCH_2{-}CH_2{-}OH} + H_2O$$

Figure 8–9 A garment made of Dacron. (Courtesy of Du Pont de Nemours and Company.)

Subsequently, the remaining acid group can react with another alcohol group, and the alcohol group can react with another acid molecule. The process continues until an extremely large polymer molecule, known as a **polyester,** is produced with a molecular weight in the range of 10,000 to 20,000.

The esterification of a dialcohol and a diacid involves two positions on each molecule.

$$HO-\overset{O}{\overset{\|}{C}}-C_6H_4-\overset{O}{\overset{\|}{C}}-O-CH_2-CH_2-O\left(\overset{O}{\overset{\|}{C}}-C_6H_4-\overset{O}{\overset{\|}{C}}-O-CH_2-CH_2-O\right)_n\overset{O}{\overset{\|}{C}}-C_6H_4-\overset{O}{\overset{\|}{C}}-O-CH_2-CH_2-OH$$

Over 2 million tons of poly(ethylene terephthalate), commonly referred to as PET, are produced in the United States each year for use in making apparel, tire cord, film for photography and magnetic recording, food packaging, beverage bottles, coatings for microwave and conventional ovens, and home furnishings. A variety of trade names are associated with the various applications. Polyester textile fibers are marketed under such names as Dacron® or Terylene® (Fig. 8–9). Films of the same polyester, when magnetically coated, are used to make audio and TV tapes. This film, Mylar®, has unusual strength and can be rolled into sheets one-thirtieth the thickness of a human hair. The inert, non-toxic, non-allergenic, non-inflammatory, and non–blood-clotting natures of Dacron polymers make Dacron tubing an excellent substitute for human blood vessels in heart bypass operations (Fig. 8–10) and as a skin substitute for burn victims.

A typical polyester is produced from a dialcohol and a diacid.

Fifty percent of all synthetic fiber is Dacron®.

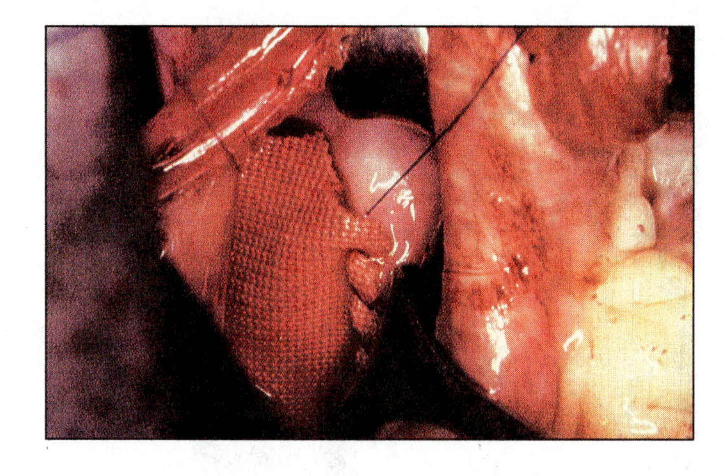

Figure 8–10 A Dacron® patch is used to close an atrial septal defect in a heart patient. (Courtesy of Drs. James L. Monro and Gerald Shore and the Wolfe Medical Publications, London, England.)

Polyamides (Nylons)

Another useful condensation reaction is that occurring between an acid and an amine to split out a water molecule to form an **amide.** Reactions of this type yield a group of polymers that perhaps have had a greater impact on society than any other type. These are the **polyamides,** or nylons.

$$\underset{\text{Acid}}{-\overset{\overset{\displaystyle O}{\|}}{C}-OH} + \underset{\text{Amine}}{H-\overset{\overset{\displaystyle H}{|}}{N}-} \longrightarrow \underset{\text{Amide}}{-\overset{\overset{\displaystyle O}{\|}}{C}-\underset{\underset{\displaystyle H}{|}}{N}-} + HOH$$

Common nylon can be made by the reaction of adipic acid and hexamethylenediamine.

In 1928, the Du Pont Company embarked on a program of basic research headed by Dr. Wallace Carothers (1896–1937), who came to Du Pont from the Harvard University faculty. His research interests were high-molecular-weight compounds, such as rubber, proteins, and resins, and the reaction mechanisms that produced these compounds. In February, 1935, his research yielded a product known as nylon 66 (Fig. 8–11) prepared from adipic acid (a diacid) and hexamethylenediamine (a diamine):

$$\underset{\text{Adipic acid}}{HO-\overset{\overset{\displaystyle O}{\|}}{C}-(CH_2)_4-\overset{\overset{\displaystyle O}{\|}}{C}-OH} + \underset{\text{Hexamethylenediamine}}{H_2N-(CH_2)_6-NH_2} \longrightarrow$$

$$-\overset{\overset{\displaystyle O}{\|}}{C}-(CH_2)_4-\boxed{\overset{\overset{\displaystyle O}{\|}}{C}-\underset{\underset{\displaystyle H}{|}}{N}}-(CH_2)_6-\boxed{\underset{\underset{\displaystyle H}{|}}{N}-\overset{\overset{\displaystyle O}{\|}}{C}}-(CH_2)_4-\boxed{\overset{\overset{\displaystyle O}{\|}}{C}-\underset{\underset{\displaystyle H}{|}}{N}}-(CH_2)_6-\overset{\overset{\displaystyle H}{|}}{N}- + xH_2O$$

Nylon 66
(The amide groups are outlined for emphasis.)

This material could easily be extruded into fibers that were stronger than natural fibers and chemically more inert. The discovery of nylon jolted the American textile industry at almost precisely the right time. Natural fibers were not meeting the needs of 20th-century Americans. Silk was not durable and was very expensive, wool was scratchy, linen crushed easily, and cotton did not lend itself to high fashion. All four had to be pressed after cleaning. As women's hemlines rose in the mid-1930s, silk stockings were in great demand, but they were very expensive and short-lived. Nylon changed all that almost overnight. It could be knitted into the sheer hosiery women wanted, and it was much more durable than silk. The first public sale of nylon hose took place in Wilmington. Delaware (the hometown of Du Pont's main office), on October 24, 1939. The stockings were so popular they had to be rationed. World War II caused all commercial use of nylon to be abandoned until 1945, as the industry turned to making parachutes and other war materials. Not until 1952 was the nylon industry able to meet the demands of the hosiery industry and to release nylon for other uses as a fiber and as a thermoplastic.

The amide linkage in nylon is the same linkage found in proteins, where it is called the peptide linkage.

Hair, wool, and silk are examples of nature's version of nylon. However, these natural polymers have only one carbon between each pair of $-\overset{O}{\overset{\|}{C}}-\underset{H}{\underset{|}{N}}-$ units instead of the half dozen or so found in synthetic nylons.

Figure 8-12 illustrates another facet of the structure of nylon—hydrogen bonding. This type of bonding explains why the nylons make such good fibers. To have good tensile strength, the chains of atoms in a polymer should be able to attract one another, but not so strongly that the plastic cannot be initially extended to form the fibers. Ordinary covalent chemical bonds linking the chains together would be too strong. Hydrogen bonds, with a strength about one tenth that of an ordinary covalent bond, link the chains in the desired manner. We shall see later that hydrogen bonding is also of great importance in protein structures.

Figure 8-11 Nylon 66. Hexamethylenediamine is dissolved in water (*bottom layer*), and a derivative of adipic acid (adipoyl chloride) is dissolved in hexane *(top layer)*. The two compounds mix at the interface between the two layers to form nylon, which is being wound onto a stirring rod. (Charles D. Winters.)

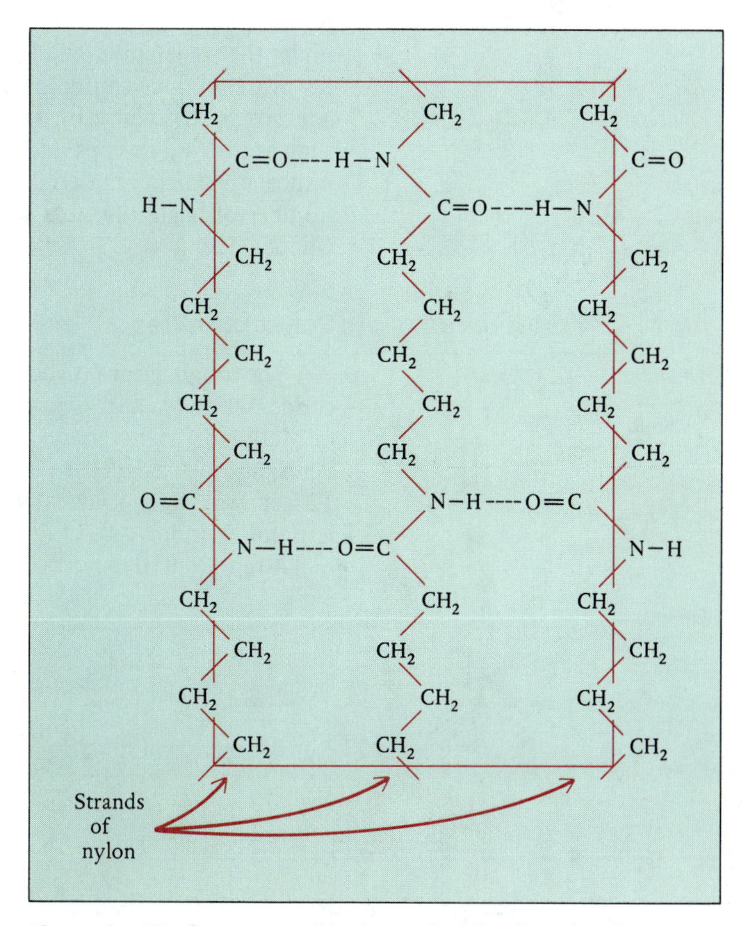

Figure 8-12 Structure and hydrogen bonding in nylon 6.

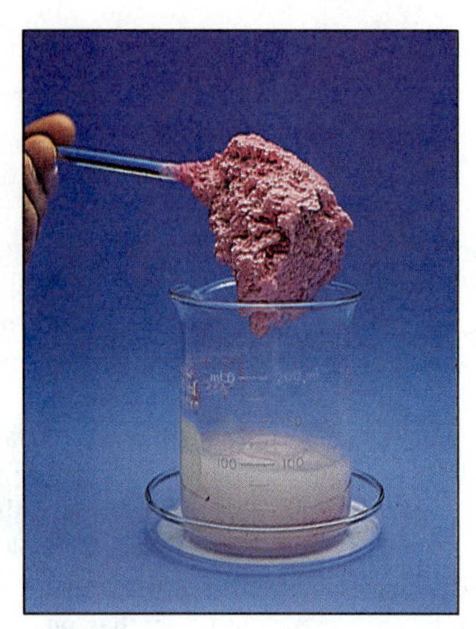

Figure 8–13 Preparation of phenol-formaldehyde polymer. (Charles D. Winters.)

Formaldehyde Resins

Formaldehyde is number 23 in chemical production in the United States primarily because of its use in synthesizing a variety of condensation polymers. The first thermosetting plastic was the *phenol-formaldehyde* copolymer (Fig. 8–13) synthesized by Leo Baekeland in 1909 and produced under the tradename Bakelite. Over 700,000 tons of phenol-formaldehyde resins are produced annually in the United States for use in making plywood adhesive, glass fiber resin, and molding compound for a variety of products such as distributor caps, radios, and buttons. Although formaldehyde is an important starting material for these condensation polymers, it presents a number of health hazards because of its toxicity and carcinogenicity (see Chapter 10).

Polycarbonates

The tough, clear polycarbonates constitute another important group of condensation plastics. One type of polycarbonate, commonly called Lexan® or Merlon®, was first made in Germany in 1953. It is as "clear as glass" and nearly as tough as steel. A 1-inch sheet can stop a .38-caliber bullet fired from 12 feet away. Such unusual properties have resulted in Lexan's use in "bulletproof" windows and as visors in astronauts' space helmets.

A representative portion of Lexan is made as follows:

$$\underbrace{Cl-\overset{\overset{\displaystyle O}{\|}}{C}}_{\text{Phosgene}}-Cl + H-O-\underbrace{C_6H_4-\overset{\overset{\displaystyle CH_3}{|}}{\underset{\underset{\displaystyle CH_3}{|}}{C}}-C_6H_4}_{R}-O-H \longrightarrow Cl\Big[\underbrace{\overset{\overset{\displaystyle O}{\|}}{C}-O-R-O}_{\text{Repeating unit}}\Big]H + HCl$$

$$\text{Repeating unit} + HO-R-OH \longrightarrow \cdots O-R-\underbrace{O-\overset{\overset{\displaystyle O}{\|}}{C}-O}_{\text{"Carbonate structure"}}-R-O\cdots$$

Moon walk. The visor in the astronaut's helmet contains Lexan, a polycarbonate. (Courtesy of NASA.)

The name *polycarbonate* comes from the linkage's similarity to an inorganic carbonate ion, CO_3^{2-}.

SILICONES

The element silicon, in the same chemical family as carbon, also forms many compounds with numerous Si—Si and Si—H bonds, analogous to C—C and C—H bonds. However, the Si—Si bonds and the Si—H bonds react with both oxygen and water; hence, there are no useful silicon counterparts to most hydrocarbons. However, silicon does form stable bonds with carbon, and especially oxygen, and this fact gives rise to an interesting group of condensation polymers containing silicon, oxygen, carbon, and hydrogen (bonded to carbon).

In 1945, at the General Electric Research Laboratory, E. G. Rochow discovered that a silicon-copper alloy reacts with organic chlorides to produce a whole class of reactive compounds, the **organosilanes.**

Silane (SiH_4) is structurally like methane (CH_4), in that both are tetrahedral.

$$\underset{\text{Methyl chloride}}{2\,CH_3Cl} + \underset{\text{Silicon-copper alloy}}{Si(Cu)} \longrightarrow \underset{\text{Dimethyldichlorosilane}}{(CH_3)_2SiCl_2} + Cu$$

The chlorosilanes readily react with water and replace the chlorine atoms with hydroxyl(—OH) groups. The resulting molecule is similar to a dialcohol.

$$(CH_3)_2SiCl_2 + 2\,H_2O \longrightarrow (CH_3)_2Si(OH)_2 + 2\,HCl$$

Two dihydroxysilane molecules undergo a condensation reaction in which a water molecule is split out. The resulting Si—O—Si linkage is very strong; the same linkage holds together all the natural silicate rocks and minerals. Continuation of this condensation process results in polymer molecules with molecular weights in the millions:

$$(CH_3)(OH)_2Si{-}CH_3 + (CH_3)(OH)_2Si{-}CH_3 \longrightarrow CH_3(CH_3)(OH)Si{-}O{-}Si(OH)(CH_3)CH_3 + HOH$$

Further reaction yields:

$$-Si(CH_3)_2{-}O{-}Si(CH_3)_2{-}O{-}\left(Si(CH_3)_2{-}O\right)_n{-}Si(CH_3)_2{-}O{-}Si(CH_3)_2{-}O{-}Si(CH_3)_2{-}O$$

Silicones are polymers held together by a series of covalent Si—O bonds.

By using different starting silanes, polymers with different properties result. For example, two methyl groups on each silicon atom result in **silicone oils,** which are more stable at high temperatures than hydrocarbon oils and also have less tendency to thicken at low temperatures.

Silicone rubbers are very high molecular weight chains cross-linked by Si—O—Si bonds. Silicone rubbers that vulcanize at room temperature are commercially available; they contain groups that readily cross-link in the presence of atmospheric moisture. The —OH groups are first produced, and then they condense in a cross-linking "cure" similar to the vulcanization of organic rubbers.

Over 3 million pounds of silicone rubber are produced each year in the United States. The uses include window gaskets; O-rings; insulation; sealants for buildings, space ships, and jet planes; and even some wearing apparel (Fig. 8-14). The first footprints on the moon were made with silicone rubber boots, which readily withstood the extreme surface temperatures.

(a)

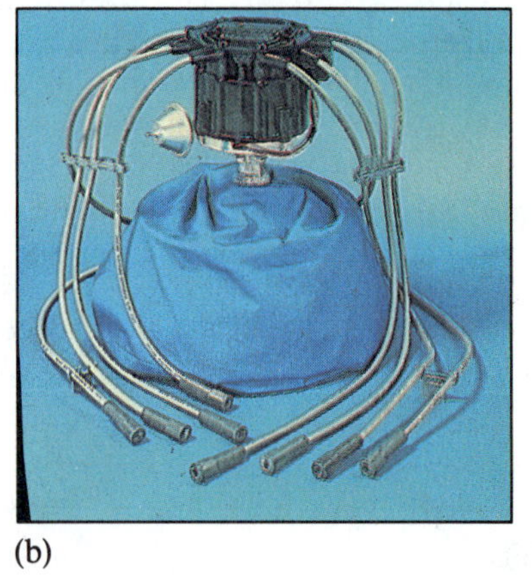

(b)

Figure 8-14 (a) Examples of some consumer products that contain silicone polymers. (b) Silicone rubber is used in automotive ignition systems because of its superior electrical properties and heat resistance. (Courtesy of Stauffer-Wacker Silicones Corporation.)

Silly Putty®, a silicone widely distributed as a toy, is intermediate between silicone oils and silicone rubber. It is an interesting material with elastic properties on sudden deformation, but its elasticity is quickly overcome by its ability to flow like a liquid when allowed to stand.

Preparation of a Substance Like Silly Putty®, an Interesting, Fun-Type Polymer

There are two receipes for making a substance that has properties similar to Silly Putty®. Try both of them and compare their properties.

1. Combine about 20 mL of white school glue with about 15 mL of liquid starch. Mix thoroughly. Rinse with water. Does the substance stretch? Roll it into a ball and drop it on a hard surface. Does it bounce? Roll it into a ball and let it sit undisturbed on a flat surface. What do you observe? Store the product in a ziplock bag.
2. Mix 25 mL of white school glue with 20 mL of water in a paper cup. Stir well. Prepare a 4% borax solution by dissolving 4 g of borax (available from the grocery store) in water to make 100 mL of solution. Measure out 5 mL of the 4% borax solution. If you want colored Silly Putty, add 1 to 5 drops of food coloring to the borax solution. Then add the 5 mL of 4% borax solution to the glue mixture and stir well. Remove the solid material from the cup, place it on a piece of waxed paper, and let it sit for a few minutes. Compare its properties to those of the substance prepared in (1). Store the Silly Putty®-type material in a ziplock bag.

RECYCLING PLASTICS

Disposal of plastics has been the subject of considerable debate in recent years as municipalities face increasing problems in locating sufficient landfill space. Americans produce 160 million tons of solid waste every year — more

The stability of plastics and their increased use contribute to the mounting problem of garbage disposal. (*The World of Chemistry*, Program 22, "The Age of Polymers.")

than 4 pounds per person every day. Table 8–3 gives the composition of this trash. Although plastics are only 7% by weight, they are estimated to make up about 20% of the volume of solid wastes. At the present time, 80% of the trash ends up in municipal landfills, 10% is incinerated, and 10% is recycled.

Only 1% of plastics waste is being recycled as compared with recycling of 60% of aluminum cans, 20% of paper, and 10% of glass. Recycling aluminum cans has been a successful money-making venture for individuals and nonprofit organizations for about 20 years, but comprehensive, community-wide recycling plans are a fairly recent development. The early success of aluminum recycling is based on economics, with scrap cans selling for 25 to 30 cents per pound. The demand for other recycled items, such as paper, has been much lower. For example, recycled newspapers often have no market. Four phases are needed for successful recycling of any waste material: collection, sorting, reclamation, and end-use. The increased public awareness of the importance of recycling has resulted in a dramatic increase in establishing these phases, particularly for plastics.

Curbside recycling and state laws requiring recycling are enhancing the collection phase. In recent years, many companies have been established to handle the sorting, reclamation, and end-use phases. Three plastics that currently have the four phases in place are polyethylene terephthalate (PET), high-density polyethylene (HDPE), and polystyrene foam.

Polyethylene terephthalate, widely used as soft-drink bottles, is the most commonly recycled plastic. The used bottles are available from retailers in states requiring refundable deposits or from curbside pickups. Over 150 million pounds of PET were recycled in the United States in 1988 (17% of annual production). Major end uses for recycled PET include fiberfill for ski jackets and sleeping bags, carpet fibers, and non-food containers. High-density polyethylene (HDPE) is the second most widely recycled resin, with 72 million pounds processed in 1988. One-gallon milk jugs are the principal source of recycled HDPE, and the base cups from PET soft-drink bottles are a second source. Recycled HDPE is used to make pipe, pails, and base cups for PET soft-drink bottles.

TABLE 8-3 Composition of Municipal Solid Wastes

Material	Percent by Weight
Paper and paperboard	36
Yard wastes	20
Food wastes	9
Metals	9
Glass	8
Plastics	7
Wood	4
Rubber and leather	3
Textiles	2
Miscellaneous	2
TOTAL	100

The use of polystyrene foam for food packaging has been a controversial issue because it symbolizes the throw-away mentality for many Americans. The decision of the McDonald Corporation to substitute paper for polystyrene foam packaging is a highly publicized example of this issue. However, recycling polystyrene packaging from fast food restaurants and school cafeterias is already big business. National Polystyrene Recycling Company is planning to recycle 250 million pounds of used polystyrene per year by 1995, which represents 25% of the annual production. Recycled polystyrene is used to make trash receptacles and plastic "lumber" for park benches, piers, and other outdoor installations.

Although recycling of plastics has shown a dramatic increase in recent years, recycling companies will not be able to increase the percentage of recycled plastics to the 50% goal by the year 2000 without significant improvement in collection and sorting. An increase in curbside recycling will help with the collection phase. Between 1989 and 1991, the number of U.S. households with curbside collection of recyclables increased from 9 million to 16 million residences, which is estimated to be 20% of U.S. households. Codes are stamped on plastic containers to help consumers identify and sort their recyclable plastics (Fig. 8–15), and several companies are developing new technology to sort plastics.

Code	Material	Percent of total bottles
1 PETE	Polyethylene terephthalate (PET)*	20-30
2 HDPE	High-density polyethylene	50-60
3 V	Vinyl polyvinyl chloride (PVC)*	5-10
4 LDPE	Low-density polyethylene	5-10
5 PP	Polypropylene	5-10
6 PS	Polystyrene	5-10
7 OTHER	All other resins and layered multi-material	5-10

*Bottle codes are different from standard industrial identification to avoid confusion with registered trademarks.

Figure 8–15 Plastic container codes.

Curbside recycling in Nashville, Tennessee.

SELF-TEST 8B

1. Nylon is an example of a ____________ polymer.
2. Polyamides are formed when ____________ is split out from the reaction of many organic acid groups and many amine groups.
3. When an acid such as terephthalic acid $HO-C(=O)-C_6H_4-C(=O)-OH$ reacts with ethylene glycol ($HOCH_2CH_2OH$), what is the structure of the resulting polymer?
4. Polyesters are formed by ( ) addition ( ) condensation reactions.
5. Many molecules of a carboxylic diacid reacting with many molecules of a dialcohol produce a ____________.
6. Formaldehyde resins are examples of ( ) thermosetting ( ) thermoplastic polymers.
7. When $(CH_3)_2SiCl_2$ reacts with water, what is a representative portion of the structure of the polymer?
8. A silicone polymer contains Si— ____ bonds.
9. Americans produce over ____________ pounds of trash per person per day.

QUESTIONS

1. In what ways is a railroad train like polystyrene?
2. Where do you suppose the first chemist who prepared a polymer got the idea for giant molecules?
3. What property does a polymer have when it is extensively cross-linked?
4. Describe on the molecular level the end result of the vulcanization process.
5. What is the origin of the word *polymer?*
6. Is polystyrene a thermoplastic or a thermosetting plastic?

7. What property of the molecular structure of rubber allows it to be stretched?
8. Explain how polymers could be prepared from each of the following compounds. (Other substances may be used.)

a. $CH_3-\overset{H}{\overset{|}{C}}=\overset{H}{\overset{|}{C}}-CH_3$

b. $HO-\overset{O}{\overset{\|}{C}}-CH_2-CH_2-\overset{O}{\overset{\|}{C}}-OH$

c. $H_2N-CH_2-C_6H_4-CH_2-NH_2$ (para-substituted benzene ring)

9. What are the monomers used to prepare the following polymers?

a. $-CH_2CH_2CH_2CH_2CH_2CH_2CH_2CH_2CH_2-$

b. $-\underset{}{\overset{CH_3}{\overset{|}{C}}}HCH_2\overset{CH_3}{\overset{|}{C}}HCH_2\overset{CH_3}{\overset{|}{C}}HCH_2-$

c. $-CH_2-\overset{H}{\overset{|}{\underset{C_6H_5}{\underset{|}{C}}}}CH_2-\overset{H}{\overset{|}{\underset{C_6H_5}{\underset{|}{C}}}}CH_2-\overset{H}{\overset{|}{\underset{C_6H_5}{\underset{|}{C}}}}CH_2-\overset{H}{\overset{|}{\underset{C_6H_5}{\underset{|}{C}}}}-$

10. What structural features must a molecule have in order to undergo addition polymerization?
11. What is meant by the term *macromolecule?*
12. Orlon has a polymeric chain structure of

$$-CH_2-\underset{CN}{\underset{|}{CH}}-CH_2-\underset{CN}{\underset{|}{CH}}-CH_2-\underset{CN}{\underset{|}{CH}}-$$

What is the monomer from which this structure can be made?
13. What feature do all condensation polymerization reactions have in common?
14. What are the starting materials for nylon 66?
15. What single property must a molecule possess in order to be a monomer?
16. Which do you think is the source of most polymers used today, green plants or petroleum? Do you think this will ever change? Explain.
17. Would isoprene make a good motor fuel? Explain.
18. A tiny sample of rubber, held in the flame of a match, burns with a small bright flame and gives a *white* flame in contrast to the black smoke of burning tires. Explain.
19. What are the four phases that must be in place for successful recycling of any solid waste?
20. Discuss what plastics are currently being recycled, and give examples of some products being made from these recycled plastics.

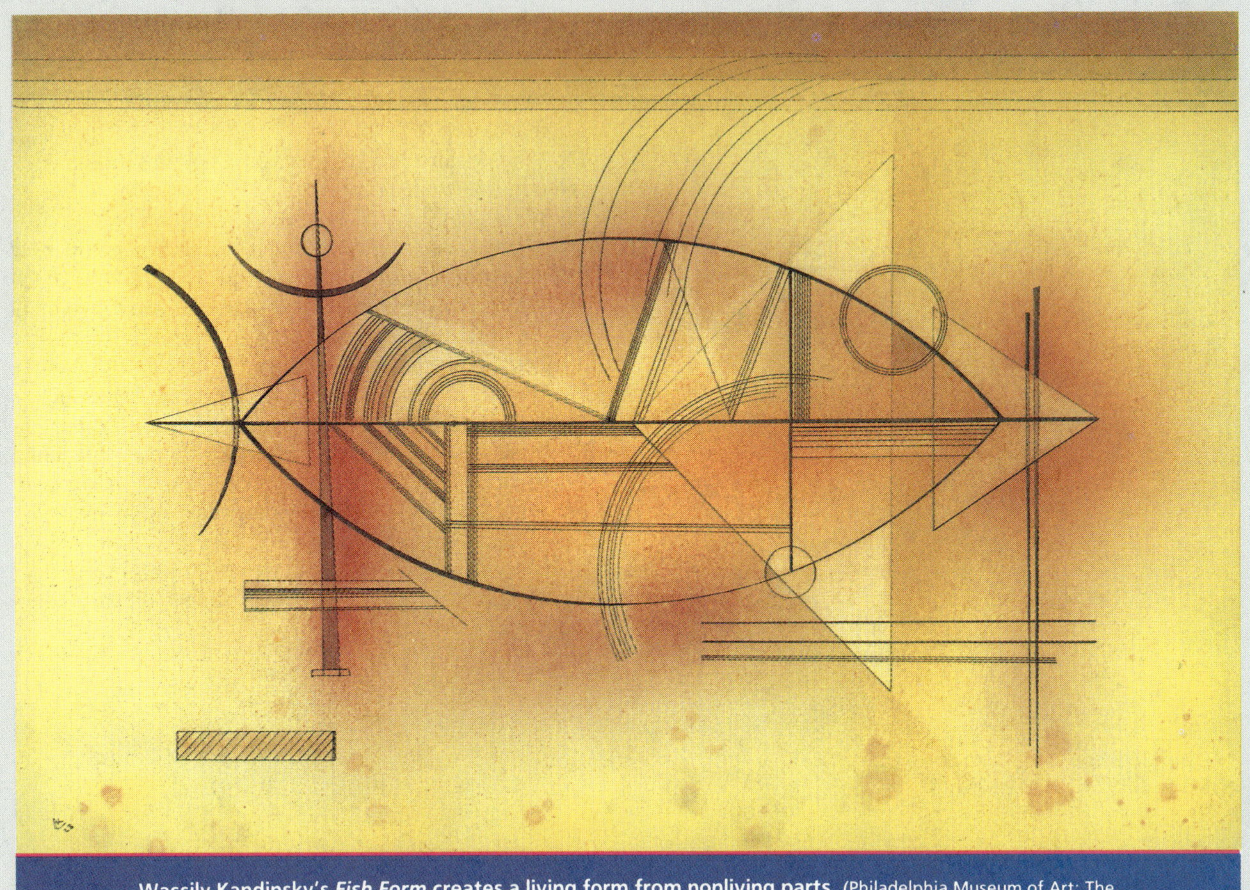

Wassily Kandinsky's *Fish Form* creates a living form from nonliving parts. (Philadelphia Museum of Art: The Louise and Walter Arensberg Collection)

CHAPTER

9

Chemistry of Life

Biochemistry is the name given to the study of the chemistry of life. Biochemicals are organic chemicals found in living things. The major classes of biochemicals include carbohydrates, lipids, proteins, enzymes, vitamins, hormones, and nucleic acids.

1. What are optical isomers, and why are they important in biochemical reactions?
2. What are the different types of sugars?
3. How does the sweetness of common sugars compare to that of artificial sweeteners?
4. What is the difference between starch and cellulose?
5. What are triglycerides?
6. What are the differences among saturated, monounsaturated, and polyunsaturated fatty acids?
7. What are steroids?
8. How are proteins formed?
9. What do enzymes do?
10. What is the human genome project?
11. What are DNA and RNA, and how do they relate to the genetic code?
12. What is biogenetic engineering?

The chemistry of life is referred to as **biochemistry,** and the organic chemicals found in living things are called biochemicals. As Wöhler's experiments demonstrated (see Chapter 7), biochemicals are not life inherent; they are simply part of living systems. Biochemicals common to all living systems are lipids, carbohydrates, proteins, enzymes, vitamins, hormones, nucleic acids, and compounds for the storage and exchange of energy, such as adenosine triphosphate (ATP). In addition to these biochemicals, certain minerals are required for proper functioning of living organisms. Vitamins and minerals are discussed in Chapter 14 on nutrition. Medicines, the chemicals frequently necessary to sustain life and to make life more bearable, are discussed in Chapter 15.

NATURAL POLYMERS

Some biochemicals are polymers. Starches are condensation polymers of simple sugars (the monomers); sucrose (table sugar) is composed of only two simple sugars. Proteins are condensation polymers of amino acids (the monomers). Nucleic acids are condensation polymers of simple sugars, nitrogenous bases, and phosphoric acid species. Other biochemicals are composed of two or more smaller molecules. For example, a fat molecule is an ester made from one glycerol and three fatty acid molecules. Enzymes are constructed of a protein alone or a protein bonded to a metal ion or a vitamin.

Many organic molecules important to the chemistry of living things exhibit "handedness," another type of isomerism. For example, all naturally occurring amino acids are "left-handed" isomers.

HANDEDNESS

Are you right-handed or left-handed? Regardless of our preference, we learn at a very early age that a right-handed glove doesn't fit the left hand and vice versa. Our hands are not identical, but they are **mirror images** of one another and are nonsuperimposable (Fig. 9–1). Many molecules also exhibit a "handedness," and right-handed or left-handed molecules are referred to as **optical isomers.** Optical isomers are possible when a molecular structure is **asymmetrical** (without symmetry). The simplest case is a tetrahedral carbon atom bonded to four *different* atoms or groups of atoms. Such a carbon atom is called a **chiral** (or asymmetrical) carbon atom. Figure 9–2 shows two ways to arrange four different atoms in the tetrahedral positions about the central carbon atom. These result in two nonsuperimposable mirror-image molecules that are called **optical isomers.**

Chiral is pronounced ki-ral and is derived from the Greek *cheir,* meaning hand.

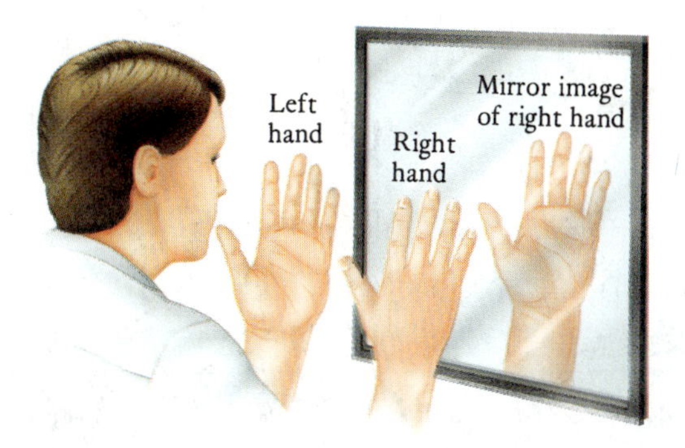

Figure 9–1 Mirror images. Your left hand is a nonsuperimposable mirror image of your right hand. For example, the mirror image of your right hand looks like your left hand. If you place one hand directly over the other, they are not identical; hence, they are nonsuperimposable mirror images.

All amino acids except glycine can exist as one of two optical isomers. In Figure 9–3, the mirror-image relationship is shown for optical isomers of alanine, an amino acid with a tetrahedral carbon atom surrounded by an amine group ($—NH_2$), a methyl group ($—CH_3$), an acid group (—COOH), and a hydrogen atom. Note that the carbon atoms in the methyl and acid groups are not asymmetrical because these carbon atoms are not bonded to four different groups.

The formula of glycine is H_2NCH_2COOH. Why doesn't glycine have optical isomers?

The "handedness" of optical isomers is represented by D for right-handed (D stands for dextro, from the Latin *dexter* meaning "right") and L for left-handed (L stands for levo, from the Latin *laevus* meaning "left"). The properties of optical isomers of a compound are almost identical—they have the same melting point, the same boiling point, the same density, and

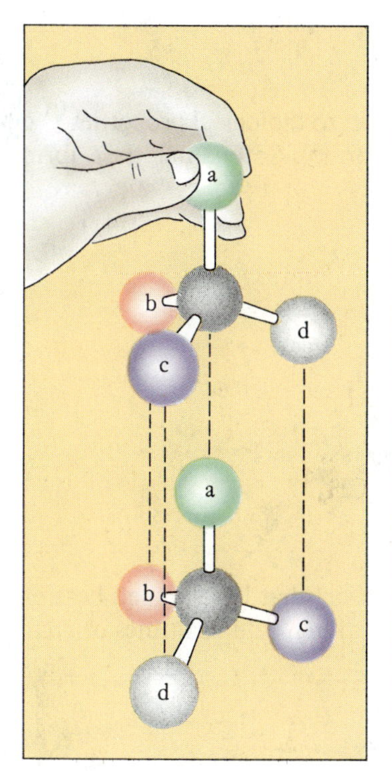

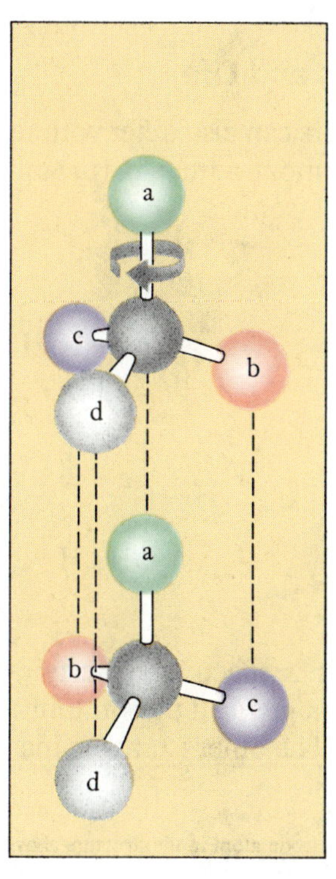

Figure 9–2 Four different atoms, or groups of atoms, are bonded to tetrahedral center atoms so that the upper isomeric form cannot be turned in any way to match exactly the lower structure. The upper structure and the lower structure are nonsuperimposable mirror images.

Figure 9–3 Optical isomers of the amino acid alanine, 2-aminopropionic acid. The D-form is the nonsuperimposable mirror image of the L-form.

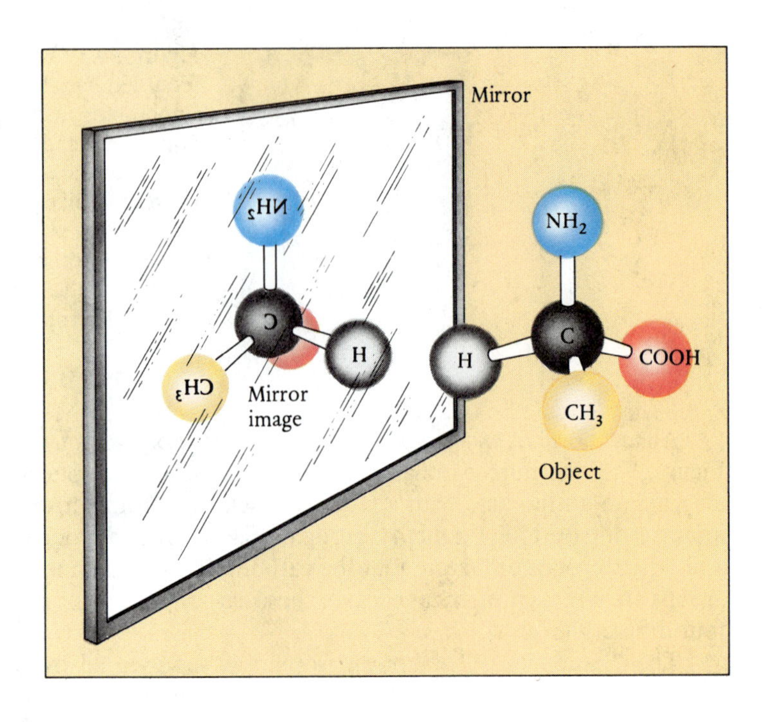

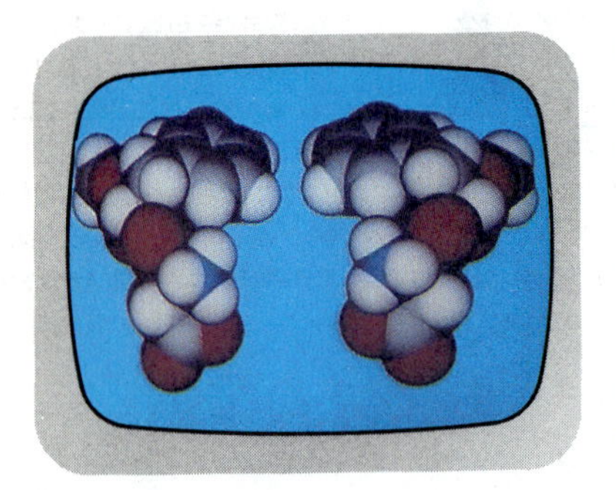

Models of the optical isomers of aspartame. (*The World of Chemistry,* Program 9, "Molecular Architecture.")

many other identical physical and chemical properties. However, they always differ with respect to one physical property: they rotate the plane of **polarized light** in opposite directions.

Optical Isomers and Life

Optical isomers can also differ with respect to biological properties. An example is the hormone adrenalin (or epinephrine). Adrenalin is the L-form of a pair of optical isomers.

Adrenalin
(Epinephrine)

Only the L-isomer is effective in starting a heart that has stopped beating momentarily, or in giving a person unusual strength during times of great emotional stress. The other isomer is inactive.

C* designates the chiral carbon atom in the structure above.

THE WORLD OF CHEMISTRY

Molecular Architecture

There are various types of isomerism common to organic chemistry. Of these, optical isomers are among the most fascinating. This is because they play such important roles in life processes. In living things, chiral molecules exist in only one form or the other but not both. How did the selection of one optical isomer over the other occur in nature? This question is addressed by Nobel laureate Christian Anfinsen who says

How this selection began in nature is anybody's guess. One assumption is that some naturally occurring minerals, for example, might have been involved in binding one form and not the other. In the process a concentration of the form we have now was built up so that when life started, it was stuck with that form. In nature we're stuck pretty much with one isomer. The world has become so evolved that living things are in general composed of one of the two possible mirror images of the basic compounds.

The World of Chemistry (Program 9) "Molecular Architecture."

Christian Anfinsen. (*The World of Chemistry,* Program 9, "Molecular Architecture.")

The optical isomers of aspartame have different properties. The L-isomer is used as an artificial sweetener (NutraSweet®), but the D-isomer is bitter.

Only L-isomers of optically active amino acids are found in proteins. Nature's preference for L-amino acids has provoked much discussion and speculation among scientists since Pasteur's discovery of optical activity in 1848 from studies of crystals of tartaric acid salts.

Enzymes, the catalysts for biochemical reactions, also have a handedness and, as a left-hand glove only fits a left hand, bind to only one of the optical isomers. For example, during contraction of muscles the body produces only the L-form of lactic acid and not the D-form.

The concentration of lactic acid in the blood is associated with the feeling of tiredness, and a period of rest is necessary to reduce the concentration of this chemical by oxidation.

COOH
C
H OH
CH_3
D-lactic acid

COOH
C
HO H
CH_3
L-lactic acid

CARBOHYDRATES

Carbohydrates are organic compounds of carbon, hydrogen, and oxygen that contain several alcohol groups. They are either monomers (simple sugars or **monosaccharides**), dimers **(disaccharides),** or condensation polymers **(polysaccharides).**

Latin *saccharum,* "sugar"; *mono-,* one; *di-,* two; *poly-,* many.

Monosaccharides

The most common simple sugar is glucose (also known as dextrose, grape sugar, and blood sugar), which is found in fruit, blood, and living cells. A solution of glucose is often given intravenously when a source of quick energy is needed to sustain life. Glucose, along with galactose and fructose, are the three common monosaccharides found in the body. They have different structures but the same molecular formula, $C_6H_{12}O_6$.

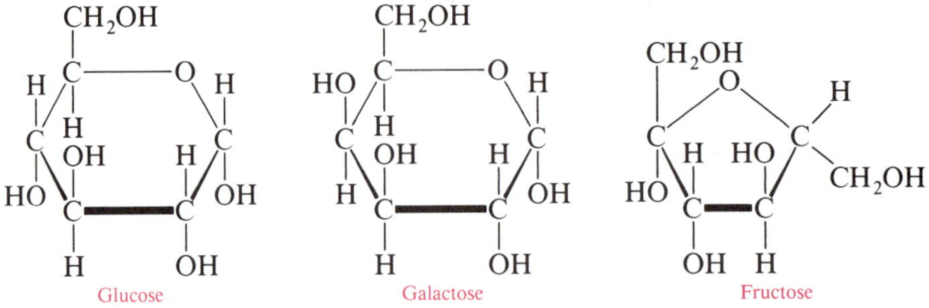

Glucose Galactose Fructose

Disaccharides

Disaccharides are two monosaccharides joined together with the elimination of water to give compounds with the general formula, $C_{12}H_{22}O_{11}$. The three most common disaccharides are **sucrose** (from sugar

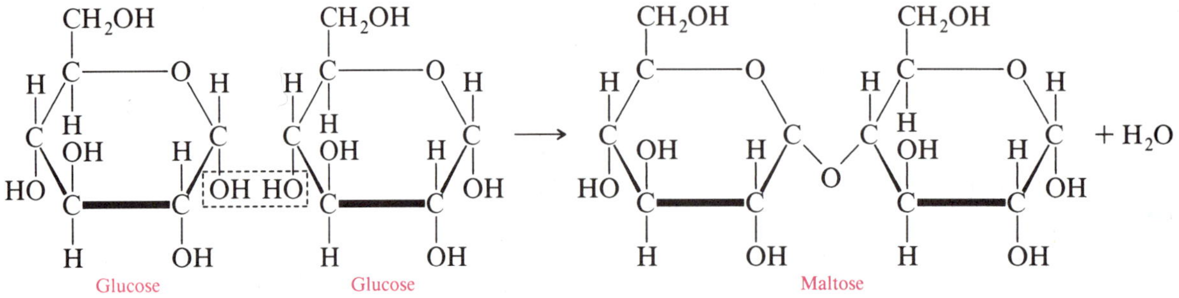

Glucose Glucose $\longrightarrow$ Maltose $+ H_2O$

cane or sugar beets), formed from a glucose monomer and a fructose monomer; **maltose** (from starch), formed from two glucose monomers; and **lactose** (from milk), formed from a glucose monomer and a galactose monomer. The structures of sucrose, maltose, and lactose and the simple sugars they form in the digestion process (hydrolysis) are given in Figure 9–4.

Hydrolysis is the term used for chemical reactions in which chemicals are decomposed through their reaction with water.

Artificial sweeteners, discussed in Chapter 14, are hundreds to thousands of times sweeter than sugars.

Sucrose is produced in a high state of purity on an enormous scale—over 80 million tons per year. About 40% of the world sucrose production comes from sugar beets and 60% from sugar cane. Although sucrose is used universally as a sweetener, it is not the sweetest sugar (Table 9–1).

See Chapter 3 for a discussion of hydrogen bonding.

The polar alcohol (—OH) groups on the disaccharides and monosaccharides cause hydrogen bonding with water. This is why table sugar dissolves readily in coffee and tea and why glucose is transported easily by the blood.

Polysaccharides

Polysaccharides are condensation polymers.

Nature's most abundant polysaccharides are starch, glycogen, and cellulose. Starch occurs in plants such as potatoes, corn, wheat, and rice. Glycogen is an energy reservoir in animals, just as starch is in plants. Glycogen is found primarily in liver and muscle tissue. Cellulose is the main structural material in plants.

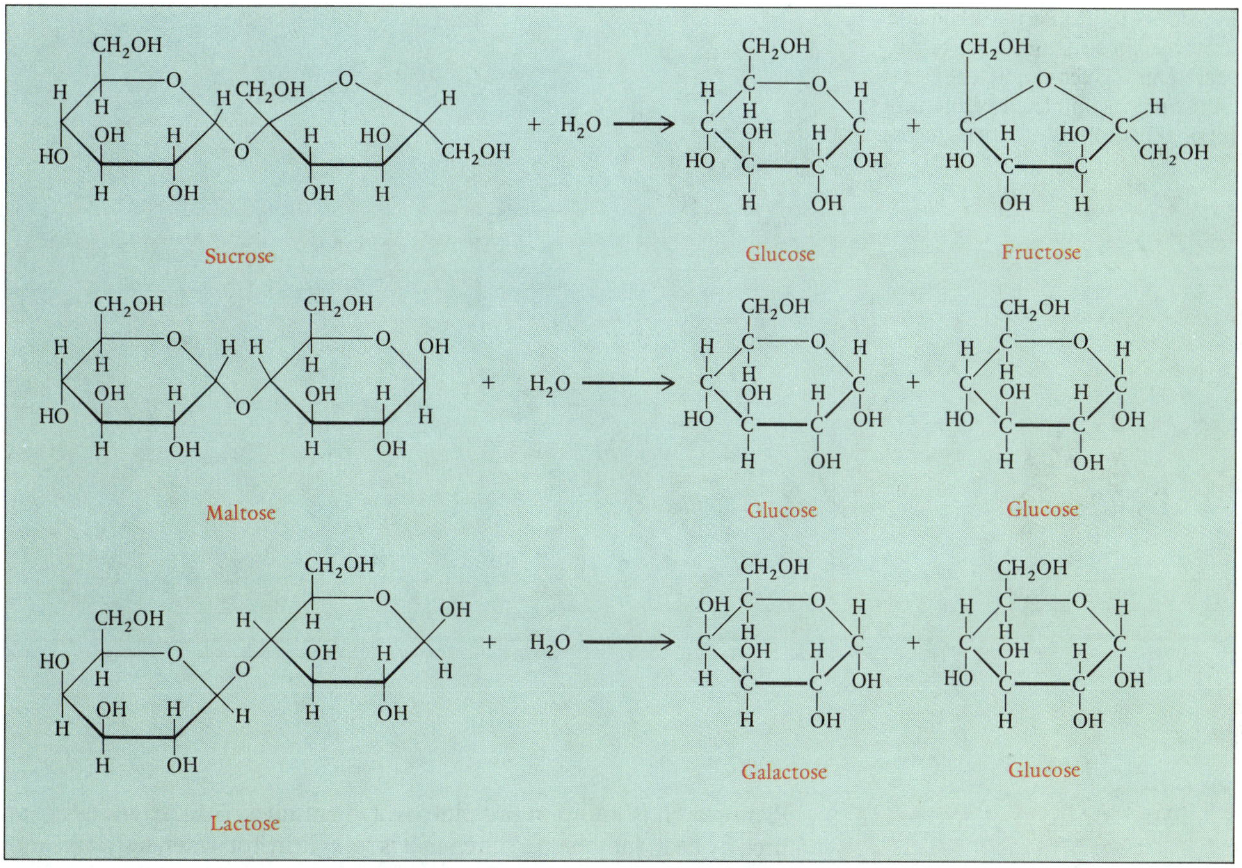

Sucrose + H_2O → Glucose + Fructose

Maltose + H_2O → Glucose + Glucose

Lactose + H_2O → Galactose + Glucose

Figure 9–4 Structures of disaccharides and the hydrolysis reactions that occur during digestion.

TABLE 9–1 Sweetness of Common Sugars and Artificial Sweeteners Relative to Sucrose

Substance	Sweetness Relative to Sucrose as 1.0
Lactose	0.16
Galactose	0.32
Maltose	0.33
Glucose	0.74
Sucrose	1.00
Fructose	1.17
Aspartame*	180
Saccharin*	300
Sucralose*	650
Alitame*	2000

* Artificial sweeteners—see Chapter 14

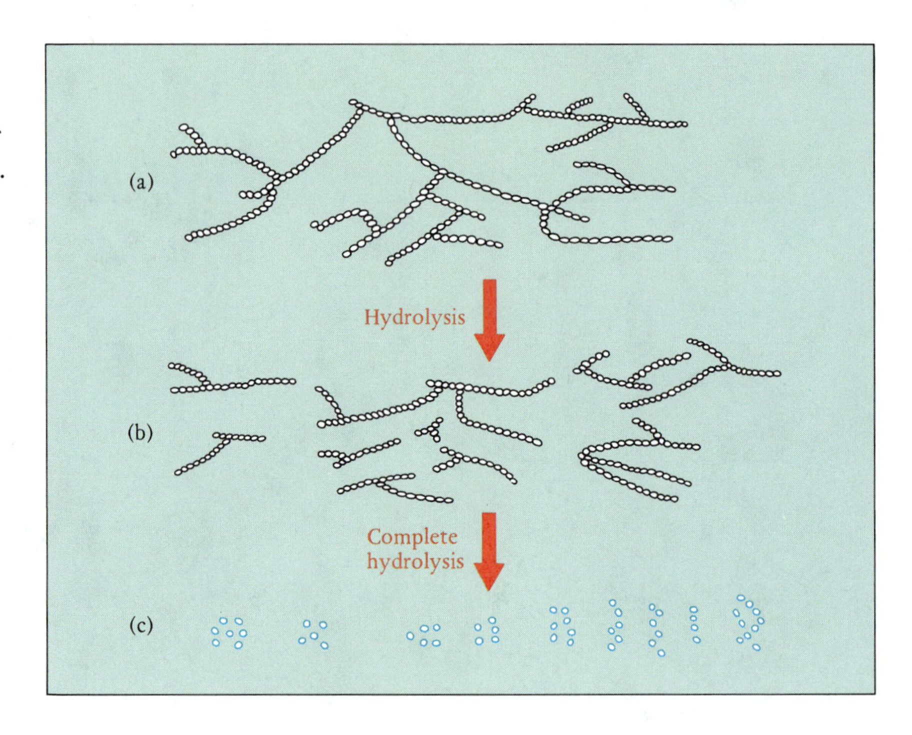

Figure 9–5 (a) Partial schematic amylopectin structure. Each circle represents a glucose unit. (b) Dextrins from incomplete hydrolysis of (c) final hydrolysis product: glucose.

Starches

Plant starch is found in protein-covered granules. If these granules are ruptured by heat, they yield a starch that is soluble in hot water, **amylose,** and an insoluble starch, **amylopectin.** Amylose constitutes about 25% of most natural starches. When tested with iodine solution, amylose turns blue-black, whereas amylopectin turns red.

Starch molecules consist of many glucose monomers bonded together.

A typical amylopectin molecule has about 1000 glucose monomers arranged into branched chains (Fig. 9–5). Complete hydrolysis yields glucose; partial hydrolysis produces mixtures called **dextrins.** Dextrins are used as food additives and in mucilage, paste, and finishes for paper and fabrics.

Structurally, amylose is a straight-chain condensation polymer with an average of about 200 glucose monomers per molecule. Each monomer is bonded to the next with the loss of a water molecule, just as the two units are bonded in maltose. A representative portion of the structure of amylose is shown in Figure 9–6.

Cellulose

Cellulose is the most abundant polysaccharide in nature. Cotton is 98% cellulose, and wood is about 50% cellulose. Like amylose, cellulose is composed of glucose units. The difference between the structures of cellulose and amylose lies in the bonding between glucose units (Fig. 9–6). The angle around the oxygen atoms connecting the glucose rings is 180 degrees in cellulose and 120 degrees in amylose. This subtle structural difference is the reason we cannot digest cellulose. Human beings do not have the necessary enzymes to hydrolyze cellulose, but grazing animals (cows, sheep, horses)

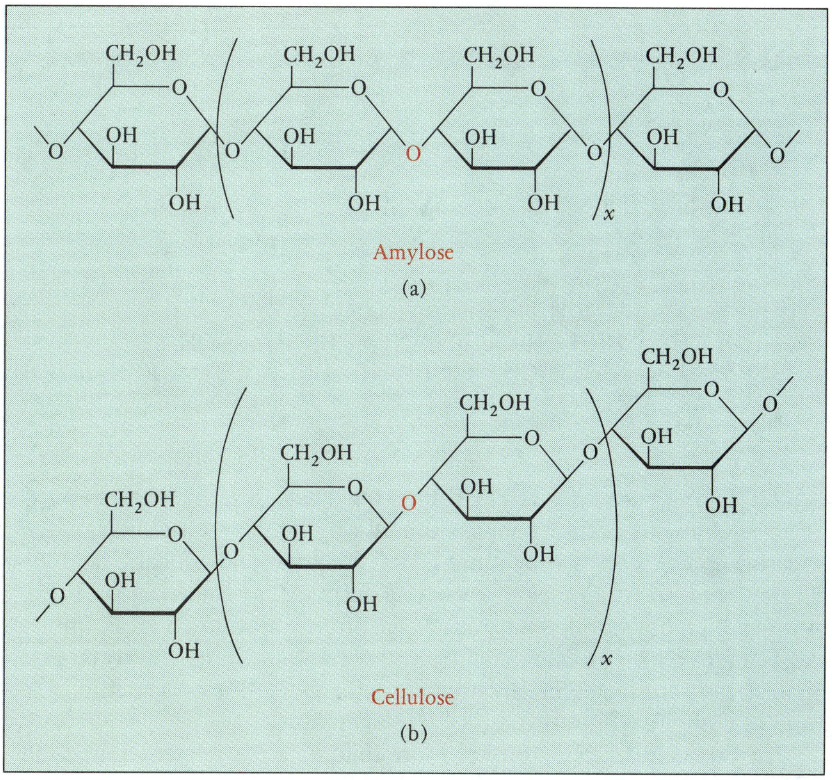

Figure 9–6 (a) Amylose structure. From 60 to 300 glucose units bond together in the manner shown. (b) Cellulose structure. About 2800 glucose units bond together in the pattern shown here.

and termites can digest cellulose because bacteria in their digestive tracts contain enzymes that break the bonds between glucose units in cellulose. Although cellulose has no nutritive value for humans, it makes up most of the dietary fiber in our diet (see Chapter 14).

In the laboratory, glucose can be obtained from cellulose by heating a suspension of cellulose in a solution of strong acid. At the present time, wood cannot be hydrolyzed to glucose economically enough to help satisfy the world's growing need for food.

LIPIDS

A lipid is an organic substance that has a greasy feel and is insoluble in water but soluble in organic solvents. Lipids include fats and oils, waxes, steroids, and phospholipids. Ninety-five percent of the lipids in our diet are triglycerides. The other 5% are phospholipids and steroids.

Fats and Oils

Fats and oils are esters of glycerol (glycerin) and fatty acids known as **triglycerides.** Fatty acids are rarely found in the free form in nature but are in the combined ester form in fats and oils. The most common carboxylic acids

Fats and oils are esters of fatty acids and glycerol. Fats are solids, and oils are liquids.

TABLE 9-2 Common Fatty Acids

Fatty Acid	Number of Carbon Atoms	Number of C=C Bonds	Condensed Formula
Lauric	12	0	$CH_3(CH_2)_{10}COOH$
Myristic	14	0	$CH_3(CH_2)_{12}COOH$
Palmitic	16	0	$CH_3(CH_2)_{14}COOH$
Stearic	18	0	$CH_3(CH_2)_{16}COOH$
Oleic	18	1	$CH_3(CH_2)_7CH{=}CH(CH_2)_7COOH$
Linoleic	18	2	$CH_3(CH_2)_4CH{=}CHCH_2CH{=}CH(CH_2)_7COOH$
Linolenic	18	3	$CH_3CH_2CH{=}CHCH_2CH{=}CHCH_2CH{=}CH(CH_2)_7COOH$
Arachidonic	20	4	$CH_3(CH_2)_4CH{=}CHCH_2CH{=}CHCH_2CH{=}CHCH_2CH{=}CH(CH_2)_3COOH$

found in fats and oils are given in Table 9-2. They are either **saturated** with only C—C single bonds, **monounsaturated** with one C=C double bond, or **polyunsaturated** with two or more C=C double bonds. Stearic acid and palmitic acid are examples of **saturated fatty acids,** and oleic acid is an example of an unsaturated fatty acid. One of the unsaturated acids, linoleic acid, is referred to as an essential fatty acid because the human body requires this acid, but cannot chemically produce it, for the synthesis of an important group of compounds known as the prostaglandins.

Prostaglandins are a group of more than a dozen related compounds with potent effects on physiological activities such as blood pressure, relaxation and contraction of smooth muscle, gastric acid secretion, body temperature, food intake, and blood platelet aggregation.

The 1990 Nobel Prize in Chemistry was awarded to Professor Elias James Corey, Harvard University, for his pioneering work in the synthesis of prostaglandins.

The formation of the esters of glycerol is shown in the following equation:

$$\begin{array}{l} H_2C{-}OH \\ \;\;| \\ HC{-}OH \\ \;\;| \\ H_2C{-}OH \end{array} + \begin{array}{l} HO{-}C({=}O){-}R \\ HO{-}C({=}O){-}R' \\ HO{-}C({=}O){-}R'' \end{array} \rightleftharpoons \begin{array}{l} H_2C{-}O{-}C({=}O){-}R \\ \;\;| \\ HC{-}O{-}C({=}O){-}R' \\ \;\;| \\ H_2C{-}O{-}C({=}O){-}R'' \end{array} + 3\,H_2O$$

Glycerol (one molecule) — Fatty acid (three molecules that may or may not be the same) — Fat or oil (one molecule) — Water (three molecules)

Hydrogenation of the double bonds in vegetable oils converts the liquid oil into a solid fat. (J. Morgenthaler.)

Saturated fatty acids are usually found in solid or semisolid fats, whereas *unsaturated* fatty acids are usually found in oils. Hydrogen can be catalytically added to the double bonds of an oil to convert it into a semisolid fat. For example, liquid soybean and other vegetable oils are **hydrogenated** to produce cooking fats and margarine.

Nutritional problems related to the amount of fat in the diet and the relative ratio of saturated and unsaturated fat are discussed in Chapter 14.

Consumers in Europe and North America have historically valued butter as a source of fat. As the population of these parts of the world increased, the advantages of a substitute for butter became apparent, and efforts to prepare such a product began about 100 years ago. One initial problem was that common fats are almost all *animal* products with very pronounced tastes of their own. Analogous compounds from vegetable oils, which are bland or have mixed flavors, were generally *unsaturated* and consequently *oils.* A solid fat could be made from the much cheaper vegetable oils if an inexpensive way could be discovered to add hydrogen across the double bonds. After extensive experiments, many catalysts were found, of which finely divided nickel is among the most effective. The nature of the process can be illustrated by the following reaction:

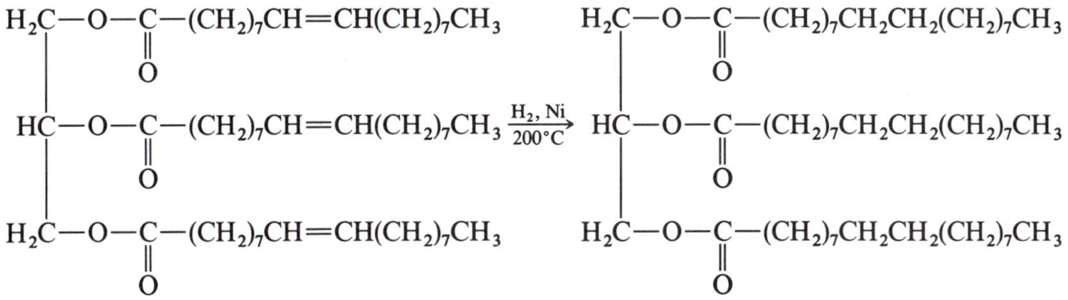

$$\begin{array}{l} H_2C{-}O{-}\underset{\displaystyle O}{\underset{\|}{C}}{-}(CH_2)_7CH{=}CH(CH_2)_7CH_3 \\ \;\; | \\ HC{-}O{-}\underset{\displaystyle O}{\underset{\|}{C}}{-}(CH_2)_7CH{=}CH(CH_2)_7CH_3 \\ \;\; | \\ H_2C{-}O{-}\underset{\displaystyle O}{\underset{\|}{C}}{-}(CH_2)_7CH{=}CH(CH_2)_7CH_3 \end{array} \xrightarrow[200^\circ C]{H_2,\ Ni} \begin{array}{l} H_2C{-}O{-}\underset{\displaystyle O}{\underset{\|}{C}}{-}(CH_2)_7CH_2CH_2(CH_2)_7CH_3 \\ \;\; | \\ HC{-}O{-}\underset{\displaystyle O}{\underset{\|}{C}}{-}(CH_2)_7CH_2CH_2(CH_2)_7CH_3 \\ \;\; | \\ H_2C{-}O{-}\underset{\displaystyle O}{\underset{\|}{C}}{-}(CH_2)_7CH_2CH_2(CH_2)_7CH_3 \end{array}$$

Triolein (a liquid oil) Tristearin (a solid fat)

Oils commonly subjected to this process include those from cottonseed, peanuts, corn germ, soybeans, coconuts, and safflower seeds. In recent years, as it has become apparent that saturated fats may encourage diseases of the heart and arteries, soft margarines and cooking oils (which still contain some of the unhydrogenated fatty acid) have been placed on the market.

Phospholipids

When glycerol forms esters that involve both phosphoric acid and fatty acids, an important class of phospholipids, the phosphoglycerides, is obtained. These molecules are found in the membranes of cells throughout the body.

Steroids

Steroids are found in all plants and animals and are derived from the following four-ring structure:

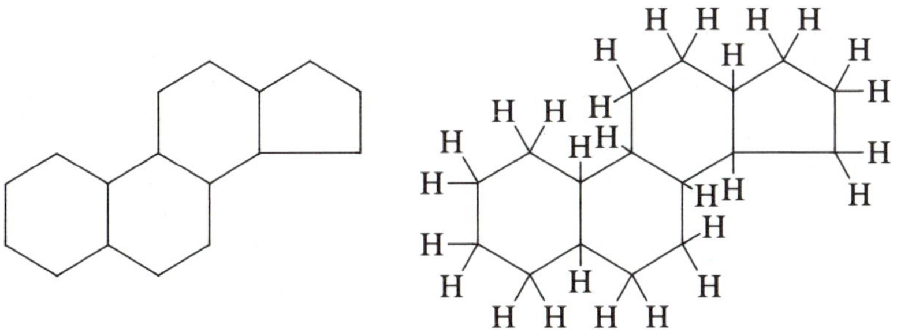

The skeletal four-ring structure drawing on the left is chemical shorthand similar to that described for cyclic hydrocarbons in Chapter 6. There is a carbon at each corner, and the lines represent C—C bonds. Since every carbon atom forms four bonds, additional bonds between carbon atoms and hydrogen atoms are understood to be present whenever the skeletal structure shows fewer than four bonds. The structure on the right shows the hydrogen atoms understood to be present in the four-ring structure shown on the left. Although all the rings in the skeletal drawing are shown as saturated rings, steroids often have one ring that is unsaturated or aromatic. For example, cholesterol has one double bond in the second ring. Note that its structure also includes alkyl groups and an alcohol group. These are substituted for hydrogen atoms in the skeletal representation shown above.

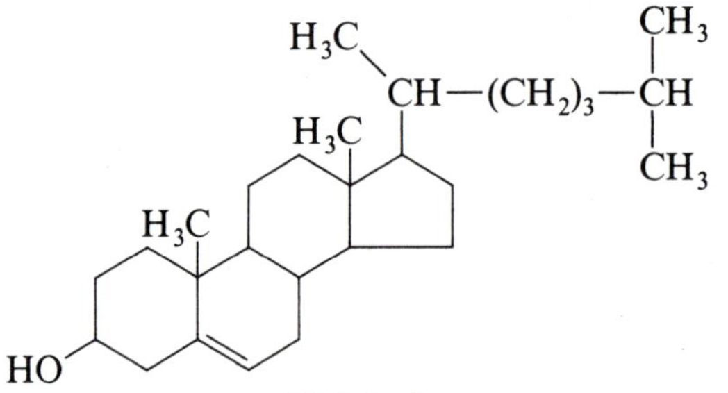

H_3C CH_3 $CH{-}(CH_2)_3{-}CH$ H_3C CH_3 H_3C HO

Cholesterol

Cholesterol is the most abundant animal steroid. The human body synthesizes cholesterol and readily absorbs dietary cholesterol through the intestinal wall. An adult human contains about 250 grams of cholesterol. Although cholesterol receives a lot of attention in connection with the correlation of blood cholesterol levels with heart disease, it is important to realize that proper amounts of cholesterol are essential to our health because cholesterol undergoes biochemical alteration or degradation to give milligram amounts of many important hormones such as vitamin D, cortisone, and the sex hormones (discussed in Chapter 15). Cholesterol combines with proteins to form lipoproteins, which transport cholesterol in the bloodstream. About 65% of the cholesterol in the blood is carried by low-density lipoproteins (LDLs), whereas 25% of the cholesterol in the blood is carried by high-density lipoproteins (HDLs). LDLs are "bad" cholesterol and HDLs are "good" cholesterol in discussions of problems relating to atherosclerotic plaque and heart disease (see Chapter 15).

Waxes

Waxes in the lipid class are esters formed from long-chain fatty acids and long-chain alcohols. The general formula of a wax is the same as that of a simple ester, $R{-}O{-}\overset{\overset{\displaystyle O}{\|}}{C}{-}R'$, with the qualification that R and R′ are limited to alkyl groups with a large number of carbon atoms. Natural waxes are usually mixtures of such esters. Animals and plants use waxes as protective coatings. Wax coatings on leaves help to protect the leaves from disease and also help the plant to conserve water. The feathers of birds are also coated with wax. Our ears are protected by wax. Several natural waxes have been used in consumer products. These include carnauba wax (from a Brazilian palm tree), which is used in floor waxes, automobile waxes, and shoe polishes; and lanolin (from lamb's wool), which is used in cosmetics and ointments. Lanolin also contains cholesterol.

SELF-TEST 9A

1. To have optical isomers in carbon compounds, a carbon atom must have __________ different groups attached.
2. In what physical property do optical isomers differ? __________ ____________________
3. The two monosaccharides released when sucrose is hydrolyzed are __________ and __________.
4. The complete hydrolysis of a polysaccharide yields __________.
5. The sugar referred to as blood sugar, grape sugar, or dextrose is actually the compound __________.
6. What kind of bonding enables sugar to dissolve in water? __________
7. Starch is a condensation polymer built of __________ monomers.
8. Cellulose is a condensation polymer built of __________ monomers.
9. Cotton is principally __________.
10. Fats and oils are esters of __________ and __________.
11. The structural difference between a saturated fat and an unsaturated fat is ____________________.

PROTEINS

Our bodies contain about 30,000 different kinds of proteins. Hair, skin, nails, muscles, enzymes—every living part of our bodies contains proteins. The close relationship between proteins and living organisms was first noted in 1835 by the Dutch chemist G. J. Mulder. He named proteins from the Greek *proteios* ("first"), thinking that proteins are the starting point for a chemical understanding of life. Each unique kind of protein is composed of several specific amino acids arranged in a definite molecular structure. In a few proteins the major fraction is only one kind of amino acid; the protein in silk, for example, is 44% glycine.

Proteins are high-molecular-weight compounds made up of amino acid monomers.

Proteins are condensation polymers of **amino acids.** The 20 different amino acids that can be found in proteins are made primarily from carbon, oxygen, hydrogen, and nitrogen. Small amounts of other elements are also found in proteins, the most common one being sulfur. As the name implies, amino acids have an amine group ($—NH_2$) and an acid (carboxyl) group ($—COOH$). Most amino acids have an amine group and an acid group bonded to the same carbon atom (Table 9–3). The general formula for an amino acid is shown below:

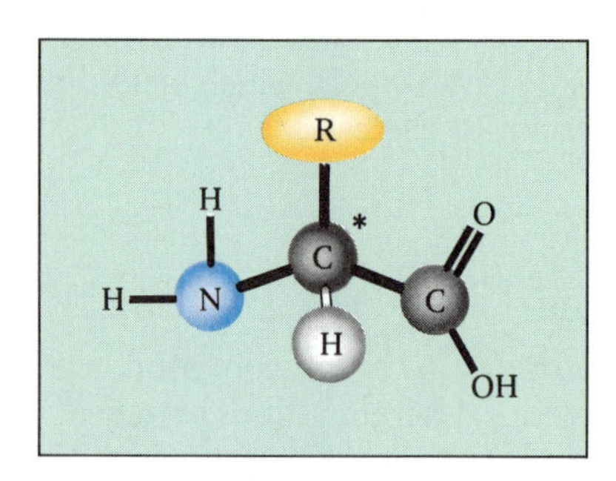

R is a characteristic group for each amino acid, and * identifies an asymmetric carbon atom. The simplest amino acid is **glycine,** in which R is a hydrogen atom. Except for glycine, the amino acids have asymmetric carbon atoms and can be optical isomers (Fig. 9–3). As mentioned at the beginning of this chapter, nature prefers the left-handed optical isomers of amino acids in protein synthesis.

Peptide bonds form polyamides like nylon 66 (Chapter 8).

Amino acid monomers are bonded together by **peptide bonds.** The chemical reaction is an acid–base reaction in which two monomers bond and water is split out. For example, when two glycine molecules react, a peptide bond is formed and a water molecule is produced:

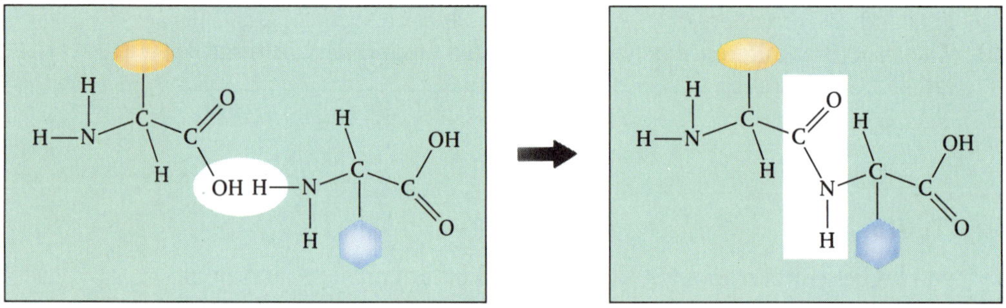

When two different amino acids are bonded, two different combinations are possible, depending on which amine reacts with which acid group. For example, when glycine and alanine react, both glycylalanine and alanylglycine can be formed.

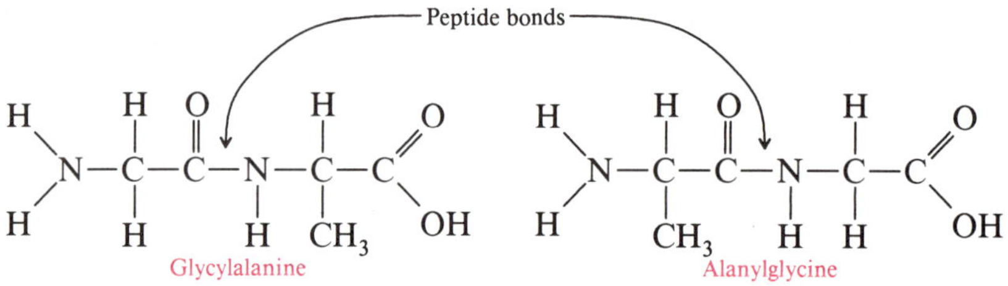

Peptide bonds

Glycylalanine

Alanylglycine

TABLE 9–3 Common L-Amino Acids found in Proteins

All of the amino acids except proline and hydroxyproline have the general formula

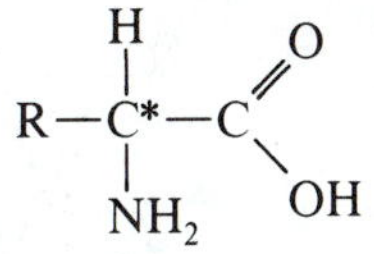

$$R{-}\underset{\substack{|\\ NH_2}}{\overset{\substack{H\\ |}}{C^*}}{-}C\begin{matrix} {/\!\!/}O \\ \backslash OH\end{matrix}$$

in which R is the characteristic group for each acid. The R groups are as follows.

1. Glycine—H
2. Alanine—CH_3
3. Serine—CH_2OH
4. Cysteine—CH_2SH
5. Cystine—$CH_2{-}S{-}S{-}CH_2{-}$
*6. Threonine—$\underset{\substack{|\\ OH}}{CH}{-}CH_3$
*7. Valine $CH_3{-}\underset{|}{CH}{-}CH_3$
*8. Leucine—$CH_2{-}\underset{\substack{|\\ CH_3}}{CH}{-}CH_3$
*9. Isoleucine—$CH\begin{matrix} {/}CH_3 \\ \backslash CH_2{-}CH_3\end{matrix}$
*10. Methionine—$CH_2{-}CH_2{-}S{-}CH_3$
11. Aspartic acid—CH_2CO_2H
12. Glutamic acid—$CH_2{-}CH_2{-}CO_2H$
*13. Lysine—$CH_2{-}CH_2{-}CH_2{-}CH_2{-}NH_2$
†14. Arginine—$CH_2{-}CH_2{-}CH_2{-}NH\overset{\substack{NH\\ \|}}{C}NH_2$
*15. Phenylalanine—$CH_2{-}C_6H_5$ (benzene ring)
16. Tyrosine—$CH_2{-}C_6H_4{-}OH$ (benzene ring)
*17. Tryptophan—CH_2— (indole ring, N—H)
18. Histidine—CH_2— (imidazole ring, N, N—H)

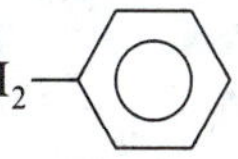

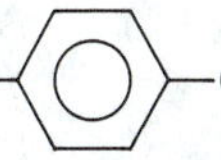

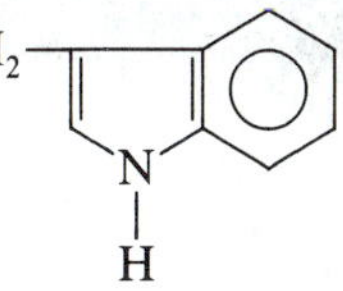

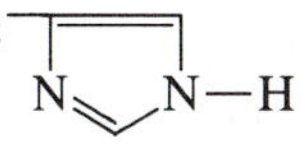

The structures for the other two are:

19. Proline

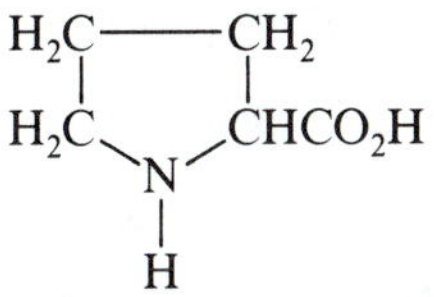

$$\begin{matrix} H_2C & — & CH_2 \\ | & & | \\ H_2C & & CHCO_2H \\ & \diagdown N \diagup & \\ & | & \\ & H & \end{matrix}$$

20. Hydroxyproline

$$\begin{matrix} HOHC & — & CH_2 \\ | & & | \\ H_2C & & CHCO_2H \\ & \diagdown N \diagup & \\ & | & \\ & H & \end{matrix}$$

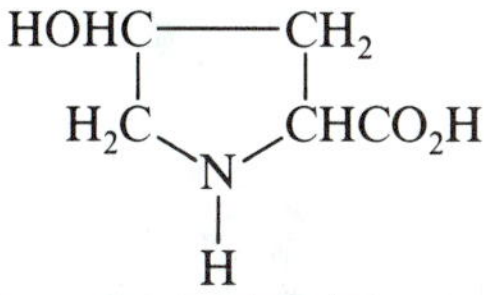

In some listings, two others are included: the amide of aspartic acid, asparagine (R group is $—CH_3CONH_2$) and the amide of glutamic acid, glutamine (R group is $—CH_2CH_2CONH_2$).

*Essential amino acids must be part of our diet. The other amino acids can be synthesized by our bodies.

†Growing children also require arginine in their diet.

Six tripeptides are possible if three amino acids (for example, glycine, Gly; alanine, Ala; serine, Ser) are linked in all possible combinations.* They are:

Gly-Ala-Ser Gly-Ser-Ala Ala-Gly-Ser Ala-Ser-Gly Ser-Ala-Gly Ser-Gly-Ala

* If the amino acids are all different, the number of arrangements is $n!$ (n factorial). For four different amino acids, the number of different arrangements is 4!, or $4 \times 3 \times 2 \times 1 = 24$. For five different amino acids, the number of different arrangements is 5!, or 120.

(a)

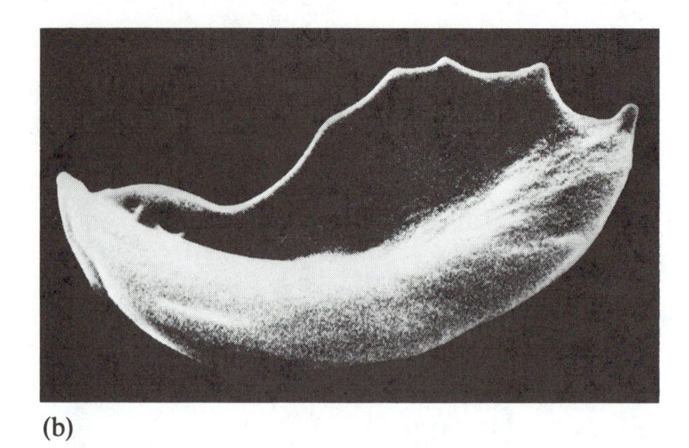

(b)

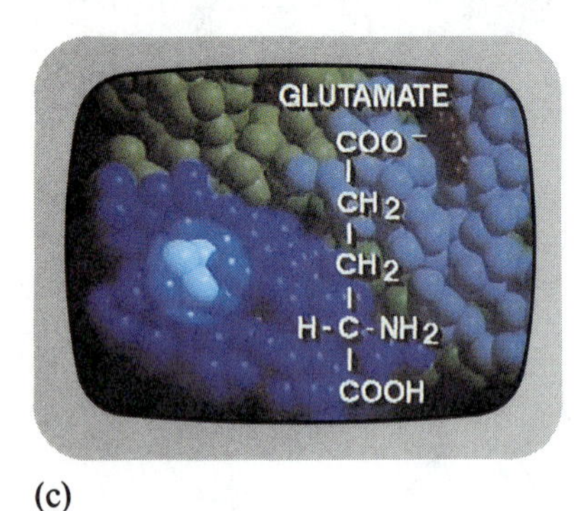

(c)

Figure 9–7 (a) Normal red blood cells. (b) Sickle red blood cells. (c) Sickle cells are caused by the substitution of the nonpolar amino acid valine for the negatively charged amino acid glutamate in the protein structure of hemoglobin. This substitution produces a crucial alteration in the tertiary structure, which causes the sickling. (a) and (b) from J. R. Holum: *Fundamentals of General, Organic, and Biological Chemistry,* 2nd ed., p. 486. New York, John Wiley & Sons, Inc., 1982. (c) is from *The World of Chemistry,* Program 24, "Genetic Code.")

As the length of the chain increases, the number of variations in the sequence of amino acids quickly increases. For example, if all 20 different amino acids are bonded, the sequences alone make 2.43×10^{18} (2.43 quintillion) uniquely different 20-monomer molecules! Since proteins can also include more than one molecule of a given amino acid, the possible combinations are essentially infinite. However, of the many different proteins that could be made from a set of amino acids, a living cell makes only the relatively small, select number it needs.

A very large number of different proteins can be prepared from a small number of different amino acids.

Protein Structure

The primary structure of a protein is the sequence of amino acids in a chain like the sequence of beads on a string. (*The World of Chemistry,* Program 23, "Proteins: Structure and Function.")

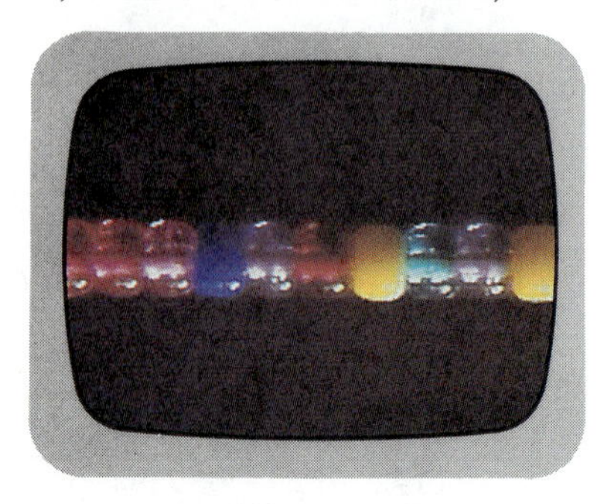

The sequence of amino acids in a protein bonded to one another by peptide bonds is called the **primary structure.** Changing the sequence alters the properties of a protein, and just one change may produce a new protein unable to function like the original one. For example, **sickle cell anemia,** a reduction in the ability of hemoglobin to transfer oxygen, is caused by the alteration of only one specific amino acid of the 146 amino acid units in a single hemoglobin chain (Fig. 9–7).

The twisting of the amino acid chain into a helical shape is an example of a **secondary structure** (Fig. 9–8). Secondary structure refers to the arrangement of chains about an axis. For example, the helical structure of proteins is caused by hydrogen bonding. Hydrogen bonds hold the helices in place as an N—H group of one amino acid hydrogen bonds with the oxygen atom in the third amino acid down the chain.

Tertiary structure refers to how a protein molecule is folded. The nature of the R groups on the amino acids in the primary structure determine the

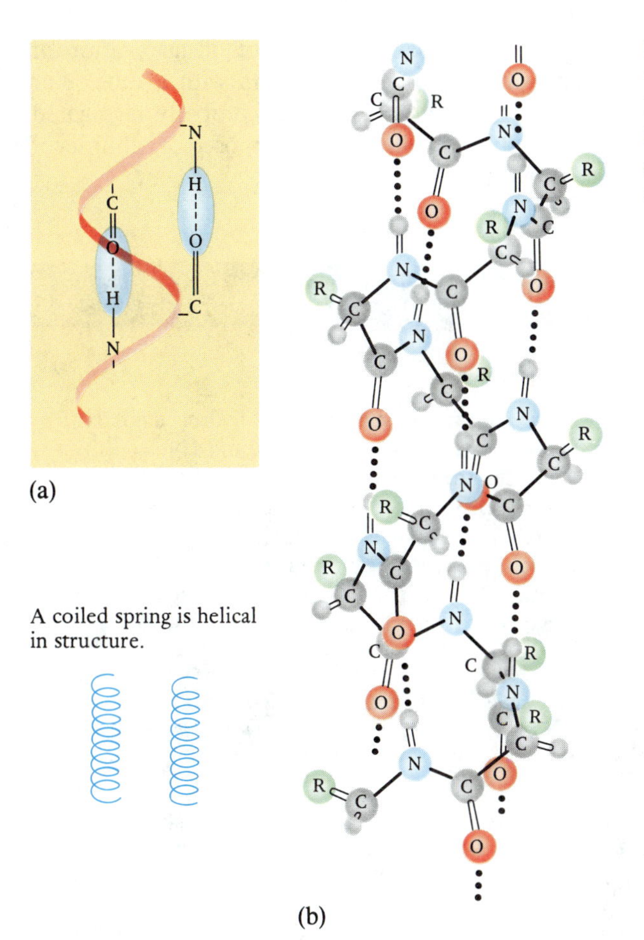

Figure 9–8 (a) Helical structure for a polypeptide in which each oxygen atom can be hydrogen-bonded to an N—H group in the third amino acid unit down the chain. (b) α-helix structure of proteins. The sketch represents the actual position of the atoms and shows where intrachain hydrogen bonds occur (*dotted line*).

A coiled spring is helical in structure.

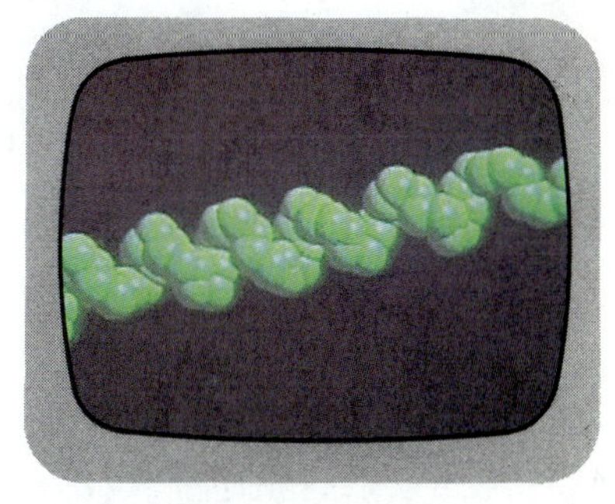

An α-helix structure of protein. (*The World of Chemistry,* Program 23, "Proteins: Structure and Function.")

tertiary structure (Fig. 9–9). Hydrogen bonding interactions, disulfide bridge bonds (—S—S—), and ionic bonds are three types of interactions that affect the folding of the protein molecule. (See Chapter 16 for a discussion of the tertiary structure of hair.)

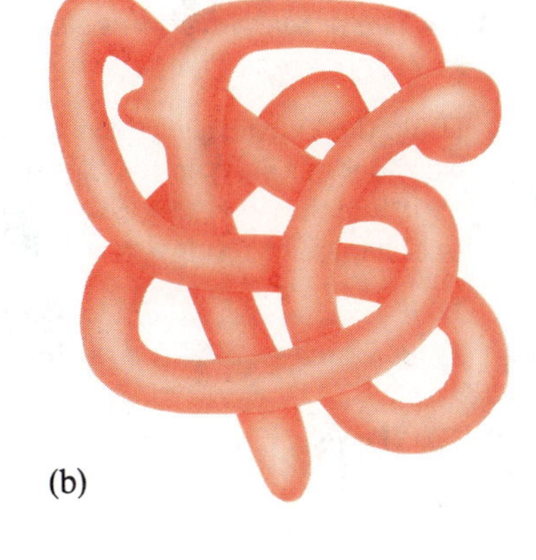

Figure 9–9 Tertiary molecular structures of proteins. (a) The imaginary twisted structure of collagen. (b) The imaginary folded structure of the helix in a globular protein.

The **quaternary structure** of proteins is the degree of aggregation of protein units. Human hemoglobin, a globular protein with a molecular weight of 68,000, must have its four amino acid chains properly aggregated in order to form active hemoglobin. Insulin is also composed of subunits of protein properly arranged into its quaternary structure.

THE WORLD OF CHEMISTRY

Unraveling The Protein Structure

One of the key steps in unraveling the mystery of hydrogen bonds in protein structure involved Linus Pauling, a cold, and a Nobel Prize. Pauling currently lives in the Big Sur region of Northern California. His living room is his office. There he spends a large part of each day at a simple desk, working on a new research interest, metals. Earlier in his career, Pauling had another interest, the structure of protein molecules. At that time there were several conflicting theories. Pauling and his colleagues thought that the first level of protein structure was a polypeptide chain. Then they asked themselves a fundamental question.

We asked: How is the polypeptide chain folded? We couldn't answer the question, but we said it's probably held together by hydrogen bonds. The conclusion we reached was that there are . . . polypeptide chains in the protein, which, far longer if they were stretched out than the diameter of the molecule, are coiled back and forth; and that they are coiled into a very well defined structure, configuration, with the different part of the chain held together by hydrogen bonds. In 1937, I spent a good bit of the summer with models for—I assumed that I knew what a polypeptide chain looks like except for the way in which it's folded. And I wanted to fold it to form the hydrogen bonds. I didn't succeed. The fact is, I thought that there was something about proteins that perhaps I didn't know.

Linus Pauling (b.1901), along with R. B. Corey, proposed the helical and sheetlike secondary structures for proteins. For his bonding theories and for his work with proteins, Pauling was awarded the Nobel Prize in 1954. For his fight against nuclear danger, he received the 1962 Nobel Peace Prize. (*The World of Chemistry*, Program 23, "Proteins: Structure and Function.")

Pauling continued to work on this problem, but the solution eluded him. Then one day he had a crucial insight in a completely different and unexpected setting.

I had a cold. I was lying in bed for two or three days, and I read detective stories, light reading, for awhile, and then I got sort of bored with that. So I said to my wife, "Bring me a sheet of paper, and I'm going to—I think I'll work on that problem of how polypeptide chains are folded in proteins. So she brought me a sheet of paper and the slide rule and pencil, and I started working.

Using the knowledge gained from his years of model building, he drew the backbone of a polypeptide chain on a piece of paper. Then it occurred to him to try to fold the paper to see how hydrogen bonds could form along the polypeptide chain. The result was a structure that twisted around like a spring.

Well, I succeeded. It only took a couple of hours of work that day, March of 1948, for me to find the structure, called the alpha helix.

The World of Chemistry (Program 23) "Proteins: Structure and Function."

ENZYMES

Enzymes function as catalysts for chemical reactions in living systems. Each enzyme performs a specific catalytic task. As we shall discuss later, most enzymes are globular proteins. Like all catalysts, enzymes increase the rate of a reaction by weakening bonds and causing a lowering of the energy of activation. The action of an enzyme on a chemical reaction is similar to the effect of a key opening a lock (Fig. 9–10). The lock can be opened without the key by using more energy (i.e., the lock can be broken). Similarly, the reaction will occur without the enzyme, but at a much slower rate. The enzyme makes the reaction go faster. For example, enzyme-catalyzed action allows a single molecule of amylase to catalyze the breaking of bonds between the glucose monomers in amylose at the rate of 4000 per second.

In 1926 at Cornell, James B. Sumner (1887–1955) separated, crystallized, identified, and characterized the first enzyme, urease. Sumner had been advised not to enter the field of chemistry because he had only one arm. In 1946 he won the Nobel Prize.

The names of most enzymes end in *-ase*.

Most enzymes are very specific. The enzyme maltase hydrolyzes maltose into two molecules of glucose. This is the only function of maltase, and no other enzyme can substitute for it. Sucrase, another enzyme, hydrolyzes only sucrose. Some enzymes are less specific. The digestive enzyme trypsin, for example, primarily hydrolyzes peptide bonds in proteins. However, the structure and polarity of trypsin are such that it can also catalyze the hydrolysis of some esters.

Some enzymes require only the protein for catalytic function, whereas other enzymes require the protein plus either a metal ion (e.g., Co^{3+}, Fe^{3+}, Mg^{2+}, or another essential mineral) or a vitamin for catalytic activity. The vitamin or the mineral is the **coenzyme,** and the protein is the **apoenzyme.** Both parts are needed for enzymatic activity, just as two keys are required to open a bank lock-box. Neither your key nor the bank's key alone will open the box; both are needed.

Why must we have minerals and vitamins? Answer: In part, because vitamins and minerals serve as coenzymes.

Figure 9–10 Lock-and-key theory for enzymatic catalysis. Although the analogy is an oversimplification, one very important point is made; the enzyme makes a difficult job easy by reducing the energy required to get the job started. The analogy also suggests that the enzyme has a particular structure at an active site that allows the enzyme to work only for certain molecules, similar to a key that fits the shape of a particular keyhole and a particular sequence of tumblers inside the lock.

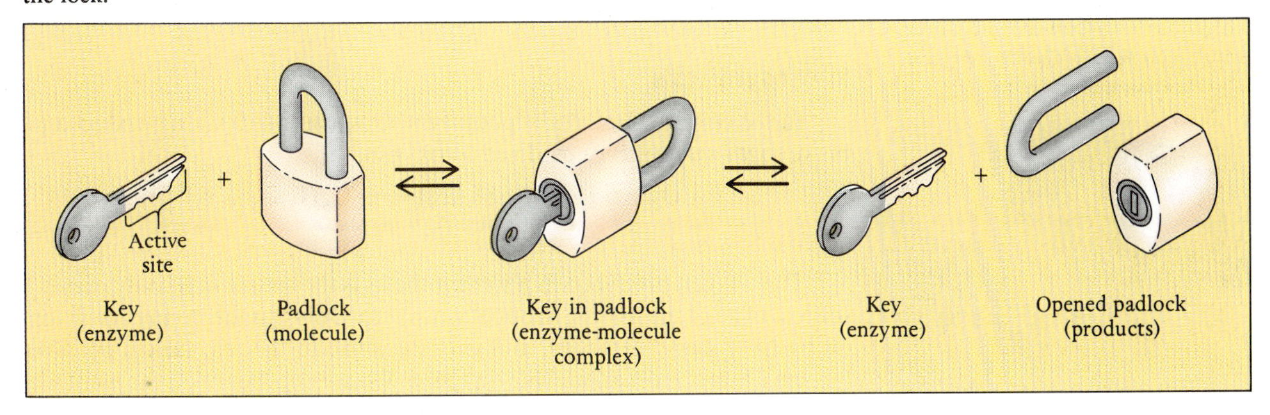

Genetic effects are often observed in the pattern of enzymes produced by individuals or races. An example of this is found in "lactose intolerance," common in certain peoples of Asia (e.g., Chinese and Japanese) and Africa (many black tribes), whose diets have traditionally contained little milk after the age of weaning. While infants, such people manufacture the enzyme **lactase** that is necessary to digest lactose, a sugar occurring in all mammals' milk. As they grow older, their bodies stop producing this enzyme because their diets normally contain no milk, and the ingestion of milk products containing lactose can lead to considerable discomfort in the form of stomach aches and diarrhea.

SELF-TEST 9B

1. The fundamental building blocks of proteins are the ____________.
2. The peptide linkage that bonds amino acids together in protein chains has the structure ____________.
3. The general structure of amino acids can be represented as ____________.
4. All amino acids except ____________ have optical isomers.
5. **a.** If we have three different amino acids and can use each one three times in any given tripeptide, we can make a total of ____________ different tripeptides.
   b. If we can use each amino acid only once, there are still ____________ possible different tripeptides.
6. In the lock-and-key analogy of enzyme activity, the enzyme functions as the ____________, and the molecule undergoing reaction serves as the ____________.
7. The helical structure of proteins is caused by ____________ bonding.

ENERGY AND BIOCHEMICAL SYSTEMS

Energy for life's processes comes from the Sun. During photosynthesis, green plants absorb energy from the Sun to make glucose and oxygen from carbon dioxide and water. The energy stored in glucose is transferred eventually to the bonds in molecules such as ATP. When needed, the ATP molecules release energy to drive other chemical reactions.

Photosynthesis requires sunlight, chlorophyll, carbon dioxide, and water. (*The World of Chemistry*, Program 17, "The Precious Envelope.")

Photosynthesis

In the complex process of photosynthesis, carbon dioxide is reduced to make sugar and water is oxidized to oxygen:

$$\underset{\text{Carbon dioxide}}{6\ CO_2} + \underset{\text{Water}}{6\ H_2O} + \underset{\text{Energy (sunlight)}}{688\ \text{kcal}} \longrightarrow \underset{\text{Glucose}}{C_6H_{12}O_6} + \underset{\text{Oxygen}}{6\ O_2}$$

The oxygen produced in photosynthesis is the source (and only present source) of all of the oxygen in our atmosphere. Only this life-giving gas, given off by trees, grass, greenery, and even by algae in the sea, makes possible human life and most animal life on Earth. We are dependent on the plant life of our planet, and we must live in balance with the oxygen output of that

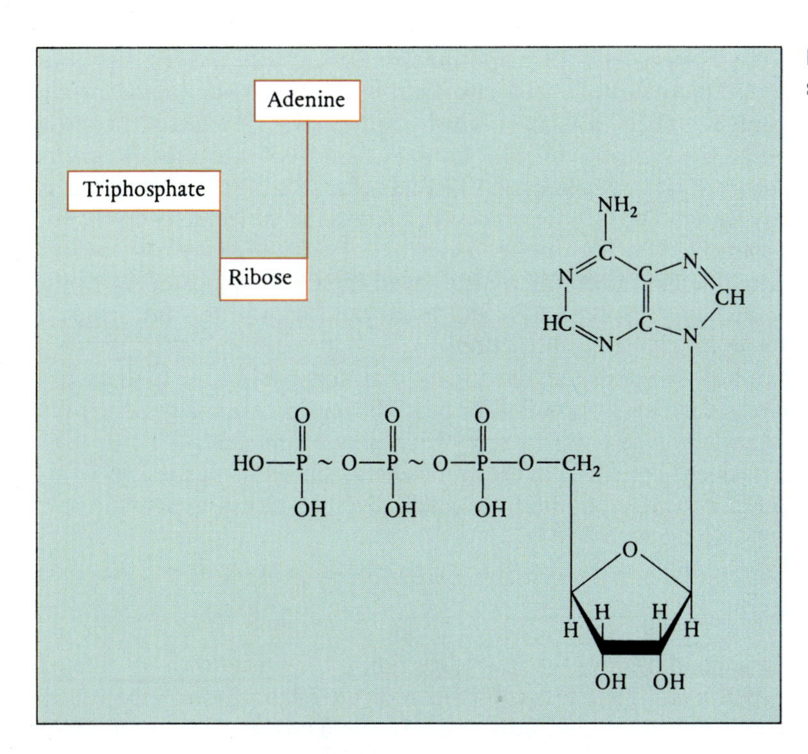

Figure 9–11 Molecular structure of adenosine triphosphate (ATP).

plant life, as well as with the food output of the same plant life. Photosynthesis is thus absolutely vital to life on Earth.

Photosynthesis involves a number of different steps and is a very complex process.

Photosynthesis is generally considered a series of **light reactions,** which occur only in the presence of light energy, and a series of **dark reactions,** which can occur in the dark. The dark reactions feed on high-energy compounds (such as ATP) produced by the light reactions. Through a series of reactions, water is oxidized to oxygen, and energy is stored in the bonds of energy-bank compounds such as ATP. ATP stores energy in two high-energy phosphate bonds, shown as wiggle lines in Figure 9–11.

In the presence of a suitable catalyst, ATP releases energy by undergoing a three-step hydrolysis. In the first step ATP is hydrolyzed to adenosine diphosphate (ADP) and releases about 12 kcal/mole (Fig. 9–12). The sec-

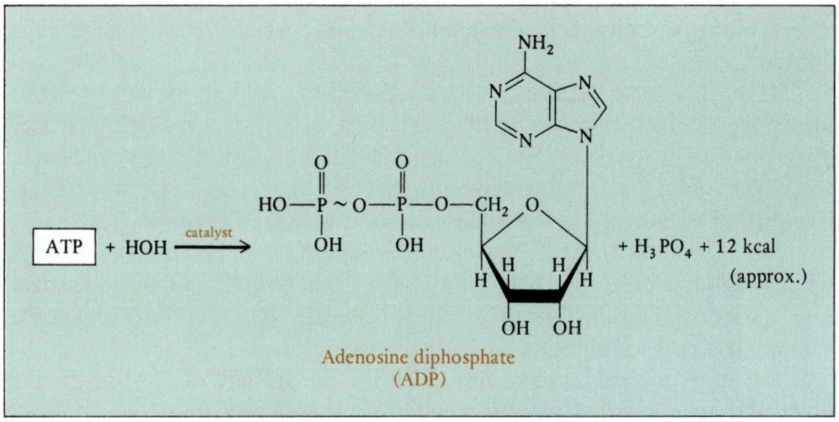

ATP + HOH —catalyst→ Adenosine diphosphate (ADP) + H_3PO_4 + 12 kcal (approx.)

Figure 9–12 Hydrolysis of ATP to ADP.

ond hydrolysis step, ADP to adenosine monophosphate (AMP), also produces about 12 kcal/mole. The last hydrolysis step, AMP to adenosine, releases only about 2.5 kcal/mole. The hydrolysis of ATP releases energy (is *exothermic*); the synthesis of ATP from AMP or ADP requires energy (is *endothermic*). It is the synthesis of ATP that occurs during the light reactions of photosynthesis, and it is this process that stores the Sun's energy in chemical compounds.

During the dark reactions, hydrolysis of the P—O bonds of ATP provides the energy to convert CO_2 and hydrogen (from water) into glucose through a series of chemical reactions.

After photosynthesis the living plant may convert glucose to disaccharides, starches, cellulose, proteins, or oils. The end-product depends on the type of plant involved and the complexity of its biochemistry.

The next steps involved in use of the energy stored in high-energy compounds are for the compounds to be eaten, digested, transported to the cells of the body, and metabolized.

Digestion

Digestion is the hydrolysis of carbohydrates, fats, and proteins to provide small molecules that can be absorbed.

From a chemical point of view, digestion is the breakdown of ingested foods by hydrolysis. The products of digestion are relatively small molecules that can be absorbed through the intestinal walls. The hydrolytic reactions of digestion are catalyzed by enzymes, there being a specific enzyme for the hydrolysis of each type of substance. The hydrolysis of carbohydrates ultimately yields simple sugars, proteins yield amino acids, and fats and oils yield fatty acids and glycerol.

In our food, carbohydrates requiring digestion are polysaccharides such as starch and disaccharides such as sucrose and lactose. The digestion process begins in the mouth with salivary amylase, or ptyalin. Starch is partially hydrolyzed into the disaccharide maltose by ptyalin, which is later rendered inactive by the high acidity of the stomach. No more digestion of carbohydrates occurs in the stomach. When the food passes from the stomach into the small intestine, the acidity is neutralized by a secretion from the pancreas. Enzymes from the pancreas complete the hydrolysis of carbohydrates into simple sugars such as glucose, fructose, and galactose. These simple sugars are then absorbed into the bloodstream. The hormone insulin (a protein) escorts simple sugars through the cell membranes and into the cells. There, in the mitochondria, these simple sugars are oxidized for their energy content.

Insulin is a protein.

If the sugar level in the bloodstream becomes too high, the simple sugars are converted into glycogen in the liver. If the sugar level is too low, stored glycogen is hydrolyzed to raise it. Malfunctions in these processes can lead to too much blood sugar, **hyperglycemia,** or too little blood sugar, **hypoglycemia.** Either condition, if sustained, produces a type of **diabetes.**

Types of diabetes are discussed in Chapter 14.

Human blood normally contains between 0.08% and 0.1% glucose.

The normal fasting level of glucose in blood occurs after 8 to 12 hours without food, which is just before most people eat breakfast. The blood sugar level for normal adults during fasting is between 70 and 100 mg of glucose for each 100 mL of blood.

The digestion of fats and oils, such as the triesters of fatty acids and glycerol, occurs primarily in the small intestine. The enzyme that catalyzes

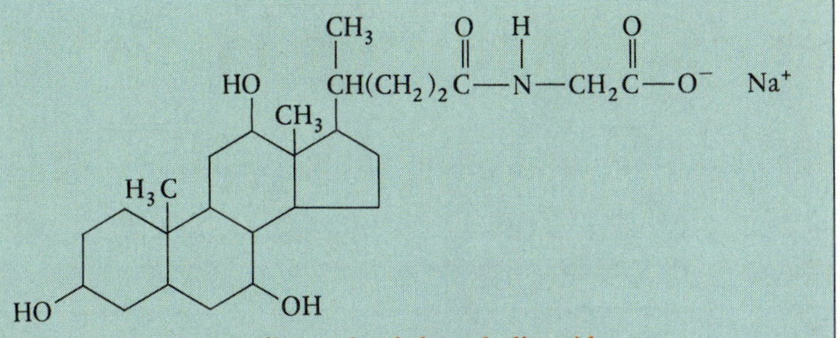

Sodium salt of glycocholic acid

Figure 9–13 The sodium salt of glycocholic acid, a bile salt made from cholesterol.

the hydrolysis of fatty acid esters is water-soluble, but the fats and oils themselves are not. Bile salts, secreted by the liver, emulsify the oil by forming an interface between the nonpolar oil and the polar water, thereby making it possible for the oil to "dissolve" in water. For a molecule to be an emulsifier between polar and nonpolar molecules, it must have both polar and nonpolar structures. The sodium salt of glycocholic acid, a bile salt synthesized from cholesterol by the body, contains the bulky nonpolar hydrocarbon groups, which are compatible with fat or oil, and the —OH and ionic groups, which attract water molecules (Fig. 9–13).

Bile salts act chemically much like detergent and soap molecules (Chapter 16).

The digestion of proteins begins in the stomach and is completed in the small intestine. Many enzymes are known to be involved. In the stomach pepsin catalyzes the hydrolysis of only about 10% of the bonds in a typical protein, leaving protein fragments with molecular weights of 600 to 3000. In the small intestine hydrolysis is completed to amino acids, which are absorbed through the intestinal wall.

Some protein enzymes are sold commercially as meat tenderizers and stain removers. Some are used to free the lens of the eye before cataract surgery.

The stomach is protected from protein-splitting enzymes by a mucous lining. The mucus is mostly protein. Although the lining is being digested slowly, it is also constantly being renewed.

The Liver: The Nutrient Bank of the Body

After digestion most food nutrients pass directly to the liver for distribution to the body. Glucose is used for energy in the liver and to prepare glucose phosphate as the first step in the preparation of glycogen (the storage carbohydrate); in addition, about one third goes on in the bloodstream to nourish the cells. From the liver a fraction of the amino acids is sent to the cells to build proteins. In the liver amino acids are used to form enzymes, and some are oxidized to obtain energy. The liver is thus the central nutrient bank, or warehouse, of the body in that it stores, converts, and classifies nutrients.

NUCLEIC ACIDS

Like polysaccharides and polypeptides, **nucleic acids** are condensation polymers. The components of the monomers are one of two simple sugars, phosphoric acid, and one of a group of ringed nitrogen compounds that have

Figure 9–14 Sugars found in RNA and DNA. The only difference between the two is indicated by the prefix *deoxy*. Deoxyribose, the sugar unit in DNA, has a hydrogen atom instead of an OH group at the position indicated in color.

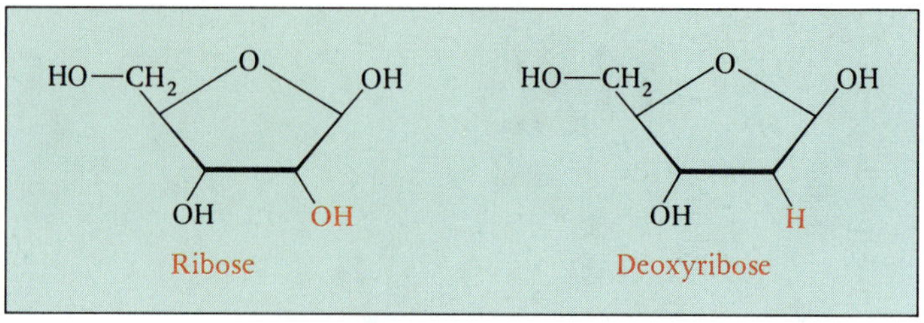

basic (alkaline) properties. The structures of the two sugars are shown in Figure 9–14. The names and formulas of the nitrogenous ring compounds are given in Figure 9–15.

Nucleic acids are **deoxyribonucleic acids (DNA)** if they contain the sugar **deoxyribose,** or **ribonucleic acids (RNA)** if they contain the sugar **ribose.** DNA is found primarily in the nucleus of the cell, whereas RNA is found mainly in the cytoplasm, outside of the nucleus. Nucleic acids are found in all living cells, with the exception of the red blood cells of mammals.

Three major types of RNA have been identified. They are messenger RNA (mRNA), transfer RNA (tRNA), and ribosomal RNA (rRNA). Each has a characteristic molecular weight and base composition. Messenger RNAs are generally the largest, with molecular weights between 25,000 and 1 million. They contain from 75 to 3000 mononucleotide units. Transfer RNAs have molecular weights in the range of 23,000 to 30,000 and contain 75 to 90 mononucleotide units. Ribosomal RNAs, which have molecular weights between those of mRNAs and tRNAs, make up as much as 80% of the total cell RNA. Besides having different molecular weights, the three types of RNA differ in function. One difference in function is described in the discussion of natural protein synthesis.

One nucleotide is joined to another by an ester-forming reaction:

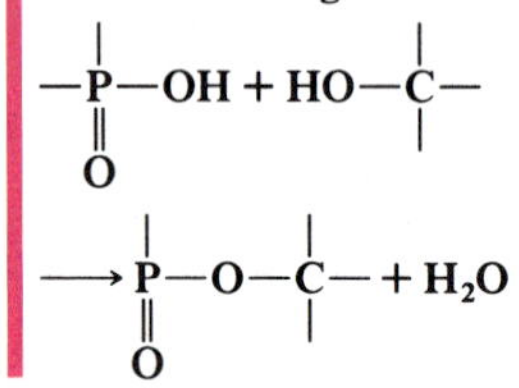

$$-\overset{|}{\underset{\|}{P}}-OH + HO-\overset{|}{\underset{|}{C}}- \longrightarrow -\overset{|}{\underset{\|}{P}}-O-\overset{|}{\underset{|}{C}}- + H_2O$$

(where each P carries a double-bonded O)

The monomers of both DNA and RNA contain a simple sugar, one of the nitrogenous bases, and one or two phosphoric acid units. The structure of a monomer, a **nucleotide,** is shown in Figure 9–16(a). The nucleotides of DNA and RNA have two structural differences: (1) the sugar (Fig. 9–14) and (2) the use of uracil base only in RNA, whereas thymine base is found only in

Figure 9–15 The five nitrogenous bases in DNA and RNA. Thymine occurs only in DNA whereas uracil is found only in RNA.

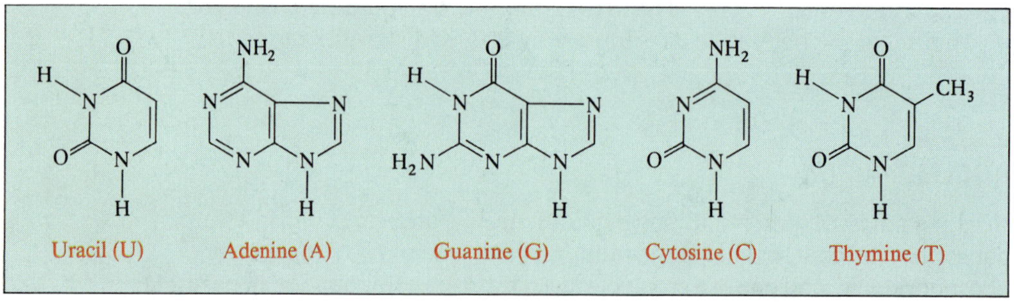

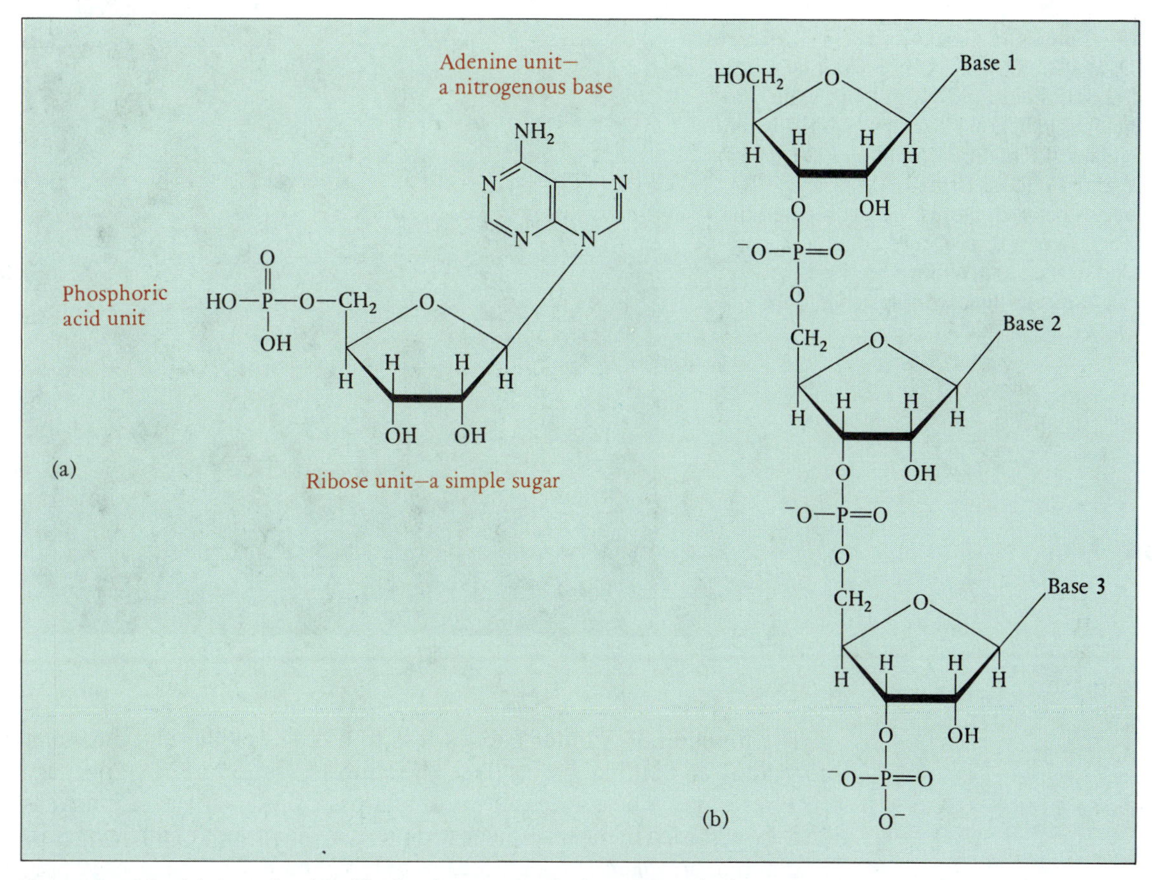

Figure 9-16 (a) A nucleotide. If other bases are substituted for adenine, several nucleotides are possible for each of the two sugars shown in Figure 9-14. (b) Bonding structure of a trinucleotide. Bases 1, 2, and 3 represent any of the nitrogenous bases obtained in the hydrolysis of DNA and RNA (Fig. 9-15). The primary structure of both DNA and RNA is an extension of this structure and produces molecular weights as high as a few million.

DNA. The other bases, adenine, guanine, and cytosine, are found in both DNA and RNA. Three monomers condensed into a trinucleotide can be seen in Figure 9-16(b).

When the phosphate group is absent, one of the sugars bonded to a nitrogenous base is a **nucleoside.**

Polynucleotides with molecular weights up to several million are known. The sequence of nucleotides in the polynucleotide chain is its **primary structure.**

In 1953, James D. Watson and Francis H. C. Crick (Fig. 9-17) proposed a **secondary structure** for DNA that has since gained wide acceptance. Figure 9-18 illustrates a small portion of the structure, in which two polynucleotides are arranged in a double helix stabilized by hydrogen bonding between the base groups opposite each other in the two chains. RNA is generally a single strand of helical polynucleotide.

Figure 9–17 Francis H. C. Crick (b. 1916) *(right)* and James D. Watson (b. 1928) *(left)*, working in the Cavendish Laboratory at Cambridge, built scale models of the double helical structure of DNA based on the X-ray data of Rosalind Franklin (1920–1958) and Maurice H. F. Wilkins (b. 1916). Knowing distances and angles between atoms, they compared the task to the working of a three-dimensional jig-saw puzzle. Watson, Crick, and Wilkins received the Nobel Prize in 1962 for their work relating to the structure of DNA.

The function of polynucleotides is to transcribe cellular and organism information so that like begets like. The almost infinite variety of primary structures of polynucleotides allows an almost infinite variety of information to be recorded in the molecular structures of the strands of nucleic acids. The different arrangements of just a few different bases give the large variety of structures. In a somewhat similar fashion, the multiple arrangements of just a few language symbols convey the many ideas in this book. The coded information in the polynucleotide is believed to control the inherited characteristics of the next generation as well as most of the continuous life processes of the organism.

The inherited traits of an organism are controlled by DNA molecules.

Double-stranded DNA forms the 46 human chromosomes. Within each chromosome are heredity areas called **genes.** Genes are segments of DNA that have as few as 1000 or as many as 100,000 base pairs such as those shown in Figure 9–18. Human DNA (the human **genome**) is estimated to have up to 100,000 genes and about 3 billion base pairs. However, genes are estimated to make up only 3% of DNA, with each gene sandwiched between "junk" or noncoding DNA sequences. There are also short segments that act as switches to signal where the coding sequence begins.

The total sequence of base pairs of a cell is called the *genome.*

Replication of DNA: Heredity

Almost all nuclei in an organism's cells contain the same chromosomal composition. This composition remains constant regardless of whether the cell is starving or has an ample supply of food materials. Each organism begins life as a single cell with this same chromosomal composition; in sexual reproduction half of a chromosome comes from each parent. These well-known biological facts, along with recent discoveries concerning polynucleotide structures, have led scientists to the conclusion that the DNA structure

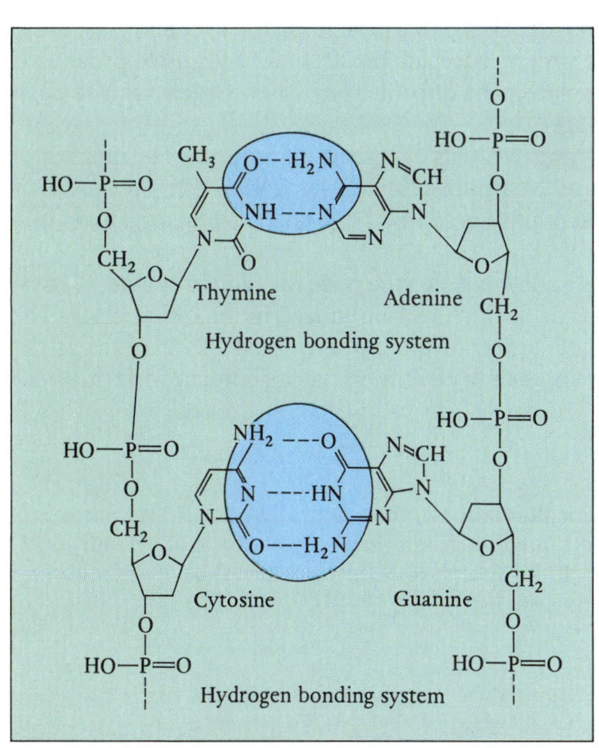

Figure 9–18 Double helix structure proposed by Watson and Crick for DNA. Hydrogen bonds in the thymine-adenine and cytosine-guanine pairs stabilize the double helix. Adenine also pairs with uracil in mRNA, which contains no thymine.

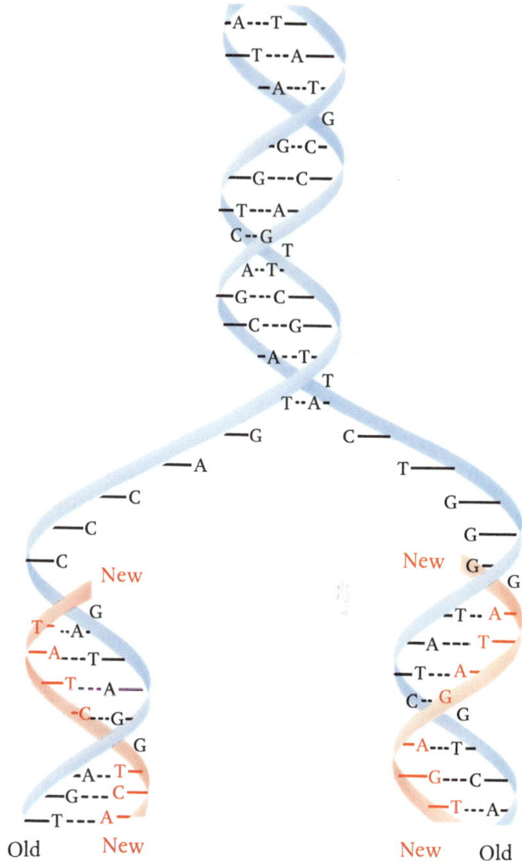

Figure 9–19 Replication of DNA structure. When the double helix of DNA *(blue)* unwinds, each half serves as a template on which to assemble subunits *(pink)* from the cell environment. A = adenine, T = thymine, G = guanine, C = cytosine.

is faithfully copied during normal cell division (**mitosis**—both strands) and that only half is copied in cell division producing reproductive cells (**meiosis**—one strand).

In replication the double helix of the DNA structure unwinds and each half of the structure serves as a template, or pattern, from which the other complementary half can be reproduced from the molecules in the cell environment (Fig. 9–19). Replication of DNA occurs in the nucleus of the cell.

The DNA molecule is capable of causing the synthesis of its duplicate.

Natural Protein Synthesis

The proteins of the body are continually being replaced and resynthesized from the amino acids available to the body.

The proteins in the human body are continually being replaced.

The use of isotopically labeled amino acids has made possible studies of the average lifetimes of amino acids as constituents in proteins—that is, the

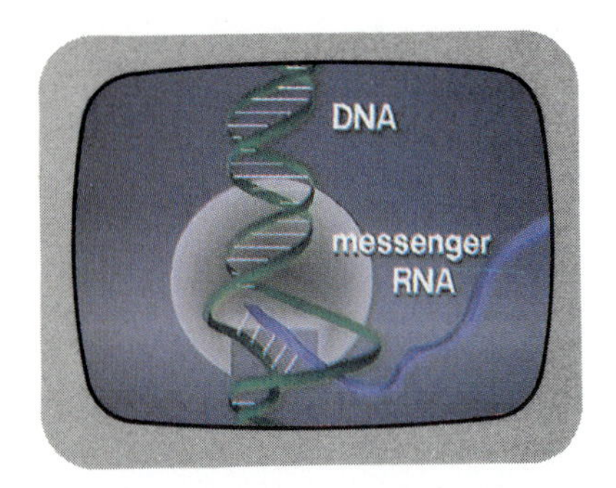

After transcription, RNA leaves DNA and the nucleus of the cell.
(*The World of Chemistry,* Program 24, "Genetic Code.")

time it takes the body to replace a protein in a tissue. For a process that must be extremely complex, replacement is very rapid. Only minutes after radioactive amino acids are injected into animals, radioactive protein can be found. Although all the proteins in the body are continually being replaced, the rates of replacement vary. Half of the proteins in the liver and plasma are replaced in 6 days; the time needed for replacement of muscle proteins is about 180 days, and replacement of protein in other tissues, such as bone collagen, takes even longer.

Recall that each organism has its own kinds of proteins. The number of possible arrangements of 20 amino acid units is 2.43×10^{18}, yet proteins characteristic of a given organism can be synthesized by the organism in a matter of a few minutes.

The DNA in the cell nucleus holds the code for protein synthesis. Messenger RNA, like all forms of RNA, is synthesized in the cell nucleus. The sequence of bases in one strand of the chromosomal DNA serves as the template from which a single strand of a messenger ribonucleotide (mRNA)

Figure 9–20 A schematic illustration of the role of DNA and RNA in protein synthesis. A, C, G, T, and U are nitrogenous bases characteristic of the individual nucleotides. See Figure 9–15 for structures of the bases and Table 9–4 for abbreviations of the amino acids used.

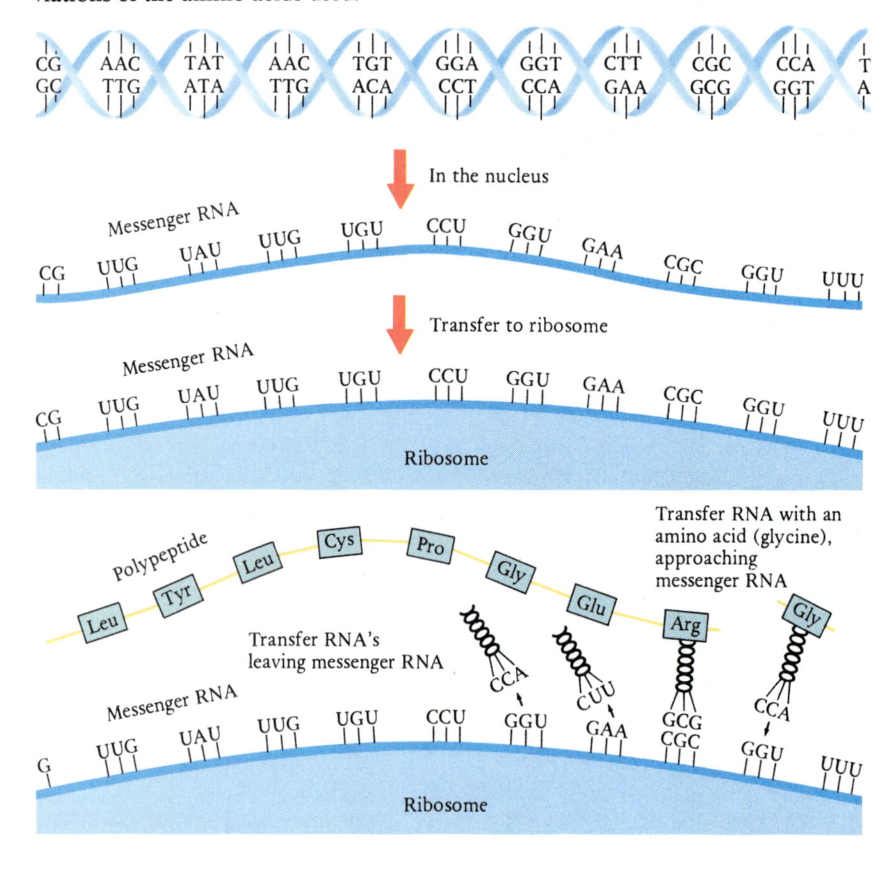

is made (Fig. 9-20). The bases of the mRNA strand complement those of the DNA strand. A pair of complementary bases is structured such that each one fits the other and forms one or more hydrogen bonds. Messenger RNA contains only the four bases adenine (A), guanine (G), cytosine (C), and uracil (U). DNA contains principally the four bases adenine (A), guanine (G), cytosine (C), and thymine (T). The base pairs are as follows:

These base pairs "fit" for hydrogen bonding.

The code on tRNA is the second genetic code; the first code is on DNA.

DNA	mRNA
A	U
G	C
C	G
T	A

This means that, provided the necessary enzymes and energy are present, wherever a DNA has an adenine base (A), the mRNA transcribes a uracil base (U).

Transfer RNAs carry the specific amino acids to the mRNA. Each of the 20 amino acids found in proteins has at least one corresponding tRNA, and some have multiple tRNAs (Table 9-4).

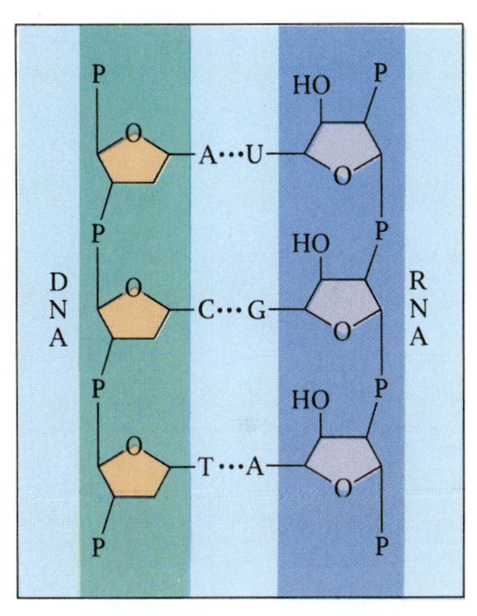

RNA is transcribed from DNA by proper base pairing. (*The World of Chemistry*, Program 24, "Genetic Code.")

TABLE 9-4 Messenger RNA Codes for Amino Acids*

Amino Acid	Shortened Notation Used for Amino Acids in Fig. 9-20	Base Code on mRNA
Alanine	Ala	GCA, GCC, GCG, GCU
Arginine	Arg	AGA, AGG, CGA, CGG, CGC, CGU
Asparagine	Asp-NH_2	AAC, AAU
Aspartic acid	Asp	GAC, GAU
Cysteine	Cys	UGC, UGU
Glutamic acid	Glu	GAA, GAG
Glutamine	Glu-NH_2	CAG, CAA
Glycine	Gly	GGA, GGC, GGG, GGU
Histidine	His	CAC, CAU
Isoleucine	Ileu	AUA, AUC, AUU
Leucine	Leu	CUA, CUC, CUG, CUU, UUA, UUG
Lysine	Lys	AAA, AAG
Methionine	Met	AUG
Phenylalanine	Phe	UUU, UUC
Proline	Pro	CCA, CCC, CCG, CCU
Serine	Ser	AGC, AGU, UCA, UCG, UCC, UCU
Threonine	Thr	ACA, ACG, ACC, ACU
Tryptophan	Try	UGG
Tyrosine	Tyr	UAC, UAU
Valine	Val	GUA, GUG, GUC, GUU

* In groups of three (called codons), bases of mRNA code the order of amino acids in a polypeptide chain. A, C, G, and U represent adenine, cytosine, guanine, and uracil, respectively. Some amino acids have more than one codon, and hence more than one tRNA can bring the amino acid to mRNA. The research on this coding was initiated by Marshall Warren Nirenberg. (Adapted from J. I. Routh, D. P. Eyman, and D. J. Burton: *Essentials of General, Organic, and Biochemistry*, 3rd ed. Philadelphia, Saunders College Publishing, 1977.)

The bases in groups of three are called codons.

BIOGENETIC ENGINEERING

Recombinant means capable of genetic recombination.

The field of biogenetic engineering started after the first successful gene-splicing and gene-cloning experiments produced **recombinant DNA** in the early 1970s. The basic idea is to use the rapidly dividing property of common bacteria, such as *Escherichia coli,* as a microbe factory for producing recombinant DNA molecules that contain the genetic information for the desired product. Bacteria have been produced that can synthesize protein, human growth hormone, and human insulin. The method of producing bacteria for a particular function involves removing a gene from the bacterium, splicing in part of a gene from a human or other organism (the part that produces human insulin, for example), placing the spliced gene back into the bacterium, and letting the bacterium make millions of other insulin-producing

THE WORLD OF CHEMISTRY

Hereditary Diseases and the Genetic Code

A change in the genetic code is passed on from generation to generation. That is why sickle cell anemia is called a hereditary disease. Sickle cell anemia is just one of almost 4000 genetic diseases caused by a change in a single gene. Hemophilia, cystic fibrosis, and Tay-Sachs disease are other familiar examples, although none is as common as sickle cell anemia. As yet, there is no cure for any genetic disease, for we have no way of correcting the defective gene, although some, like sickle cell anemia, can be successfully treated. But the future holds great promise, thanks to a new technique called **recombinant DNA technology.** Scientists have discovered a special group of enzymes, restriction enzymes, which recognize specific base sequences in DNA and cut the strands at that point. This has made it possible to remove a gene from one DNA molecule and insert it into another. The genetic code can now be edited. This technology has already made important contributions to medical science. Genes coding for human insulin have been inserted into the DNA of bacteria, which are then grown in huge fermenting tanks. Here billions of bacteria act like human insulin factories, producing a virtually limitless supply of this important hormone. More and more diabetics are now injecting themselves with this human insulin rather than the animal insulin alternative, which can produce undesirable side effects. There are many who view recombinant DNA technology as the beginning of a new age in science. Laboratories throughout the world are already genetically modifying crop species, introducing genes that code for proteins that improve growth and protect the plant against insect pests. And scientists are hopeful that children with life-threatening genetic diseases may someday be cured by introducing normal copies of the defective gene into their cells. But recombinant DNA technology has even more to offer. It is allowing scientists to unravel chemistry's most remarkable secret, the molecular basis of life.

The World of Chemistry (Program 24) "Genetic Code."

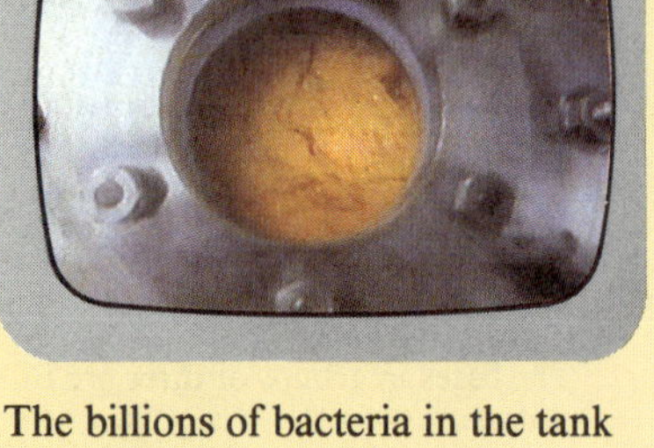

The billions of bacteria in the tank have DNA coded to produce human insulin. The bacteria act like human insulin factories and produce a virtually limitless supply of this hormone. (*The World of Chemistry,* Program 24, "Genetic Code.")

bacteria. This process of splicing and recombining genes is referred to as **recombinant DNA technology** or **biogenetic engineering.** The implications of gene splicing are tremendous—for both good and bad—and demand responsible human decision making for guidance toward the common good.

See Chapter 13 for a discussion of applications of biogenetic engineering to agriculture.

A **mutation** occurs whenever an individual characteristic appears that has not been inherited but is duly passed along as an inherited factor to the next generation. A mutation can readily be accounted for in terms of an alteration in the DNA genetic code; that is, some force alters the nucleotide structure in a reproductive cell. Some sources of energy, such as gamma radiation, are known to produce mutations. This is entirely reasonable because certain kinds of energy can disrupt some bonds, which can re-form in another sequence.

A mutation results when there has been an alteration of the genetic code contained within the DNA molecule.

If scientists can control the genetic code, can they control hereditary diseases such as sickle cell anemia, gout, some forms of diabetes, and mental retardation? If our understanding of detailed DNA structure and the enzymatic activity required to build these structures continues to grow, it is reasonable to believe that some detailed relationships between structure and gross properties will emerge. If this happens, it may be possible to build compounds that, when introduced into living cells, can combat or block inherited characteristics.

Ethics and risks were discussed in Chapter 1.

HUMAN GENOME PROJECT

Until 1986, the experimental determination of the sequence of base pairs in a DNA strand was laboriously slow, taking place at a speed of about 200,000 base pairs per year. With the invention of an automatic DNA sequencer by a team headed by Professor Leroy Hood at California Institute of Technology in Pasadena, the number of base pairs that can be determined per day increased dramatically (over 10,000 base pairs per day). This made complete sequencing of genomes a possibility. In January, 1989, the National Institutes of Health (NIH) announced plans to determine the complete sequence of the human genome, estimated to contain 3 billion base pairs. The Director of the Human Genome Project at NIH is James D. Watson, one of the team that received the Nobel Prize in 1962 for determining the structure of DNA (see Fig. 9–17). The project is estimated to take 15 years at a cost of $3 billion. The first five-year phase of the project involves the study of some model genome systems, and research groups are working on sequencing the base pairs in *E. coli,* estimated to contain 4.5 million base pairs; a yeast genome estimated to contain 12.5 million base pairs; and a nematode genome with 100 million base pairs. It is hoped that the techniques and knowledge gained from the study of these model systems, along with continued improvement in the number of base pairs that can be sequenced per day, will make it possible to sequence the base pairs in human DNA in 15 years. A complete mapping of the human genome will improve the knowledge of the estimated 4000 hereditary diseases and lead to better diagnosis and treatment of these diseases.

SELF-TEST 9C

1. The reactants in the photosynthesis process are ____________ and ____________; ____________ must also be supplied.
2. Most of the energy obtained by the oxidation of food is used immediately to synthesize the molecule ____________.
3. The hydrolysis of ATP produces ____________ and phosphoric acid. ____________ is also released.
4. Digestion is the breakdown of foodstuffs by ____________, with the help of enzymes.
5. The basic code for the synthesis of protein is contained in the ____________ molecule.
6. The sugar in RNA is ____________, whereas the one in DNA is ____________.
7. A nucleotide contains ____________, ____________, and ____________.
8. The secondary structure of DNA is in the shape of a(n) ________________.
9. When DNA replicates itself, each nitrogenous base in the chain is matched to another one via ____________ bonds.
10. The synthesis of a protein is carried out when ____________ molecules bring up the required amino acids to mRNA.
11. Base pairs are formed between A and ____, G and ____, T and ____, U and ____.
12. Human DNA is estimated to have ____________ base pairs.

QUESTIONS

1. Which arrangement has a mirror image that is nonsuperimposable?

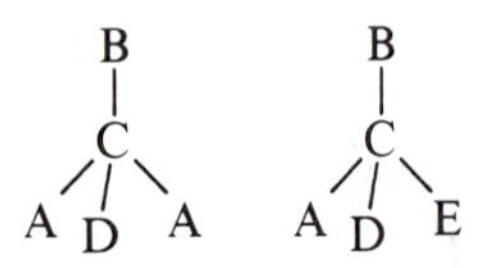

2. Give two examples that illustrate the importance of optical isomers in our body.
3. What is the structural difference between glucose and sucrose? Between sucrose and maltose?
4. Explain the basic difference between amylose and cellulose.
5. Why can't we digest cellulose?
6. What is the purpose of ATP?
7. Show the structure of the product that would be obtained if two alanine molecules (Table 9–3) were to react to form a dipeptide.
8. Sugars are soluble in water, but fats and oils are not soluble in water. Explain.
9. What is the structural difference between saturated and unsaturated fats?
10. Why are fats and oils known as triglycerides?
11. What is the difference between HDLs and LDLs?
12. Our body needs cholesterol, yet we are told to have a low cholesterol diet. Explain.
13. In a protein, what type of bond holds the helical structure in place?
14. Explain the difference between the primary structure and the secondary structure of a protein.
15. What three molecular units are found in nucleotides?
16. What are the basic differences between DNA and RNA structures?
17. a. Which of the following biochemicals are polymers: starch, cellulose, glucose, fats, proteins, DNA, and RNA?

    b. What are the monomer units for those that are polymers?
18. What stabilizing forces hold the double helix together in the secondary structure of DNA proposed by Watson and Crick?
19. What is meant by a base pair in protein synthesis? What type of bonds holds base pairs together?
20. What is recombinant DNA?

21. The molecular structures of enzymes are most closely related to which of the following structures: proteins, fats, carbohydrates, or polynucleotides?
22. A mutation can be explained in terms of a change in which chemical in the cell?
23. Does a strand of DNA actually duplicate itself base for base in the formation of a strand of mRNA? Explain.
24. Give an example of the application of biogenetic engineering techniques.
25. What is the human genome?

Jacques Louis David's *The Death of Socrates* depicts Socrates' death by hemlock poisoning. (The Metropolitan Museum of Art, Wolfe Fund, 1931. Catharine Lorillard Wolfe Collection)

CHAPTER 10

Toxic Substances

Toxic substances surround us. They affect our lives by causing us to be on guard against them. In addition, our bodies are always fighting the effects of toxic substances that are found in the air we breathe, the water we drink, and the food we eat.

1. What is meant by the term "dose"?
2. How do corrosive poisons work?
3. Why is carbon monoxide toxic?
4. How does cyanide act as a poison?
5. Why are heavy metals like arsenic, mercury, and lead toxic?
6. How do nerve poisons like nicotine and the war gases act?
7. How do chemicals cause birth defects and mutations?
8. How do some chemicals cause cancer?
9. How are we exposed to cancer-causing chemicals in our diets?

Toxic substances upset the incredibly complex system of chemical reactions occurring in the human body. Sometimes toxic substances cause mere discomfort; sometimes they cause illness, disability, or even death. Toxic symptoms can be caused by very small amounts of extremely toxic materials (an example is sodium cyanide) or larger amounts of a less toxic substance. The term ***toxic substances*** usually is limited to materials that are dangerous in small amounts. However, as most of us know, ill effects can be caused by excessive intake of substances normally considered harmless (eating too much candy, for example). Fortunately, in most cases the human body is capable of recognizing "foreign" chemicals and ridding itself of them. In this chapter, we focus on the chemistry of toxic substances.

A large enough dose of any compound can result in poisoning.

DOSE

"Dosis sola facit venenum"—the dose makes the poison.

Lethal doses of toxic substances are customarily expressed in milligrams (mg) of substance per kilogram (kg) of body weight of the subject. For example, the cyanide ion (CN^-) is generally fatal to human beings in a dose of 1 mg of CN^- per kg of body weight. For a 200-lb (90.7 kg) person, about 0.1 g of cyanide is a lethal dose. Examples of somewhat less toxic substances and the range of lethal doses for human beings follow:

Morphine	1–50 mg per kg
Aspirin	50–500 mg per kg
Methyl alcohol	500–5000 mg per kg
Ethyl alcohol	5000–15,000 mg per kg

A quantitative measure of toxicity is obtained by introducing into laboratory animals (such as rats) various dosages of substances to be tested. The dosage found to be lethal in 50% of a large number of the animals under controlled conditions is called the LD_{50} (lethal dosage—50%) and is reported in milligrams of poison per kilogram of body weight. Thus, if a statistical analysis of data on a large population of rats showed that a dosage of 1 mg/kg was lethal to 50% of the population tested, the LD_{50} for this poison would be 1 mg/kg. Obviously, metabolic variations and other differences between species produce different LD_{50} values for a given poison in different kinds of animals. For some toxic chemicals, the effect between animal species can be very great. For example, the toxicity of dioxin, a multi-ring chlorinated compound produced when chlorinated compounds are incinerated, and also found as an impurity in some defoliants, such as Agent Orange, varies over an extremely large range (Table 10–1). The extremely low LD_{50} for the guinea pig is unusual. Some have classed dioxin as one of the most toxic compounds known. Dioxin certainly is toxic to animals, but it does not rank as a potent human toxin.

TABLE 10-1 LD_{50} Values for Dioxin

Species*	LD_{50} (mg/kg)
Guinea pig	0.0006
Rat	0.04
Monkey	0.07
Rabbit	0.115
Dog	0.150
Mouse	0.200
Hamster	3.5
Bullfrog	>1.0

* No known human deaths have been reported for this compound.

TABLE 10-2 Approximate Comparison of LD_{50} Values with Lethal Doses for Human Adults

Oral LD_{50} for Any Animal (mg/kg)	Probable Lethal Oral Dose for Human Adult
<5	A few drops
5-50	"A pinch" to 1 tsp
50-500	1 tsp-2 T
500-5000	1 oz-1 pt (1 lb)
5000-15,000	1pt-1 qt (2 lb)

Owing to species differences in LD_{50} values, toxicity cannot be extrapolated to human beings with any assurance, but it is generally safe to assume that a chemical with a low LD_{50} value for several species is also quite toxic to human beings (Table 10-2).

Toxic substances can be classified according to the way in which they disrupt the chemistry of the body. Some modes of action of toxic substances can be described as **corrosive, metabolic, neurotoxic, mutagenic, teratogenic,** and **carcinogenic,** and these will serve as the bases of our discussion.

Metabolism (from the Greek, *metaballein,* meaning "to change or alter") is the sum of all the physical and chemical changes by which living organisms are produced and maintained.

CORROSIVE POISONS

Toxic substances that actually destroy tissues are corrosive poisons. Examples include strong acids and alkalies and many oxidants, such as those found in laundry products, which can destroy tissues. Sulfuric acid (found in auto batteries) and hydrochloric acid (also called muriatic acid, used for cleaning purposes) are very dangerous corrosive poisons. So is sodium hydroxide, used in cleaning clogged drains. Death has resulted from the swallowing of 1 ounce of concentrated (98%) sulfuric acid, and much smaller amounts can cause extensive damage and severe pain.

Concentrated mineral acids such as sulfuric acid act by first dehydrating cellular structures. The cell dies because its protein structures are destroyed by the acid-catalyzed hydrolysis of the peptide bonds.

Strong acids and bases destroy cell protoplasm.

$$R-\overset{O}{\overset{\|}{C}}-\overset{H}{\overset{|}{N}}-R' + H_2O \xrightarrow[\text{From acid}]{H^+} R-\overset{O}{\overset{\|}{C}}-OH + H-\overset{H}{\overset{|}{N}}-R'$$

Peptide link (in protein) — Carboxyl end of smaller peptide or amino acid — Amine end of smaller peptide or amino acid

In the early stages of this process there is a large proportion of larger fragments present. Subsequently, as more bonds are broken, smaller and smaller fragments result, leading to the ultimate disintegration of the tissue.

Chemical "warfare gases," such as phosgene, were outlawed by an international conference in 1925.

Some poisons act by undergoing chemical reaction in the body to produce corrosive poisons. Phosgene, the deadly gas used during World War I, is an example. When inhaled, it is hydrolyzed (broken down by water) in the lungs to hydrochloric acid, which causes pulmonary edema (a collection of fluid in the lungs) owing to the dehydrating effect of the strong acid on tissues. The victim "drowns" because oxygen cannot be absorbed effectively by the flooded and damaged tissues.

$$\underset{\text{Phosgene}}{\mathrm{Cl{-}C({=}O){-}Cl}} + H_2O \longrightarrow \underset{\text{Hydrochloric acid}}{2\,HCl} + \underset{\text{Carbon dioxide}}{CO_2}$$

Sodium hydroxide, NaOH (caustic soda—a component of drain cleaners), is a very strongly alkaline, or basic, substance that can be just as corrosive to tissue as strong acids. The hydroxide ion also catalyzes the splitting of peptide linkages:

$$\mathrm{R{-}C({=}O){-}N(H){-}R'} + H_2O \xrightarrow[\text{Base}]{OH^-} \mathrm{R{-}C({=}O){-}OH} + \mathrm{H{-}N(H){-}R'}$$

Both acids and bases, as well as other types of corrosive poisons, continue their action until they are consumed in chemical reactions.

TABLE 10-3 Some Corrosive Poisons

Substance	Formula	Toxic Action	Possible Contact
Hydrochloric acid	HCl	Acid hydrolysis	Tile and concrete floor cleaner; concentrated acid used to adjust acidity of swimming pools
Sulfuric acid	H_2SO_4	Acid hydrolysis, dehydrates tissue—oxidizes tissue	Auto batteries
Phosgene	$ClCOCl$	Acid hydrolysis	Combustion of chlorine-containing plastics (PVC or Saran)
Sodium hydroxide	$NaOH$	Base hydrolysis	Caustic soda, drain cleaners
Trisodium phosphate	Na_3PO_4	Base hydrolysis	Detergents, household cleaners
Sodium perborate	$NaBO_3 \cdot 4\,H_2O$	Base hydrolysis—oxidizing agent	Laundry detergents, denture cleaners
Ozone	O_3	Oxidizing agent	Air, electric motors
Nitrogen dioxide	NO_2	Oxidizing agent	Polluted air, automobile exhaust
Iodine	I_2	Oxidizing agent	Antiseptic
Hypochlorite ion	OCl^-	Oxidizing agent	Bleach
Peroxide ion	O_2^{2-}	Oxidizing agent	Bleach, antiseptic
Oxalic acid	$H_2C_2O_4$	Reducing agent, precipitates Ca^{2+}	Bleach, ink eradicator, leather tanning, rhubarb, spinach, tea
Sulfite ion	SO_3^{2-}	Reducing agent	Bleach
Chloramine	NH_2Cl	Oxidizing agent	Produced when household ammonia and chlorinated bleach are mixed
Nitrosyl chloride	$NOCl$	Oxidizing agent	Mixing household ammonia and bleach

Some corrosive poisons destroy tissue by oxidizing it. This is characteristic of substances such as ozone, nitrogen dioxide, and possibly iodine, which destroy enzymes by oxidizing their functional groups. Specific groups, such as the —SH and —S—S— groups in the enzyme, are believed to be converted by oxidation to nonfunctioning groups; alternatively, the oxidizing agents may break chemical bonds in the enzyme, leading to its inactivation.

A summary of some common corrosive poisons is presented in Table 10-3.

METABOLIC POISONS

Metabolic poisons are more subtle than the tissue-destroying corrosive poisons. In fact, many of them do their work without actually indicating their presence until it is too late. Metabolic poisons can cause illness or death by interfering with a vital biochemical mechanism to such an extent that it ceases to function or is prevented from functioning efficiently.

Carbon Monoxide

The interference of carbon monoxide with extracellular oxygen transport is one of the best understood processes of metabolic poisoning. As early as 1895, it was noted that carbon monoxide deprives body cells of oxygen (asphyxiation), but it was much later before it was known that carbon monoxide, like oxygen, combines with hemoglobin:

$$O_2 + \text{hemoglobin} \longrightarrow \text{oxyhemoglobin}$$

$$CO + \text{hemoglobin} \longrightarrow \text{carboxyhemoglobin}$$

Laboratory tests show that carbon monoxide reacts with hemoglobin to give a compound (carboxyhemoglobin) that is 140 times more stable than the compound of hemoglobin and oxygen (oxyhemoglobin) (Fig. 10-1). Because hemoglobin is so effectively tied up by carbon monoxide, it cannot perform its vital function of transporting oxygen.

An organic material that undergoes incomplete combustion always liberates carbon monoxide. Sources include auto exhausts, smoldering leaves, lighted cigars or cigarettes, and charcoal burners. In the United States alone, combustion sources of all types dump about 200 million tons of carbon monoxide per year into the atmosphere.

Although the best estimates of the maximum global background level of carbon monoxide are of the order of 0.1 ppm, the background concentration in cities is higher. In heavy traffic, sustained levels of 100 ppm or more are common; for offstreet sites an average of about 7 ppm is typical for large cities. A concentration of 30 ppm for 8 hr is sufficient to cause headache and nausea. Breathing an atmosphere that is 0.1% (1000 ppm) carbon monoxide for 4 hr converts approximately 60% of the hemoglobin of an average adult to carboxyhemoglobin, and death is likely to result (Fig. 10-2).

ppm—parts per million—a measure expressing concentration. 50 ppm CO means 50 mL CO for every million mL of air.

Since both the carbon monoxide and oxygen reactions with hemoglobin involve easily reversed reactions, the concentrations, as well as relative

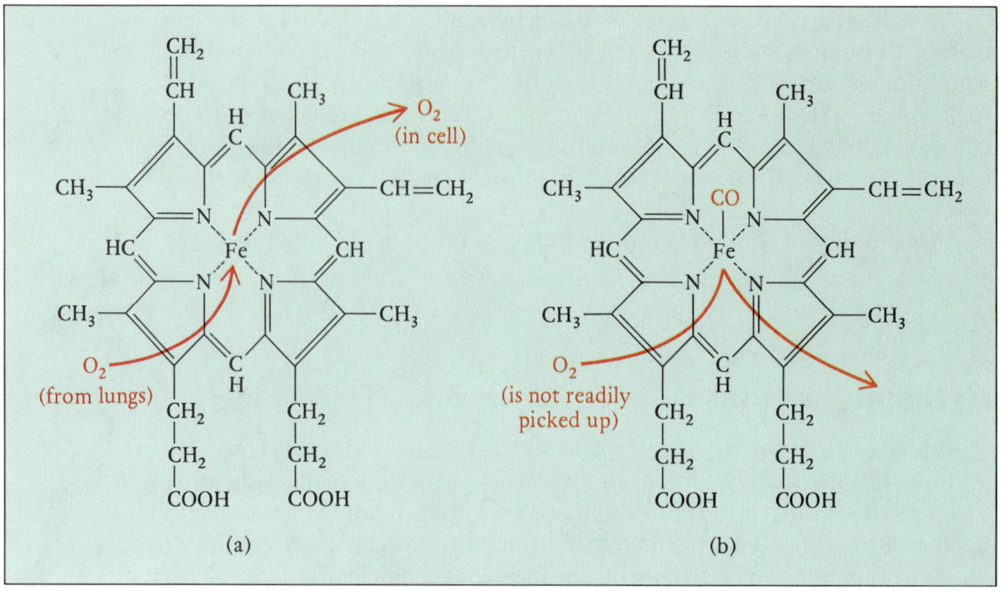

Figure 10–1 Structure of the heme portion of hemoglobin. (a) Normal acceptance and release of oxygen. (b) Oxygen blocked by carbon monoxide.

strengths of bonds, affect the direction of the reaction. In air that contains 0.1% CO, oxygen molecules outnumber CO molecules 200 to 1. The larger concentration of oxygen helps to counteract the greater combining power of CO with hemoglobin by shifting the reaction equilibrium (see Chapter 2) to the right. Consequently, if a carbon monoxide victim is exposed to fresh air

To convert ppm to percent, divide by 10,000.

Air is 21% O_2 by volume; in 1 million "air molecules" there would be 210,000 O_2 molecules.

Figure 10–2 A healthy adult can tolerate 100 ppm carbon monoxide in air without suffering ill effect. A 1-hr exposure to 1000 ppm causes a mild headache and a reddish coloration of the skin. A 1-hr exposure to 1300 ppm turns the skin cherry red and a throbbing headache develops. A 1-hr exposure to concentrations greater than 2000 ppm will likely cause death.

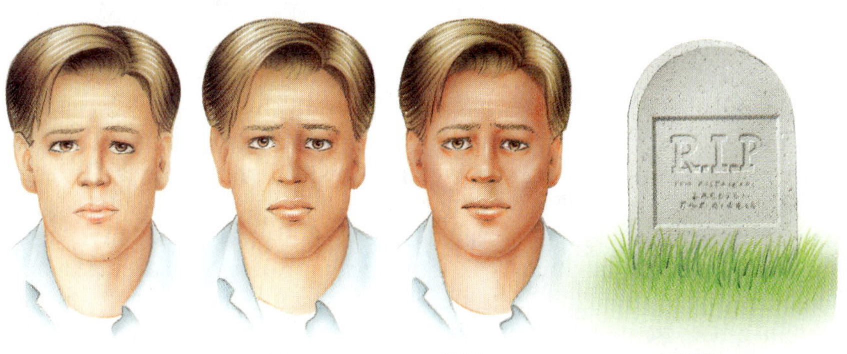

or, still better, pure oxygen (provided that he or she is still breathing), the carboxyhemoglobin (HbCO) is gradually decomposed, owing to the greater concentration of oxygen:

$$HbCO + O_2 \leftrightarrows HbO_2 + CO$$

Equilibrium shifted to right because of greater concentration of oxygen

See Chapters 2 and 4 for more discussions on these unequal length arrows.

Although carbon monoxide is not a cumulative poison, permanent damage can occur if certain vital cells (e.g., brain cells) are deprived of oxygen for more than a few minutes.

Individuals differ in their tolerance to carbon monoxide, but generally those with anemia or an otherwise low reserve of hemoglobin (e.g., children) are more susceptible. No one is helped by carbon monoxide, and smokers suffer chronically from its effects. There is also a strong relationship between low infant birth weight and mothers who smoke.

Cigarette smoke contains carbon monoxide.

Finally, it should be noted that carbon monoxide is a subtle poison because it is colorless, odorless, and tasteless.

Cyanide

The cyanide ion (CN^-) is the toxic agent in cyanide salts such as sodium cyanide used in electroplating. Because the cyanide is a relatively strong base, it reacts easily with many acids (weak and strong) to form volatile hydrogen cyanide (HCN):

Natural sources of cyanide include the seeds of the cherry, plum, peach, apple, and apricot.

$$\underset{\text{Acetic acid}}{CH_3COOH} + \underset{\text{Sodium cyanide}}{Na^+CN^-} \rightleftarrows \underset{\text{Hydrogen cyanide}}{HCN} + \underset{\text{Sodium acetate}}{Na^+CH_3COO^-}$$

Because HCN boils at a relatively low temperature (26°C), it is a gas at temperatures slightly above room temperature. It is often used as a fumigant in storage bins and holds of ships because it is toxic to most forms of life and, in gaseous form, can penetrate into tiny openings, even into insect eggs.

The cyanide ion is one of the most rapidly working poisons. Lethal doses taken orally act in minutes. Cyanide poisons by asphyxiation, as does carbon monoxide, but the mechanism of cyanide poisoning is different (Fig. 10–3). Instead of preventing the cells from getting oxygen, cyanide interferes with oxidative enzymes, such as cytochrome oxidase. Oxidases are enzymes con-

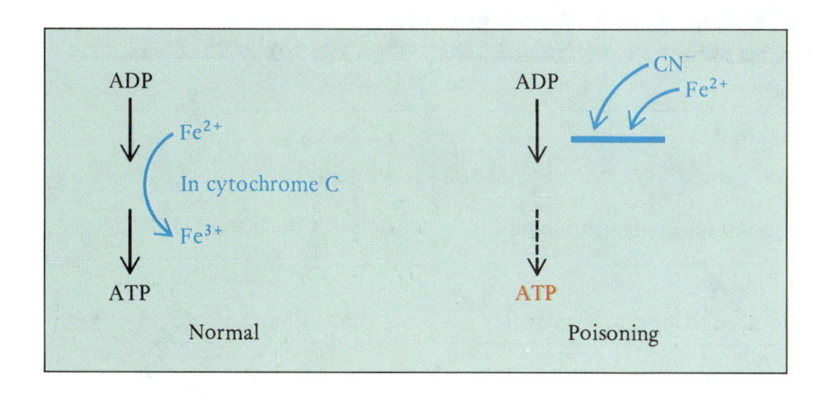

Figure 10–3 The mechanism of cyanide (CN^-) poisoning. Cyanide binds tightly to the enzyme cytochrome C, an iron compound, thus blocking the vital ADP–ATP reaction in cells.

taining a metal, usually iron or copper. They catalyze the oxidation of substances such as glucose:

A metabolite results from natural chemical change in organisms, i.e., glucose to starch.

$$\text{Metabolite } (H)_2 + \tfrac{1}{2}\,O_2 \xrightarrow{\text{Oxidase}} \text{Oxidized metabolite} + H_2O + \text{energy}$$

The iron atom in cytochrome oxidase is oxidized from Fe^{2+} to Fe^{3+} to provide electrons for the reduction of O_2. The iron regains electrons from other steps in the process. The cyanide ion forms stable cyanide complexes with the metal ion of the oxidase and renders the enzyme incapable of reducing oxygen or oxidizing the metabolite.

$$\text{Cytochrome oxidase (Fe)} + CN^- \longrightarrow \text{Cytochrome oxidase } \underbrace{\text{(Fe)} \cdots CN^-}_{\text{Complex}}$$

In essence, the electrons of the iron ion are "frozen"—they cannot participate in the oxidation–reduction processes. Plenty of oxygen gets to the cells, but the mechanism by which the oxygen is used in the support of life is stopped. Hence, the cell dies, and if this occurs fast enough in the vital centers, the victim dies.

HEAVY METALS

Most heavy metals are cumulative poisons.

Heavy metals are perhaps the most common of all the metabolic poisons. These include such common elements as lead and mercury, as well as many less common ones such as cadmium, chromium, and thallium. In this group we should also include the infamous poison, arsenic, which is really not a metal but is metal-like in many of its properties, including its toxic action.

Arsenic

Arsenic, a classic homicidal poison, occurs naturally in small amounts in many foods. Shrimp, for example, contain about 19 ppm arsenic; corn may contain 0.4 ppm arsenic. Some agricultural insecticides contain arsenic (Table 10–4), and so some arsenic is observed in very small amounts on

TABLE 10–4 Some Arsenic-Containing Insecticides

Name	Formula
Lead arsenate	$Pb_3(AsO_4)_2$
Monosodium methanearsenate	$CH_3{-}As(=O)(OH){-}O^-Na^+$
Paris green (copper acetoarsenite)	$3\,CuO \cdot 3\,As_2O_3 \cdot Cu(C_2H_3O_2)_2$

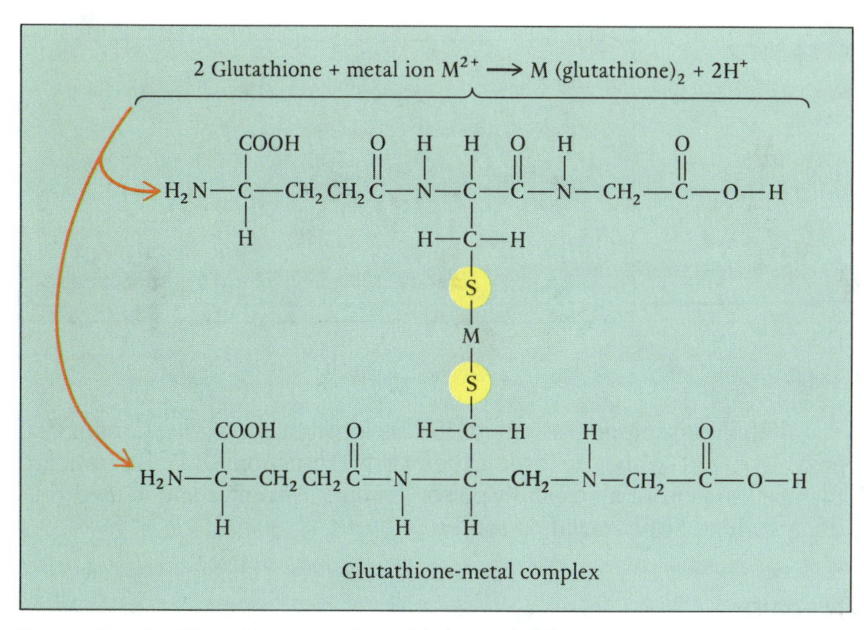

Figure 10–4 Glutathione reaction with a metal (M).

some fruits and vegetables. The Federal Food and Drug Administration (FDA) has set a limit of 0.15 mg of arsenic per pound of food, and this amount apparently causes no harm. Several drugs, such as arsphenamine, which has found some use in treating syphilis, contain covalently bonded arsenic. In its ionic forms, arsenic is much more toxic.

See Chapter 16 for a discussion of arsphenamine.

Arsenic and heavy metals owe their toxicity primarily to their ability to react with and inhibit sulfhydryl (—SH) enzyme systems, such as those involved in the production of cellular energy. For example, glutathione (a tripeptide of glutamic acid, cysteine, and glycine) occurs in most tissues; its behavior with metals illustrates the interaction of a metal with sulfhydryl groups. The metal replaces the hydrogen on two sulfhydryl groups on adjacent molecules (Fig. 10–4), and the strong bond that results effectively eliminates the two glutathione molecules from further reaction. Glutathione is involved in maintaining healthy red blood cells.

The problem of developing a compound to counteract *Lewisite,* an arsenic-containing poison gas used in World War I, led to an understanding of how arsenic acts as a poison and subsequently to the development of an antidote. Once it was understood that Lewisite poisoned people by the reaction of arsenic with protein sulfhydryl groups, British scientists set out to find a suitable compound that contained highly reactive sulfhydryl groups that could compete with sulfhydryl groups in the natural substrate for the arsenic, and thus render the poison ineffective. Out of this research came a compound now known as British anti-Lewisite (BAL).

$CH_2-CH-CH_2$ (with OH, SH, SH attached respectively)

BAL
(British anti-Lewisite)

The BAL, which bonds to the metal at several sites, is called a **chelating agent** (Greek, *chela,* meaning "claw"), a term applied to a reacting agent that envelops a species such as a metal ion. BAL is one of many compounds that can act as chelating agents for metals (Fig. 10–5).

A chelating agent encases an atom or ion like a crab or an octopus surrounds a bit of food.

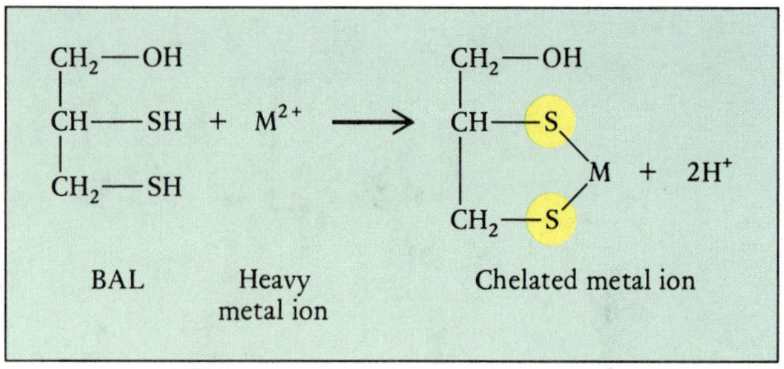

Figure 10–5 BAL chelation of arsenic or a heavy metal ion such as lead.

$$\begin{array}{l} CH_2{-}OH \\ | \\ CH{-}SH \\ | \\ CH_2{-}SH \end{array} + M^{2+} \longrightarrow \begin{array}{l} CH_2{-}OH \\ | \\ CH{-}S \\ | \qquad\quad \backslash M \\ CH_2{-}S \end{array} + 2H^+$$

BAL — Heavy metal ion — Chelated metal ion

With the arsenic or heavy metal ion tied up, the sulfhydryl groups in vital enzymes are freed and can resume their normal functions. BAL is a standard therapeutic item in a hospital's poison emergency center and is used routinely to treat heavy-metal poisoning.

Mercury

A vivid description of the psychic changes produced by mercury poisoning can be found in the Mad Hatter, a character in Lewis Carroll's *Alice in Wonderland.* The fur felt industry once used mercury (II) nitrate, $Hg(NO_3)_2$, to stiffen the felt. Chronic mercury poisoning accounted for the Mad Hatter's odd behavior; it also gave the workers in hat factories symptoms known as "hatter's shakes."

Mercury deserves some special attention because it has a rather peculiar fascination for some people, especially children, who love to touch it. It is poisonous and, to make matters worse, mercury and its salts accumulate in the body. This means the body has no quick means of ridding itself of this element, and there tends to be a buildup of the toxic effects leading to *chronic* poisoning.

Although mercury is rather unreactive compared with other metals, it is quite volatile and easily absorbed through the skin. In the body, the metal atoms are oxidized to Hg_2^{2+} [mercury (I) ion] and Hg^{2+} [mercury (II) ion]. Compounds of both Hg_2^{2+} and Hg^{2+} are known to be toxic.

Amalgam: Any mixture or alloy of metals of which mercury is a constitutent.

Today mercury poisoning is a potential hazard to those working with or near this metal or its salts, such as dentists (who use it in making amalgams for fillings), various medical and scientific laboratory personnel (who routinely use mercury compounds or mercury pressure gauges), and some agricultural workers (who employ mercury salts as fungicides).

Incinerated trash containing old mercury batteries releases the mercury to be washed into streams and lakes.

Mercury can also be a hazard when it is present in food. It is generally believed that mercury enters the food chain through small organisms that feed at the bottom of bodies of water that contain mercury from industrial waste or mercury minerals in the sediment. These in turn are food for bottom-feeding fish. Game fish in turn eat these fish and accumulate the largest concentration of mercury, the accumulation of poison building up as the food chain progresses.

Lead

$1\ \mu g$ (microgram) $= 10^{-6}$ g

1 mg (milligram) $= 10^{-3}$ g $= 1000\ \mu g$

Lead is another widely encountered heavy-metal poison. The body's method of handling lead provides an interesting example of a "metal equilibrium" (Fig. 10–6). Lead often occurs in foods (100–300 $\mu g/kg$), beverages (20–30 $\mu g/L$), public water supplies (100 $\mu g/L$, from lead-sealed pipes), and even air (up to 2.5 $\mu g/m^3$ from lead compounds). With this many sources and contacts per day, it is obvious that the body must be able to rid itself of this poison; otherwise everyone would have died long ago of lead poisoning! The average person can excrete about 2 mg of lead a day through the kidneys and intestinal tract; the daily intake is normally less than this. However, if intake exceeds this amount, accumulation and storage result. In the body lead not only resides in soft tissues but also is deposited in bone. In the bones lead acts on the bone marrow, in tissues it behaves like other heavy-metal poisons, such as mercury and arsenic. Lead, like mercury and arsenic, can also affect the central nervous system.

At the height of the Roman empire, lead production worldwide was about 80,000 tons per year. Today it is about 3 million tons per year.

Unless they are very insoluble, lead salts are always toxic, and their toxicity is directly related to the salt's solubility. But even metallic lead can be absorbed through the skin; cases of lead poisoning have resulted from repeated handling of lead foil, bullets, and other lead objects.

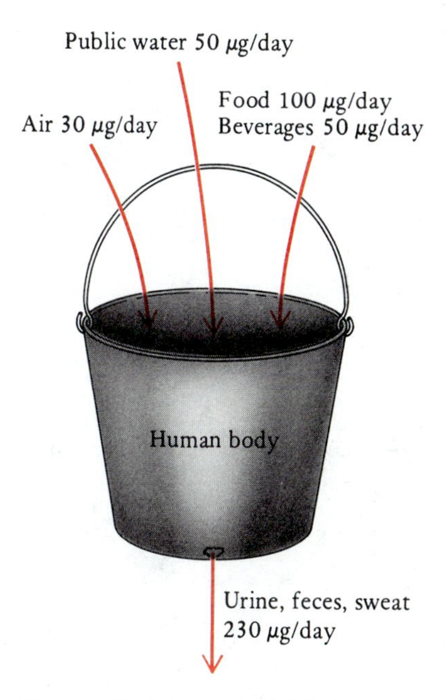

Figure 10–6 Lead equilibrium in humans. Here, the body is likened to a bucket with a hole in it. Figures chosen for intake are probable upper limits.

Health experts estimate that up to 225,000 children become ill from lead poisoning each year (Fig. 10–7), with many experiencing mental retardation or other neurological problems. The reason for this is twofold. Lead-based paints still cover the walls of many older dwellings. Coupled with this is the fact that many children in poverty-stricken areas are ill fed and anemic. These children develop a peculiar appetite trait called **pica,** and among the items that satisfy their cravings are pieces of flaking paint, which may contain lead. Lead salts also have a sweet taste, which may contribute to this consumption of lead-based paint.

If a child consumes as little as 200 mg (200,000 μg) of older, lead-based paint, he or she may ingest as much as 2600 μg of lead, of which 550 μg will be absorbed. Since 1977, lead in paints has been limited to 600 ppm (dry weight). For every gallon of paint, about 8 lb, there should be no more than 0.0048 lb, or 2,100,000 μg, of lead. If this paint is spread uniformly over 400 ft^2 or about 371,600 cm^2, a paint chip 1 cm square will contain less than 6 μg of lead. Some older paints contained up to 50,000 ppm lead (or about

Figure 10–7 The office of the Lead Poisoning Prevention Program in Chatham County (Savannah), Georgia.

Gamma ray fluorescent device for determining lead in paint. (*The World of Chemistry,* Program 6, "The Atom.")

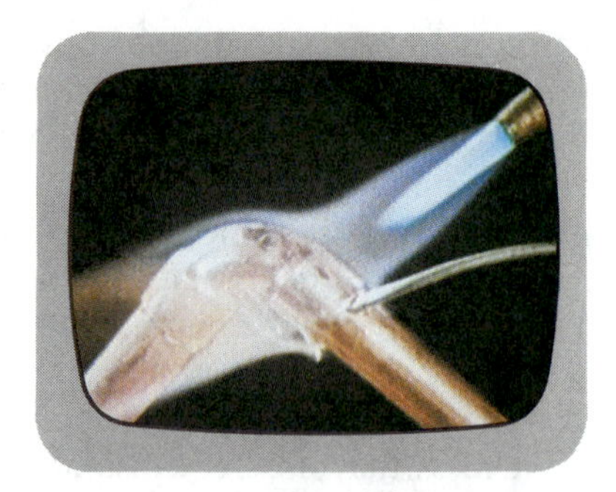

Soldering copper water pipes together with lead-based solder has been the cause of lead in drinking water. (*The World of Chemistry,* Program 19, "Metals.")

In 1991 the EPA lowered the standard for lead in drinking water. The new standard is an average level of about 5 ppb (parts per billion).

175 μg of lead/cm² of paint). Lead compounds have been used in paints as pigments, such as yellow lead chromate, and as drying agents, such as lead naphthenate.

Children retain a larger fraction of absorbed lead than do adults, and children do not immediately tie up absorbed lead in their bones as adults do. This inability to quickly absorb lead into their bones means the lead stays in a child's blood longer, where it can exert its toxic effects on various organs. Table 10–5 shows the effects of lead in children's blood.

Studies in the mid-1980s by the U.S. Department of Health showed that about 1.5 million children have blood levels above 15 μg/dL. About 900,000 children have blood levels above 20 μg/dL.

TABLE 10–5 Effects of Lead in Children's Blood

Pb Levels (μg/dL)*	Effect
~5	Elevated blood pressure
~10	Lowered intelligence
15–25	Decreased heme synthesis, decreased vitamin D and calcium metabolism
25–40	Impaired central nervous system functions, delayed cognitive development, reduced IQ scores, impaired hearing, reduced hemoglobin formation
40–80	Peripheral neuropathy, anemia
80+	Coma, convulsions, irreversible mental retardation, possible death

* μg/dL = micrograms per deciliter of blood. A deciliter is 1/10 of a liter.

Source: Agency for Toxic Substance and Disease Registry, U.S. Department of Health and Human Services, Atlanta, 1988.

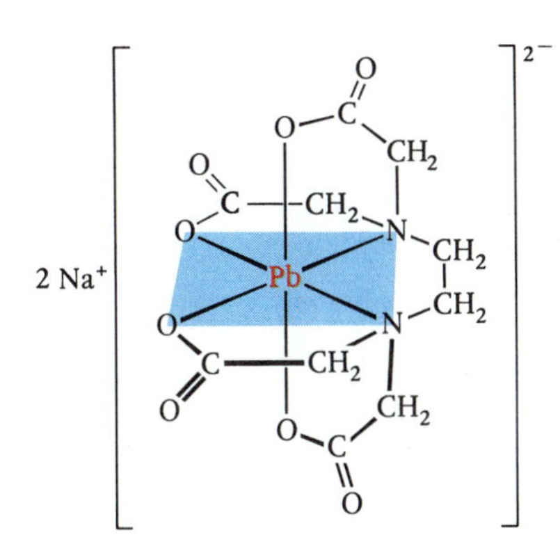

Figure 10–8 The structure of the chelate formed when the anion of EDTA envelopes a lead (II) ion.

Toxicologists have discovered an effective chelating agent for removing lead from the human body—ethylenediaminetetraacetic acid, also called EDTA (Fig. 10–8).

$$\begin{array}{l} HOOCCH_2 \\ \quad\quad \searrow \\ \quad\quad\quad N-CH_2-CH_2-N \\ \quad\quad \nearrow \\ HOOCCH_2 \end{array} \begin{array}{l} CH_2COOH \\ \swarrow \\ \\ \nwarrow \\ CH_2COOH \end{array}$$

EDTA
(ethylenediaminetetraacetic acid)

The calcium disodium salt of EDTA is used in the treatment of lead poisoning because EDTA by itself would remove too much of the blood serum's calcium. In solution, EDTA has a greater tendency to complex with lead (Pb^{2+}) than with calcium (Ca^{2+}). As a result, the calcium is released and the lead is tied up in the complex:

$$[CaEDTA]^{2-} + Pb^{2+} \longrightarrow [PbEDTA]^{2-} + Ca^{2+}$$

The lead chelate is then excreted in the urine.

The definition of lead poisoning by the U.S. Department of Health is 25 μg/dL (micrograms per deciliter of blood). This number has been downwardly revised in the past and may be again in the future as more is learned about lead toxicity.

SELF-TEST 10A

1. Corrosive poisons such as sulfuric acid destroy tissue by ______________ followed by ______________ of proteins.
2. Corrosive poisons, such as ozone, nitrogen dioxide, and iodine, destroy tissue by ______________ it.
3. Carbon monoxide poisons by forming a strong bond with iron in ______________ and thus preventing the transport of ______________ from the lungs to the cells throughout the body.
4. CO is a cumulative poison: True ( ) False ( )
5. The cyanide ion has the formula ______________. It poisons by complexing with iron in the enzyme ______________, thus preventing the use of ______________ in the oxidative processes in the cells.
6. Arsenic (a) is a corrosive poison, (b) attacks hemoglobin and robs cells of oxygen, (c) attacks enzymes by bonding with sulfhydryl groups.

7. BAL is an antidote for ______________. BAL is effective because its sulfhydryl (—SH) groups ______________ arsenic and heavy metals and render them ineffective toward enzymes.
8. Mercury is a cumulative poison: True ( ) False ( )
9. Name three ways children can be exposed to lead.

______________________________, ______________________________, and ______________________________.

NEUROTOXINS

Anticholinesterase Poisons

Some metabolic poisons are known to limit their action to the nervous system. These include poisons such as strychnine and curare (a South American Indian dart poison), as well as the dreaded nerve gases developed for chemical warfare. The exact modes of action of most neurotoxins are not known for certain, but investigations have discovered the action of a few.

Investigations of the actions of neurotoxins have provided insight into how the nervous system works.

A nerve impulse or stimulus is transmitted along a nerve fiber by electric impulses. The nerve fiber connects either with another nerve fiber or with some other cell (such as a gland or cardiac, smooth, or skeletal muscle) capable of being stimulated by the nerve impulse (Fig. 10–9). Neurotoxins

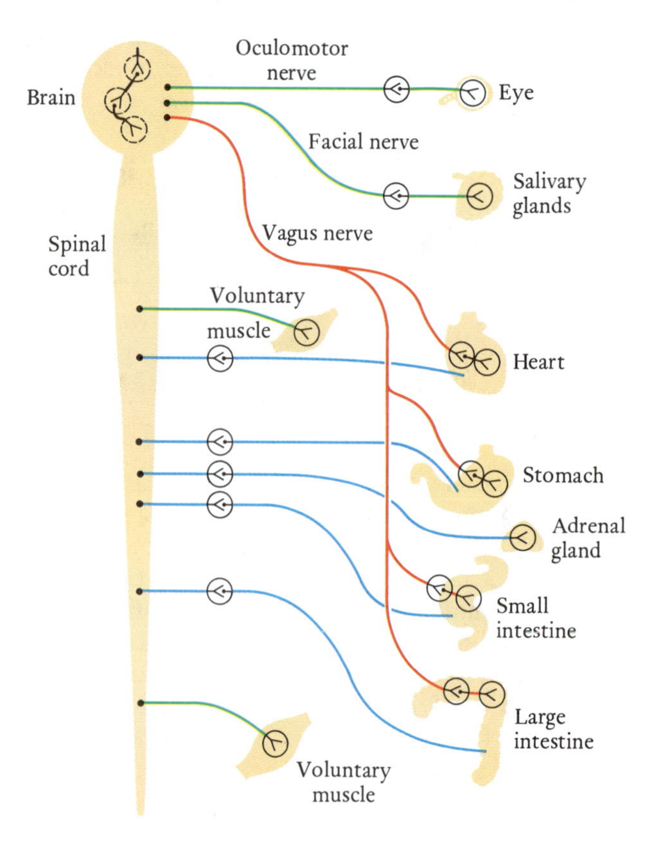

Figure 10–9 "Cholinergic" nerves, which transmit impulses by means of acetylcholine, include nerves controlling both voluntary and involuntary activities. Exceptions are parts of the "sympathetic" nervous system that utilize norepinephrine instead of acetylcholine. Sites of acetylcholine secretion are circled in color, poisons that disrupt the acetylcholine cycle can interrupt the body's communications at any of these points. The role of acetylcholine in the brain is uncertain, as is indicated by the broken circles.

often act at the point where two nerve fibers come together, called a **synapse.** When the impulse reaches the end of certain nerves, a small quantity of **acetylcholine** is liberated. This activates a receptor on an adjacent nerve or organ. The acetylcholine is thought to activate a nerve ending by changing the permeability of the nerve cell membrane. The method of increasing membrane permeability is not clear, but it may be related to an ability to dissociate fat–protein complexes or to penetrate the surface films of fats. Such effects can be brought about by as little as 10^{-6} mole of acetylcholine, which could alter the permeability of a cell so that ions can cross the cell membrane more freely.

Permeability: The ability of a membrane to let chemicals pass through it.

10^{-6} of a mole of acetylcholine is 6×10^{17} molecules.

To enable the receptor to receive further electrical impulses, the enzyme **cholinesterase** breaks down acetylcholine into acetic acid and choline (Fig. 10–10):

$$\underset{\text{Acetylcholine}}{CH_3\overset{\displaystyle O}{\overset{\|}{C}}OCH_2CH_2\overset{\displaystyle CH_3}{\overset{|}{\underset{\underset{\displaystyle CH_3}{|}}{N^+}}}{-}CH_3,\ OH^-} + \underset{\text{Water}}{H_2O} \xrightarrow{\text{Cholinesterase}} \underset{\text{Acetic acid}}{CH_3\overset{\displaystyle O}{\overset{\|}{C}}OH} + \underset{\text{Choline}}{HOCH_2CH_2\overset{\displaystyle CH_3}{\overset{|}{\underset{\underset{\displaystyle CH_3}{|}}{N^+}}}{-}CH_3,\ OH^-}$$

In the presence of potassium and magnesium ions, other enzymes such as acetylase resynthesize new acetylcholine from the acetic acid and the choline within the incoming nerve ending:

$$\text{Acetic acid} + \text{Choline} \xrightarrow{\text{Acetylase}} \text{Acetylcholine} + H_2O$$

The new acetylcholine is available for transmitting another impulse across the gap (see Fig. 10–10a).

Neurotoxins can affect the transmission of nerve impulses at nerve endings in a variety of ways. The **anticholinesterase poisons** prevent the breakdown of acetylcholine by deactivating cholinesterase. These poisons are usually structurally analogous to acetylcholine, so they bond to the enzyme cholinesterase and deactivate it (Fig. 10–10b). The cholinesterase molecules bound by the poison are held so effectively that the restoration of proper nerve function must await the manufacture of new cholinesterase. In the meantime, the excess acetylcholine overstimulates nerves, glands, and muscles, producing irregular heart rhythms, convulsions, and death. Many of the organic phosphates widely used as insecticides are metabolized in the body to produce anticholinesterase poisons. For this reason, they should be treated with extreme care. Some poisonous mushrooms also contain an anticholinesterase poison. Figure 10–11 contains the structures of some anticholinesterase poisons.

Neurotoxins such as **atropine** and **curare** (Table 10–6) are able to occupy the receptor sites on nerve endings of organs that are normally occupied by the impulse-carrying acetylcholine. When atropine or curare occupies the receptor site, no stimulus is transmitted to the organ (Fig. 10–10b). Acetylcholine in excess causes a slowing of the heartbeat, a decrease in blood pressure, and excessive saliva, whereas atropine and curare produce excessive thirst and dryness of the mouth and throat, a rapid heartbeat, and an increase in blood pressure. The normal responses to acetylcholine activation

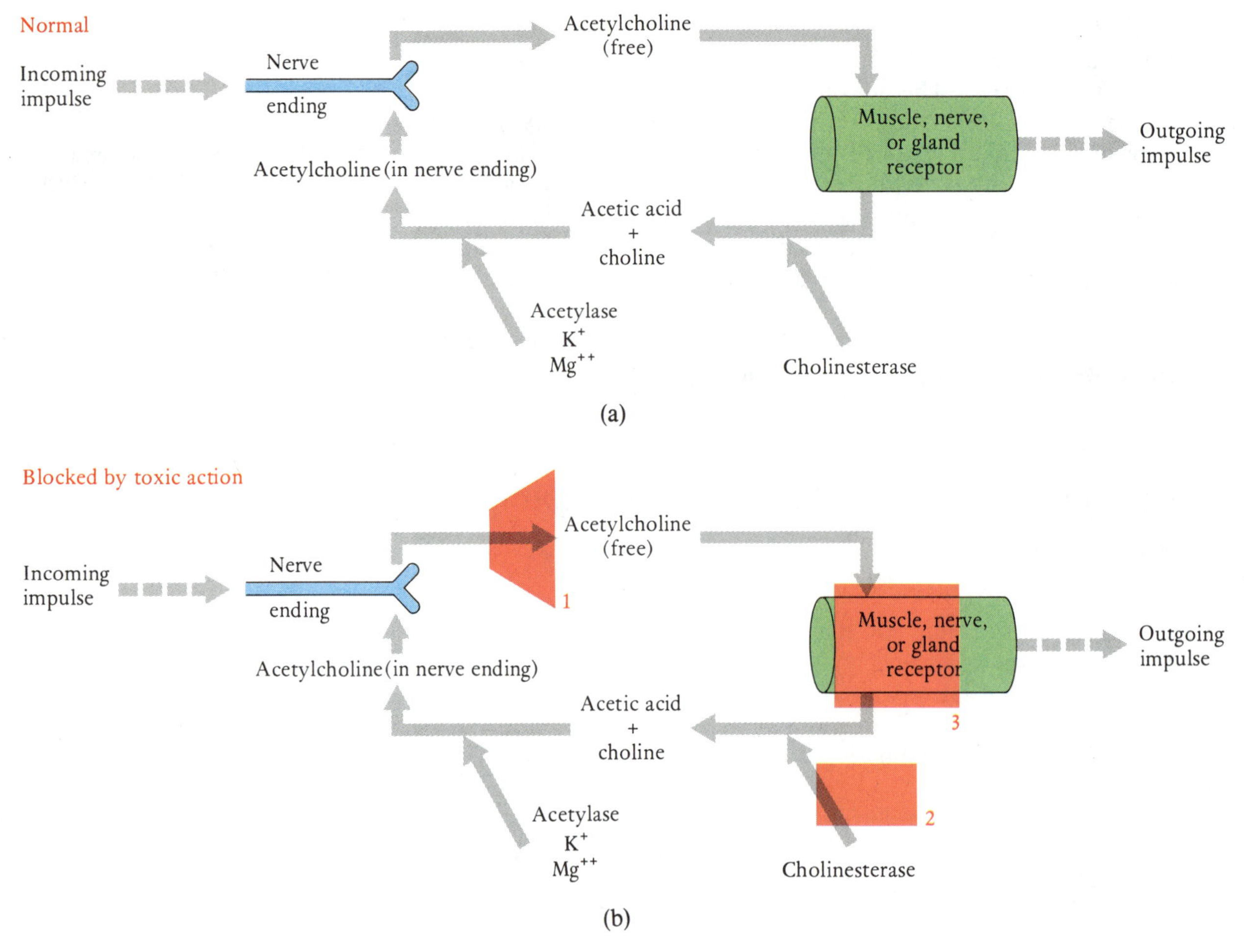

Figure 10–10 The acetylcholine cycle, a fundamental mechanism in nerve impulse transmission, is affected by many poisons. An impulse reaching a nerve ending in the normal cycle (a) liberates acetylcholine, which then stimulates a receptor. To enable the receptor to receive further impulses, the enzyme cholinesterase breaks down acetylcholine into acetic acid and choline; other enzymes resynthesize these into more acetylcholine. (b) Botulinus and dinoflagellate toxins inhibit the synthesis, or the release, of acetylcholine(1). The "anticholinesterase" poisons inactivate cholinesterase and therefore prevent the breakdown of acetylcholine(2). Curare and atropine desensitize the receptor to the chemical stimulus.(3).

are absent, and the opposite responses occur when there is sufficient atropine present to block the receptor sites.

Neurotoxins of this kind can be extremely useful in medicine. For example, atropine is used to dilate the pupil of the eye to facilitate examination of its interior. Applied to the skin, atropine sulfate and other atropine salts relieve pain by deactivating sensory nerve endings on the skin. Atropine is also used as an antidote for anticholinesterase poisons. Curare has long been used as a muscle relaxant.

Name	Structure	LD_{50} (rat;oral), mg/kg	Use
Sarin	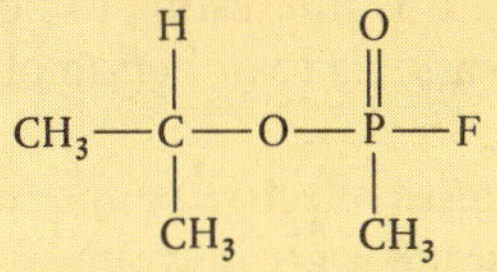 $CH_3-CH(CH_3)-O-P(=O)(CH_3)-F$	0.55	Chemical warfare agent first produced in World War II
Tabun	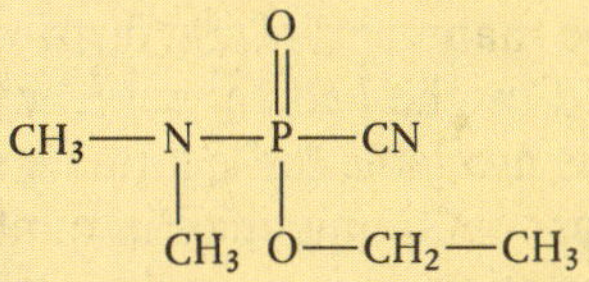 $CH_3-N(CH_3)-P(=O)(O-CH_2-CH_3)-CN$	3.7	Chemical warfare agent first produced in World War II
Parathion	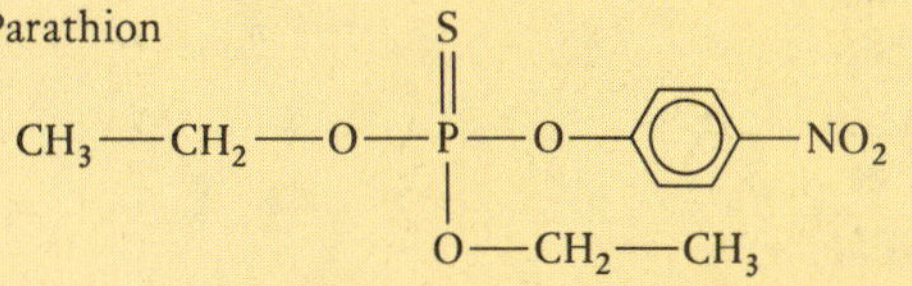 $CH_3-CH_2-O-P(=S)(O-CH_2-CH_3)-O-C_6H_4-NO_2$	20	Insecticide
Malathion	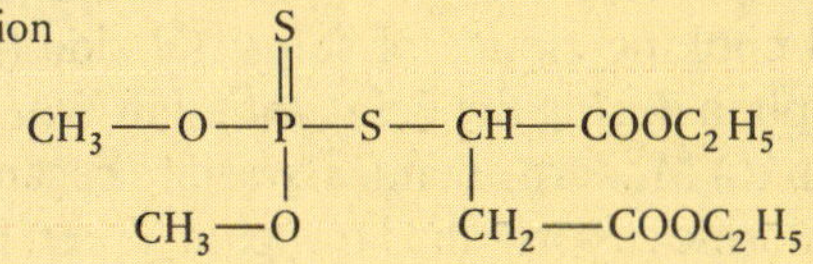 $(CH_3-O)_2P(=S)-S-CH(COOC_2H_5)-CH_2-COOC_2H_5$	885	Insecticide
Carbaryl (sevin)	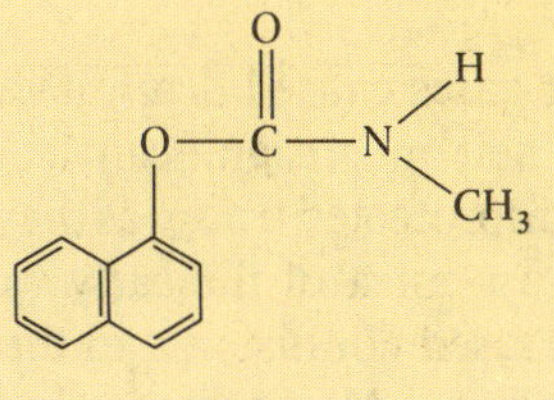 $C_{10}H_7-O-C(=O)-NH-CH_3$	400	Insecticide

Figure 10-11 Some anticholinesterase poisons. In animals, parathion is converted into paraoxon in the liver. Carbaryl and malathion do not bind to cholinesterase as strongly. Malathion was the insecticide used in California in July 1981 to eradicate Medflies.

TABLE 10-6 Neurotoxins That Compete with Acetylcholine for the Receptor Site

Name	Normal Contact	Lethal Dose (for a 70-kg human)
Atropine	Dilation of the pupil of the eye	100 mg
Curare	Muscle relaxant	20 mg
Nicotine	Tobacco, insecticides	75 mg
Caffeine	Coffee, tea, cola drinks	13.4 g (one cup of coffee contains about 40 mg caffeine)
Morphine	Opium—pain killer	100 mg
Codeine	Opium—pain killer	300 mg
Cocaine	Leaves of *Erythroxylon coca* plant in South America	1 g

Alkaloids are discussed in Chapter 7.

A well-known alkaloid that blocks receptor sites in a manner similar to that of curare and atropine is **nicotine.** This powerful poison causes stimulation and then depression of the central nervous system. The probable lethal dose for a 70-kg person is less than 0.3 g. It is interesting to note that pure nicotine was first extracted from tobacco and its toxic action observed *after* tobacco use was established as a habit.

Morphine is the most effective pain killer known.

Natural or synthetic **morphine** is the most effective pain reliever known. It is widely used to relieve short-term acute pain resulting from surgery, fractures, burns, and so on, as well as to reduce suffering in the latter stages of terminal illnesses such as cancer. The manufacture and distribution of narcotic drugs are stringently controlled by the Federal government through laws designed to keep these products available for legitimate medical use. Under Federal law, some preparations containing small amounts of narcotic drugs may be sold without a prescription (for example, cough mixtures containing codeine), but not many.

CHEMICAL WARFARE AGENTS

Chemical warfare is the use of toxic chemicals to kill and incapacitate the enemy. The Greeks used choking clouds of sulfur dioxide (SO_2) gas caused by burning sulfur and pitch during the Peloponnesian War between Sparta and Athens (413–404 B.C.) Modern chemical warfare began in 1915 when the Germans released chlorine gas on Allied troops at Ypres, Belgium, during World War I. After the initial use of chlorine, various other gases were developed and used (Table 10–7).

In general, the World War I war gases caused death if the victim was exposed to high enough doses, but their most significant contribution to warfare was their effect on dispersing unprotected troops as they ran from the areas of highest concentration. Because most of the early war gases were strongly irritating, their use always caused confusion and disorder among troop concentrations. The actual number of deaths due to chemical warfare agents during World War I was fairly small. This was probably due to technical problems of delivering the toxic chemical so as to produce consistently a lethal concentration exactly where the enemy troops were located. In addi-

TABLE 10–7 Some Chemical Warfare Agents

Type	Example	Action	History
Mustard gas	Bis (2-chloroethyl) sulfide $(Cl{-}CH_2{-}CH_2)_2S$	Strong blisters, strong irritant	WW I
Choking gas	Phosgene Cl_2CO Chlorine Cl_2	Lung damage	WWI
Blood gas	Hydrogen cyanide HCN	Cell death	WW I
Nerve gas	Tabun (see text)	Anticholinesterase poisons	WW II
	Sarin (see text)		WW II

tion, gas masks were quickly issued to troops of all belligerent nations. These gas masks offered sufficient protection to prevent death from exposures except in cases where wounded troops could not put on their masks as the toxic cloud approached.

After World War I, most nations agreed to never use toxic chemicals in warfare—yet development of these agents continued. In its war with Ethiopia in 1938, Italy used both nerve gases and mustard gas. During World War II, the Germans developed Tabun and Sarin (Table 10–7), two nerve gases that are anticholinesterase poisons. Their discovery led to our present-day phosphate ester insecticides such as Parathion® and Malathion®. Throughout World War II, war gases were available but were never used.

Recently, in the 1980s, chemical agents were used in the Iran–Iraq war against both troops and civilians. Against civilians, chemical warfare agents are especially devastating because civilians are not only untrained and uninformed about the effects of these chemicals, but are unprepared to protect themselves. Modern concern regarding chemical warfare agents centers on protecting civilians, especially against terrorist attacks using weapons of this sort.

SELF-TEST 10B

1. Substances that poison the nervous system are called ____________.
2. The junction between two nerve cells where information is passed is called a(n) ____________.
3. The electrical impulse at the junction between two nerve cells is passed by a chemical called ____________.
4. Which chemical is not a neurotoxin? (a) curare, (b) nicotine, (c) Sarin (a war gas), (d) sulfuric acid, (e) cocaine.
5. Which war gas causes severe skin blisters? (a) Sarin, (b) mustard gas, (c) phosgene.
6. Just prior to which war were the so-called nerve gases discovered? (a) WW I, (b) WW II, (c) Vietnam War, (d) Gulf War.

TERATOGENS

The effects of chemicals on human reproduction are a frightening aspect of toxicity. The study of birth defects produced by chemical agents is the discipline of **teratology.** The root word *terat* comes from the Greek word meaning "monster." There are three known classes of **teratogens:** radiation, viral agents, and chemical substances.

Birth defects occur in 2% to 3% of all births. About 25% of these occur from genetic causes, some possibly due to contact with mutagens, and 5% to 10% are the result of teratogens. The remaining 65% to 70% result from unknown causes.

In the development of the newborn, there are three basic periods during which the fetus is at risk. For a period of about 17 days between conception

Cell death by chemical "insult" is thought to be the cause of cancer (see text).

and implantation in the uterine wall, a chemical "insult" results in cell death. The rapidly multiplying cells often recover, but if a lethal dose is administered, death of the organism occurs followed by spontaneous abortion or reabsorption. The chemical in the so-called abortion pill, RU-486, developed in France in 1988 and being tested in the United States, works in this way.

During the critical embryonic stage (18 to 55 days) organogenesis, or development of the organs, occurs. At this time the fetus is extremely sensitive to teratogens. During the fetal period (56 days to term), the fetus is less sensitive. Contact with teratogens results in reduction of cell size and number. This is manifested in growth retardation and failure of vital organs to reach maturity.

The horrible thalidomide disaster in 1961 focused worldwide attention on chemically induced birth defects. Thalidomide®, a tranquilizer and sleeping pill, caused gross deformities (flipper-like arms, shortened arms, no arms or legs, and other defects) in children whose mothers used this drug during the first 2 months of pregnancy. The use of this drug resulted in more than 4000 surviving malformed babies in West Germany, more than 1000 in Great Britain, and about 20 in the United States. With shattering impact, this incident demonstrated that a compound can appear to be remarkably safe on the basis of animal studies (so safe, in fact, that thalidomide was sold in West Germany without prescription) and yet cause catastrophic effects in

H N O O C O C O N

Thalidomide (a teratogen)

TABLE 10-8 Teratogenic Substances

Substances	Species	Effects on Fetus
Metals		
Arsenic	Mice Hamsters	Increase in males born with eye defects, renal damage
Cadmium	Mice	Abortions
	Rats	Abortions
Cobalt	Chickens	Eye, lower limb defects
Gallium	Hamsters	Spinal defects
Lead	Humans Rats Chickens	Low birth weights, brain damage, stillbirth, early and late deaths
Lithium	Primates	Heart defects
Mercury	Humans	Minamata disease (Japan)
	Mice	Fetal death, cleft palate
	Rats	Brain damage
Thallium	Chickens	Growth retardation, abortions
Zinc	Hamsters	Abortions
Organic Compounds		
DES (diethylstilbestrol)	Humans	Uterine anomalies
Caffeine (15 cups per day equivalent)	Rats	Skeletal defects, growth retardation
PCBs (polychlorinated biphenyls)	Chickens	Central nervous system and eye defects
	Humans	Growth retardations, stillbirths

humans. Although the tragedy focused attention on chemical mutagens, thalidomide presumably does not cause genetic damage in germinal cells and is really not mutagenic. Rather, thalidomide, when taken by a woman during early pregnancy, causes direct injury to the developing embryo.

Any chemical substance that can cross the placenta is a potential teratogen, and any activity resulting in the uptake of these chemicals into the mother's blood might prove a dangerous act for the health and well-being of the fetus. Smoking a cigarette results in higher than normal blood levels of such substances as carbon monoxide, hydrogen cyanide, cadmium, nicotine, and benzo(α)pyrene. Of course, many of these substances are present in polluted air as well. Table 10–8 lists a number of chemical substances known to be teratogenic in humans and laboratory animals.

A pregnant woman should always be advised to limit her exposure throughout the term of her pregnancy to chemicals of unknown toxicity, any of which could be harmful to the developing child. This is especially true during the 18th through 55th days. She should take no drugs or medicines except on the advice of her physician, and she should avoid the use of alcohol and tobacco.

Fetal alcohol syndrome and cocaine addiction are observed in many newborns delivered by mothers who consume alcohol or use cocaine.

MUTAGENS

Mutagens are chemicals capable of altering the genes and chromosomes sufficiently to cause abnormalities in offspring. Chemically, mutagens alter the structures of DNA and RNA, which compose the genes (and, in turn, the chromosomes) that transmit the traits of parent to offspring. Mature sex, or germinal, cells of humans normally have 23 chromosomes; body, or somatic, cells have 23 *pairs* of chromosomes.

A mutagen is a chemical that can change the hereditary pattern of a cell. See Chapter 9.

Although many chemicals are under suspicion because of their mutagenic effects on laboratory animals, it should be emphasized that no one has yet shown conclusively that any chemical induces mutations in human germinal cells. Part of the difficulty of determining the effects of mutagenic chemicals in humans is the extreme rarity of mutation. A specific genetic disorder may occur as infrequently as only once in 10,000 to 100,000 births. Therefore, to obtain meaningful statistical data, a carefully controlled study of the entire population of the United States would be required. In addition, the very long time between generations presents great difficulties, and there is also the problem of tracing a medical disorder to a single specific chemical out of the tens of thousands of chemicals with which we come in contact.

If there is no direct evidence for specific mutagenic effects in human beings, why, then, the interest in the subject? The possibility of a deranged, deformed human race is frightening; the chance for an improved human body is hopeful; and the evidence for chemical mutation in plants and lower animals is established. A wide variety of chemicals is known to alter chromosomes and to produce mutations in rats, worms, bacteria, fruit flies, and other plants and animals. Some of these are listed in Table 10–9.

In the early 1980s, Professor Bruce Ames and his colleagues at the University of California, Berkeley, developed a simple test that can identify chemicals capable of causing mutations in sensitive strains of bacteria. In

TABLE 10-9 Mutagenic Substances as Indicated by Experimental Studies on Plants and Animals

Substance	Experimental Results
Aflatoxin (from mold, *Aspergillus flavus*)	Mutations in bacteria, viruses, fungi, parasitic wasps, human cell cultures, mice
Benzo(α)pyrene (from cigarette and coal smoke)	Mutations in mice
Caffeine (coffee, soft drinks)	Chromosome changes in bacteria, fungi, onion root tips, fruit flies, human tissue cultures
Captan (a fungicide)	Mutagenic in bacteria and molds; chromosome breaks in rats and human tissue cultures
Chloroprene (industrial chemical)	Mutagenic in male sex cells; results in spontaneous abortions
Dimethyl sulfate (used extensively in chemical industry to methylate amines, phenols, and other compounds)	Methylates DNA base guanine; potent mutagen in bacteria, viruses, fungi, higher plants, fruit flies
LSD (lysergic acid diethylamide)	Chromosome breaks in somatic cells of rats, mice, hamsters, white blood cells of humans and monkeys
Maleic hydrazide (plant growth inhibitor; trade names Slo-Gro®, MH-30®)	Chromosome breaks in many plants and in cultured mouse cells
Nitrous acid (HNO_2) (from cured meats)	Mutations in bacteria, viruses, fungi
Ozone (O_3) (atmospheric pollution, electric sparks)	Chromosome breaks in root cells of broadleaf plants
Solvents in glue (glue sniffing) (toluene, acetone, hexane, cyclohexane, ethyl acetate)	4% more human white blood cells showed breaks and abnormalities (6% versus 2% normal)
TEM (triethylenemelamine) (anticancer drug, insect chemosterilants)	Mutagenic in fruit flies, mice

this test, about 100 million bacteria unable to synthesize the amino acid histidine are mixed in an agar suspension along with a suspected mutagenic chemical. This mixture is then added to a hard agar gel containing salts and glucose, and incubated in a Petri dish for several days. If the suspected chemical is a mutagen, some of the histidine-requiring cells mutate and the biosynthesis of histidine resumes. The growth of these bacterial colonies can be seen in the Petri dish (Fig. 10-12).

The Ames test has utility in identifying not only mutagenic chemicals, but potential carcinogenic chemicals as well, because mutagenic chemicals are often carcinogenic. In studies involving hundreds of chemicals, nearly four of every five animal carcinogens have been found to be mutagenic in the Ames test.

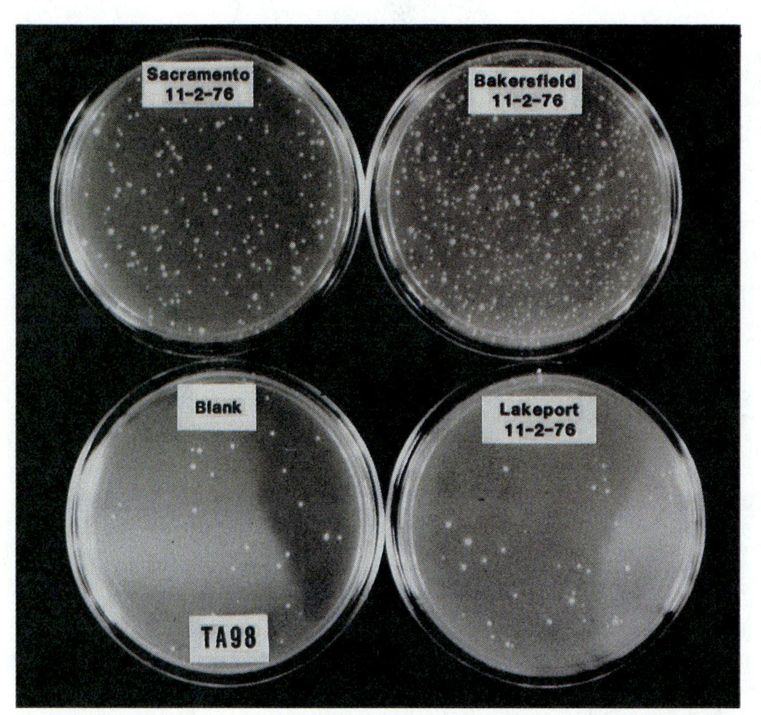

Figure 10–12 The effect of mutagens in the air in various California cities (© American Chemical Society, From P. Flessel, et al.: "Ames Testing for Mutagens and Carcinogens in Air," *J. Chemical Education,* 64:391–395, 1987. Reproduced by permission.)

CARCINOGENS

Carcinogens are chemicals that cause cancer. Just how they cause cancer is not clearly understood. **Cancer** is an abnormal growth condition in an organism that manifests itself in at least three ways. The rate of cell growth (that is, the rate of cellular multiplication) in cancerous tissue differs from the rate in normal tissue. Cancerous cells may divide more rapidly or more slowly than normal cells. Cancerous cells spread to other tissues; they know no bounds. Normal liver cells divide and remain a part of the liver. Cancerous liver cells may leave the liver and be found, for example, in the lung. Most cancer cells show partial or complete loss of specialized functions. Although located in the liver, cancer cells no longer perform the functions of the liver.

Attempts to determine the cause of cancer have evolved from early studies in which the disease was linked to a person's occupation. Dr. Percivall Pott, an English physician, first noticed in 1775 that persons employed as chimney sweeps had a higher rate of skin cancer than the general population. It was not until 1933 that **benzo(α)pyrene,** $C_{20}H_{12}$ (a five-ringed aromatic hydrocarbon), was isolated from coal dust and shown to be metabolized in the body to produce one or more carcinogens. In 1895, the German physician Rehn noted three cases of bladder cancer, not in a random population, but in employees of a factory that manufactured dye intermediates in the Rhine Valley. Rehn attributed these cancers to his patients' occupation. These and other cases confirmed that at times as many as 30 years passed between the time of the initial employment and the occurrence of bladder

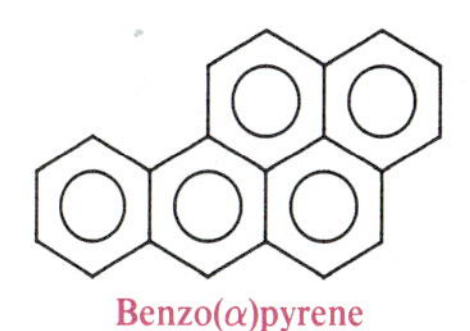

Benzo(α)pyrene

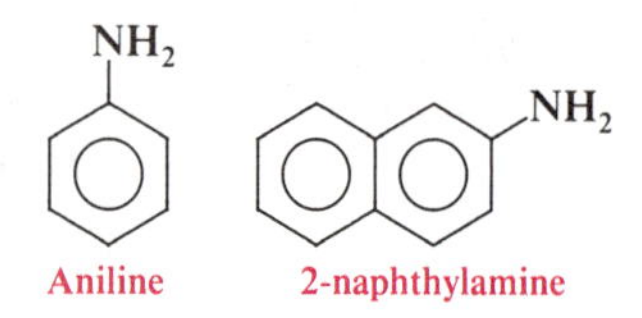

cancer. The principal product of these factories was aniline. Although aniline was first thought to be the carcinogenic agent, it was later shown to be noncarcinogenic. It was not until 1937 that continuous long-term treatment with **2-naphthylamine,** one of the suspected dye intermediates, in dosages of up to 0.5 g per day, produced bladder cancer in dogs. Since then other dye intermediates have been shown to be carcinogenic.

A vast amount of research has verified the carcinogenic behavior of a large number of diverse chemicals. Some of these are listed in Table 10–10. Some carcinogens are relatively nontoxic in a single, large dose, but may be quite toxic, often increasingly so, when administered continuously. Thus,

TABLE 10–10 Some Industrial Chemicals That Are Carcinogenic for Humans

Compound	Use or Source	Site Affected	Confirming Animal Tests*
Inorganic Compounds			
Arsenic (and compounds)	Insecticides, alloys	Skin, lungs, liver	−
Asbestos	Brake linings, insulation	Respiratory tract	+
Beryllium	Alloy with copper	Bone, lungs	+
Cadmium	Metal plating	Kidney, lungs	+
Chromium	Metal plating	Lungs	+
Nickel	Metal plating	Lungs, nasal sinus	+
Organic Compounds			
Benzene	Solvent, chemical intermediate in syntheses	Blood (leukemia)	+
Acrylonitrile	Monomer	Colon, lungs	+
Carbon tetrachloride	Solvent	Liver	+
Diethylstilbestrol	Hormone	Female genital tract	+
Benzo(α)pyrene	Cigarette and other smoke	Skin, lungs	+
Benzidine	Dye manufacture, rubber compounding	Bladder	+
Ethylene oxide	Chemical intermediate used to make ethylene glycol, surfactants	Gastrointestinal tract	±
Soots, tar, and mineral oils	Roofing tar, chimney soot, oils of hydrocarbon nature	Skin, lungs, bladder	+
Vinyl chloride	Monomer for making PVC	Liver, brain, lungs, lymphatic system	+

* For animal tests, (+) means positive supporting data, (−) means a lack of supporting data, (±) means conflicting data.

much patience, time, and money must be expended in carcinogen studies. The development of a sarcoma in humans, from the activation of the first cell to the clinical manifestation of the cancer, takes from 20 to 30 years. With life expectancy of an average person in the United States now set at about 70 years, it is not surprising that the number of deaths due to cancer is increasing.

Sarcoma—cancer of connective tissue.

An abnormal growth is classified as cancerous or malignant when examination shows it is invading neighboring tissue. A growth is benign if it is localized at its original site.

Recently, Professor Bruce Ames, whose mutagen test was mentioned earlier, and others have suggested that the way chemicals are tested for carcinogenicity in animals needs to be changed. Animal cancer tests, usually on rats or mice specially bred for genetic uniformity, are carried out with nearly toxic doses (the maximum tolerated dose, or MTD) for the test animal. It has been observed that the MTD over a long period of time can cause chronic mitogenesis (cell production). This chronic mitogenesis gives rise to a higher probability of mutagenesis (owing to the higher number of occurrences of cell production) and hence carcinogenesis. In the eyes of these researchers, chronic dosing of chemicals at the MTD is like chronic wounding, which is known to be a promoter of cancer in animals and a cancer risk factor for humans.

Some compounds cause cancer at the point of contact. Other compounds cause cancer in an area remote from the point of contact. The liver, the site at which most toxic chemicals are removed from the blood, is particularly susceptible to such compounds. Again, this is consistent with the chronic wounding theory. Because the original compound does not cause cancer on contact, some other compound made from it must be the cause of cancer. For example, it appears that the substitution of an $>$NOH group for an $>$NH group in an aromatic amine derivative produces at least one of the active intermediates for carcinogenic amines. If R denotes a two- or three-ring aromatic system, then the process can be represented as follows:

$$\underset{\text{Inactive on contact}}{\mathrm{R\overset{H}{\overset{|}{N}}COCH_3}} \longrightarrow \underset{\text{Active on contact}}{\mathrm{R\overset{OH}{\overset{|}{N}}COCH_3}} \longrightarrow \underbrace{\mathrm{RX?} \longrightarrow \mathrm{RY?}}_{\text{Other unknown intermediates}} \xrightarrow{\text{Tissue}} \text{Tumor cell}$$

As indicated by the variety of chemicals in Table 10–10, many molecular structures produce cancer, whereas some closely related to them do not. The 2-naphthylamine mentioned earlier is carcinogenic, but repeated testing gives negative results for 1-naphthylamine.

NH_2 NH_2

1-naphthylamine (noncarcinogenic) 2-naphthylamine (carcinogenic)

Almost all human cancers caused by chemicals have a long induction period, which makes it extremely difficult for researchers to obtain meaningful interpretations of exposure data. For example, asbestos particles are known to cause cancer, but only those particles in a narrow size range—that is, 5 to 100 μm long and less than 2 μm wide—seem to be capable of actually

Asbestos sample as seen under a polarizing light microscope. This sample contains 60% chrysotile asbestos. (Courtesy of Particle Data Laboratories.)

Asbestos is thought to be responsible for over 10,000 cancer deaths per year—second only to tobacco-caused cancer deaths.

causing cancer under controlled conditions involving laboratory animals. Neither lung cancer nor mesothelioma (a very malignant tumor of the linings of the chest and abdominal cavities) commonly occurs less than 20 years after the first exposure. Because very precise measurements of asbestos in the air at workplaces were made prior to 1950 and it wasn't until 1965 that the smaller-sized asbestos particles could be measured, it is not surprising that we know very little about the concentrations of asbestos particles that caused cancer in so many factory and construction workers.

Cancer does not occur with the same frequency in all parts of the world. Breast cancer occurs less frequently in Japan than in the United States or Europe. Cancer of the stomach, especially in males, is more common in Japan than in the United States. Cancer of the liver is not widespread in the western hemisphere but accounts for a high proportion of the cancers among the Bantu in Africa and in certain populations in the Far East. The widely publicized incidence of lung cancer is higher in the industrialized world and is increasing at an appreciable rate.

DIETARY CARCINOGENS

Apart from industrial chemicals with which we may come in contact in the workplace and which may contaminate our atmosphere (see Chapter 12) and our drinking water (see Chapter 11), our everyday diets contain a great variety of natural carcinogens. Some of these chemicals are also mutagens and teratogens. Plants produce these chemicals as defense mechanisms, or natural pesticides, and often produce more of them when diseased or stressed. Celery, for example, contains isoimpinellin, a member of a chemical family called psoralens, at a level of 100 μg/100 g. This level increases by 100-fold if the celery is diseased. This compound is a potent light-activated carcinogen. Psoralens, when activated by sunlight, damage DNA. Oil of bergamont, found in citrus fruits, contains a psoralen that was once used by a French manufacturer of suntan oil. Sunlight caused the psoralens to enhance tanning.

Isoimpinellin

Hydrazine

Sym-dimethylhydrazine

Agaritine

$CH_2{=}CH{-}CH_2N{=}C{=}S$

Allyl isothiocyanate

Black pepper contains small amounts of saffrole, a known carcinogen, and large amounts (up to 10% by weight) of piperine, a compound related to saffrole. Black pepper extracts cause tumors in mice at a variety of sites at dose rates equivalent to 4 mg of dried pepper per day for three months. Many people consume over 140 mg of black pepper per day for life.

Most hydrazines that have been tested are carcinogenic. This includes industrial products such as rocket fuels, and hydrazines produced by plants. The widely eaten false morel *(Gyromitra esculenta)* contains eleven hydrazines, three of which are known carcinogens. One of these is *N*-methyl-*N*-formylhydrazine, which is present at concentrations of 50 mg/100 g and causes lung tumors in mice at very low dietary doses of 20 μg per mouse per day. The common mushroom, *Agaricus bisporus,* contains about 300 mg of agaritine per 100 g. Agaritine is a mutagen and is metabolized to a compound that is extremely carcinogenic.

Allyl isothiocyanate, the main flavor ingredient in oil of mustard and horseradish, has been shown to cause cancers in rats. It also causes chromosome damage in hamster cells at low concentration.

THE WORLD OF CHEMISTRY

Toxic Substances

How do scientists perceive the risk from chemicals and the environment? Interestingly most chemicals, even vitamins, can be toxic at large doses. Most chemicals, in fact, have a threshold dose of toxicity. A dose above the threshold is dangerous, a dose below it is not. But there's a great deal of debate over whether potential carcinogens have such thresholds, that is, concentrations below which they won't cause cancer. Thus, given conflicting and imprecise evidence, some scientists talk about carcinogens in terms of risk or probabilities. Risk is relative. For example, some think the cancer risk from hazardous waste is negligible compared with the naturally occurring carcinogens we successfully resist every day.

The originator of a test for determining whether a chemical has the potential to cause cancer is Dr. Bruce Ames, biochemist at the University of California, Berkeley. He claims:

People have been very worried about toxic waste dumps but, in fact, the evidence that they're really causing any harm is really minimal, there's not very much evidence. And the levels of chemicals are very tiny, so we don't really know whether there's no hazard or a little bit of hazard.

The world is full of carcinogens, because half the natural chemicals they've tested have come out as carcinogens. Some plants have toxic chemicals to keep off insects, and we are eating those every time we eat a tomato or potato. Mushrooms have carcinogens, celery has carcinogens, and an apple has formaldehyde in it. So there is an incredible number of carcinogens in nature; we're getting much more of those than man-made chemicals.

Dr. Halina Brown, Professor of Toxicology, Clark University, has stated:

If we accept those risks, why can't we accept small risks from chemical carcinogens in the environment? It's a valid argument. But then there is, of course, the counter argument. The counter argument is we cannot do much about trace amounts of carcinogens that are present in food, why should we add to this burden that we already have by increasing the amount of exposure to carcinogens. But then it boils down to money. Unfortunately, it takes tremendous resources to reduce the levels of exposure in the environment to carcinogens, especially when you get to very low levels. Reducing it by another order of magnitude may take millions of dollars at one hazardous waste site. And the pie is not unlimited. Even those who don't consider hazardous waste dumps a health threat think the money should be spent to clean them up.

Dr. Bruce Ames also states:

I mean if Congress has put $10 billion for cleaning up toxic waste dumps, you might as well find the worst ones and clean them up. Now, whether you're getting anything—whether you're gaining much in public health for cleaning them up is something one could argue about. I think probably very little. But, in any case, you can—you might as well spend the money cleaning up the worst dumps.

The World of Chemistry (Program 25) "Chemistry and the Environment."

Dr. Bruce Ames. (*The World of Chemistry,* Program 25, "Chemistry and the Environment.")

The average intake of these and other natural carcinogens is about 1500 mg per person per day. By comparison, synthetic pesticide residues on foodstuffs amount to about 0.09 mg per person per day. This means that many of the cancers we may get as we grow older may have been caused not by the chemicals getting all of the publicity, but by the very foods that we have been eating all along. We are not likely to eliminate all carcinogens either from our diets or from our general environment.

SELF-TEST 10C

1. Which class of chemical (teratogen/mutagen) causes deformities at birth?
2. Which class of chemical (teratogen/mutagen) causes damage to DNA?
3. When is a chemical more likely to harm a developing fetus? (a) during the last three months just prior to birth, (b) during the embryonic stage (18 to 55 days), (c) during the second three months of pregnancy.
4. Name two very common chemicals that can cause mutations in laboratory animals or human cells: ____________ and ____________.
5. The "Ames test" tests for (a) mutagenicity, (b) carcinogenicity.
6. A chemical that can cause cancer is called a(n) ____________.
7. Of the chemicals listed below, which one chemical is not a known or suspected carcinogen? Beside each chemical name, indicate where you might encounter this chemical.
   a. benzene ____________
   b. mineral oil ____________
   c. sucrose ____________
   d. asbestos ____________
   e. arsenic ____________
8. A dose of a chemical which is just below the dose that is toxic is called the ____________ ____________ ____________.
9. There are cancer-causing chemicals in the foods we eat: True ( ) False ( )

QUESTIONS

1. Discuss the saying "the dose makes the poison."
2. What does the term LD_{50} mean? Would a given chemical have the same LD_{50} for all species?
3. Name one chemical that has gained quite a bit of notoriety as one of the most toxic chemicals known. How does its LD_{50} vary among species?
4. How do corrosive poisons work? Name one war gas that is a corrosive poison.
5. How does carbon monoxide work as a poison? Name two sources where we are most likely to encounter carbon monoxide.
6. How does cyanide work as a poison? How is cyanide similar to carbon monoxide and how is it different as they act as poisons?
7. Name three heavy metals known for their toxicity. In general terms, how do heavy metals act as poisons?
8. Which heavy metal is most likely to cause poisoning in small children, particularly those living in older housing?
9. Name a common source for the metal in Question 8 which can give almost everyone a daily dose of this metal. Why do most people show little if any harmful effects from this dosage?
10. What is the name of the enzyme that helps break apart acetylcholine which is formed after a nerve synapse has fired?
11. How does anticholinesterase work? Describe the effect an anticholinesterase poison would have on the heart.

12. Name two common chemicals that are anticholinesterase poisons, although we do not generally think of them as poisons.
13. Several common pesticides used both on farms and around the home are chemically related to nerve gases that have been used in various wars. Name two ways in which these pesticides are different from the nerve gases.
14. Phosgene (Cl_2CO) is a gas that was used in WW I. It has other chemical uses as well. What is the target organ of phosgene? Explain how phosgene can cause injury and death upon exposure.
15. Explain how teratogens work. Name one well-known teratogen.
16. Explain how mutagens work. Name one well-known mutagen.
17. Explain how carcinogens work. Fit the concept of maximum tolerated dose (MTD) into your explanation. Name one common carcinogen.
18. If a chemical is a carcinogen to mice, does this mean it is a carcinogen to humans? Explain.
19. Approximately what time period elapses between exposure to a carcinogen and observation of a cancer?
20. Many animal tests on the carcinogenicity of a chemical involve doses that are very near the MTD. Define the MTD and discuss the theory of MTD, cell damage, and cancer.
21. According to tests carried out on various foodstuffs, our average daily intake of natural carcinogens is about 1500 mg. What is the estimate for our daily intake of synthetic pesticides, many of which can cause cancer in laboratory animals (an approximate number will do)? What additional questions need to be answered regarding these two numbers as their impact on our dietary habits is considered?

The importance of water is rendered serenely in Seurat's *Bathers.* (The National Gallery, London)

CHAPTER

11

Water—Plenty of It, But of What Quality?

Water is the most abundant compound on earth, and it is necessary for all forms of life. Yet, because water is so accessible and has excellent solvent properties, it ends up polluted.

1. What are the properties of water that make it so special?
2. How much water is there, and what are the sources of water?
3. What is the difference between clean water and polluted water?
4. How does Biochemical Oxygen Demand affect water quality?
5. How do hazardous industrial and household wastes impact water quality?
6. How is water purified?
7. How can fresh water be taken from the sea?

Water is an unusually unique substance and without it life on this planet would not be possible. Certainly there are other media on earth and in the universe wherein much chemistry occurs. However, on Earth the chemistry in water solutions and the chemistry of water dominate. Water plays an important role as a reactant, a product, or a coordinating chemical in most of the chemical reactions in our environment. What are water's unique properties, and what are their effects on life as we know it?

Some Physical Properties of Water

1. The density of solid water (ice) is less than that of liquid water. Put another way, water expands when it freezes. If ice were a normal solid, it would be denser than liquid water, and lakes would freeze from the bottom up. This would have disastrous consequences for marine life, which could not survive in areas with winter seasons.
2. Water is a liquid at room temperature in spite of being composed of very small molecules. In contrast, the hydrogen compounds of the other nonmetals around oxygen in the periodic table are toxic, corrosive gases such as NH_3, H_2S, and HF.
3. Water has a relatively high heat capacity per unit of weight. This means it can absorb large quantities of heat without large changes in temperature. For comparison, the heat capacity of water is about ten times that of copper or iron for equal weights. This property accounts for the moderating influence of lakes and oceans on the climate. Huge bodies of water absorb heat from the Sun and release the heat at night or in cooler seasons. The Earth would have extreme temperature variations if it were not for this property of water. By contrast, the temperatures on the surface of the Moon and the planet Mercury vary by hundreds of degrees through the light and dark cycle.

Heat capacity is defined as the amount of heat required to raise the temperature of a sample of matter 1°C.

4. Water has a relatively high heat of vaporization. The heat needed to vaporize 1 g of water at 100°C is 540 cal. A consequence of this is the cooling effect that occurs when water evaporates from moist skin.
5. Water has a large surface tension. The large surface tension of water and its ability to wet surfaces are the bases for capillary action, which carries water to leaves in plants and trees.
6. Water is an excellent solvent, often referred to as the universal solvent. As a result, water from natural sources is not pure water but a solution of substances dissolved by contact with water.

Surface molecules of a liquid are pulled inward by the intermolecular interactions with molecules below the surface. Surface tension is a measure of this force.

The high boiling point, high heat of vaporization, and large heat capacity of water are partially the result of the energy needed to break the hydrogen bonds in liquid water as it is heated or vaporized, as described in Chapter 3.

Hydrogen bonding also accounts for the large surface tension of liquid water. The water molecules on the surface are pulled inward by hydrogen bonding to water molecules below the surface. This unbalanced force at the surface causes the surface layer to contract, and energy is required to break this surface. Insects can walk on water because their weight is not sufficient to break through the surface tension.

The water strider with little weight has an easy time walking on water. (The insect does not provide enough force per unit area to break through the surface tension.) Note that the insect does not walk on the sharp ends of its "toes." (Manfred Danegger/Peter Arnold, Inc.)

Interaction of Water with Other Chemicals

The chemical properties of water are also a function of the polar nature of the water molecule—its ability to engage in hydrogen bonding and the strength of the covalent hydrogen-oxygen bond within the molecule. Water tends to react with any negatively charged species by surrounding it with the positive ends (hydrogen) of the water dipole.

$$\ominus \text{ charge} + x\, H_2O \longrightarrow \ominus\cdot(H_2O)_x \qquad (x = \text{no. of water molecules})$$

Conversely, water reacts with any positively charged species by surrounding it with the negative end of the water dipole.

$$\oplus \text{ charge} + y\, H_2O \longrightarrow \oplus\cdot(H_2O)_y \qquad (y = \text{no. of water molecules})$$

(See Fig. 4-3 for a visual image of dissolving ions being hydrated by water molecules.)

The number of water molecules involved (x or y) depends on the intensity of the charge and on the size of the charged species. Water's unique properties result from its molecular polarity and hydrogen bonding, which enable it to carry dissolved chemicals, some of them quite toxic, as well as harmful bacteria and viruses. In developed countries throughout the world, pure water is often taken for granted, but producing water that is both clean and pure enough for use by man, animals, and plants is not easy. The job of purifying water is becoming more difficult as water becomes contaminated by the chemical byproducts of mining, farming, industry, and household activities. Serious water supply problems exist in some states and communities, and this has resulted in recent years in the rationing of all types of water use. For nearly two decades, the U.S. Environmental Protection Agency (EPA) has issued standards for toxic contaminants commonly found in our drinking water. Therefore, we often do not have enough water, and what we have is at risk of being contaminated by chemicals that can cause harm. In this chapter we examine these problems by looking at some of their causes and the possible solutions to them.

The gathering of water molecules around a charged center is somewhat analogous to a magnet picking up a group of paper clips.

WATER—THE MOST ABUNDANT COMPOUND

Water is the most abundant substance on the Earth's surface. Oceans (with an average depth of 2.5 miles) cover about 72% of the Earth. They are the reservoir of 97.2% of the Earth's water. The rest consists of 2.16% in glaciers, 0.0197% fresh water in lakes and rivers, 0.61% groundwater (water underground), 0.01% in brine wells and brackish waters, and 0.001% in atmospheric water.

Water is the major component of all living things. For example, the water content of human adults is 70%—the same proportion as the Earth's surface (Table 11-1).

TABLE 11-1 Water Content

Marine invertebrates	97%
Human fetus (1 month)	93%
Adult human	70%
Body fluids	95%
Nerve tissue	84%
Muscle	77%
Skin	71%
Connective tissue	60%
Vegetables	89%
Milk	88%
Fish	82%
Fruit	80%
Lean meat	76%
Potatoes	75%
Cheese	35%

Brackish water contains dissolved salts but at a lower level than sea water.

Transpiration is the release of water by leaves of plants. An acre of corn is estimated to release 3000 gal per day, while a large oak tree releases 110 gal per day.

It is estimated that an average of 4350 billion gal of rain and snow fall on the contiguous United States each day. Of this amount, 3100 billion gal return to the atmosphere by evaporation and transpiration. The discharge to the sea and to underground reserves amounts to 800 billion gal daily, leaving 450 billion gal of surface water each day for domestic and commercial use. The 48 contiguous states withdrew 40 billion gal per day from natural sources in 1900 and 394 billion gal in 1985, and it is estimated that the demand will be at least 900 billion gal per day by the year 2000. The demand for water by our growing population is already greater than the resupply by natural resources in many parts of the country.

SOURCES OF WATER

The two sources of usable water are **surface water** (lakes, rivers) and **groundwater.** Groundwater is that part of underground water that is below the water table. Figure 11-1 shows the various parts of the water cycle and the flow of groundwater. About 90 billion gal of the 394 billion gal per day of water usage come from groundwater supplies. These groundwater supplies are from wells drilled into the **aquifers.** The supply and demand for surface water and groundwater are uneven across the country, and in many areas the

An aquifer is a water-bearing stratum of permeable rock, sand, or gravel, as illustrated in Figure 11-1.

Figure 11-1 How groundwater gets polluted.

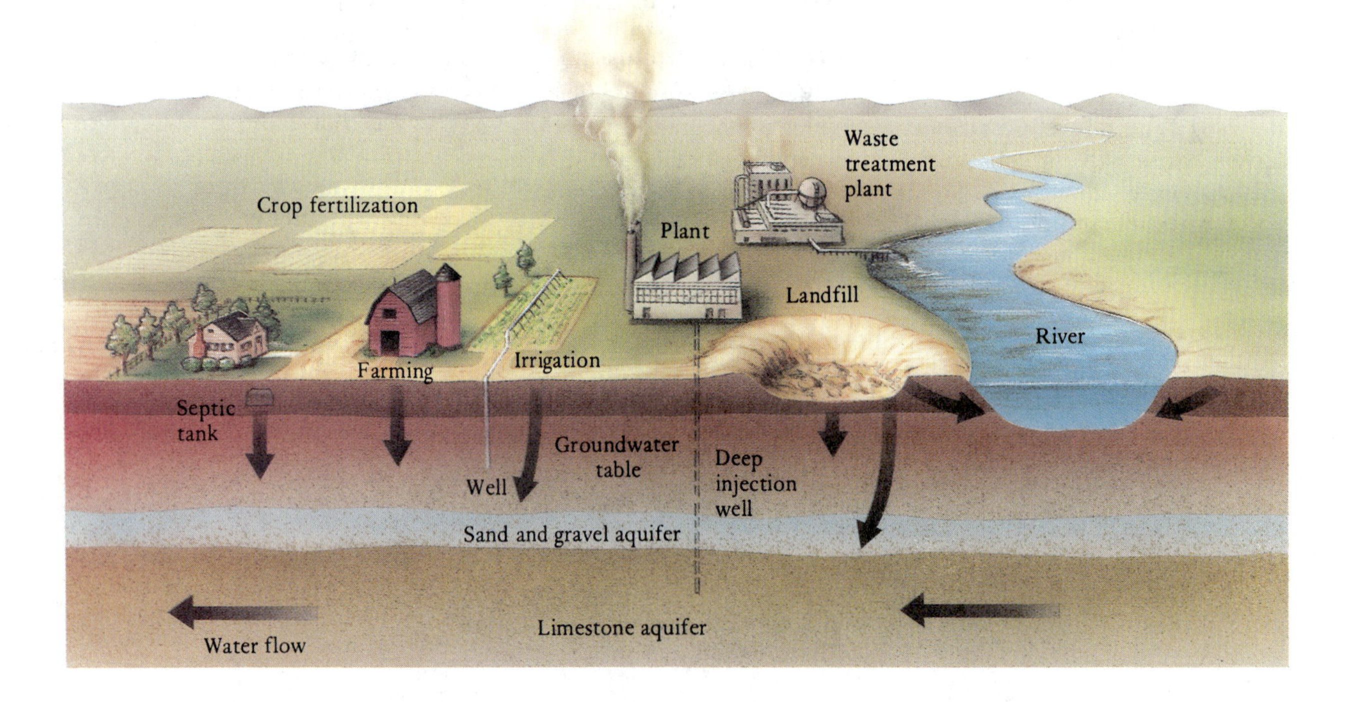

quantity and quality of the withdrawn water are not being resupplied to the lakes, rivers, and aquifers at the rate needed.

In the arid West, wells used to pump water for irrigation either are going dry or are requiring drilling so deep that irrigation is no longer economically feasible. The huge Ogallala aquifer that stretches from South Dakota to Texas has 150,000 wells tapping it for irrigation of 10 million acres. As a result, the Ogallala aquifer is being drawn down at a rate that has reduced the average thickness of the aquifer from 58 ft in 1930 to 8 ft today. At the current rate, the Ogallala aquifer will be used up as a source of groundwater in 20 to 30 years!

Even drinking water has enough impurities to build up in water pipes over time. (*The World of Chemistry,* Program 12, "Water")

The depletion of the Ocala aquifer along the eastern seaboard has caused large sinkholes in Georgia and Florida when the limestone rock strata of the aquifer collapse as the water is withdrawn. Many coastal cities are also experiencing problems with brackish drinking water that comes from aquifers where the fresh water has been drawn off, causing sea water to flow into the depleted aquifer.

Depletion of underground sources has also caused sinkholes in Texas. Houston has sunk several feet as the result of extensive use of the underground water sources in that area! Figure 11–2 is a graphic illustration of the change in surface level in California's San Joaquin Valley as a result of groundwater depletion.

WHAT IS CLEAN WATER?—WHAT IS POLLUTED WATER?

Water that is judged unsuitable for drinking, washing, irrigation, or industrial uses is **polluted water.** The pollution may be heat, radioisotopes, toxic metal ions, organic solvent molecules, acids, or alkalies. Water suitable for some uses might be considered polluted and therefore unsuitable for other uses. Water that is unsuitable for use has generally been polluted by human activity, although natural leaching of some metal ions from rocks and soil, organic substances like tannins from decaying leaves and animal wastes, and silt can pollute clean water. As human activities have continued to pollute water, various governments have passed laws designed to cause us to keep our waters clean and unpolluted.

The Clean Water Act of 1977 shifted the burden of producing water suitable for reuse from the user to the wastewater discharger. This action was a crucial step in improving the quality of our rivers and lakes, because it is easier to clean the wastewater prior to dumping than to clean the river water after the untreated waste has been discharged. In addition, the quality of the wastewater effluent is often high enough to be used as a resource of water for other purposes, such as irrigation or cooling towers.

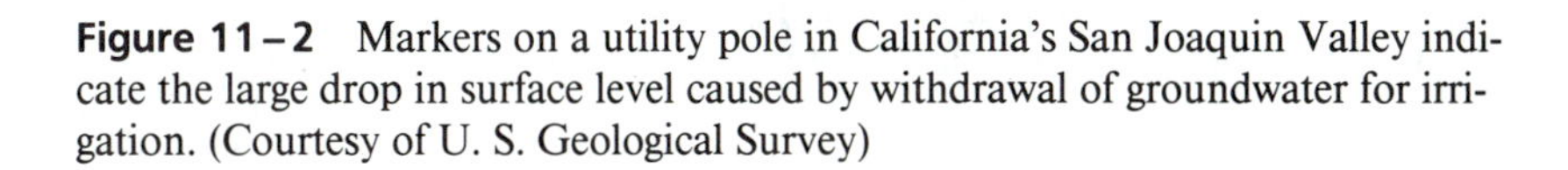

Figure 11–2 Markers on a utility pole in California's San Joaquin Valley indicate the large drop in surface level caused by withdrawal of groundwater for irrigation. (Courtesy of U. S. Geological Survey)

As our industrial and commercial use of water has increased, water pollution has become more diversified. The U.S. Public Health Service now classifies water pollutants into eight broad categories (Table 11-2). The very fact that the EPA has published limits for chemical contaminants in drinking water (Table 11-3) is evidence enough that our water supplies are polluted.

TABLE 11-2 Classes of Water Pollutants, with Some Examples

Oxygen-demanding wastes	Plant and animal material
Infectious agents	Bacteria, viruses
Plant nutrients	Fertilizers, such as nitrates and phosphates
Organic chemicals	Pesticides such as DDT,* detergent molecules
Other minerals and chemicals	Acids from coal mine drainage, inorganic chemicals such as iron from steel plants
Sediment from land erosion	Clay silt on stream bed, which may reduce or even destroy life forms living at the solid-liquid interface
Radioactive substances	Waste products from mining and processing of radioactive material, used radioactive isotopes
Heat from industry	Cooling water used in steam generation of electricity

* Banned in the United States, but still produced and exported.

TABLE 11-3 Maximum Contaminant Levels for Drinking Water Supplies

Contaminant	Maximum Concentration (mg/L)
Inorganics	
Arsenic	0.05
Barium	1
Cadmium	0.01
Chromium	0.05
Lead	0.05
Mercury	0.002
Nitrate	10
Selenium	0.01
Silver	0.05
Organics	
Endrin	0.0002
Lindane	0.004
Methoxychlor	0.1
Toxaphene	0.005
2,4-D	0.1
2,4,5-TP (Silvex)	0.01
Total trihalomethanes (includes bromotrichloromethane, dibromochloromethane, tribromomethane, trichloromethane (chloroform)	0.1

Source: EPA, 1988.

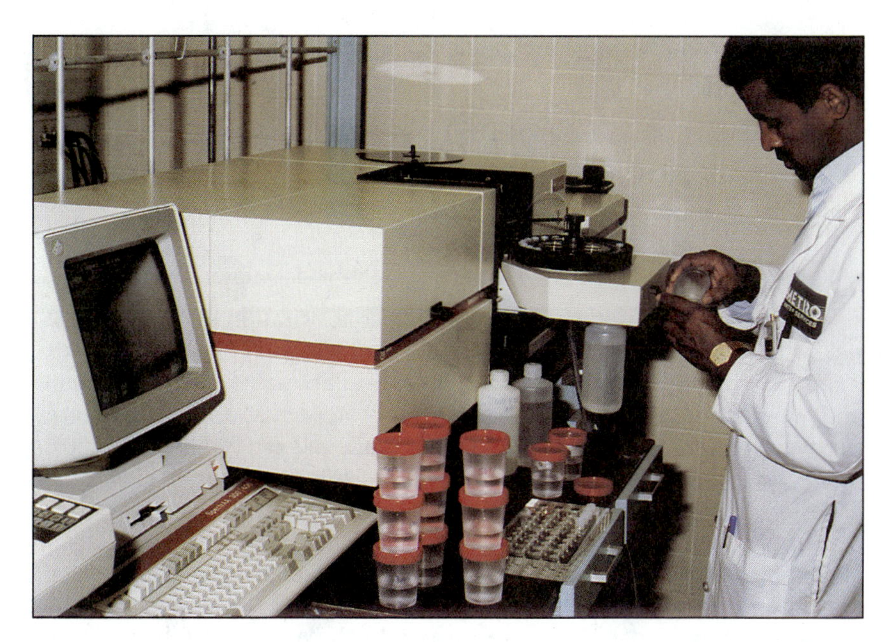

Figure 11–3 A water quality laboratory. This atomic absorption instrument can analyze both drinking water and treated waste water samples for trace amounts of various metals. (Courtesy of Metro Water Services, Nashville, Tennessee)

The EPA requires that municipal water supplies be monitored constantly (Fig. 11–3).

SELF-TEST 11A

1. Approximately what percentage of the human body is water? ________
2. What is the major reservoir of water on the planet Earth? __________
3. Three common water pollutants are ______________, ______________, and ______________.
4. Water held in a stratum of porous rock is called ______________.
5. A water-bearing stratum of porous rock, sand, or gravel is called a(n) ______________.
6. What happens to most of the water that falls on the United States each day? ________________________________

BIOCHEMICAL OXYGEN DEMAND

The way in which organic materials are oxidized in the natural purification of water deserves special attention. Even in the natural state, living organisms found in natural waters are constantly discharging organic debris into the water. To change this organic material into simple inorganic substances (such as CO_2 and H_2O) requires oxygen. The amount of oxygen required to oxidize a given amount of organic material is called the **biochemical oxygen demand (BOD).** The oxygen is required by microorganisms, such as many forms of bacteria, to metabolize the organic matter that constitutes their food. Ultimately, given nearly normal conditions and enough

A quantitative relationship exists between oxygen needs and organic pollutants to be destroyed. This is BOD.

Low available oxygen in a stream results in dead fish. (*The World of Chemistry,* Program 12, "Water")

Fish cannot live in water that has less than 0.004 g_{O_2}/L (4 ppm).

Characteristic BOD Levels

	g_{O_2}/L
Untreated municipal sewage	0.1–0.4
Runoff from barnyards and feed lots	0.1–10
Food-processing wastes	0.1–10

time, the microorganisms convert huge quantities of organic matter into the following end products:

$$\text{Organic carbon} \longrightarrow CO_2$$
$$\text{Organic hydrogen} \longrightarrow H_2O$$
$$\text{Organic oxygen} \longrightarrow H_2O$$
$$\text{Organic nitrogen} \longrightarrow NO_3^- \text{ or } N_2$$

Highly polluted water often has a high concentration of organic material, with resultant large biochemical oxygen demand (Fig. 11–4). In extreme cases, more oxygen is required than is available from the environment, and putrefaction results. Fish and other freshwater aquatic life can no longer survive. The aerobic bacteria (those that require oxygen for the decomposition process) die. As a result of the death of these organisms, even more lifeless organic matter results and the BOD soars. Nature, however, has a backup system for such conditions. A whole new set of microorganisms (anaerobic bacteria) takes over; these organisms take oxygen from oxygen-containing compounds to convert organic matter to CO_2 and water. Organic nitrogen is converted to elemental nitrogen by these bacteria. Given enough time, enough oxygen may become available, and aerobic oxidation then returns.

A stream containing 10 ppm by weight (just 0.001%) of an organic material, the formula of which can be represented by $C_6H_{10}O_5$, contains 0.01 g of this material per liter. It can be calculated that the BOD for this stream is 0.012 g of oxygen per liter. At 68°F (20°C), the solubility of oxygen in water under normal atmospheric conditions is 0.0092 g of oxygen per liter.

Because the BOD (0.012 g/L) is greater than the equilibrium concentration of dissolved oxygen (0.0092 g/L), as the bacteria utilize the dissolved oxygen in this stream, the oxygen concentration of the water soon drops too

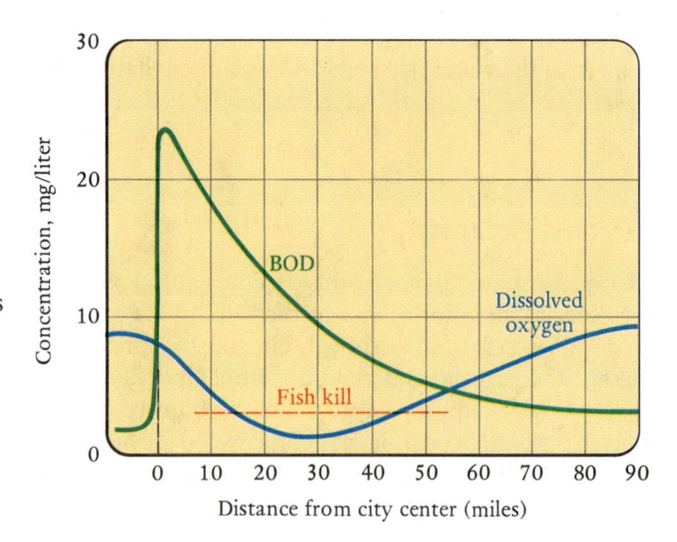

Figure 11–4 Graph showing oxygen content and oxidizable nutrients (BOD) as a result of sewage introduced by a city. The results are approximated on the basis of a river flow of 750 gal/sec. Note that it takes 70 miles for the stream to recover from a BOD of 0.023 g oxygen per liter.

low to sustain any form of fish life. Life forms can survive in water where the BOD exceeds the dissolved oxygen if the water is flowing vigorously in a shallow stream (this facilitates the absorption of more oxygen from the air via aeration).

BOD values can be greatly reduced by treating industrial wastes and sewage with oxygen and/or ozone. Numerous commercial cleanup operations now being developed and used employ this type of "burning" of the organic wastes. Another benefit of treating waste water with oxygen is that some of the nonbiodegradable material becomes biodegradable as a result of partial oxidation.

High concentration of organic pollutants
↓
Low oxygen concentration
↓
Dead organisms
↓
Higher concentration of organic pollutants
↓
Lower oxygen concentration
↓
Anaerobic conditions

IMPACT OF HAZARDOUS WASTES ON WATER QUALITY

Industrial wastes can be an especially vexing sort of pollution problem because often they either are not removed or are removed very slowly by naturally occurring purification processes and are generally not removed at all by a typical municipal water treatment plant. Table 11–4 lists some of the industrial pollutants that result from products important to us.

Disposal of hazardous wastes in landfills has been the principal method of disposal for industries, agriculture, and municipalities for decades. Incidents such as the Love Canal disaster drew attention to the serious contamination of groundwater by hazardous wastes. Action on local, state, and Federal levels began in the 1970s to solve problems caused by past disposal and to develop workable methods for future disposal of hazardous wastes. In 1980, Congress established the "Superfund," a $1.6 billion program designed to clean up hazardous waste sites that were threatening to contaminate the nation's underground water supplies. In 1985, the Office of Tech-

TABLE 11–4 Important Industrial Products and Consequent Hazardous Wastes

The Products We Use	The Potentially Hazardous Waste They Generate
Plastics	Organic chlorine compounds
Pesticides	Organic chlorine compounds, organic phosphate compounds
Medicines	Organic solvents and residues, heavy metals like mercury and zinc, for example
Paints	Heavy metals, pigments, solvents, organic residues
Oil, gasoline, and other petroleum products	Oils, phenols, and other organic compounds, lead, salts, acids, alkalies
Metals	Heavy metals, fluorides, cyanides, acids, and alkaline cleaners, solvents, pigments, abrasives, plating salts, oils, phenols
Leather	Chromium, zinc
Textiles	Heavy metals, dyes, organic chlorine compounds, organic solvents

Field sampling of industrial waste water. (Courtesy of Isco, Inc.)

nology Assessment estimated that the number of hazardous waste sites requiring cleanup will increase, perhaps to as many as 10,000, and the cost of cleanup may reach $100 billion. By April, 1991, the EPA had placed 1189 hazardous waste sites on its National Priorities List for cleanup under the Superfund Law (Fig. 11–5).

The EPA has listed over 2000 sites in the United States where toxic wastes have been stored and should be cleaned up (Fig. 11–5). This cleanup will cost billions of dollars.

Although only 1% or 2% of the aquifers are known to be polluted by hazardous wastes, many of these aquifers are near large population centers, so the problem is a serious one. The basic problem with land disposal of hazardous wastes is the contamination of groundwater as it moves through the disposal area (see Fig. 11–1). Water pollution from these sites generally occurs as seepage into an underlying aquifer.

In 1976, the Federal government passed the Resource Conservation and Recovery Act (RCRA). This law is designed to give "cradle-to-grave" (origin to disposal) responsibility to generators of hazardous wastes. The RCRA regulations cover generation, transportation, storage treatment, and disposal of hazardous wastes.

Considerable attention has been given to safe disposal of hazardous wastes, monitoring groundwater near hazardous waste sites and reducing the quantity of hazardous wastes by recycling chemicals. The technology for safe disposal exists, but the costs are high (Fig. 11–6). Data reported by the EPA in 1988 (Table 11–5) indicate that 7% of hazardous wastes are being disposed of by environmentally unsound methods. The effect of the present government regulations and the greater public awareness are making current disposal methods safer, but the cleanup of Superfund sites and other landfills that are contaminating groundwater will take time.

Figure 11–5 Hazardous waste sites in the United States as designated by the EPA Superfund.

Number of hazardous waste sites in each state:

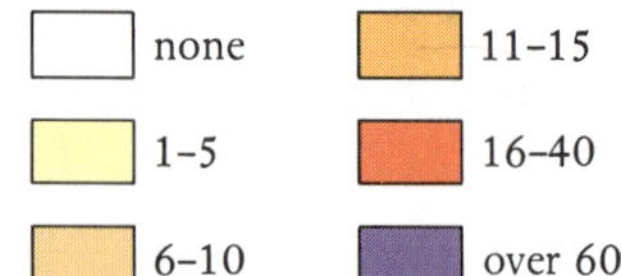

Hawaii: 6
Alaska: none
Puerto Rico: 8
Guam: 1

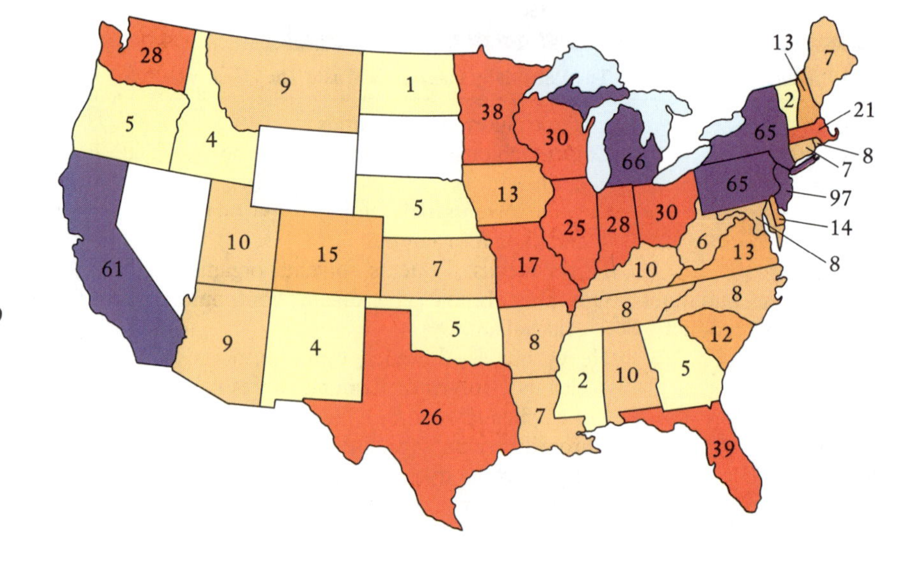

Figure 11–6 Waste drums containing hazardous wastes. Their cleanup is both time-consuming and costly. (John D. Cunningham/Visuals Unlimited)

TABLE 11–5 Hazardous Waste Disposal Methods in 1988

Method	Percentage of Total
Unacceptable	
Unlined surface impoundment	5
Land disposal	1
Uncontrolled incineration	1
Acceptable	
Controlled incineration	55
Secure landfills	20
Recovered	25

Source: EPA, 1989.

It was once considered good engineering practice to put all wastes into landfills. There, many of these wastes leached into the groundwater, causing serious water pollution problems. Today, industrial wastes categorized as **hazardous wastes** (Table 11–6) must be placed into secure landfills, incinerated, or treated in some way to render them nonhazardous. Secure landfills generally have plastic linings (Fig. 11–7) and carefully spaced monitoring wells so any leaching from the landfill's contents may be detected. Other

TABLE 11–6 Hazardous Wastes as Defined by EPA

Wastes containing the following metals and pesticides:
  Arsenic, barium, cadmium, chromium, lead, mercury, selenium, silver, endrin, lindane, methoxychlor, toxaphene, 2,4-D, 2,4,5-TP (Silvex)
Wastes that have the following characteristic properties*:
  Ignitible, corrosive, reactive, acutely toxic
Twenty-one wastes from nonspecific sources: (such as)
  Wastes containing the cyanide ion, distillation residues, used halogenated solvents like carbon tetrachloride
Eighty-nine wastes from specific sources: (such as)
  Wastewater sludges from chloride production, wastewater from pesticide manufacture, sludges from the production of petroleum products
A large number of various discarded and off-specification chemicals, many of which are used in the chemical industry to manufacture pharmaceuticals, polymers, paints, dyes, automotive products, cosmetics, etc.

Note: Shipments of hazardous wastes are carefully monitored by EPA and state governments. In addition, all facilities receiving these wastes are permitted and licensed.

Source: EPA.

* Detailed definitions apply to these waste characteristics.

Workers in protective suits at a waste site. (*The World of Chemistry,* Program 25, "Chemistry and the Environment")

Figure 11–7 Pits for holding containers of hazardous wastes must be lined with thick polymer liners that help prevent the escape of wastes into the groundwater. (Courtesy of Chemical Waste Management, Inc.)

wastes from industry, not considered hazardous but still containing potentially harmful chemicals (Table 11–7), may go into public landfills or may go to sewers or receiving bodies of water after some form of pretreatment. The amounts of wastes disposed of this way are very large, and the potential for polluting ground and surface waters is considerable.

In 1988, new reporting regulations regarding releases of certain listed hazardous chemicals went into effect for manufacturing industries. These releases must be reported each year. Data for 1989 (Table 11–8) indicate that the state of Louisiana had the largest total amount of hazardous chemicals released into water.

TABLE 11–7 Some Additional Hazardous Chemicals Not Considered Hazardous Wastes

Chemical	Use
Ammonium nitrate (solution)	Fertilizer, explosives
Beryllium nitrate	Chemical manufacture
Biphenyl	Fungistat for citrus fruit
Cobalt oxide	Pigment
Copper nitrate	Electroplating, light-sensitive papers
Ethylene glycol	Antifreeze
Manganese dioxide	Battery manufacture
Nickel nitrate	Metal plating
Quinoline	Pharmaceuticals, flavorings

Note: Chemicals on this list must be reported when released into the environment by any means according to the Superfund Reauthorization and Amendments Act of 1986.

Source: EPA.

TABLE 11-8 Releases of Hazardous Chemicals by Industry to Water: The Top 10 States for 1989

State	Quantity Released (millions of pounds)
1. Louisiana	46.2
2. Illinois	16.7
3. Washington	15.6
4. Virginia	11.1
5. California	10.7
6. Arkansas	9.1
7. Alabama	7.5
8. Florida	6.6
9. Ohio	6.1
10. Texas	6.1

Source: EPA, 1991.

Note: These release data are now available to the public from the National Library of Medicine as an on-line database. This release reporting is a part of the "Community Right-to-Know" provisions of the Superfund Reauthorization and Amendments Act of 1986.

An industrial waste discharge. (Karen Roeder)

Because of known releases of harmful chemicals into water, states like California have taken drastic steps to protect their ground and surface waters. In California, Proposition 65, the Safe Drinking Water and Toxic Enforcement Act of 1986, lists approximately 200 chemicals or classes of chemicals known to cause cancer or reproductive toxicity and prohibits their discharge into drinking water supplies.

Proposition 65:
"No person in the course of doing business shall knowingly discharge or release a chemical known to cause cancer or reproductive toxicity into water or onto or into land where such chemical passes or probably will pass into any source of drinking water."

HOUSEHOLD WASTES AS HAZARDOUS WASTES

Often we do not think about the things we discard in our garbage, but what we throw away and how we do it can affect our groundwater purity. Household wastes that are incinerated can contribute to air pollution (see Chapter 12), but since the bulk of our household waste goes to landfills, we too can be responsible for causing pollution of groundwater. Table 11-9 lists some common household products and the kinds of chemicals they contain. Because we are the consumers of industrial products, we can put the very same chemicals into our groundwater as industry can. Although the individual amounts of harmful chemicals used in a household are less than those used in a large industry, the total amounts of harmful chemicals disposed of daily by all households can be very large, even for a medium-sized city.

Households have a greater problem disposing of chemicals that are potentially harmful to the groundwater than industry does. Even if a city has

TABLE 11–9 Some Household Hazardous Wastes and Recommended Disposal

Type of Product	Harmful Ingredients	Disposal*
Bug sprays	Pesticides, organic solvents	Special
Oven cleaner	Caustics	Drain
Bathroom cleaners	Acids or caustics	Drain
Furniture polish	Organic solvents	Special
Aerosol cans (empty)	Solvents, propellants	Trash
Nail-polish remover	Organic solvents	Special
Nail polish	Solvents	Trash
Antifreeze	Organic solvents, metals	Special
Insecticides	Pesticides, solvents	Special
Auto battery	Sulfuric acid, lead	Special
Medicine (expired)	Organic compounds	Drain
Paint (latex)	Organic polymers	Drain
Gasoline	Organic solvents	Special
Motor oil	Organic compounds, metals	Special
Drain cleaners	Caustics	Drain
Shoe polish	Waxes, solvents	Trash
Paints (oil-based)	Organic solvents	Special
Mercury batteries	Mercury	Special
Moth balls	Chlorinated organic compound	Special
Batteries	Heavy metals such as Hg	Special

* Special: Professional disposal as a hazardous waste. Drain: disposal down the kitchen or bathroom drain. Trash: Treat as normal trash—no harm to the groundwater. In most households, the items marked special are disposed of as normal trash, which results in groundwater pollution.

Source: "Household Hazardous Waste: What You Should and Shouldn't Do," Water Pollution Control Federation, 1986.

The EPA estimates that each year 350 million gallons of waste oil are poured on the ground or flushed down the drain. That's 35 times more oil than the Exxon Valdez spilled.

Recycling of materials like glass, paper, metals, and plastics (see Chapter 8) helps conserve resources like raw materials and energy. Recycling also conserves valuable landfill space and keeps some otherwise harmful chemicals from the groundwater.

Ordinary garbage costs about $27/ton for disposal, whereas hazardous waste costs about $1000/ton for proper disposal.

an active recycling project for glass, paper, metals, and plastics, most municipalities provide no means of pickup of those chemicals separated from the ordinary trash destined for the landfill. If these chemicals are mixed with ordinary garbage, all of it goes to the city landfill or incinerator. Professional hazardous waste disposers seldom offer their services to homeowners at a cut rate, so their high prices deter most households from disposing of these chemicals in the proper way.

How can we dispose of hazardous household wastes without danger to the groundwater supply? We can ask our city's municipal waste authorities to provide disposal sites for these wastes. As an example of how this can be done, in some U.S. cities and some European countries (such as the Netherlands) special trucks pick up paints, oils, batteries, and other products for disposal. Another approach is to purchase products with their ingredients in mind. Ordinary alkaline batteries often work just as well as mercury batteries, for example.

It is important to remember that when metals and plastics are mixed with incinerated wastes, they add pollutants to the atmosphere (see Chapter 12). Used motor oil and discarded automobile batteries can add metals, acids, and unwanted hydrocarbons to groundwater and therefore should be recycled whenever possible. Most communities have companies that specialize in recycling one or more of these household waste products, but it is up to each homeowner to seek out alternatives to indiscriminate mixing of garbage and other household wastes.

SELF-TEST 11B

1. The amount of oxygen required to oxidize a given amount of organic material is called the ______________________, which is abbreviated __________.
2. Name two industrial products whose manufacture introduces heavy metals into groundwater. __________, __________
3. Name two industrial products whose manufacture introduces chlorinated organic compounds into groundwater. __________, __________
4. Name three household wastes that can contaminate groundwater with the same harmful chemicals as industrial wastes. Beside each list the harmful chemical.

Household waste	Harmful chemical
__________	__________
__________	__________
__________	__________

5. List four household waste types that lend themselves to recycling. __________, __________, __________, __________

WATER PURIFICATION: CLASSICAL AND MODERN PROCESSES

The outhouses of some rural dwellers had their counterparts in city cesspools. The terrible job of cleaning led to the development of cesspools that could be flush-cleaned with water, followed by a connecting series of such pools that could be flushed from time to time. City sewer systems with no holding of the wastes were the next step.

Cesspools were an early and crude form of the modern activated sludge process.

Sewage is still 99.9% water!

Since there were not enough pure wells and springs to serve the growing population, water purification techniques were developed. The classical method, which is now termed **primary water treatment,** involved settling and filtration (Fig. 11–8).

In the settling stage, calcium hydroxide and aluminum sulfate are added to produce aluminum hydroxide. Aluminum hydroxide is a sticky, gelatinous precipitate that settles out slowly, carrying suspended dirt particles and bacteria with it.

$$3\ Ca(OH)_2 + Al_2(SO_4)_3 \longrightarrow 2\ Al(OH)_3 + 3\ CaSO_4$$

If the intake water is polluted enough with biological wastes, the primary treatment, even with chlorination, cannot render the water safe. To be sure,

PLANT FOR PRIMARY SEWAGE TREATMENT

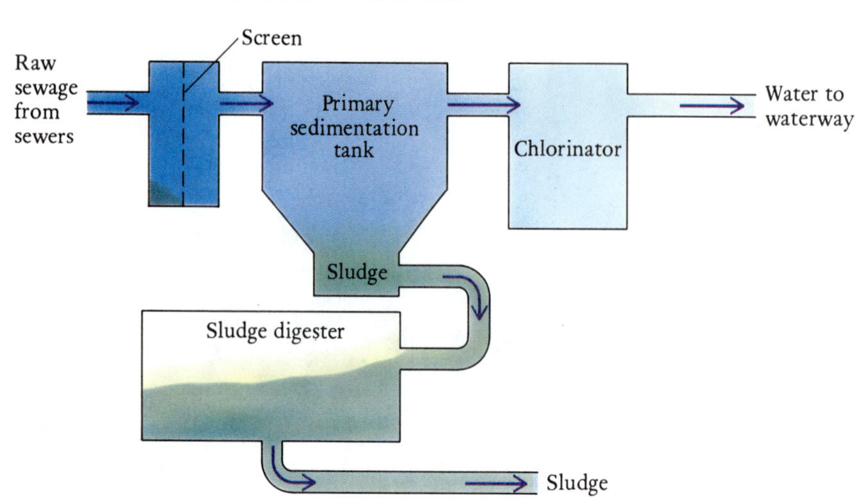

PLANT FOR SECONDARY SEWAGE TREATMENT

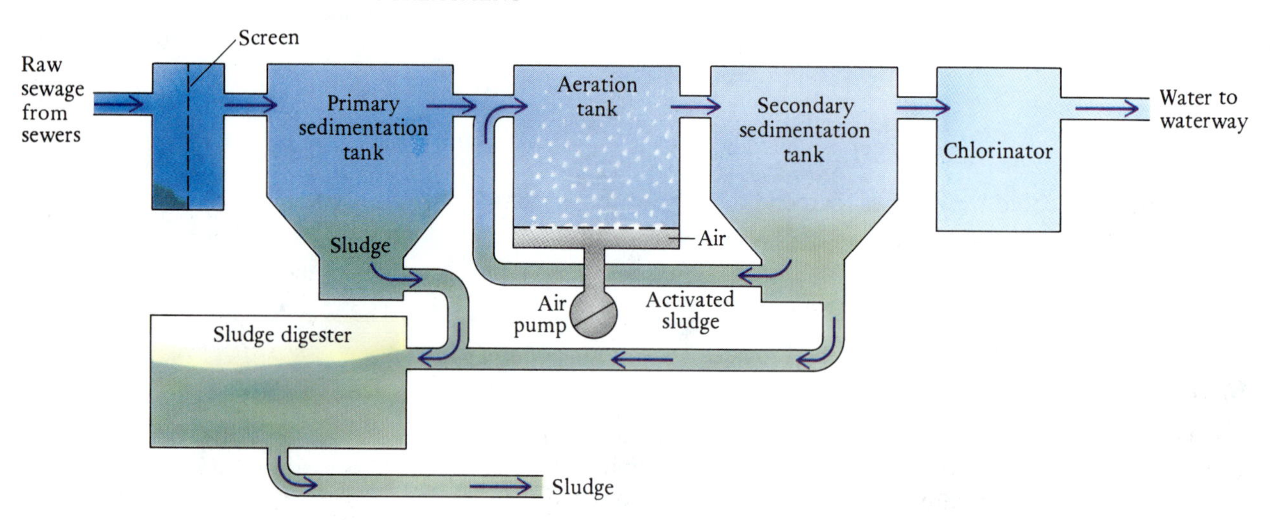

Figure 11–8 Sewage plant schematic, showing facilities for primary and secondary treatment. (Redrawn from *The Living Waters.* U.S. Public Health Service Publication No. 382)

Americans spend about $350 million a year on bottled water. Buyer beware: A very wide variety of standards exist for bottled water.

Primary, secondary, and tertiary water treatment methods can be used in both the purification of water to be consumed and the preparation of sewage to be sent back into a stream.

enough chlorine or other oxidizing agents could be added to kill all life forms, but the result would be water loaded with a wide variety of noxious chemicals, especially chlorinated organics, many of which are suspected carcinogens. Some way had to be found to coagulate and separate out the organic material that passed through the primary filters.

Secondary water treatment revives the old cesspool idea under a more controlled set of conditions and acts only on the material that will not settle or cannot be filtered. Modern secondary treatment operates in an oxygen-rich environment (aerobic), whereas the cesspool operates in an oxygen-poor environment (anaerobic) (Fig. 11–9). The results are the same: The organic molecules that do not settle are consumed by organisms; the resulting sludge does settle. Bacteria and even protozoa are introduced into the oxygen-rich

(a)

(b)

Figure 11–9 Water treatment. (a) Aerobic. (b) Anaerobic. (*The World of Chemistry,* Programs 12 and 16, "Water" and "The Proton in Chemistry")

environment for this purpose. Two techniques, the trickle filter and the activated sludge method, have been widely used in secondary water treatment.

Primary and secondary water treatment systems do not remove dissolved inorganic materials such as poisonous metal ions or even residual amounts of organic materials. These materials are removed by a variety of **tertiary water treatments.**

Two technologies are now being used for the removal of toxic materials from wastewater; these are carbon adsorption and activated sludge. Carbon black has been used for many years for adsorbing vapors and solute materials from liquid streams. Many toxic organic materials can be removed with activated or baked carbon granules that have been activated by high-temperature baking. This activated carbon has a high surface area that readily adsorbs chemicals from the wastewater. Activated sludge is a hurry-up version of natural stream purification. Bacteria and other microorganisms degrade the water pollutants in the sludge medium.

CHLORINATION

With the advent of chlorination of water supplies in the early 1900s, the number of deaths in the United States caused by typhoid and other water-borne diseases dropped from 35/100,000 population in 1900 to 3/100,000 population in 1930.

Chlorine is introduced into water as the gaseous free element (Cl_2) (Fig. 11–10) and it acts as a powerful oxidizing agent for the purpose of killing bacteria that remain in water after preliminary purification. The principal water-borne diseases spread by bacteria include cholera, typhoid, paratyphoid, and dysentery.

Most city water supplies are not bacteria-free. Surviving bacteria usually produce counts numbering in the tens of thousands, but only rarely do these

Figure 11–10 A chlorinator apparatus for a 60 million gal/day water plant. (Courtesy of the Robert L. Lawrence Jr. filtration plant)

Ton cylinders of chlorine. Each cylinder holds 2000 lb of gaseous chlorine for final treatment of drinking water. (Courtesy of the Robert L. Lawrence Jr. filtration plant)

surviving bacteria cause disease. Today the most common water-borne bacterial disease is giardiasis, a gastrointestinal disorder. Most often this disease comes from surface water but, on occasion, it can be traced to city water systems.

Chlorination of industrial wastes and city water supplies presents a potential threat because of the reaction of chlorine with residual concentrations of organic compounds. Traditional purification methods do not remove chlorinated hydrocarbons or, for that matter, most organics. The chlorinated hydrocarbons, which may be present at levels of a few parts per million or less, include dichloromethane, chloroform, trichloroethylene, and chlorobenzene, all suspected carcinogens.

A number of these chemicals in the same concentration range have been shown to be mutagenic to salmonella bacteria. Studies show an increased risk of 50% to 100% in rectal, colon, and bladder cancers in individuals who drink chlorinated water. According to the EPA, mutagenic or carcinogenic chemicals have been found in 14 major river basins in the United States. It is estimated that more than 500 water systems in the United States exceed EPA's maximum of 0.1 ppm for chlorinated hydrocarbons. The presence of these chlorinated hydrocarbons can be prevented either by using another disinfectant or by removing the low-level organic compounds before chlorination.

An efficient process for reducing the level of organic compounds in water is to pass the water through biologically activated carbon.

FRESH WATER FROM THE SEA

Because sea water covers 72% of the Earth, it is not surprising that this source would be a major consideration for areas where fresh water supplies aren't sufficient to meet the demand. The oceans contain an average 3.5% dissolved salts by weight, a concentration too high for most uses. The solvent properties of water are illustrated by the average composition of sea water in Table 11–10. If you add these up in terms of the number 0.001 g/kg, you have over 35,000 ppm of dissolved ions. The total must be reduced to below 500 ppm before the water is suitable for human consumption.

The technology has been developed for the conversion of sea water to fresh water. The extent to which this technology is actually put to use depends on the availability of fresh water and the cost of the energy for the conversion. Over 2200 desalination plants were in operation throughout the world in the 1980s. One method used to purify sea water is reverse osmosis.

Reverse Osmosis

When a membrane is permeable to water molecules but not to ions or molecules larger than water, it is called a **semipermeable membrane.** If a semipermeable membrane is placed between sea water and pure water, the pure water passes through the membrane to dilute the sea water. This is **osmosis.** The liquid level on the sea water side rises as more water molecules

TABLE 11-10 Ions Present in Sea Water at Concentrations Greater Than 0.001 g/kg

Ion	g/kg Sea Water
Cl^-	19.35
Na^+	10.76
SO_4^{2-}	2.71
Mg^{2+}	1.29
Ca^{2+}	0.41
K^+	0.40
HCO_3^-, CO_3^{2-}	0.106
Br^-	0.067
$H_2BO_3^-$	0.027
Sr^{2+}	0.008
F^-	0.001
Total	35.129

enter than leave, and pressure is exerted on the membrane until the rates of diffusion of water molecules in both directions are equal. **Osmotic pressure** is defined as the external pressure required to prevent osmosis. Figure 11-11 illustrates the concept of osmosis and osmotic pressure.

Reverse osmosis is the application of pressure to cause water to pass through the membrane to the pure-water side (Fig. 11-12). The osmotic pressure of normal sea water is 24.8 atm (atmospheres). As a result, pressures greater than 24.8 atm must be applied to cause reverse osmosis. Pressures up to 100 atm are used to provide a reasonable rate of filtration and to account for the increase in salt concentration that occurs as the process proceeds.

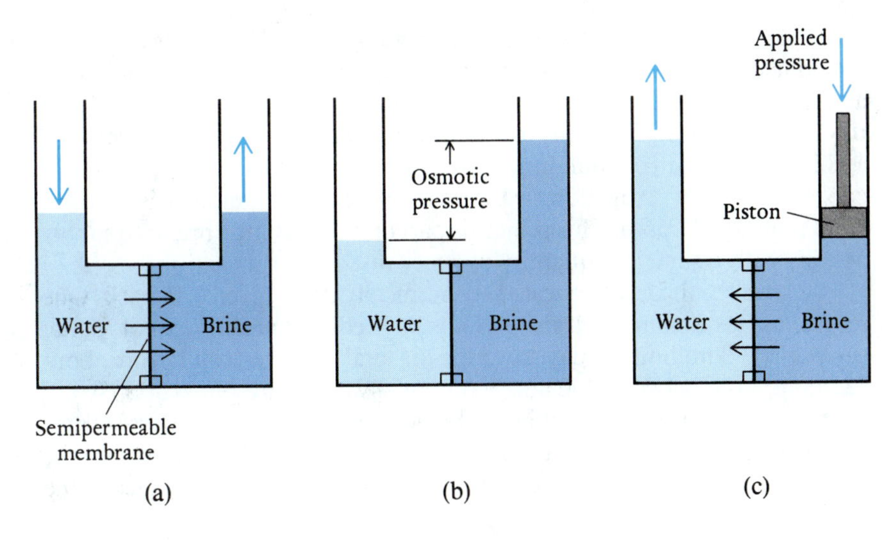

Figure 11-11 Normal osmosis is represented by (a) and (b). Water molecules create osmotic pressure by passing through the semipermeable membrane to dilute the brine solution. Reverse osmosis, represented in (c), is the application of an external pressure in excess of osmotic pressure to force water molecules to the pure water side.

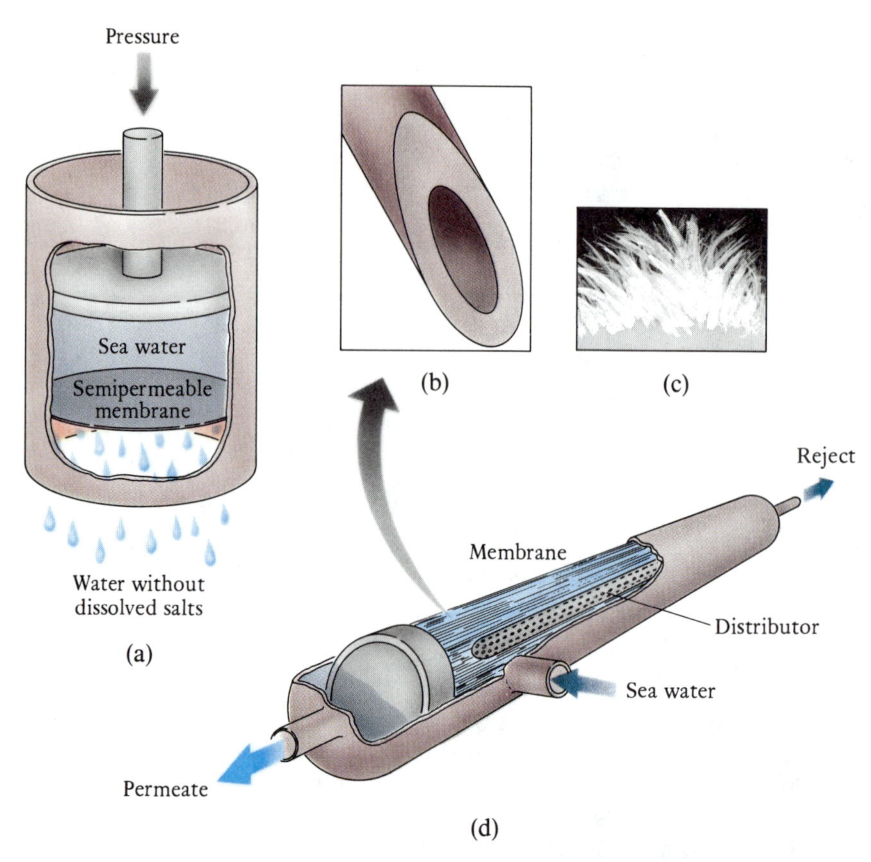

Figure 11–12 Commercial reverse osmosis units. (a) Mechanical pressure forces water against osmotic pressure to the region of pure water. (b) Enlargement of individual membrane. (c) Mass of many membranes. (d) Industrial unit; feed water (salt) that passes through membranes collects at the left end (permeate). The more concentrated salt solution flows out to the right as the reject.

The most common semipermeable membrane used in reverse osmosis is a modified cellulose acetate polymer, although several polyamide polymers also have been used. The largest reverse osmosis plant in operation today is the Yuma Desalting Plant in Arizona. This plant, which began operation in 1982, can produce 100 million gal/day. The plant was built to reduce the salt concentration of irrigation wastewater in the Colorado River from 3200 ppm to 283 ppm. This project is part of a U.S. commitment to supply Mexico with a sufficient quantity of water suitable for irrigation.

Irrigation water of desert fields dissolves about 2 tons of salt per acre per year. Irrigation wastewater carries the salt back to the Colorado River.

Sea water can also be purified economically using reverse osmosis. One reverse osmosis plant, built in 1983 on the Mediterranean island of Malta, can purify 5.3 million gal/day, lowering the total dissolved solids from about 38,000 ppm to between 400 and 500 ppm—well within the World Health Organization's limits for drinking water. Malta now uses four reverse osmosis plants that produce a total of 12 million gal of fresh water per day from the sea. On Florida's Sanibel Island, increasing salinity in its wellwater led to the installation of a reverse osmosis system. This facility has a design capacity of 3.6 million gal/day and has one of the lowest energy consumptions per 1000 gal of potable water of any comparably sized system in commercial use.

SELF-TEST 11C

1. Primary water treatment involves settling followed by ____________.
2. An oxygen-rich environment is called ____________.
3. An oxygen-poor environment is called ____________.
4. What chemical element is used to purify water? ____________.
5. What is the most abundant ion in sea water? ____________.
6. One water purification method that is the opposite of osmosis is called ________________________.

QUESTIONS

1. Approximately how many billions of gallons of water fall on the contiguous United States every day?
2. Distinguish between surface water and groundwater.
3. What is an aquifer?
4. Define the term *polluted water.*
5. Name four classes of water pollutants and give an example of each as it pollutes water in your community.
6. What is BOD and how can a high BOD level be harmful? Describe the kinds of pollutants that can cause the BOD to be high.
7. How does industry contribute to water pollution? List some ways this type of pollution can be prevented.
8. Name three of the top ten states with the largest releases of hazardous chemicals into water.
9. Name ten household wastes that can be hazardous wastes. By each waste, indicate how they are disposed of in your household. Make an estimate of how much (in pounds) of each waste your household disposes of annually. Then multiply by your estimate of how many households there are in the United States, and indicate the total pounds disposed of per year in the United States.
10. Explain why primary water treatment alone is insufficient to purify polluted water.
11. Explain how secondary water treatment involving air oxidation is like chlorination of water.
12. Name two diseases that can be effectively controlled by chlorination of waste water.
13. What is an undesirable side effect of the chlorination of waste water?
14. Explain how reverse osmosis works to purify water.
15. What is the "Superfund" and what is it applied to?
16. Name one state with no Superfund sites.
17. What do the initials RCRA relate to?
18. A community discovers drums containing waste solvent buried in a landfill of household wastes. Describe what problems this waste might cause.
19. Assume that you have several drums of a sludge containing chromium and lead. Call local landfill authorities and ask them how you can dispose of it. Record their answers, and be prepared to discuss them in class.
20. Explain how a chemical with hazardous properties can be considered not a hazardous waste.

Urban air pollution is not a recent phenomenon as shown in this painting by Claude Monet, entitled *The Gare St-Lazare, Paris: Arrival of a Train.* (Collection of Maurice Wertheim, Harvard University Art Museum)

CHAPTER

12 Clean Air—Should It Be Taken for Granted?

The air we breathe is often taken for granted, and yet this vast ocean of nitrogen, oxygen, and other chemicals often gets polluted by nature and by mankind's activities.

1. What is the composition of unpolluted air?
2. What is air pollution, and what chemical reactions cause air pollution?
3. What is smog?
4. What is the difference between "good" ozone and "bad" ozone?
5. What causes acid rain, and what damage does it do?
6. What is causing a hole in the ozone layer?
7. What is the greenhouse effect, and what can be done about it?
8. How does industry pollute the atmosphere?
9. What causes indoor air pollution?
10. What are the dangers of radon pollution in the home?

Planet Earth is enveloped by a few vertical miles of chemicals that compose the gaseous medium in which we exist—the atmosphere. Close to the Earth's surface and near sea level, the atmosphere is mostly nitrogen (80%) and life-sustaining oxygen (20%). It is the few little fractions of a percentage point of other chemicals that make a difference in the quality of life in various spots on Earth. Extra water in the atmosphere can mean a rain forest; a little less water produces a balanced rainfall; and practically no water results in a desert.

The atmosphere of the Earth is a fantastically large source of the elements nitrogen (N_2) and oxygen (O_2) and much smaller amounts of certain of the noble gases, including argon (Ar), neon (Ne), and xenon (Xe) (see Table 12–1). Figure 12–1 presents some of the basic facts about our atmosphere, including the naming of the stratified layers, the chemical species present in the layers, atmospheric pressures, and human activity. Our main concerns are with the layers called the troposphere (the air we breathe and where our weather takes place) and the stratosphere, where the UV-protective ozone layer is found.

Urbanization created an unhealthful, unpleasant medium for the existence of human life. With its consequent vast number of vehicles and increase in industrialization, urbanization produced an unwanted (and for a while ignored) increase in some of the pesky, naturally occurring "minor" chemicals in the atmosphere (nitrogen oxides, sulfur dioxide, carbon monoxide, carbon dioxide, and ozone). An atmosphere containing such unwanted and harmful ingredients is called **polluted.**

TABLE 12–1 The Dry Atomspheric Composition of the Earth at Sea Level

Gas	Percentage by Volume
Nitrogen	78.084
Oxygen	20.948
Argon	0.934
Carbon dioxide	0.033
Neon	0.00182
Hydrogen	0.0010
Helium	0.00052
Methane	0.0002
Krypton	0.0001
Xenon	0.000008

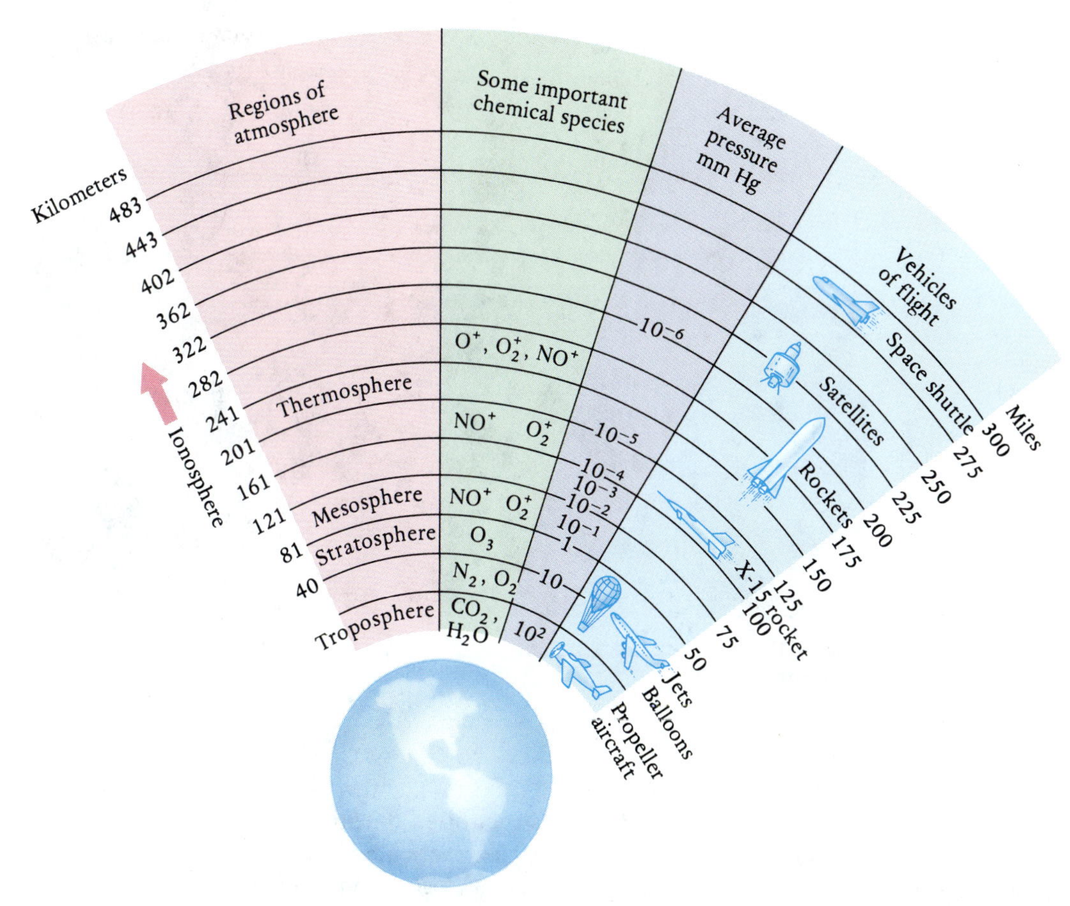

Figure 12–1 Some facts about our limited atmosphere. The troposphere was named by British meteorologist Sir Napier Shaw from the Greek word *tropos,* meaning "turning." The stratosphere was discovered by the French meteorologist Leon Philippe Teisserenc de Bort, who believed that this region consisted of an orderly arrangement of layers with no turbulence or mixing. The word *stratosphere* comes from the Latin word *stratum,* meaning "layer."

Air pollution is nothing new. Shakespeare wrote about it in the 17th century.

> *this most excellent canopy, the air,*
> *look you, this excellent o'erhanging*
> *firmament, this magestical roof fretted*
> *with golden fire, why, it appears*
> *no other thing to me but a foul*
> *and pestilent congregation of vapors.*

Hamlet *(act II, scene 2)*

Nature pollutes the air on a massive scale with volcanic ash, mercury vapor, and hydrogen sulfide from volcanoes and reactive and odorous organic compounds from coniferous plants such as pine trees. But automobiles, power plants, smelting plants, other metallurgical plants, and petroleum refineries add significant quantities of toxic chemicals to the atmosphere, especially in heavily populated areas. Atmospheric pollutants

Figure 12–2 The New York City skyline obscured by air pollution. (John D. Cunningham/Visuals Unlimited)

cause burning eyes, coughing, acid rain, smog, the destruction of ancient monuments, and even the destruction of the atmosphere itself.

A few decades ago, we operated on the principle that "Dilution is the solution to pollution."

Prior to 1960, there was little concern about air pollution. Most smoke, carbon monoxide, sulfur dioxide, nitrogen oxides, and organic vapors were emitted into the air with little apparent thought of their harmful nature as long as they were scattered into the atmosphere and away from human smell and sight (Fig. 12–2).

Early in the 1960s, air pollution became generally recognized as a problem and caused widespread concern, although devastating air pollution was prevalent earlier in certain geographical areas such as London, England, and where volcanic eruptions and burning of large areas occurred.

THE CLEAN AIR ACT OF 1990

On November 15, 1990, the President signed into law the 1990 Clean Air Act (CAA) amendments. These were the first major overhaul of U.S. air laws since 1977. The 1990 CAA affects almost everything we manufacture and consume in this country, all in the name of cleaner, safer air. The 1990 CAA speaks of the very chemicals discussed in this chapter—particulates, ozone, carbon monoxide, oxides of nitrogen and sulfur, hydrocarbons, carbon dioxide, and stratospheric ozone-depleting chemicals. Let's begin by looking at the particles that obscure our vision.

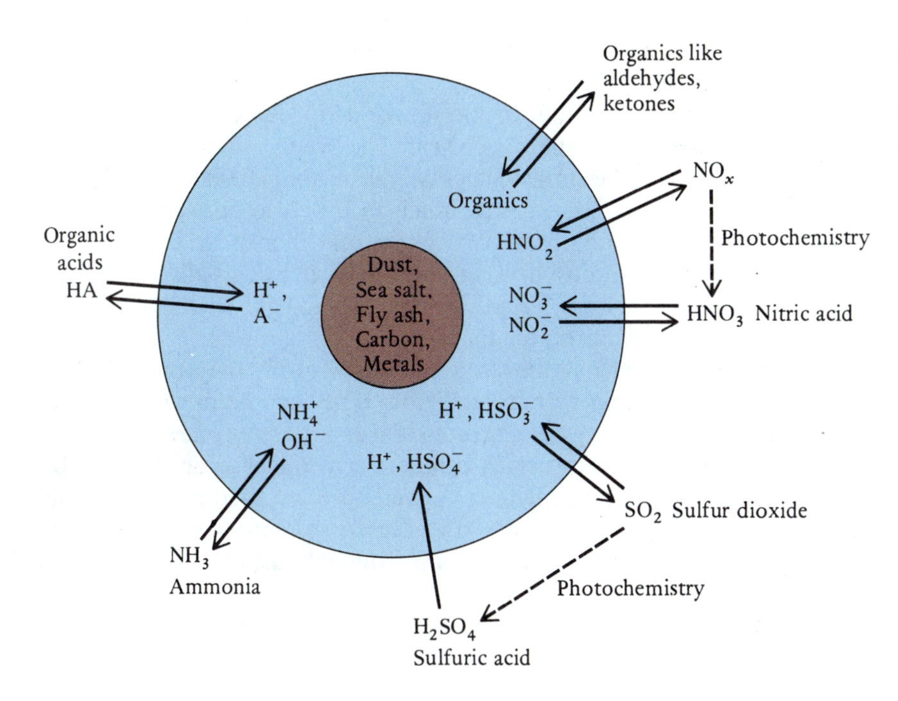

Figure 12–3 Schematic of an aerosol particle and some of its chemical reactions involving urban air pollutants.

AIR POLLUTANTS—PARTICLE SIZE MAKES A DIFFERENCE

Pollutants may exist in particle sizes from fly ash particles big enough to see down to single, isolated molecules, ions, or atoms. Often, because of their polar nature, many pollutants are attracted into water droplets and form **aerosols,** or onto larger particles called **particulates.** The solids in an aerosol or particulate may be various dusts, consisting of metal oxides and soil particles, sea salt, fly ash from electric generating plants and incinerators, elemental carbon, or even small metal particles. Aerosols range upward from a diameter of 1 nm (nanometer) to about 10,000 nm and may contain as many as a trillion atoms, ions, or small molecules. Aerosol particles are small enough to remain suspended in the atmosphere for long periods. Such particles are easily breathable and can cause lung diseases. They may also contain mutagenic or carcinogenic compounds. Smoke, dust, clouds, fog, mist, and sprays are typical aerosols. Because of their vast combined surface area, aerosol particles have great capacities to *adsorb* and concentrate chemicals on the surfaces of the particles. Liquid aerosols or particles covered with a thin coating of water may *absorb* air pollutants, thereby concentrating them and providing a medium in which reactions may occur. A typical urban aerosol is shown schematically in Figure 12–3.

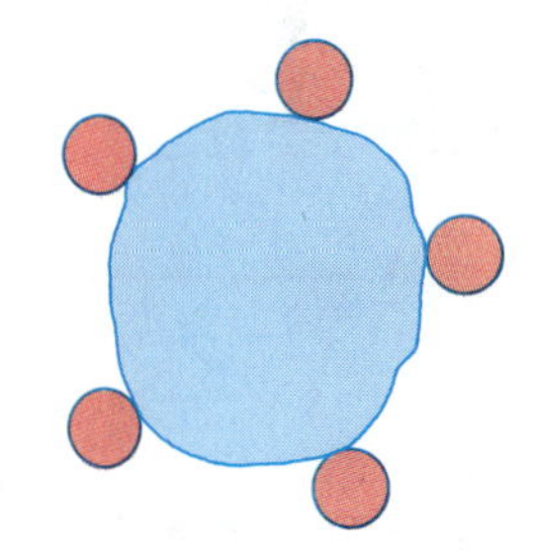

Adsorption is the attachment of particles to a surface.

1 nanometer = 10^{-9} m.

Absorption is pulling particles inside.

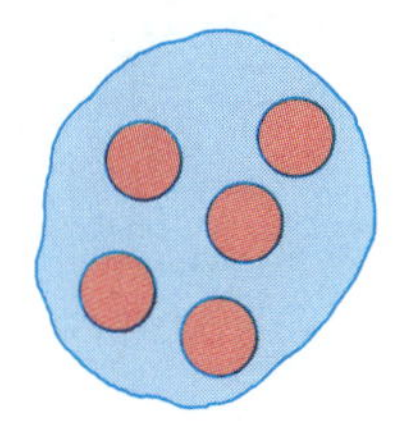

Ammonia may concentrate in this aerosol as ammonium hydroxide, sulfur dioxide may react to form sulfurous acid, nitric oxide may form nitrous acid, and many other reactions may occur.

Many aerosols are large enough to be seen; these are the particulates. Particles of sizes below 2 μm (microns) in diameter are largely responsible

1 μm = 10^{-6} m, or 1000 nm.

for the deterioration of visibility often observed in highly populated cities like Los Angeles and New York.

Major contributors to the amount of atmospheric particulates are volcanic eruptions by Krakatoa, Indonesia, 1883; Mt. Katmai, Alaska, 1912; Hekla, Iceland, 1947, Mt. Spurr, Alaska, 1953; Bezymyannaya, U.S.S.R., 1956; Mt. St. Helens, Washington, 1980; Mt. Pinatubo, 1991.

Millions of tons of soot, dust, and smoke particles are emitted into the atmosphere of the United States each year. The average suspended particulate concentrations in the United States vary from about 0.00001 g/m^3 of air in rural areas to about six times as much in urban locations. In heavily polluted areas, concentrations of particulates may increase to 0.002 g/m^3.

Particulates in the atmosphere can cool the Earth by partially shielding it from the Sun. Large volcanic eruptions such as that from Mt. St. Helens in 1980 have a cooling effect on the Earth.

Particulates and aerosols are removed naturally from the atmosphere by gravitational settling and by rain and snow. They can be prevented from entering the atmosphere by treating industrial emissions by one or more of a variety of physical methods such as filtration, centrifugal separation, scrubbing, and electrostatic precipitation. A method often used is electrostatic precipitation, which is better than 98% effective in removing aerosols and dust particulates even smaller than 1 μm from exhaust gases of industrial plants.

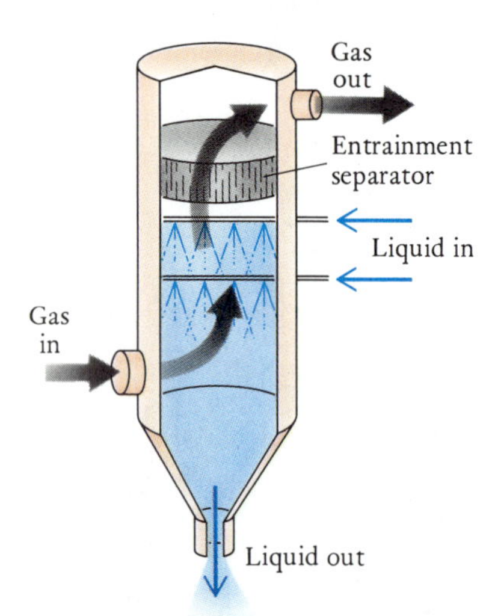

Removing particulates and aerosols by scrubbing. Schematic drawing of a spray collector, or scrubber.

SMOG

The poisonous mixture of smoke, fog, air, and other chemicals was first called **smog** in 1911 by Dr. Harold de Voeux in his report on a London air pollution disaster that caused the deaths of 1150 people. Through the years, smog has been a technological plague in many communities and industrial regions.

What general conditions are necessary to produce smog? Although the chemical ingredients of smogs often vary, depending on the unique sources of the pollutants, certain geographical and meteorological conditions exist in nearly every instance of smog.

Thermal inversion: mass of warmer air over a mass of cooler air.

There must be a period of windlessness so that pollutants can collect without being dispersed vertically or horizontally. This lack of movement in the ground air can occur when a layer of warm air rests on top of a layer of cooler air. This sets the conditions for a **thermal inversion,** which is an abnormal temperature arrangement for air masses (Fig. 12–4). Normally the warmer air is on the bottom nearer the warm Earth, and this warmer, less dense air rises and transports most of the pollutants to the upper troposphere where they are dispersed. In a thermal inversion the warmer air is on top, and the cooler, more dense air retains its position nearer the Earth. The air becomes stagnated. If the land is bowl shaped (surrounded by mountains, cliffs, or the like), this stagnant air mass can remain in place for quite some time.

When these natural conditions exist, humans supply the pollutants by combustion and evaporation in automobiles, electric power plants, space heating, and industrial plants. The chief pollutants are sulfur dioxide (from burning coal and some oils), nitrogen oxides, carbon monoxide, and hydrocarbons (chiefly from the automobile). Add to these ingredients the radiation from the Sun, and a massive smog is in the offing.

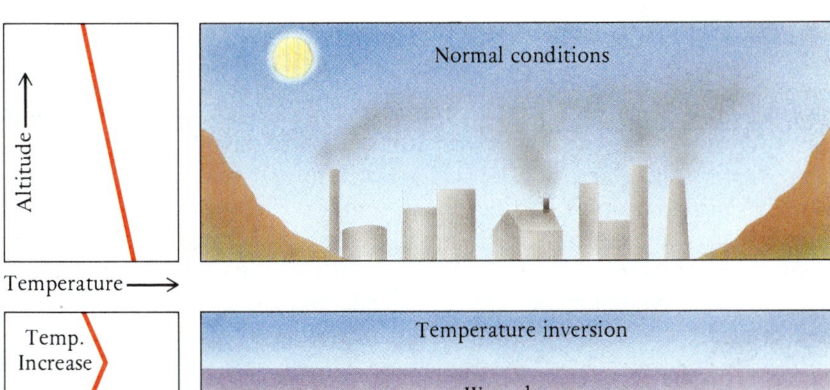

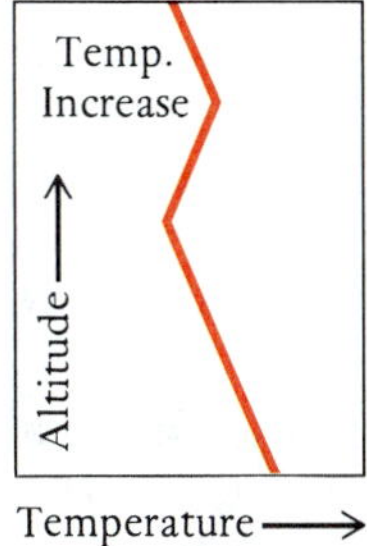

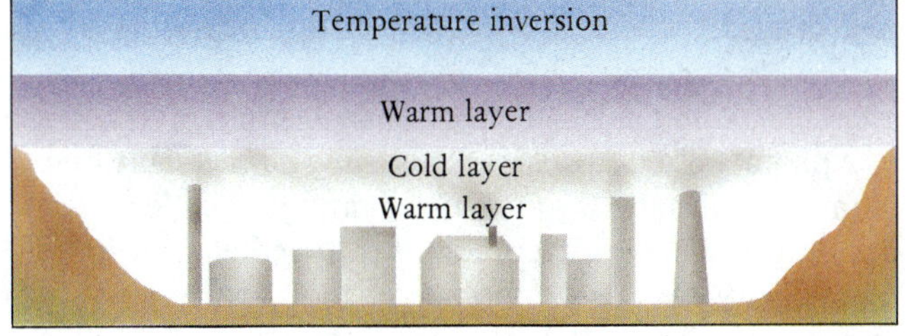

Figure 12–4 A diagram of a temperature inversion over a city. Warm air over a polluted air mass effectively acts as a lid, holding the polluted air over the city until the atmospheric conditions change. The line on the left of the diagram indicates the relative air temperature.

Two general kinds of smog have been identified. One is the chemically reducing type that is derived largely from the combustion of coal and oil and contains sulfur dioxide mixed with soot, fly ash, smoke, and partially oxidized organic compounds. This is the **industrial type** (sometimes called London smog), which is diminishing in intensity and frequency as less coal is burned and more controls are installed.

The type of smog formed around some industrial centers and power plants is thought to be caused by sulfur dioxide. Laboratory experiments have shown that sulfur dioxide increases aerosol formation, particularly in the presence of mixtures of hydrocarbons, nitrogen oxides, and air energized by sunlight (Fig. 12–5). For example, mixtures of 3 ppm olefin, 1 ppm NO_2,

"Olefin" is another name for an unsaturated hydrocarbon.

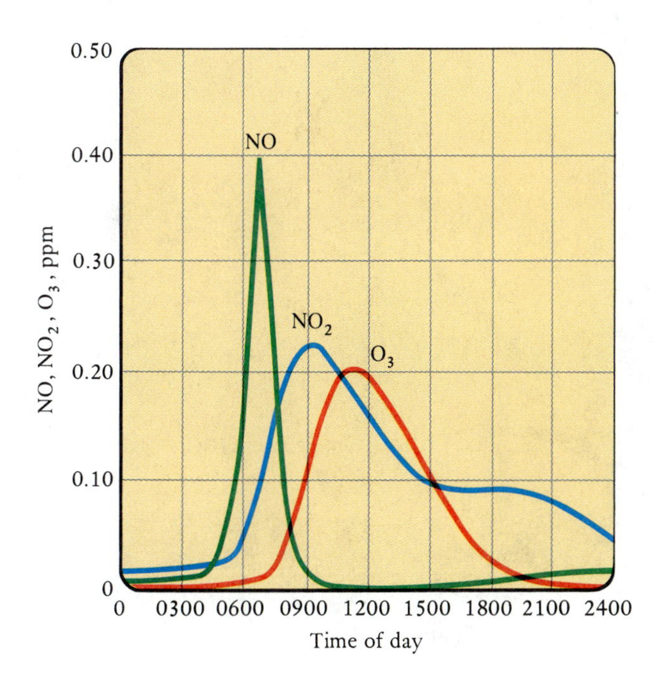

Figure 12–5 Average concentrations of the pollutants NO, NO_2, and O_3 on a smoggy day in Los Angeles, California. The NO concentration builds up first during the morning rush hour. Later in the day the concentrations of NO_2 and O_3 build up.

and 0.5 ppm SO_2 at 50% relative humidity form aerosols that have sulfuric acid as a major product. Even with 10% to 20% relative humidity, sulfuric acid is a major product. Sulfuric acid, which is formed in this kind of smog, is very harmful to people suffering from respiratory diseases such as asthma or emphysema. At a concentration of 5 ppm for 1 hr, SO_2 can cause constriction of bronchial tubes. A level of 10 ppm for 1 hr can cause severe distress.

Relative humidity is a measure of the amount of water vapor air contains compared with the maximum amount it can contain.

Industrial smog: fog + SO_2.

Photochemical smog: fog + NO_x + hydrocarbons.

A second type of smog is the chemically oxidizing type, typical of Los Angeles and other cities where exhausts from internal combustion engines are highly concentrated in the atmosphere. This type is called **photochemical smog** because light—in this instance sunlight—is important in initiating the photochemical process. This smog is practically free of sulfur dioxide but contains substantial amounts of nitrogen oxides, ozone, ozonated olefins, and organic peroxide compounds, together with hydrocarbons of varying complexity. Detailed laboratory studies have revealed that the chemical reactions in the smog-making process involve aerosols that serve to keep the reactants together long enough to form **secondary pollutants.** Ultraviolet

Organic peroxides contain the R—O—O—R′ structure and are produced by ozone reacting with organic molecules. Hydrogen peroxide is H—O—O—H.

Figure 12–6 The formation of photochemical smog. The details of these processes are discussed in this chapter.

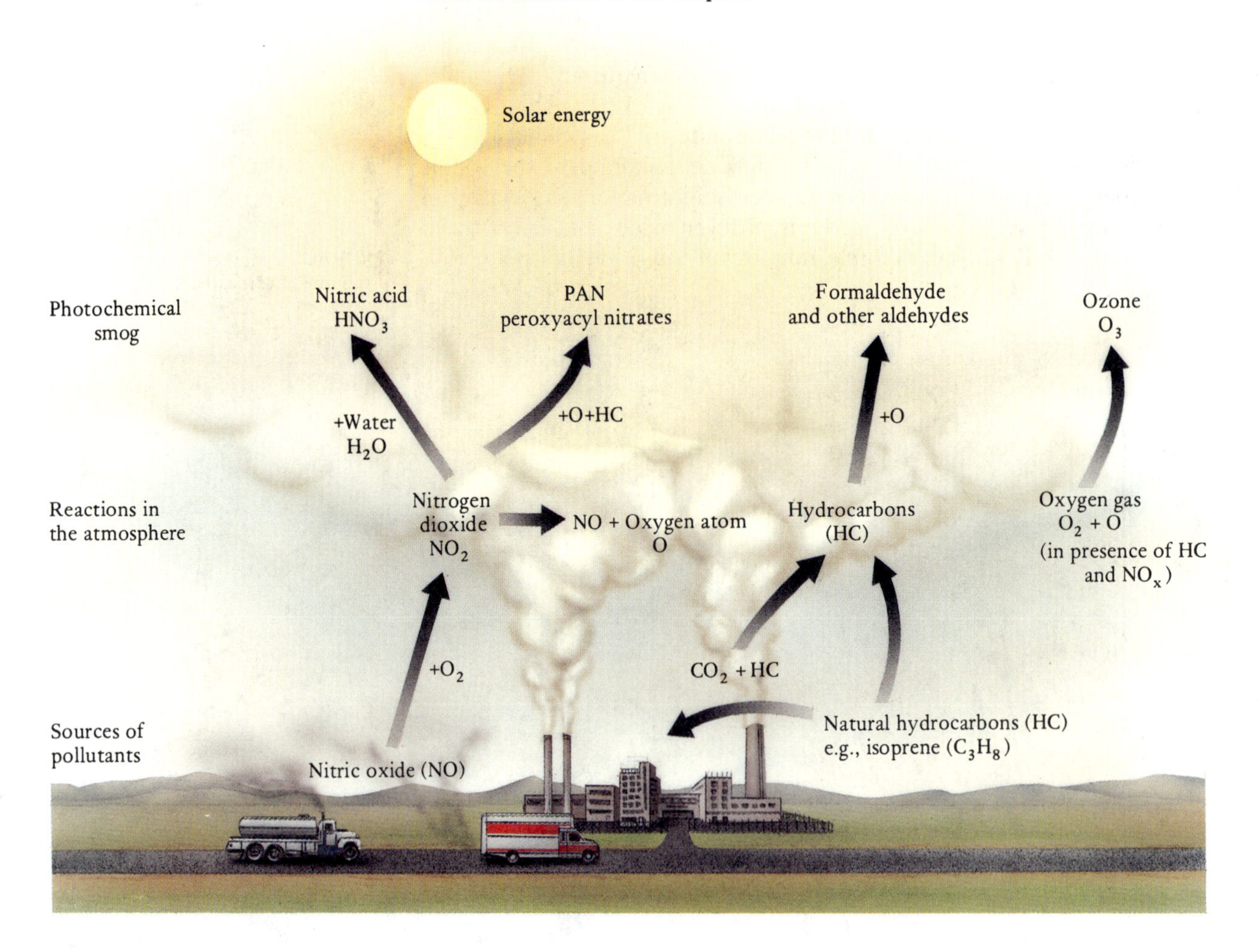

radiation from the sun is the energy source for the formation of this photochemical smog.

The exact reaction scheme by which primary pollutants are converted into the secondary pollutants found in smog is still not completely understood (Fig. 12–6). The process is thought to begin with the absorption of a quantum of light by nitrogen dioxide, which causes its breakdown into nitrogen oxide and atomic oxygen, a chemical radical. The very reactive atomic oxygen reacts with molecular oxygen to form ozone (O_3), which is then consumed by reacting with nitrogen oxide to form the original reactants—nitrogen dioxide and molecular oxygen. Atomic oxygen, however, also reacts with reactive hydrocarbons—olefins and aromatics—to form other chemical radicals. These radicals, in turn, react to form other radicals and secondary pollutants such as aldehydes (e.g., formaldehyde). About 0.2 ppm of nitrogen oxides and 1 ppm of reactive hydrocarbons are sufficient to initiate these reactions. The hydrocarbons involved come mostly from unburned petroleum products like gasoline.

Primary pollutants: pollutants emitted into the air.

Secondary pollutants: pollutants formed in the air by chemical reaction.

A chemical radical is a species with an unpaired valence electron. They are usually very reactive and short lived.

In the following sections, we shall look at the major ingredients of photochemical smog, the primary pollutants, the oxides of nitrogen and hydrocarbons, and the secondary pollutant ozone, and see how they interact with oxygen in our atmosphere to produce urban pollution.

NITROGEN OXIDES

There are eight known oxides of nitrogen, two of which are recognized as important components of the atmosphere: dinitrogen oxide (N_2O) and nitrogen dioxide (NO_2) (see Figure 2–5 for a photo of NO_2).

About 97% of the nitrogen oxides in the atmosphere are naturally produced and only 3% result from human activity.

Most of the nitrogen oxides emitted are in the form of NO, a colorless reactive gas. In a combustion process involving air, some of the atmospheric nitrogen reacts with oxygen to produce NO (nitric oxide):

$$N_2 + O_2 + \text{heat} \longrightarrow 2\,NO$$

Nitric oxide is formed in this manner during electrical storms. Because the formation of nitrogen oxide requires heat, it follows that a higher combustion temperature would produce relatively more NO.

In the atmosphere NO reacts rapidly with atmospheric oxygen to produce NO_2:

$$2\,NO + O_2 \longrightarrow \underset{\text{Nitrogen dioxide}}{2\,NO_2}$$

Normally the atmospheric concentration of NO_2 is a few parts per billion (ppb) or less. In the United States, most oxides of nitrogen (NO_x) are produced from fossil fuel combustion such as automobile engines burning gasoline, with significantly less coming from natural sources like lightning (Table 12–2).

In a concentration of 3 ppm for 1 hr, nitrogen dioxide causes bronchioconstriction in humans, and short exposures at high levels (150–220 ppm) cause changes in the lungs that produce fatal results. A seemingly harmless exposure one day can cause death a few days later.

The brown haze shown above this city is caused by NO_2. (National Center for Atmospheric Research)

TABLE 12-2 Emissions of NO_x

	Emissions (millions of tons)	
Source	*United States*	*Global*
Fossil fuel combustion	66	231
Biomass burning	1.1	132
Lightning	3.3	88
Microbial activity in soil	3.3	88
Input from the stratosphere	0.3	5.5
Total (uncertainty in estimates)	74 (±1)	544.5 (±275)

Note: The large uncertainty for global emissions is due to incomplete data for much of the world.
Source: Stanford Research Institute, 1983.

The symbol hν represents a photon of light. ν is the frequency. The larger the frequency, the greater the energy of the photon.

Ozone is discussed in the next section.

Nitrogen dioxide reacts with photons of light, hν, with a wavelength between 280 and 430 nm, in a **photodissociation** reaction that produces nitric oxide and free oxygen atoms (O, oxygen radicals) that can react with molecular oxygen to produce ozone.

$$NO_2 + h\nu \longrightarrow NO + O$$

$$O_2 + O \longrightarrow O_3$$

The nitric oxide then reacts with an ozone molecule to regenerate NO_2:

$$NO + O_3 \longrightarrow NO_2 + O_2$$

Nitric acid is a source of acid rain, which is discussed later in this chapter.

Nitrogen dioxide can also react with water, such as in aerosol particles, to form nitric acid and nitrous acid.

$$2\ NO_2 + H_2O \longrightarrow \underset{\text{Nitric acid}}{HNO_3} + \underset{\text{Nitrous acid}}{HNO_2}$$

Nitrates are important components of fertilizers.

These acids in turn can react with ammonia or metallic particles in the atmosphere to produce nitrate or nitrite salts. For example,

$$\underset{\text{Ammonia}}{NH_3} + HNO_3 \longrightarrow \underset{\text{Ammonium nitrate (a salt)}}{NH_4NO_3}$$

The acids or the salts, or both, ultimately form aerosols, which eventually settle from the air or dissolve in raindrops. Nitrogen dioxide, then, is a primary cause of haze in urban or industrial atmospheres because of its participation in the process of aerosol formation. Normally nitrogen dioxide has a lifetime of about three days in the atmosphere.

OZONE AND ITS ROLE IN AIR POLLUTION

Ozone is an allotropic form of oxygen consisting of three oxygen atoms bound together in a molecule with the formula O_3. Ozone has a pungent odor that can be detected at concentrations as low as 0.02 ppm. We often smell the ozone produced by sparking electric appliances, or after a thunderstorm when lightning-caused ozone washes out with the rainfall.

As we shall see in this chapter, there is "good" and "bad" ozone. The bad ozone is that found in the air we breathe, whereas the good ozone is found in the stratosphere, where it forms a protective blanket, absorbing harmful ultraviolet radiation.

The only significant chemical reaction producing ozone in the atmosphere is one involving molecular oxygen and atomic oxygen.

$$O_2 + O + M \longrightarrow O_3 + M$$

In the reaction above, M stands for a third molecule, like a nitrogen or an oxygen molecule. This third molecule takes away some of the energy of the reaction and keeps the ozone molecule from dissociating immediately. At high altitudes oxygen atoms are produced by ultraviolet photons striking oxygen molecules, and most of these high-energy photons are absorbed before they get to the lower atmosphere (see the discussion of the protective "good" ozone layer later in this chapter), where only photons with wavelengths greater than 280 nm are present. The low-energy photons are sufficiently energetic to react with nitric oxide,

$$NO_2 + h\nu \longrightarrow NO + O$$

so there can be plenty of oxygen atoms, especially if there is a ready source of nitric oxide, like automobile exhaust or other high-temperature combustion sources.

Ozone is a secondary air pollutant and is the most difficult pollutant to control. According to the EPA in 1990 the standard for ozone of 0.12 ppm was exceeded in 96 urban areas of the United States. This high ozone was primarily related to excess nitrogen oxides emissions from automobiles, buses, and trucks. Most major urban areas have vehicle inspection centers for passenger automobiles in an effort to control nitrogen oxide emissions as well as those of carbon monoxide and unburned hydrocarbons (Fig. 12–7).

Allotropes are different forms of the same element that differ significantly in physical and chemical properties. Diatomic oxygen (O_2) and triatomic ozone (O_3) are allotropes. Carbon has three common allotropes: graphite, diamond, and carbon black. A new allotropic form, C_{60}, is discussed in Chapter 6.

Figure 12–7 At this test station, automobiles are tested for carbon monoxide, nitrogen oxides, and unburned hydrocarbons. Those failing the standards established by the EPA must be repaired.

Figure 12–8 Ozone can affect the breathing of children at play in urban environments. (*The World of Chemistry,* Program 25, "Chemistry and the Environment")

As difficult as it is to attain, the ozone standard may not be low enough for good health. Exposure to concentrations of ozone at or near 0.12 ppm lowers the volume of air a person breaths out in 1 sec (the forced expiratory volume, or FEV_1). Children studied who were exposed to ozone concentrations close to the EPA standard, but not exceeding it, showed a 16% decrease in the FEV_1. Some scientists have been urging the EPA to lower the standard to 0.08 ppm. If that is done even some midsized cities would probably fail to meet the EPA standards.

Lower FEV_1 accelerates the aging of the lungs.

No matter what the standard becomes, present ozone concentrations in many urban areas represent health hazards to children at play (Fig. 12–8), joggers, others doing outdoor exercise, and older persons who may have diminished respiratory capabilities.

HYDROCARBONS AND AIR POLLUTION

Hydrocarbons occur in the atmosphere from both natural sources and human activities. Isoprene and α-pinene are produced in large quantities by both coniferous and deciduous trees. Methane gas is produced by such diverse sources as ruminant animals, termites, ants, and decay-causing bacteria acting on dead plants and animals. Human activities such as the use of industrial solvents, petroleum refining and its distribution, and unburned gasoline and diesel fuel components account for a large amount of hydrocarbons in the atmosphere. The types of hydrocarbons found in urban air (Table 12–3) read like something you saw in the organic chemistry chapter (Chapter 6).

In September, 1988, William Chameides, of Georgia Tech in Atlanta, published a report in *Science* magazine, in which he stated that in some cities trees may account for more hydrocarbons in the atmosphere than that produced from human activities. The EPA has since found this to be true. This fact is causing a rethinking about how to control urban pollution.

In addition to simpler hydrocarbons like alkanes, alkenes (olefins), and alkynes, a large number of polynuclear aromatic hydrocarbons (PAH) are released into the atmosphere, primarily from motor vehicle exhaust. These chemicals can react with hydroxyl radicals and oxygen much like simpler hydrocarbons, but their greatest danger is their toxic properties. One PAH, benzo(α)pyrene (BAP), is a known carcinogen. Concentrations as high as

TABLE 12-3 The Ten Most Abundant Ambient Air Hydrocarbons Found in Cities. Results from 800 Air Samples Taken in 39 Different Cities

Compound	Concentration Median	Maximum ppb of Carbon
Isopentane	45	3393
n-Butane	40	5448
Toluene	34	1299
Propane	23	399
Ethane	23	475
n-Pentane	22	1450
Ethylene	21	1001
m-Xylene, *p*-Xylene	18	338
2-Methylpentane	15	647
Isobutane	15	1433

Source: Air Pollution Control Association, 1988.

Trees in urban environments may emit as many reactive hydrocarbons as automobiles do.

60 $\mu g/m^3$ of air have been found in urban air. Coal smoke contains about 300 ppm benzo(α)pyrene. Measurements have shown that for every million tons of coal burned, about 750 tons of benzo(α)pyrene can be produced. British researchers reported that a typical London resident in the 1950s inhaled about 200 mg of BAP a year. Heavy smokers (those who smoke about two packs a day without filters) receive an additional 150 mg a year. This is about 40,000 times the amount necessary to produce cancer in mice. Extracts of urban air taken at various times during the past decade have in fact produced cancer in mice, but not all of these cancers were caused by PAHs like benzo(α)pyrene; other organic chemicals were present as well (see later in this chapter).

Benzo(α)pyrene, a carcinogenic polynuclear aromatic hydrocarbon found in smoke.

SELF-TEST 12A

1. Parts per million is abbreviated ______________. To change from ppm to percent, divide by ______________. Thus, 10 ppm would be ______________ percent.
2. What has been the general trend in air pollutants for approximately the past decade? ( ) increase ( ) decrease.
3. Name a chemical that is considered both an air pollutant and a beneficial chemical. ______________
4. Because of their large surface areas, aerosol particles can ( ) absorb ( ) adsorb chemicals onto their surfaces.
5. A liquid aerosol particle will likely ( ) adsorb ( ) absorb a chemical.
6. A thermal inversion occurs when ( ) warm ( ) cool air is above ( ) warm ( ) cool air below.
7. Industrial type smog is often associated with ( ) coal burning ( ) sunlight.

8. In all combustion processes in air, some nitrogen ______________ are formed.

9. What are the products of the photodissociation of nitrogen dioxide?

____

10. What species reacts with molecular oxygen to form ozone?

____

SULFUR DIOXIDE—A MAJOR PRIMARY POLLUTANT

Sulfur dioxide is produced when sulfur or sulfur-containing compounds are burned in air.

$$S + O_2 \longrightarrow SO_2$$

Most of the coal burned in the United States contains sulfur in the form of the mineral pyrite (FeS_2). The weight percent of sulfur in this coal ranges from 1% to 4%. The pyrite is oxidized as the coal is burned.

$$4\ FeS_2 + 11\ O_2 \longrightarrow 2\ Fe_2O_3 + 8\ SO_2$$

Electric power plants are discussed in Chapter 5.

—SH
Mercaptan group.

Large amounts of coal are burned in this country to generate electricity. A 1000 megawatt coal-fired generating plant can burn about 700 tons of coal an hour. If the coal contains 4% sulfur, that equals 56 tons of SO_2 an hour, or 490,560 tons of SO_2 every year. Oil-burning electric generating plants can also produce comparable amounts of SO_2 because fuel oils can contain up to 4% sulfur. The sulfur in the oil is in the form of mercaptan compounds in which sulfur atoms are bound to carbon and hydrogen atoms. Eight states have the highest SO_2 emissions in the United States. Table 12–4 shows the number of coal-fired power plants and the amounts of SO_2 emitted in those eight states. Operators of these facilities are under EPA orders to eliminate most of the SO_2 before it reaches the stack. The 1990 Clean Air Act requires that by the year 2000, SO_2 emissions from all power-generating sources will be no greater than 8.9 million tons per year. That's a 10 million ton per year reduction from 1980 levels.

TABLE 12–4 Characteristics of Coal-Fired Power Plants in Eight States

State	Plants	SO_2 Emissions (thousand tons/yr)	Capacity (gigawatts)
Ohio	99	2,221	22.31
Indiana	66	1,588	14.58
Pennsylvania	70	1,427	17.93
Missouri	41	1,214	9.97
Illinois	59	1,136	15.75
West Virginia	33	966	14.46
Tennessee	37	950	9.41
Kentucky	54	947	11.82

Note: Some of these plants have emissions controls installed; others do not.
Source: EPA, 1988.

Most low-sulfur coals are mined far from the major metropolitan areas where they are most needed for power generation. The cleansing of sulfur from closer, high-sulfur coal is costly and incomplete. One method is to pulverize the coal to the consistency of talcum powder and remove the pyrite (FeS_2) by magnetic separation. Technology is available to decrease the sulfur content of fuel oil to 0.5%, but this process, too, is costly. It involves the formation of hydrogen sulfide (H_2S) by bubbling hydrogen through the oil in the presence of metallic catalysts, such as a platinum-palladium catalyst.

Several efficient methods are available to trap SO_2. In one method, limestone is heated to produce lime. The lime reacts with SO_2 to form calcium sulfite, a solid particulate, which can be removed from an exhaust stack by an electrostatic precipitator.

$$\underset{\text{Limestone}}{CaCO_3} \xrightarrow{\text{Heat}} \underset{\text{Lime}}{CaO} + CO_2$$

$$CaO + SO_2 \longrightarrow \underset{\text{Calcium sulfite}}{CaSO_3(s)}$$

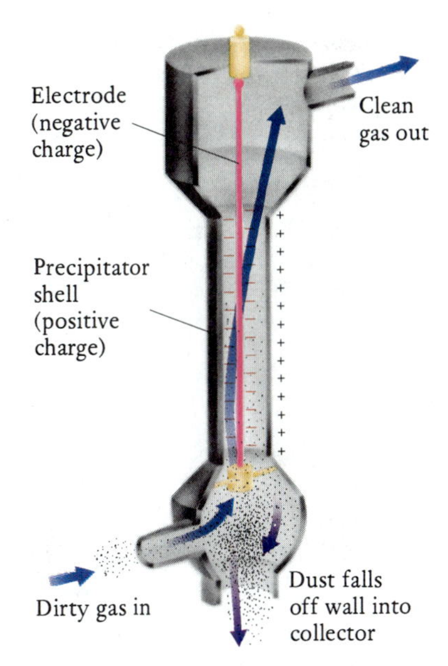

An electrostatic precipitator. The central electrode is negatively charged and imparts a negative charge to particles in the gases. These are attracted to the positively charged walls.

Another trapping method involves the passage of SO_2 through molten sodium carbonate. Solid sodium sulfite is formed.

$$SO_2 + \underset{\text{Sodium carbonate}}{Na_2CO_3} \xrightarrow{800°C} \underset{\text{Sodium sulfite}}{Na_2SO_3} + CO_2$$

The less desirable method of dissipating SO_2 is by tall stacks. Although tall stacks emit SO_2 into the upper atmosphere away from the immediate vicinity and give SO_2 a chance to dilute itself on the way down, the fact remains that SO_2 does come down, and the longer it stays up the greater chance it has to become sulfuric acid. A 10-year study in Great Britain showed that although SO_2 emissions from power plants increased by 35%, the construction of tall stacks decreased the ground level concentrations of SO_2 by as much as 30%. The question is, who got the SO_2? In this case, Britain's solution was others' pollution. In the United States, the EPA may have added to a pollution problem unwittingly with rules in 1970 that caused plants to increase the height of smokestacks and caused pollutants to be carried longer distances by winds. There are about 179 stacks in the United States that are 500 ft or higher, and 20 stacks are 1000 ft or more tall.

Most of the SO_2 in the atmosphere reacts with oxygen to form sulfur trioxide (SO_3). Several reactions are possible. SO_2 may react with atomic oxygen:

$$SO_2 + O \longrightarrow SO_3$$

It may also react with molecular oxygen:

$$2\ SO_2 + O_2 \longrightarrow 2\ SO_3$$

The SO_3 formed has a strong affinity for water and dissolves in aqueous aerosol particles, forming sulfuric acid, a strong acid.

$$SO_3 + H_2O \longrightarrow H_2SO_4$$

Sulfur dioxide in the air is harmful to people, animals, plants, and buildings.

Sulfur dioxide can corrode metals and decay building stones, in particular marble and limestone. Both marble and limestone are forms of calcium carbonate ($CaCO_3$), which reacts readily with acid (H^+) and with SO_2 and water.

$$CaCO_3 + 2\,H^+ \longrightarrow Ca^{2+} + H_2O + CO_2$$

$$CaCO_3 + SO_2 + 2\,H_2O \longrightarrow \underset{\text{Calcium sulfite (soluble)}}{CaSO_3 \cdot 2H_2O} + CO_2$$

An alarming example is the disintegration of marble statues and buildings on the Acropolis in Athens, Greece. As all coatings have failed to protect the marble adequately, the only known solution is to bring the prized objects into air-conditioned museums protected from SO_2 and other corroding chemicals.

Sulfur dioxide is physiologically harmful to both plants and animals. Most healthy adults can tolerate fairly high levels of SO_2 without apparent lasting ill effects. Individuals with chronic respiratory difficulties such as bronchitis or asthma tend to be much more sensitive to SO_2.

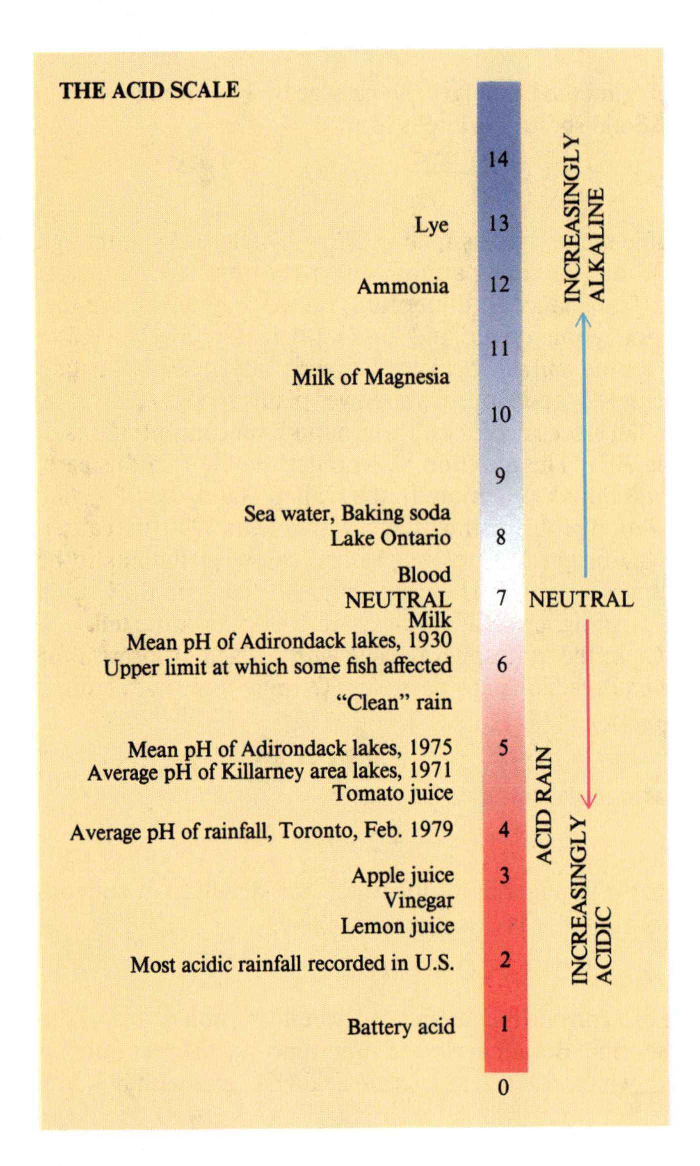

Figure 12–9 The pH of acid rain as compared with the pH of other mixtures.

ACID RAIN

The term **acid rain** was first used in 1872 by Robert Angus Smith, an English chemist and climatologist. He used the term to describe the acidic precipitation that fell on Manchester just at the start of the Industrial Revolution. Although neutral water has a pH of 7, rainwater becomes naturally acidified from dissolved carbon dioxide, a normal component of the atmosphere. The carbon dioxide reacts reversibly with water to form a solution of the weak acid, carbonic acid.

Reversibility and equilibria are discussed in Chapter 2. Weak acids are discussed in Chapter 4. The longer double arrow to the left indicates that very little CO_2 is converted to carbonic acid.

$$H_2O + CO_2 \leftleftharpoons H^+ + HCO_3^-$$

At equilibrium, the pH of this solution is 5.6. Thus, natural rainfall is slightly acidic from carbon dioxide (Fig. 12–9). Any precipitation with a pH below 5.6 is considered excessively acidic.

As we have seen earlier in this chapter, NO_2 and SO_2 can react with chemicals in the atmosphere to produce acids. These gases can dissolve in water droplets or aerosol particles where they greatly lower the pH, and when conditions are favorable, this moisture precipitates as rain or snow. NO_2 yields nitric acid (HNO_3) and nitrous acid (HNO_2); SO_2 yields sulfuric acid (H_2SO_4) and sulfurous acid (H_2SO_3). Ice core samples taken in Greenland and dating back to 1900 contain sulfate (SO_4^{2-}) and nitrate (NO_3^-) ions. This indicates that at least from 1900 onward, acid rain has been commonplace, and more importantly, has been deposited far from where the oxides of nitrogen and sulfur were formed.

A more adequate term for acid rain might be acid precipitation. Some scientists use *acid deposition.*

Acid rain is a problem today due to the large annual amounts of these acidic oxides being produced by human activities and put into the atmosphere (Fig. 12–10). When this acidic precipitation falls on natural areas that cannot easily tolerate all of this abnormal acidity, serious environmental problems occur. The average annual pH of precipitation falling on much of the northeastern United States and northeastern Europe is between 4 and

The March 1991 eruption of Mt. Pinatubo in the Philippines injected over 10^8 kg of SO_2 into the stratosphere. The SO_2 will come down as acid rain. Aerosols of particles containing sulfuric acid are presently enhancing the beauty of sunrises and sunsets.

Figure 12–10 Major source and components of acid rain.

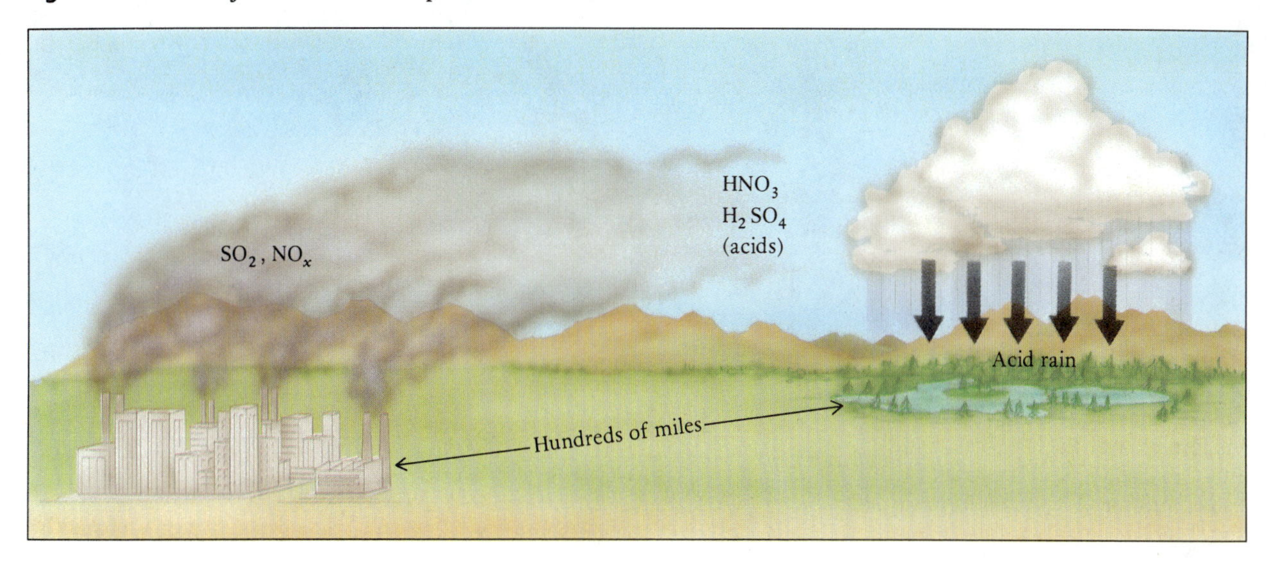

4.5. Specific storms in some areas like West Virginia have been recorded with rain having pH values as low as 1.5. Further complicating matters is the fact that acid rain is an international problem—rain and snow don't observe borders. Many Canadian residents are offended by the government of the United States because some of the acid rain produced in the United States falls on Canadian cities and forests (Fig. 12–11).

Acid rain destroys lakes.

The extent of the problems with acid rain can be seen in dead (fishless) ponds and lakes, dying or dead forests, and crumbling buildings. Because of wind patterns, Norway and Sweden have received the brunt of western Europe's emission of sulfur oxides and nitrogen oxides as acid rain. As a result, of the 100,000 lakes in Sweden, 4000 have become fishless, and 14,000 other lakes have been acidified to some degree. In the United States, 6% of all ponds and lakes in the Adirondack Mountains of New York are now fishless, and 200 lakes in Michigan are dead. For the most part, these "dead" lakes are still picturesque, but no fish can live in the acidified water. Lake trout and yellow perch die at pH below 5.0, and smallmouth bass die at pH below 6.0. Mussels die at pH below 6.5.

Dead trees caused by acid rain. (*The World of Chemistry*, Program 16, "The Proton in Chemistry")

Acid rain damages trees in several other ways. It disturbs the stomata (openings) in tree leaves and causes increased transpiration and a water deficit in the tree. Acid rain can acidify the soil, damaging fine root hairs, and thus diminish nutrient and water uptake. The acid can leach out needed minerals in the soil, and the minerals are carried off in the groundwater. The surface structures of the bark and the leaves can be destroyed by the acid in the rain. The effects of acid rain have been so severe on some forests already that experts predict that they will be lost, possibly forever. This is the case with Germany's famous Black Forest.

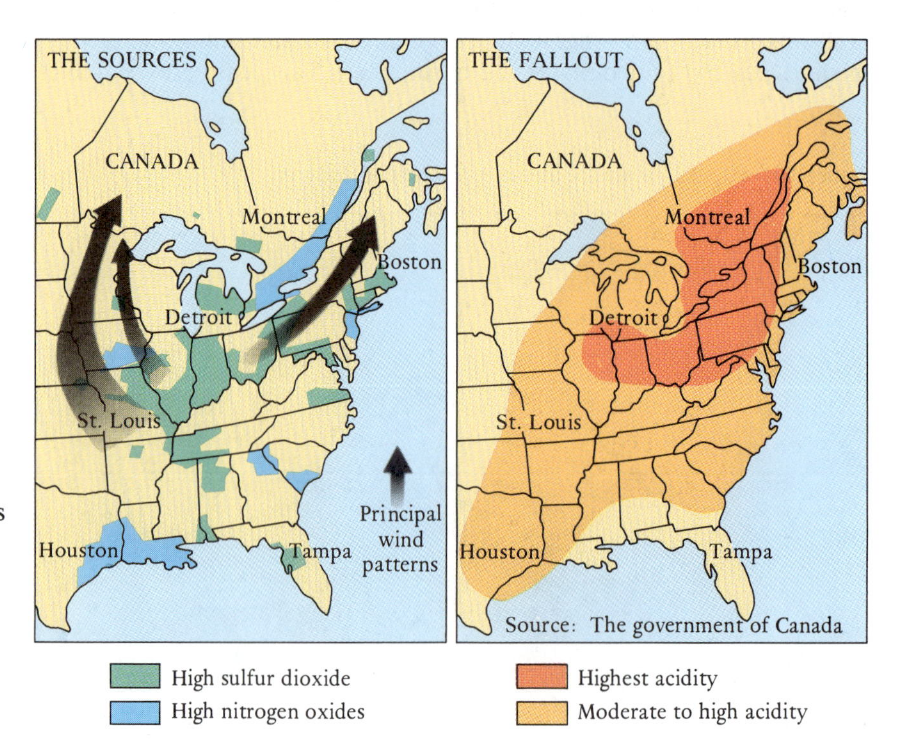

Figure 12–11 Most of the oxides of sulfur responsible for acid rain come from Midwestern states. Prevailing winds carry the acid droplets formed over the Northeast and into Canada. Oxides of nitrogen also contribute to acid rain formation.

Aerial view of dead trees at a lake acidified by acid rain. (*The World of Chemistry,* Program 16, "The Proton in Chemistry")

Acid rain also damages forests by releasing harmful minerals in the soil. Many of these minerals contain metal ions that are toxic to plant life. These minerals can exist in natural soil indefinitely because they are insoluble in groundwater and surface waters of normal pH. However, acid solutions can increase the solubility of many minerals. For example, protons in acid rain can react with insoluble aluminum hydroxide found in the soil, causing aluminum ions (Al^{3+}) to be taken up by the roots of plants where they have toxic effects.

The leaching of toxic metal ions into groundwater by acid rain may also increase groundwater pollution.

$$Al(OH)_3 + 3\ H^+ \longrightarrow Al^{3+} + 3\ H_2O$$

The effects of acid rain and other pollution on stone and metal structures are more subtle. These effects are especially devastating because of their irreversibility. By damaging stone buildings in Europe, acid rain is slowly but surely dissolving the continent's historical heritage. The bas-reliefs on the Cologne (West Germany) cathedral are barely recognizable. The Tower of London, St. Paul's Cathedral, and the Lincoln Cathedral in London (Fig. 12-12) have suffered the same fate. Other beautifully carved statues and

Acid rain crumbles buildings.

(a)

(b)

Figure 12-12 The photo on the left was taken at the Lincoln Cathedral in London in 1910. The photo on the right was taken in 1984. (Dean and Chapter of London)

Loading a statue onto a boat in Venice, Italy, to take it to a cleaner environment. (*The World of Chemistry*, Program 18, "The Chemistry of Earth")

bas-reliefs on buildings throughout Europe and the eastern part of the United States and Canada are slowly passing into oblivion by the action of pollutants, in particular, acid rain.

What can be done about acid rain? Some stopgap measures are being taken, such as spraying lime, $Ca(OH)_2$, into acidified lakes to neutralize at least some of the acid and raise the pH toward 7 (Fig. 12–13).

$$Ca(OH)_2 + 2\ H^+ \longrightarrow Ca^{2+} + 2\ H_2O$$

Sweden is spending $40 million a year to neutralize the acid in some of its lakes. Some lakes in the problem areas have their own safeguard against acid rain by having limestone-lined bottoms, which supply calcium carbonate ($CaCO_3$) for neutralizing the acid from acid rain (just as an antacid tablet relieves indigestion). Statues and bas-reliefs have been coated with a variety of plastics and other materials. None of these materials appears to be a long-range protector.

The ultimate answers to acid rain problems lie with governments. Twenty-one European countries agreed in 1985 to lower their SO_2 emissions by 30% or more over a 10-year period. By 1989, more than half of those

Figure 12–13 Calcium carbonate (limestone) is being dispersed over forests affected by acid rain. (Courtesy of Ohio Edison)

countries had already reached that goal. In 1988, the Canadian government set a goal of lowering its SO_2 emissions by half by 1994. In the United States, the 1990 Clean Air Act requires a reduction of 10 million tons annually from the levels of 1980, approximately a 40% reduction.

CARBON MONOXIDE

Carbon monoxide (CO) is the most abundant and widely distributed air pollutant found in the atmosphere. Like ozone, carbon monoxide is one of the most difficult pollutants to control. Cities like Los Angeles with high densities of automobiles tend to be repeatedly cited by the EPA for not attaining the required ambient air quality for carbon monoxide.

Heavy traffic contributes to high CO levels. (*The World of Chemistry,* Program 2, "Color")

Carbon monoxide is always produced when carbon or carbon-containing compounds are oxidized by insufficient oxygen.

$$2\,C + O_2 \longrightarrow 2\,CO$$

For every 1000 gal of gasoline burned, 2300 lb of carbon monoxide are formed in the combustion gases. Modern catalytic converters on car mufflers convert much of this carbon monoxide to carbon dioxide, but the amounts that are not converted make being near a heavily traveled street dangerous because of the carbon monoxide concentrations. At peak traffic times, concentrations as high as 50 ppm are common (refer to Chapter 10 on toxic substances). In the countryside, carbon monoxide levels are closer to the global average of 0.1 ppm.

At least ten times more carbon monoxide enters the atmosphere from natural sources than from all industrial and automotive sources combined. Of the 3.8 billion tons of carbon monoxide emitted every year, about 3 billion tons are emitted by the oxidation of decaying organic matter in the topsoil.

About 10^{14} g of CO are released each year in the United States, about 1000 lb for every person.

A bit of a mystery concerning carbon monoxide is that its global level does not seem to be changing as is the case with some pollutants. Although polar carbon monoxide molecules dissolve readily in water, they react very slowly with oxygen to form carbon dioxide. Where carbon monoxide goes is the subject of ongoing research in atmospheric chemistry.

CHLOROFLUOROCARBONS AND THE OZONE LAYER

Most pollutants are adsorbed onto surfaces, absorbed into water droplets and react, or they react in the gas phase with other pollutants in the lower atmosphere (troposphere) and eventually wash out in precipitation. There is one class of industrial pollutants, the halogenated hydrocarbons collectively called **chlorofluorocarbons,** or **CFCs,** which are relatively unreactive and are not eliminated in the troposphere. Instead, many of them eventually mix with air in the stratosphere where they reside for many years. As we shall see, it is the series of reactions between CFCs and ozone that cause so much concern about the presence of CFCs in our atmosphere.

The chemistry of CFCs is discussed in Chapter 7.

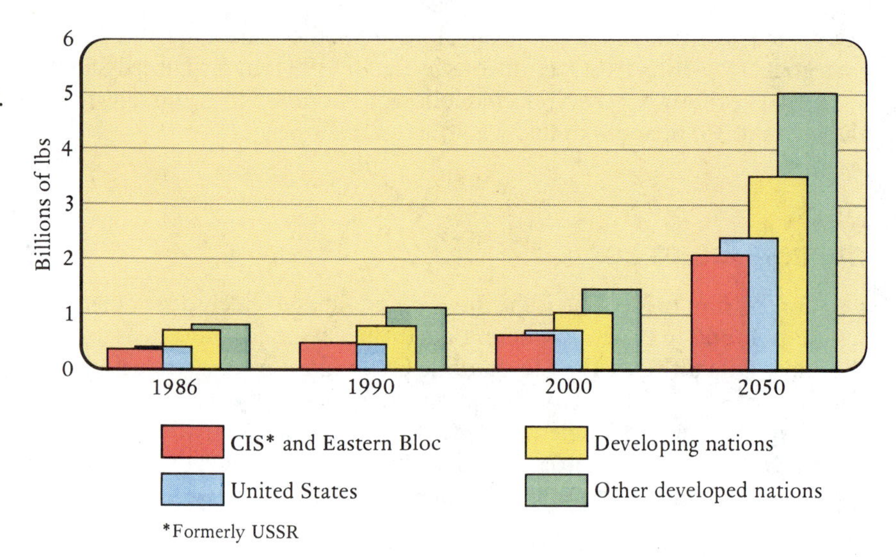

Figure 12–14 CFC growth projected by geographic region, 1986–2050 (absent international controls). (Environmental Protection Agency)

Over the years, the use of CFCs has increased dramatically in both industralized and third-world countries (Fig. 12–14). Because of the strength and stability of the carbon-fluorine and carbon-chlorine bonds, CFCs are generally nonreactive toward things like oxidizers, acids, and bases. This nonreactivity is a desirable property for many applications, but it means that CFCs, once released into the troposphere, do not react, but rather eventually end up in the stratosphere where they cause harm.

The dangers of CFCs were announced in 1974, when M. J. Molina and F. S. Rowland of the University of California, Irvine, published a scientific paper in which they predicted that continued use of CFCs would lead to a serious depletion of the Earth's protective stratospheric ozone layer. The reason the depletion of the ozone layer is important is that for every 1% decrease in stratospheric ozone, an additional 2% of the sun's most damaging ultraviolet radiation reaches the Earth's surface, resulting in increased skin cancer, damage to plants, and possibly other effects we know little about now. Let's examine how these CFCs destroy the ozone layer.

In the stratosphere there is an abundance of ozone being produced because high-energy ultraviolet photons with wavelengths below 280 nm readily photodissociate oxygen molecules.

$$O_2 + h\nu \longrightarrow 2\ O$$

$$O + O_2 \longrightarrow O_3$$

Recall that these same reactions take place to a lesser extent in the troposphere, with the ozone becoming involved in producing much of the urban secondary pollutants.

This stratospheric ozone is so abundant (about 10 ppm) that it absorbs between 95 and 99% of the sunlight in the 200 to 300 nm wavelength range (the ultraviolet range). Light in this wavelength range is especially damaging to living organisms, so it is important that this protective ozone layer remain intact if we are to avoid drastic harmful effects on our planet.

TABLE 12-5 Halons—Halogenated Hydrocarbons Containing Carbon-Bromine Bonds

Name	Formula	Uses	Atmospheric Lifetime (yrs)
Halon-1211	CF_2BrCl	Portable fire extinguishers	25
Halon-1301	CF_3Br	Fire extinguisher systems	110

The carbon-chlorine bond in a CFC molecule is broken easily by an ultraviolet photon, especially those available in the stratosphere. When the CFC molecule photodissociates, an active chlorine atom (Cl) is produced. The reaction below shows one of the most common CFCs, CFC-11, undergoing photodissociation.

CFC-11 is commonly used in automotive air conditioners.

$$CFCl_3 + h\nu \longrightarrow CFCl_2 + Cl$$

The chlorine atom then combines with an ozone molecule, producing a chlorine oxide (ClO) radical and an oxygen molecule.

$$Cl + O_3 \longrightarrow ClO + O_2$$

Thus, an ozone molecule has been destroyed. If this were the only reaction the single CFC molecule caused, there would be little danger to the ozone layer. However, the ClO radical can react with oxygen atoms and produce the chlorine atom again, which is ready to react with yet another ozone molecule.

$$ClO + O \longrightarrow O_2 + Cl$$

$$Cl + O_3 \longrightarrow O_2 + ClO$$

The overall reaction from these steps is the destruction of an ozone molecule for every reaction cycle in which the chlorine atom participates.

$$O_3 + O \longrightarrow 2\ O_2$$

A single chlorine atom may undergo reaction thousands of times. Eventually the chlorine atom reacts with a water molecule to form HCl, which mixes into the troposphere and washes out in acidic rainfall. This chlorine atom chain is thought to account for about 80% of the loss of ozone observed.

The bromine oxide radical can cause ozone depletion in a manner similar to chlorine oxide. Bromine oxide (BrO) is produced from halons, a class of compounds structurally similar to the CFCs (Table 12-5) but containing a carbon-bromine bond. When a halon like Halon-1301 reacts with a photon, a bromine atom is produced.

$$CF_3Br + h\nu \longrightarrow CF_3 + Br$$

The bromine atom then reacts with ozone to form BrO and O_2.

$$Br + O_3 \longrightarrow BrO + O_2$$

Nitric oxide can also deplete ozone. Nitric oxide, also a radical species and written as NO, is formed from microbially produced nitrous oxide (N_2O) reacting with oxygen atoms.

$$O + N_2O \longrightarrow NO + NO$$

For NO the reactions are

$$NO + O_3 \longrightarrow NO_2 + O_2$$

$$NO_2 + O \longrightarrow NO + O_2$$

At one time it was feared that high-flying supersonic aircraft might destroy ozone by the nitric oxides they produced, but this has not proved to be the case. Molina and Rowland's warning regarding CFCs has proven correct, however. Satellite and ground-based measurements since 1978 indicate that global concentrations of ozone in the stratosphere have been decreasing.

Recently NASA scientists published measurements that show an average of about 2.5% decrease in the ozone layer worldwide in the decade 1978 to 1988. Local decreases in ozone may be even larger. In a latitude band that includes Dublin, Ireland; Moscow, Russia; and Anchorage, Alaska, ozone decreased 8% from January 1969 to January 1986 (Fig. 12–15).

Figure 12–15 Atmospheric ozone levels—a global decline. (Ozone Trends Panel)

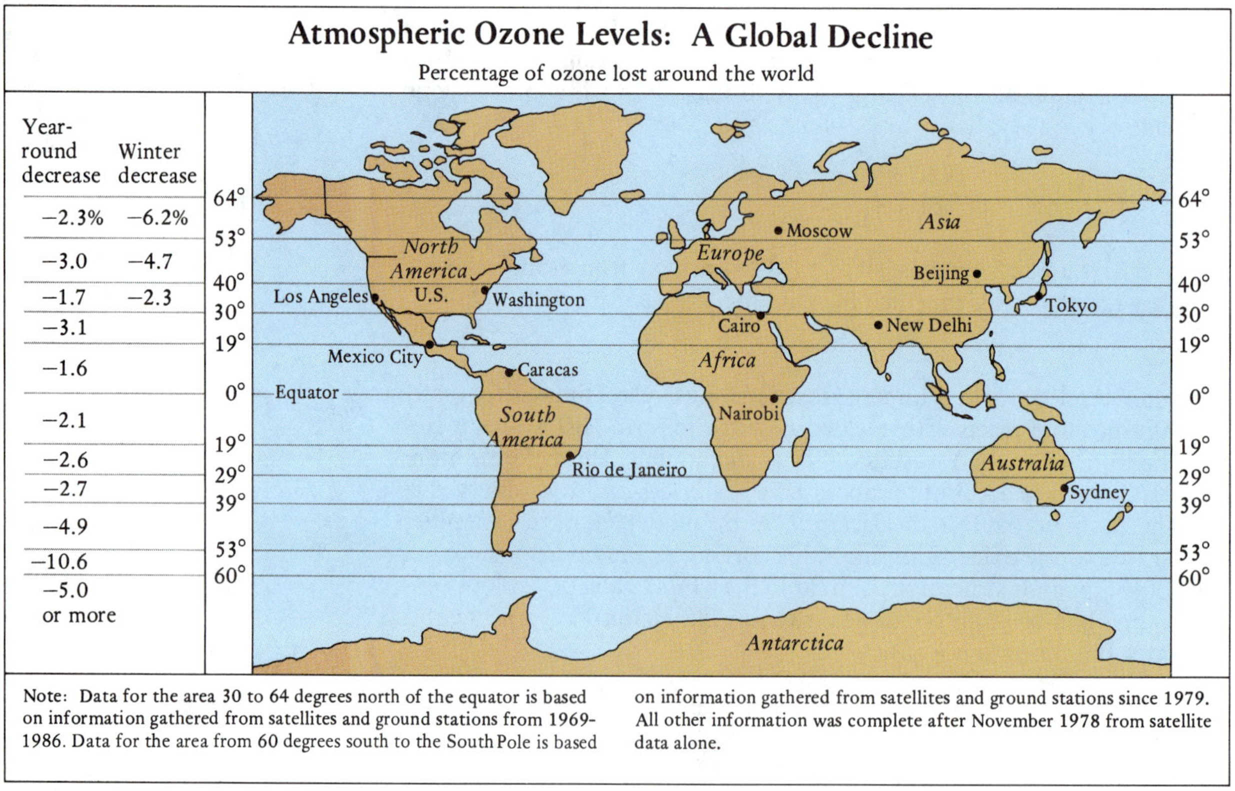

Atmospheric Ozone Levels: A Global Decline

Percentage of ozone lost around the world

Year-round decrease	Winter decrease
−2.3%	−6.2%
−3.0	−4.7
−1.7	−2.3
−3.1	
−1.6	
−2.1	
−2.6	
−2.7	
−4.9	
−10.6	
−5.0 or more	

Note: Data for the area 30 to 64 degrees north of the equator is based on information gathered from satellites and ground stations from 1969-1986. Data for the area from 60 degrees south to the South Pole is based on information gathered from satellites and ground stations since 1979. All other information was complete after November 1978 from satellite data alone.

Near the North and South Poles ozone losses have been between 1 and 2.5% per year (Fig. 12–16). Recently, massive losses, termed "holes," have been observed. These ozone holes are of special concern because many scientists believe that they may happen at the mid-latitudes in the future.

Intense studies since 1987 have shown that other factors lower the ozone concentration in addition to the gas-phase reactions mentioned earlier. In an Antarctic winter, a vortex of intensely cold air containing ice crystals builds up. On the surfaces of these crystals additional reactions take place that produce two chemical species not ordinarily involved in chlorine atom production. These species are hydrogen chloride (HCl) and chlorine nitrate ($ClONO_2$).

On the surface of these ice crystals, HCl and $ClONO_2$ react to form chlorine molecules (Cl_2). These are readily photodissociated into chlorine atoms, which become involved in the ozone destruction reactions.

$$HCl + ClONO_2 \longrightarrow Cl_2 + HNO_3$$

$$Cl_2 + h\nu \longrightarrow 2\ Cl$$

In an effort to reduce the use of CFCs and other compounds that harm the stratospheric ozone layer, some state legislatures have begun to pass laws restricting CFC use. Some states have passed laws requiring automobile service centers to reclaim automobile air-conditioning refrigerant gases.

Key Dates

Dec 1973:	Rowland and Molina discovery
Oct 1978:	Use of CFCs in aerosols banned in U.S.
Oct 1984:	British team reports 40% loss of ozone over Antarctica during austral spring
Sep 1987:	Montreal Protocol-Representatives from 43 nations agree to CFC reductions of 50% by 2000
Oct 1987:	Antarctic expedition verifies huge losses of ozone over Antarctica during austral spring
Mar 1988:	U.S. ratifies Montreal Protocol, DuPont announces it will cease production of CFCs
Apr 1988:	Plastic foam manufacturers announce they will stop using CFCs
Mar 1989:	700 representatives from 124 countries attend London conference on saving the ozone layer
Jun 1990:	Environment ministers from 93 nations agree to strengthen Montreal Protocol with complete phaseout of CFCs by 2000 (and HCFCs by 2040)
Oct 1990:	U.S. Congress passes revised Clean Air Act that includes phaseout of CFCs by 2000
Jan 1991:	Environment ministers of European Community agree to complete CFC ban by 1997
Jan 1992:	Increased concentrations of ozone-depleting chemicals found over populated areas in the Northern Hemisphere.
Feb 1992:	U.S. president moves target date for phase-out of CFCs from the year 2000 to 1995.

Figure 12–16 The hole in the ozone layer over the Antarctic continent. The purple region has the lowest ozone concentration. Dobson unit is a measure of the thickness of the ozone layer. 300 DU equals a 3 mm thick layer of ozone at 1 atmosphere pressure. (Photo courtesy of NASA)

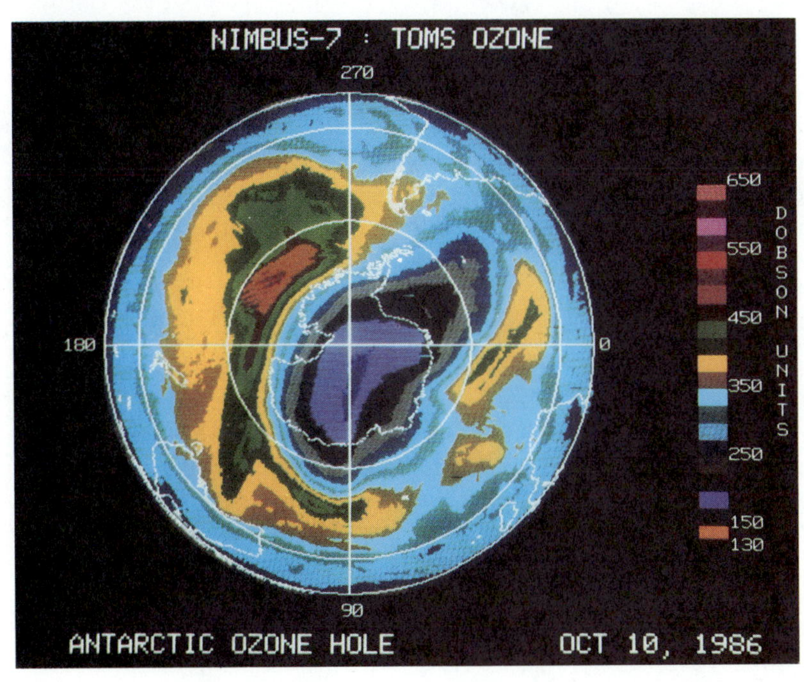

Figure 12–17 CFCs like those used in automotive air conditioners can be recycled during repairs. At many service centers the CFCs are simply vented into the atmosphere prior to repair work. Many states and localities are beginning to require recycling as a means of lowering the amounts of CFCs escaping into the atmosphere. (Courtesy of Robinair)

Vermont banned CFC refrigerants in 1990, and the state of Connecticut will ban CFCs from automobile air conditioners by 1993 (Fig. 12–17). Nations are also acting. In January, 1989, the Montreal Protocol on Substances That Deplete the Ozone Layer went into effect. Signed by 24 nations, the protocol calls for reductions in production and consumption of several of the long-lived CFCs. There will be a total ban on CFCs by the year 2000. In addition, the use of two halons is to remain at 1986 levels. CFC and halon manufacturers are seeking alternatives to these compounds. Two such possible substitutes are HCFC-22 and HCFC-141b.

$$\begin{array}{ccc} & \text{Cl} & \\ & | & \\ \text{H}- & \text{C} & -\text{F} \\ & | & \\ & \text{F} & \end{array} \qquad \begin{array}{cccc} & \text{H} & \text{Cl} & \\ & | & | & \\ \text{H}- & \text{C} & -\text{C} & -\text{F} \\ & | & | & \\ & \text{H} & \text{F} & \end{array}$$

HCFC-22 HCFC-141b

Replacing one or more C—Cl bonds with C—H bonds in CFC molecules makes the molecules less stable in the lower atmosphere so that they do not get into the stratosphere to react and form ozone-depleting Cl atoms.

SELF-TEST 12B

1. When coal and fuel oil are burned, what two primary pollutants are formed? __________ and __________
2. When sulfur dioxide reacts with oxygen, what oxide of sulfur is formed? __________ When this oxide reacts with water, what acid is formed? __________
3. Which chemical would most likely react with sulfur dioxide and remove it from combustion gases? ( ) sodium chloride (NaCl), ( ) lime (CaO), ( ) nitric oxide (NO).
4. Name two acids found in acid rain. One must contain sulfur, the other nitrogen. __________ and __________

5. What is the pH of normal rainfall? ______ What dissolved chemical causes this pH to be below pH 7? ______
6. Approximately when was acid rainfall first observed? ______
7. Name the chemical bond broken by a photon of light in a typical CFC molecule. ______
8. Write the reaction producing ozone from oxygen. ______
______
9. What is the chemical species containing chlorine that destroys ozone molecules? ______
10. What type of chemical bond does a typical halon molecule contain?
11. Over what continent have scientists found an ozone "hole"?
______
12. If ozone in the stratosphere is destroyed, what form of radiation will pass through to the Earth below? ______

CARBON DIOXIDE AND THE GREENHOUSE EFFECT

How can carbon dioxide (CO_2) be considered a pollutant when it is a natural product of respiration and fossil fuel burning and is a required reactant for photosynthesis? Actually, CO_2 is not a pollutant, but the fact that it is increasing in the Earth's atmosphere is cause for deep concern. Consequently, it is treated as a pollutant. Without human influences, the flow of carbon between the air, plants, animals, and the oceans would be roughly balanced. However, between 1900 and 1970, the global concentration of CO_2 increased from 296 ppm to 318 ppm, an increase of 7.4%. By 1985, the concentration was 350 ppm and expectations are that the CO_2 concentration will continue to increase (Fig. 12–18). For example, since the end of World War II, a world energy growth rate of about 5.3% per year took place until the OPEC oil embargo in the mid-1970s. Rates of energy use have actually decreased since then.

The ocean dissolves carbon dioxide to form bicarbonates and carbonates. (Charles D. Winters)

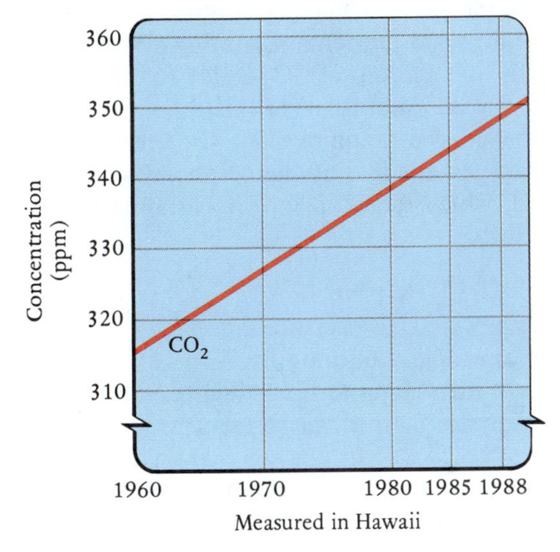

Figure 12–18 Atmospheric carbon dioxide is up more than 10% since 1960. (Adapted from *Chemical and Engineering News,* March 13, 1989)

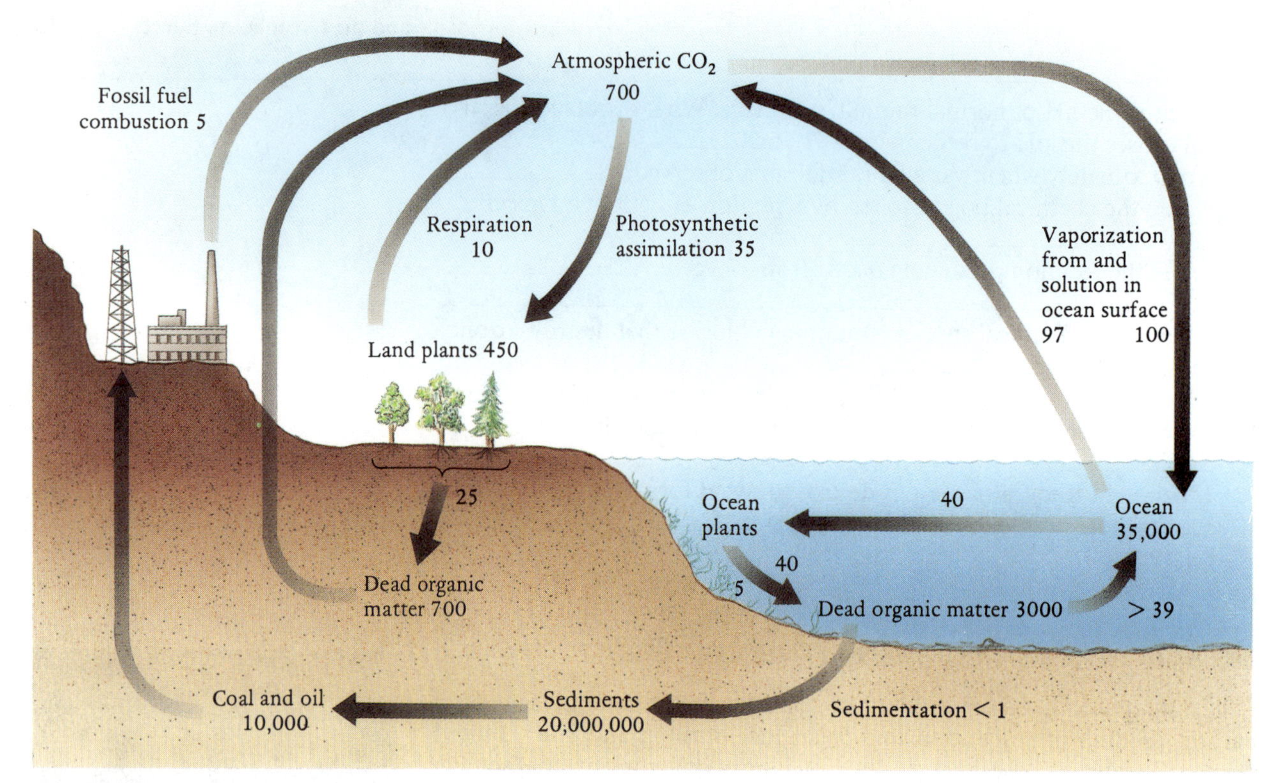

Figure 12–19 The carbon cycle in the biosphere. The numbers are in units of 10^9 tons.

To see how easily our everyday activities affect the amount of CO_2 being put into the atmosphere, consider a round-trip flight from New York to Los Angeles. Each passenger pays for about 200 gal of jet fuel, which weighs 1400 lb. When burned, each pound of jet fuel produces about 3.14 lb of carbon dioxide. So 4400 lb, or 2.2 tons of carbon dioxide are produced per passenger during that trip.

Population pressures are also contributing heavily to increased CO_2 concentrations. In the Amazon region of Brazil, for example, extensive cut-and-burn practices are being used to create cropland. This is causing a tremendous burden on the natural CO_2 cycle, because CO_2 is being added to the atmosphere during burning while there are fewer trees present to photosynthesize this additional CO_2 into plant nutrients. Counting all forms of fossil fuel combustion worldwide, the amount of CO_2 added to the atmosphere is about 50 billion tons a year. About half of this remains in the atmosphere to increase the global concentration of CO_2. The other half is taken up by plants during photosynthesis and by the oceans, where CO_2 dissolves to form carbonic acid, which then can form bicarbonates and carbonates (Fig. 12–19).

$$CO_2 + H_2O \rightleftharpoons H^+ + \underset{\text{Bicarbonate}}{HCO_3^-} \rightleftharpoons H^+ + \underset{\text{Carbonate}}{CO_3^{2-}}$$

It seems reasonable that if we are rapidly burning fossil fuels that took millions of years to form, we are then going to be adding back into the atmosphere CO_2 at a more rapid pace than it can be used up in natural processes.

How does increasing global CO_2 concentrations produce a greenhouse effect? Consider solar radiation as it arrives at the Earth's atmosphere. About

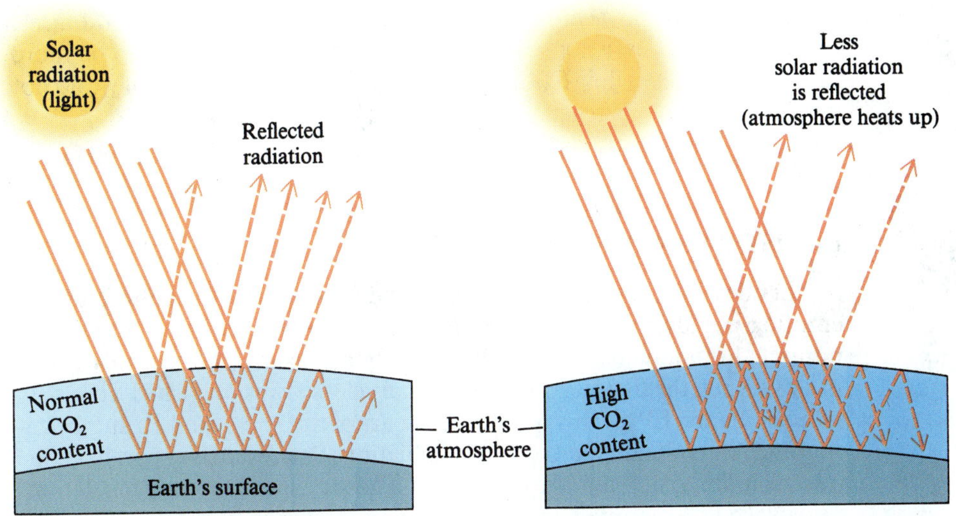

Figure 12–20 The greenhouse effect. Owing to a balance of incoming and outgoing energy in Earth's atmosphere, the mean temperature of Earth is 14.4°C (58°F). Carbon dioxide permits the passage of visible radiation from the Sun to Earth but traps some of the heat radiation attempting to leave Earth.

half of the visible light (400–700 nm) striking the Earth is reflected back into space. The remainder reaches the Earth's surface and causes warming (Fig. 12–20). These warmed surfaces then re-radiate this energy as heat. Water vapor and CO_2 readily absorb some of this radiated energy coming from the Earth's surface and in turn warm the atmosphere. A botanical greenhouse works on the same principle. Glass transmits visible light, but blocks infrared radiation trying to leave. The effect is a warming of the air inside the greenhouse. In warm weather, the windows of a greenhouse must be opened or the plants inside will overheat and die.

The Earth, with an average surface temperature of about 300 K (27°C), radiates energy in the infrared region of the spectrum. CO_2, water vapor, and ozone all absorb in various portions of this infrared region. So all three are "greenhouse gases" and act as absorbing blankets that prevent radiation losses and keep the Earth's atmospheric temperature comfortable (although not in all locations at the same time!). Water vapor in the atmosphere is subject to such vast cycles that human activity doesn't seem to bother it. Because ozone is present in relatively low concentrations, our attention is focused on CO_2.

Recently, Russian scientists took ice core samples dating back 160,000 years. In these ice samples were tiny pockets of air that could be analyzed for CO_2 content. They found a direct correlation between CO_2 and geologic temperatures known by other means. As the CO_2 increased, the global temperature increased and as the CO_2 decreased, the global temperature decreased. It is generally agreed that rising CO_2 concentrations will probably lead to increasing global temperatures and corresponding changes in climates. If predictions by the National Academy of Sciences prove correct, when and if the global concentration of CO_2 reaches 600 ppm, the average global temperature could rise by 1.5 to 4.5°C (2.7–8.1°F). Even a 1.5°C warming would produce the warmest climate seen on Earth in the past 6000 years, and a 4.5°C warming would produce world temperatures higher than any since the Mesozoic era—the time of dinosaurs.

A chemical plant. Possible source for release of toxic chemicals into the atmosphere. (*The World of Chemistry,* Program 17, "The Precious Envelope")

Clearly, global warming is a major potential problem the world faces. Controlling CO_2 emissions worldwide will undoubtedly prove to be more difficult than controlling CFCs or the precursors to acid rain, but if we do not do something in this regard, the future may be bleak indeed.

INDUSTRIAL AIR POLLUTION

Besides CO_2, CFCs, NO_x, and SO_2, industry pollutes the atmosphere by emitting a wide variety of solvents, metal particulates, acid vapors, and unreacted monomers. The extent to which this takes place became evident in 1989 when the first summary of annual releases was published from data received by the EPA. This report was a part of the Superfund Reauthorization and Amendments Act of 1986, regulations that resulted in part from the tragedy in Bhopal, India (see Chapter 1). These release-reporting regulations were placed on manufacturers who use any of a group of about 320 chemicals and classes of compounds representing special health hazards. The reporting was divided into releases to air, water, and land. (Water and land releases directly affect surface and groundwater purity and are discussed in Chapter 11.)

These regulations have been called "Community Right-to-Know" regulations because they inform communities about releases of harmful chemicals in their areas.

In the report, an industrial facility was asked to count all releases to the atmosphere, regardless of type. This meant that leaky valves, fittings, accidental spills, vapor losses while filling tank trucks and rail tank cars, emissions at stacks, and so forth, were all added together. As expected, heavily industrialized states and states with a lot of chemical industry had high releases (Table 12–6), but the amounts of some chemicals released were also surprisingly large (Table 12–7). These data are now available by means of publicly accessible computerized databases. Interested persons may call the EPA at 1-800-535-0202 for more information.

This form of release reporting could be termed a pollution *"glasnost."*

TABLE 12–6
The Top Ten States by Toxic Chemicals Released to Air in 1989

State	Emissions (millions of lbs)
1. Texas	189.7
2. Tennessee	152.5
3. Ohio	133.1
4. Utah	124.9
5. Indiana	122.9
6. Louisiana	101.5
7. Illinois	94.4
8. North Carolina	89.8
9. Alabama	86.1
10. Michigan	69.1

Note: The total emissions for all 50 states in 1989 was 2,428,000,000 lb.

Source: U.S. EPA, 1991.

Testing for leaks at a valve in a chemical plant. (*The World of Chemistry,* Program 25, "Chemistry and the Environment")

TABLE 12-7 Top Ten Chemicals Released into Air in 1989

Chemical	Emissions (millions of lbs)	Uses
1. Toluene	255	Gasoline, solvent
2. Ammonia	244	Refrigerant, reactant*
3. Acetone	198	Solvent in paints
4. Methanol	198	Solvent, reactant
5. 1,1,1-Trichloroethane	169	Degreasing operations
6. Chlorine	130	Reactant, bleach†
7. Methyl ethyl ketone	127	Solvent in paints
8. Xylene (mixed isomers)	113	Gasoline, solvents
9. Dichloromethane	109	Solvent, reactant‡
10. Carbon disulfide	96	Solvent, reactant

* Ammonia's use as a fertilizer not reported.

† Chlorine's use to disinfect water and wastewater not reported.

‡ A known carcinogen.

Source: U.S. EPA, 1991.

INDOOR AIR POLLUTION

As if the data about pollutants in the outside air were not enough to concern us, the air inside our homes and workplaces is also contaminated and usually by the same chemicals emitted by industry.

We shouldn't be surprised that air in our homes is contaminated by industrial chemicals—after all, we bring industry's products into our homes.

At Home

Some scientists have concluded that air in our homes may be more harmful than the air outdoors, even in heavily industrialized areas. A study by the EPA indicated that indoor pollution levels in rural homes were about the same as for homes in industrialized areas. One cause for this is the emphasis on tighter, more energy-efficient homes, which tend to trap air inside for long periods.

What are the sources of home air pollution (see Fig. 12–21)? Tobacco smoke, if present, is an obvious source. Benzene, a known carcinogen, occurs at 30% to 50% higher levels in homes of smokers than homes of nonsmokers. Building materials and other consumer products are also sources of pollutants (Table 12–8). Entire buildings can acquire a "sick building syndrome" when a particular chemical or group of chemicals is found in sufficiently high concentration to cause headaches, nausea, stinging eyes, itching nose, or some combination of these symptoms. Usually, the best cure for all forms of indoor air pollution is to limit the introduction of the offending chemicals and to have better exchange with the outside air.

Radon is also a major potential air pollutant inside the home. Radon (Rn) is the heaviest member of the noble-gas family of elements. Radon-222, the most common isotope of radon, is radioactive, with a half-life of 3.82 days. Radon-222 is a direct result of the decay of radium-226, a naturally

One curie (Ci) is 37 billion disintegrations per second. One picocurie (pCi) is one disintegration about every 0.037 seconds.

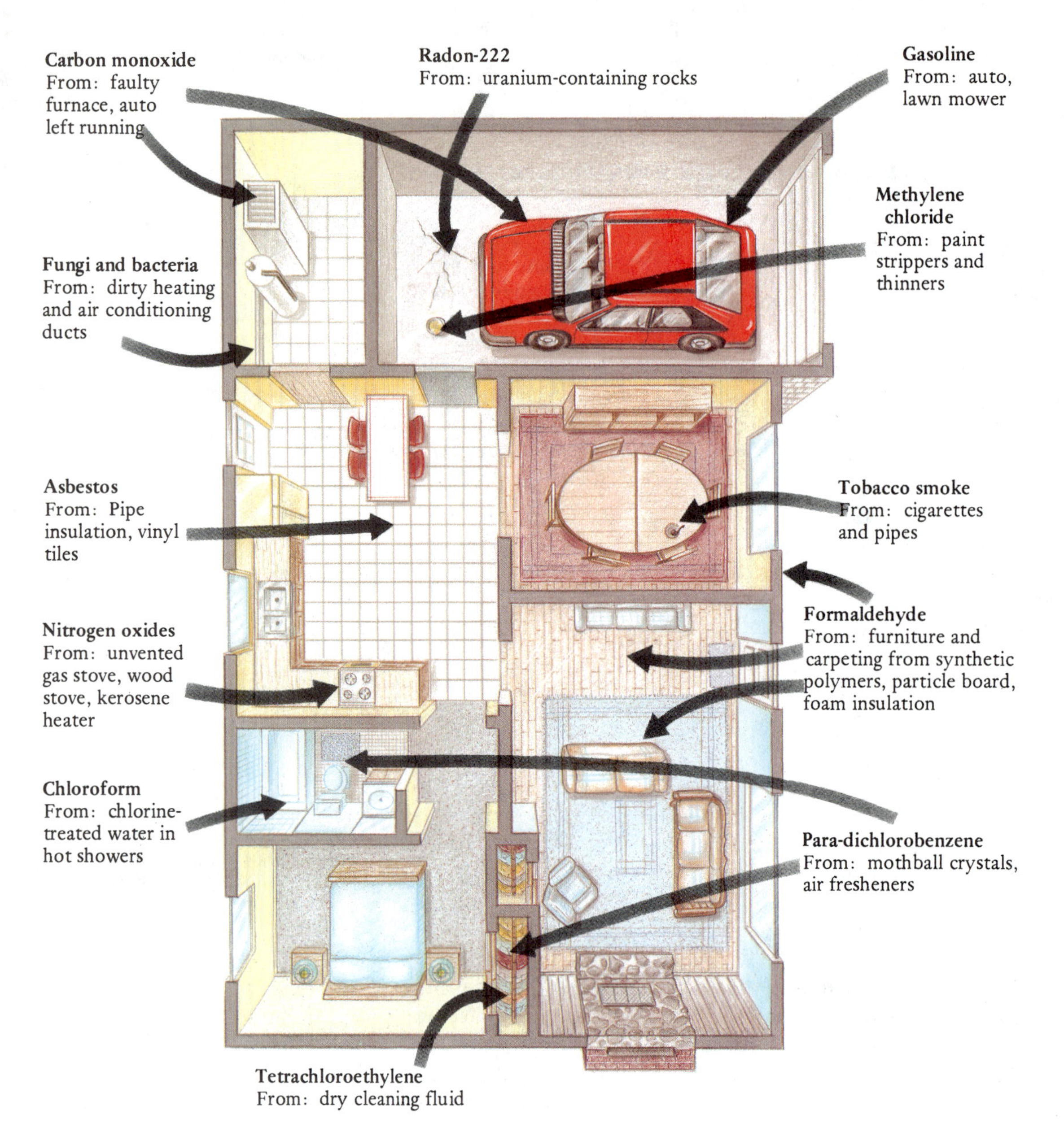

Figure 12–21 Some indoor air pollutants and their sources.

occurring isotope that is present at levels of about 1 pCi (picocurie: 1×10^{-12} Ci) per gram in ordinary soil and rocks.

$$^{226}_{88}\text{Ra} \longrightarrow {}^{222}_{86}\text{Rn} + {}^{4}_{2}\text{He}$$

When Rn decays, it produces alpha particles and another short-lived radioisotope, polonium-218.

$$^{222}_{86}\text{Rn} \longrightarrow {}^{218}_{84}\text{Po} + {}^{4}_{2}\text{He}$$

Polonium-218 (half-life 3.05 min) also decays, producing an alpha particle, to lead-214 (half-life 26.8 min).

$$^{218}_{84}\text{Po} \longrightarrow {}^{214}_{82}\text{Pb} + {}^{4}_{2}\text{He}$$

Lead-214 decays through several other radioisotopes (Table 12–9) to lead-206. The stable isotope lead-206 is the end result of the decay of all these

TABLE 12-8 Some Common Household Products and the Chemicals They Contribute to Indoor Air Pollution

Product	Major Organic Chemicals
Silicone caulk	Methyl ethyl ketone, butyl propionate, 2-butoxyethanol, butanol, benzene, toluene
Floor adhesive	Nonane, decane, undecane, xylene
Particleboard	Formaldehyde, acetone, hexanal, propanal, butanone, benzaldehyde, benzene
Moth crystals	Naphthalene, *p*-dichlorobenzene
Flood wax	Nonane, decane, dimethyloctane, ethylmethylbenzene
Wood stain	Nonane, decane, methyloctane, trimethylbenzene
Latex paint	2-propanol, butanone, ethylbenzene, toluene
Furniture polish	Trimethylpentane, dimethylhexane, ethylbenzene, limonene
Room freshener	Nonane, limonene, ethylheptane, various substituted aromatics (as fragrances)

Source: EPA, 1988.

radioisotopes formed from the decaying radon-222 atom. Collectively, these radioisotopes are called *radon daughters*.

Being a member of the noble-gas family, radon exists as a gas and is nonreactive chemically. Radon atoms in the air we breathe are inhaled and exhaled without any reaction, although some may dissolve in the fluids found in the lungs. If the radon atom happens to decay while inside our lungs (recall the half-life of radon-222 is 3.82 days, so half of a sample of these atoms decay every 3.82 days), atoms of nongaseous and reactive metallic elements such as polonium, bismuth, and lead form. These atoms quickly react to form ionic compounds. The ionic compounds, for the most part, remain inside the lung. All of the radon daughters up to lead-210 have short half-lives and emit damaging radiation in close contact to delicate lung tissue. The result is a higher than normal risk of lung cancer.

Miners in deep mines are exposed to much higher than normal radon levels, and as early as 1950, government agencies began monitoring radon

TABLE 12-9 Radon Daughters from the Decay of Radon-222

Isotope	Decay Particle	Half-Life
$^{218}_{84}Po$	$^{4}_{2}He$ (alpha)	3.05 min
$^{214}_{82}Pb$	$^{0}_{-1}e$ (beta)	26.8 min
$^{214}_{83}Bi$	$^{0}_{-1}e$ (beta)	19.7 min
$^{214}_{84}Po$	$^{4}_{2}He$ (alpha)	1.6×10^{-4} sec
$^{210}_{82}Pb$	$^{0}_{-1}e$ (beta)	22 years
$^{210}_{83}Bi$	$^{0}_{-1}e$ (beta)	5.0 days
$^{210}_{84}Po$	$^{4}_{2}He$ (alpha)	46 sec
$^{206}_{82}Pb$	stable	

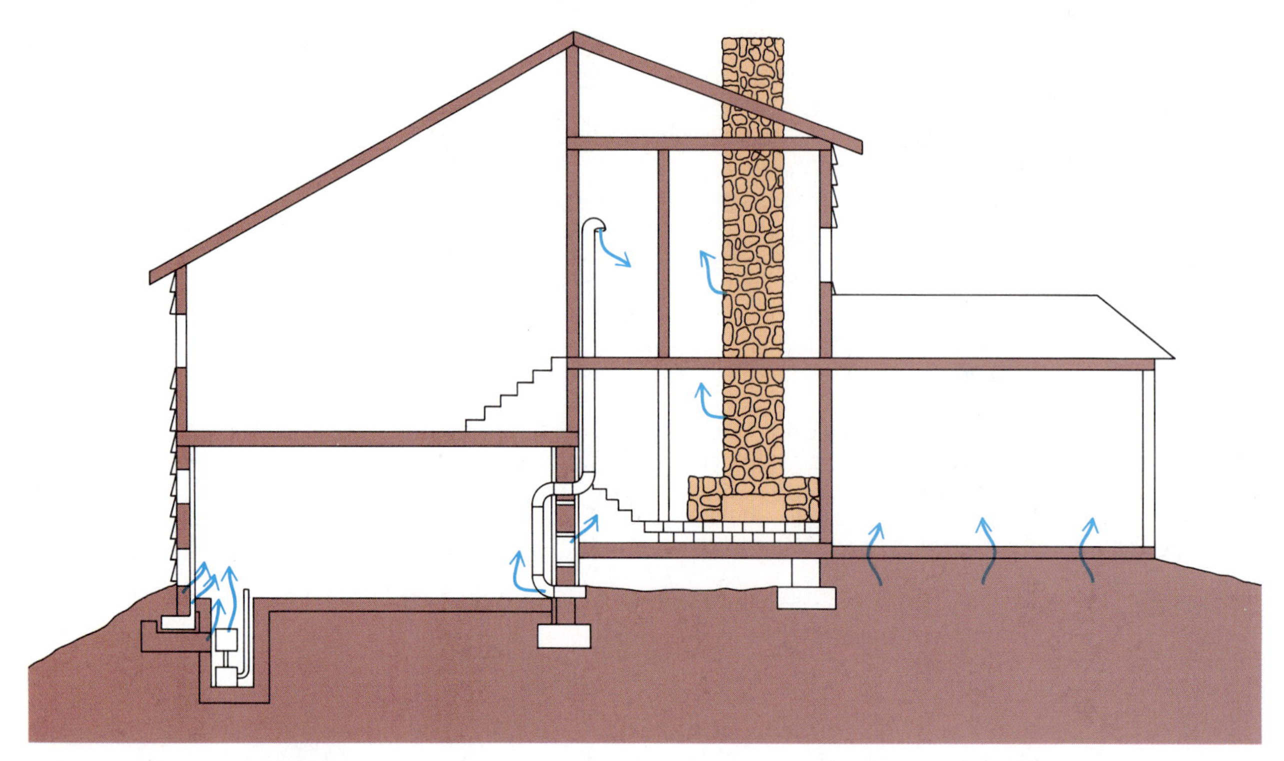

Figure 12–22 Some of the ways radon can get into buildings.

exposures and incidences of lung cancer. Today, it is well known that radon exposure increases one's chance of developing lung cancer. If you smoke, the chances are even greater, because there seems to be a synergistic effect between smoking and radon levels in causing lung cancer. Both alpha and beta particles from the decay of radon daughters can cause disruption of chemical bonds in DNA strands in lung cells, which may cause cancerous growth in these cells. Smoking introduces additional chemicals capable of disrupting DNA. It is estimated that about 10,000 of the annual 140,000 lung cancer deaths in the United States are caused by radon exposure.

When buildings are built over soil containing the heavy radioactive elements that decay to radium-226, some of the radon gas produced seeps through minute fissures in the soil or rock and migrates into the air in these buildings (Fig. 12–22). The building literally funnels the radon through it by "chimney effects" from such sources as clothes dryers, fireplaces, furnaces, and warm air rising and leaving through openings near the roof. If a building is built on a foundation over a crushed stone ballast, holes may be drilled into the ballast. Then a suction pump is attached to produce a negative pressure. For buildings not built on concrete foundations, increasing ventilation both in the basement and inside the building is effective in removing the radon from the building. Of course this means energy losses, but these might be acceptable when the alternatives of the radiation damage from the radon daughters are considered.

It has been estimated that perhaps 8 million homes in the United States are affected by radon contamination. Radon has been detected in homes in almost every state. Some areas have higher soil radon than others.

In 1986, the EPA surveyed over 11,000 homes for radon contamination. Almost 39% of the homes in Colorado had levels above the EPA action level of 4 pCi/L. Some homes in the states with the lowest percentage of homes

4 pCi/L is a little less than 1.5 disintegrations every 10 sec.

exceeding the action level had the highest measured readings. One home in Alabama had a radon level of 180 pCi/L, although the state as a whole had only 6% of its homes with over 4 pCi/L of radon.

This EPA action level is important for two reasons. First, levels of radon above 4 pCi/L should be reduced because levels above this lead to unacceptable risks of lung cancer. Second, it is difficult to lower the level of radon in most contaminated homes below 4 pCi/L. This last point is basically an acceptance of the fact that radiation exposure will always be with us.

Outdoor radon concentrations are approximately 0.1 to 0.15 pCi/L worldwide.

SELF-TEST 12C

1. Name three different human activities that produce large amounts (millions of tons/year) of CO_2. ______________________________
____
2. What are two principal processes whereby CO_2 is consumed? ____________, ____________
3. Name three major greenhouse gases found in the atmosphere. ____________, ____________, ____________ Which is most closely associated with human activities? ____________
4. Global temperature seems to follow carbon dioxide concentrations: True ( ) False ( ).
5. Approximately what is the current global CO_2 concentration? ______
6. What state had the highest air releases of toxic chemicals in 1989, the latest year for which such data are available? ____________
7. What chemical was released in greatest amount nationwide in 1989, according to the EPA release report? ____________
8. What single activity inside the home can account for increased concentrations of benzene, a known human carcinogen? ____________
9. What is the source of radon in the home? ____________

QUESTIONS

1. Define *polluted air.*
2. Name three major air pollutants and give their sources.
3. What role do aerosols play in air pollution?
4. How is an industrial-type smog different from a photochemical smog?
5. What is the main source of nitrogen oxides in the atmosphere?
6. What causes photodissociation to occur?
7. What is meant by "bad" ozone? Where does this ozone occur?
8. What is meant by "good" ozone? Where does this ozone occur?
9. Describe an air pollution problem in your community. How can it be solved?
10. Name two sources of hydrocarbons in the atmosphere.
11. How is benzo(α)pyrene formed? Why is it a dangerous air pollutant?
12. What is the principal source of sulfur dioxide as an air pollutant? How is sulfur dioxide harmful?
13. Define the term *acid rain.* What would be the approximate pH of acidic rainfall? Name two ways acid rain can be harmful.
14. Give details about the regions in the world where acid rain is a problem.
15. How can acid rain be controlled?
16. What is the primary source of carbon monoxide as an air pollutant?
17. What is a CFC? What is an HCFC? How are they used?
18. How are CFCs harmful to the atmosphere? Give some details.
19. How can CFCs be controlled?
20. How is carbon dioxide released into the atmosphere? Describe the trend observed in atmospheric carbon dioxide? What is the implied danger in this trend?
21. Define the greenhouse effect.
22. Describe the term *indoor air pollution.* What are the major sources of indoor air pollution?
23. What human activity increases the chances that radon exposure will lead to lung cancer?

J. F. Millet's *The Gleaners* portrays our dependence on farmers to feed the world's population.
(GIRAUDON/ART RESOURCE, NY. Paris: Louvre)

CHAPTER

13

Agricultural Chemistry

When hunting and gathering from nature's bounty were the primary means of obtaining food, only catastrophic events could void the fundamental relationship between effort and a satisfied stomach. Population increases and the concentration of people in cities forced the development of basic agriculture, which has mostly succeeded but sometimes dramatically failed. Now with a world population in excess of 5 billion people, scientific and technological advances in agriculture are absolutely necessary.

1. What is the structure of a typical soil?
2. What is humus, and what is its origin?
3. What are the plant nutrients, and into what four categories are they organized?
4. What three elements are of primary concern in crop fertilization?
5. What resources are necessary for the production of fertilizers?
6. How is nitrogen from the air "fixed" for crop utilization?
7. How dependent is modern agriculture on insecticides and herbicides?
8. What is the potential of genetic engineering in the production of food?

Through the 8,000 to 10,000 years of recorded human history, food production techniques have developed enormously. It is generally estimated that 90% of the U.S. population worked to provide food and fiber during most of the 19th century. Sophistication in agriculture now allows one farm worker to feed 80 people, and farming efficiency is likely to increase! However, we are still playing catchup in supplying the human appetite for food. Today's scientific farming methods are required and will need to be improved if we are to feed adequately the approximately 5.4 billion people who are now alive in the expanding world population (Fig. 13–1). Some demographic experts predict the human population at 6 billion in the year 2000 and 10 billion by 2050!

The British Isles recorded 201 famines between A.D.10 and 1846, with none since. In China there were 1846 famines between 108 B.C. and A.D.1828.

To help grow the enormous amount of food we need, the chemical industry supplies modern scientific agriculture with a large assortment of chemicals—the **agrichemicals**—including fertilizers, medicine for livestock, chemicals to destroy unwanted pests and plant diseases, food supplements, and many others.

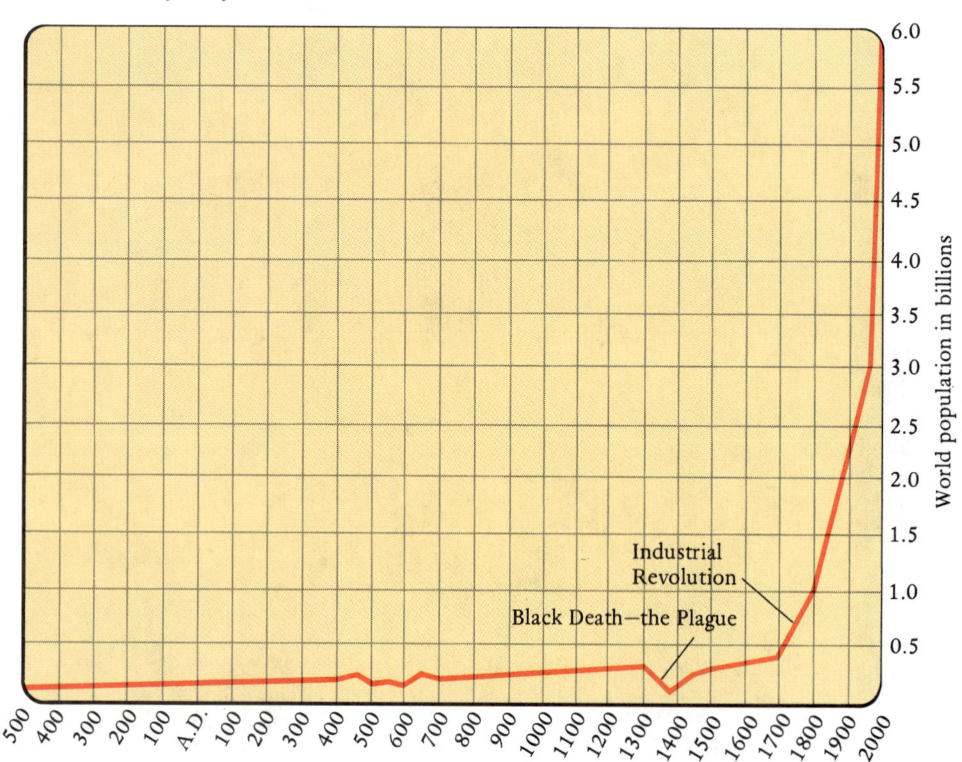

Figure 13–1 World populations for 500 B. C. to A. D. 2000. The curve is shaped like a J because slow growth over a long period of time was followed by rapid growth over a relatively short period. It is projected that 6 billion humans will inhabit the earth by the year 2000.

The use of too much of a chemical or the use of the wrong chemical produces unwanted side effects that negate effective food production. Two notable examples are the phosphate runoff from fertilized fields, which can pollute streams, and the inclusion of DDT in bird eggs, which interferes with calcium metabolism and causes eggshells to become thin and to crack prematurely.

Our fundamental food is plants. Either we eat plants or animals that eat plants. In order to grow, plants require the proper temperature, nutrients, air, water, and freedom from disease, weeds, and harmful pests. Chemistry has provided chemicals to assist nature in giving plants the proper nutrients and freedom from disease and competitive life forms; these chemicals are the subject of this chapter. However, we must point out that the use of agrichemicals involves risks to the environment and human health; it is important to measure the risks versus benefits in the use of these chemicals.

The United States exported $41.3 billion while importing $22.0 billion worth of agricultural commodities in 1989.

George Washington had marl, an alkaline mixture of limestone and clay, dug from the Potomac River bed for application to his fields.

"No man qualifies as a statesman who is entirely ignorant of the problems of wheat."—Socrates

Benefits and problems related to DDT are discussed later in this chapter.

The best estimate is that two thirds of the world's agriculture is "backwards."

NATURAL SOILS SUPPLY NUTRIENTS TO PLANTS

Layers within the soil are called **horizons** (Fig. 13–2). The **topsoil** contains most of the presently living material and humus from dead organisms. It is not uncommon to find as much as 5% organic matter in topsoil. Topsoil is usually several inches thick, and in some locations morc than 3 ft of topsoil can be found. The **subsoil,** up to several feet in thickness, contains the inorganic materials from the parent rocks as well as organic matter, salts, and clay particles washed out of the topsoil.

Since healthy topsoil has abundant life forms and their remains, it must contain an abundant supply of oxygen. Soil that supports a rich vegetative growth and serves as a host for insects, worms, and microbes is typically full of pores; such soil is likely to have as much as 25% of its volume occupied by air. The ability of soil to hold air depends on soil particle size and how well

Through photosynthesis powered by energy from the Sun, Earth's green plant population provides the food to sustain life on Earth. (*The World of Chemistry,* Program 23, "Proteins: Structure and Function")

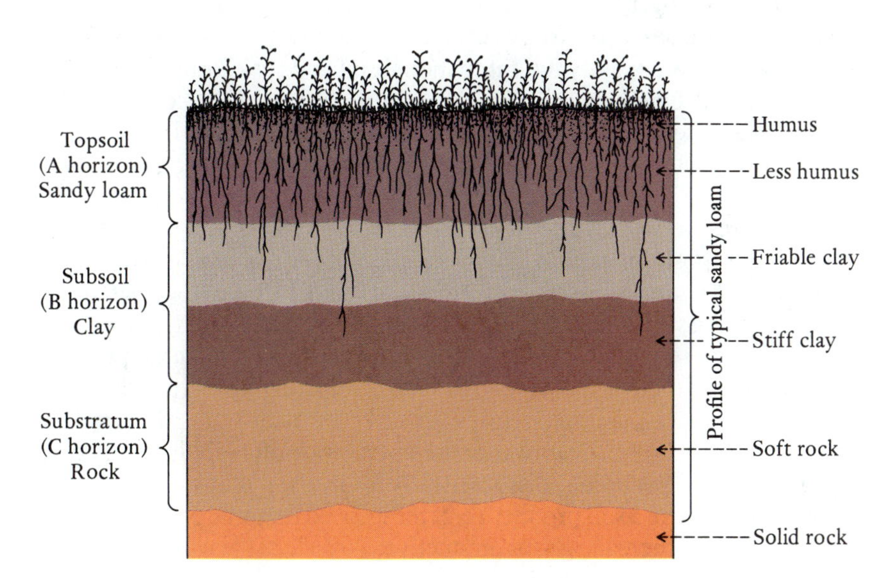

Figure 13–2 Structure of a typical sandy soil. Note that the soil is composed of layers or horizons that can be built up through the weathering of the parent rock below and the interaction of the weathered material with air and plant life from above. (U.S. Department of Agriculture)

the particles pack and cling together to form a solid mass. The particle size groups in soils, called **separates,** vary from clays (the finest) through silt and sand to gravel (the coarsest). The particle size of a clay is 0.005 mm or less. The small particles in a clay deposit pack closely together to eliminate essentially all air and thus support little or no life. A typical soil horizon is composed of several separates. A **loam,** for example, is a soil consisting of a friable mixture of varying proportions of clay, sand, and organic matter; a loam is rich in air content.

Friable material crumbles easily under slight pressure.

Air in soil has a different composition from the air we breathe. Normal dry air at sea level contains about 21% oxygen (O_2) and 0.03% carbon dioxide (CO_2). In soil the percentage of oxygen may drop to as low as 15%, and the percentage of carbon dioxide may rise above 5%. This results from the partial oxidation of organic matter in the closed space. The carbon in the organic material uses oxygen to form carbon dioxide. This increased concentration of carbon dioxide tends to cause groundwater to become acidic; acidic soils are described as *sour* soils because of the sour taste of aqueous acids.

$$CO_2 + H_2O \rightleftharpoons H^+ + HCO_3^-$$

A slightly basic soil is a *sweet* soil.

Crushed limestone ($CaCO_3$) applied to soil combines with hydrogen ions to form bicarbonate ions, thus raising the pH.

$$H^+ + CO_3^{2-} \rightleftharpoons HCO_3^-$$

If enough limestone is added to neutralize the acid in the soil and leave an excess of limestone, the pH of the soil becomes alkaline.

WATER IN SOIL: TOO MUCH, TOO LITTLE, OR JUST RIGHT

Water can be held in soil in three ways: it can be *absorbed* into the structure of the particulate material, it can be *adsorbed* onto the surface of the soil particles, and it can occupy the pores ordinarily filled with air.

Water is removed from soil in four ways: plants transpire it while carrying on life processes, soil surfaces evaporate water, water is carried away in plant products, and gravity pulls it to the subsoil and rock formations below in a process called **percolation.** Percolation is the ability of a solid material to drain a liquid from the spaces between the solid particles. Soils with good percolation drain water from all but the small pores in the natural flow of the water.

It takes several hundred pounds of water for the typical food crop to make 1 lb of food.

The percolation of a soil depends on the soil particle size and on the chemical composition of the soil material. Because of the small particle sizes involved, clays, and to a lesser degree silts, tend to pack together in an impervious mass with little or no percolation. Of course, sand, gravel, and rock pass water readily. Waterlogged soils that do not percolate support few crops because of their lack of air and oxygen. Rice is an important exception. A negative aspect of the massive flow of water through soil is the **leaching effect.** Water, known as the universal solvent because of its ability to dissolve so many different materials, dissolves away, or leaches, many of the chemicals needed to make a soil productive. If the leached material is not replaced, the soil becomes increasingly unproductive.

In addition to the food manufacturing process, a leaf must transport large masses of materials from one place to another and exchange gases and water vapor with the surrounding air. All of these processes depend on an abundant supply of water. Desert plants have protective mechanisms to protect a high water content for chemically active cells. (*The World of Chemistry*, Program 12, "Water")

It is estimated that 2 million gal of water per person per year are used in the United States and that 80% of this water is for agriculture. We want to grow crops where rainwater is not sufficient and as we become more adept at using fresh groundwater, it is becoming increasingly evident that the fresh water supply will continue to be the limiting factor in food production. Note the prospects for using salt water for agriculture at the end of Chapter 11. (*The World of Chemistry*, Program 12, "Water")

Soils become acidic, or sour, not only because of the oxidation of organic matter but also because of *selective leaching* by the passing groundwater. Salts of the IA and IIA metals are more soluble than salts of the group IIIA and transition metals. For example, a soil containing calcium, magnesium, iron, and aluminum is likely to be slightly alkaline, or sweet, before leaching with water. If calcium and magnesium are removed and iron and aluminum salts remain, the soil becomes acidic. The iron and aluminum ions each tie up hydroxide ions from water and release an excess of hydrogen ions:

In some arid regions, calcium collects as calcium carbonate just under the solum, or true soil.

$$Fe^{3+} + H_2O \longrightarrow FeOH^{2+} + H^+$$

$$Al^{3+} + H_2O \longrightarrow AlOH^{2+} + H^+$$

Acids and bases are discussed in Chapter 4.

HUMUS

Organic matter varies in soil from the relatively fresh remains of leaves, twigs, and other plant and animal parts to peat, which results from old, decayed animal and vegetable matter. Peat is the precursor of coal and oil. Humus is not far removed in time from the living debris. However, it is well decomposed, dark-colored, and rather resistant to further decomposition. As a source of nutrients for plants, humus is almost like a time-release

Humus releases its nutrients to plants slowly.

capsule, taking considerable time to release its contents while holding them in an insoluble form. Eventually, humus is decomposed into minerals and inorganic oxides.

In addition to being a source of plant nutrients, humus is important in maintaining good soil structure, often keeping the soil friable in a soil rich in clay. Soil rich in humus may contain as much as 5% organic matter. Soils in the grasslands of North America are rich in humus to a considerable depth, in contrast to rain-forest regions, where there is only a thin film of humus on the ground surface.

NUTRIENTS

At least 18 known elemental nutrients are required for normal green plant growth (Table 13–1). Three of these, the **nonmineral nutrients,** carbon, hydrogen, and oxygen, are obtained from air and water. The **mineral nutrients** must be absorbed through the plant root system as solutes in water. The 15 known mineral nutrients fall into three groups: **primary nutrients, secondary nutrients,** and **micronutrients,** depending on the amounts necessary for healthy plant growth.

Nonmineral Nutrients

Sir Humphrey Davies argued the humus theory. "Carbon for plants came from humus." A Swiss, de Saussure, showed the carbon came from carbon dioxide.

Carbon, hydrogen, and oxygen are available from the air and water. Carbon comes to plants as carbon dioxide, and hydrogen and oxygen come as water; in addition, plant roots absorb some free oxygen dissolved in water. During photosynthesis, green plants produce an excess of oxygen, which is released through the leaves and other green tissue.

Primary Nutrients

The air above each acre of earth's surface contains 36,000 tons of nitrogen.

The primary nutrients are nitrogen, phosphorus, and potassium. Although bathed in an atmosphere of nitrogen, most plants are unable to use the air as a supply of this vital element. **Nitrogen fixation** is the process of

TABLE 13–1 Essential Plant Nutrients

Nonmineral	Primary	Secondary	Micronutrients
Carbon	Nitrogen	Calcium	Boron
Hydrogen	Phosphorus	Magnesium	Chlorine
Oxygen	Potassium	Sulfur	Copper
			Iron
			Manganese
			Molybdenum
			Sodium
			Vanadium
			Zinc

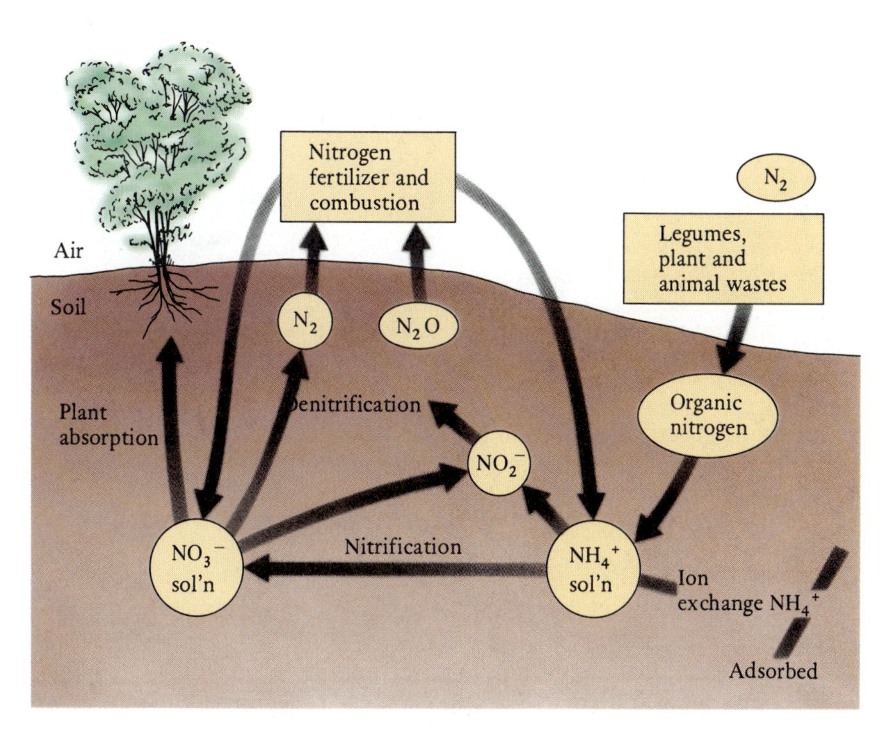

Figure 13–3 Nitrogen pathways through the soil.

changing atmospheric nitrogen into the compounds of this element that can be dissolved in water, absorbed through the plant's roots, and assimilated by the plant (Fig. 13–3). Most plants thrive on soils rich in nitrates, but many plants that grow in swamps, where there is a lack of oxidized materials, can use reduced forms of nitrogen such as the ammonium ion. The nitrate ion is the most highly oxidized form of combined nitrogen, and the ammonium ion is the most reduced form of nitrogen.

Nitrogen species ranked by degree of oxidation: NO_3^-, NO_2, NO, N_2O, N_2, NH_4^+.

Nitrogen is an important element in amino acids and proteins.

Nature fixes nitrogen on a massive scale in two ways. Nitrogen is oxidized under highly energetic conditions, such as in the discharge of lightning, or to a lesser extent, such as in a fire. The initial reaction is the reaction of nitrogen and oxygen to form nitric oxide (NO):

$$N_2 + O_2 \longrightarrow 2\ NO$$

A vast amount of atmospheric nitrogen is fixed as a result of natural electric discharges in the atmosphere. The energy in a bolt of lightning is sufficient to disrupt the very stable triple bond in a nitrogen molecule. Oxidation of nitrogen results. (Gary Ladd)

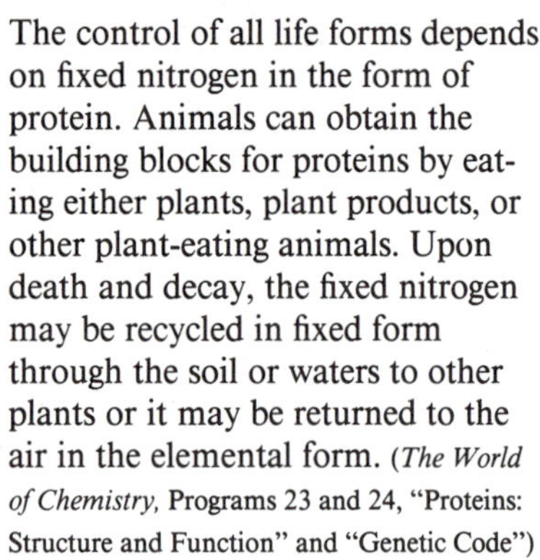

The control of all life forms depends on fixed nitrogen in the form of protein. Animals can obtain the building blocks for proteins by eating either plants, plant products, or other plant-eating animals. Upon death and decay, the fixed nitrogen may be recycled in fixed form through the soil or waters to other plants or it may be returned to the air in the elemental form. (*The World of Chemistry,* Programs 23 and 24, "Proteins: Structure and Function" and "Genetic Code")

Nitric oxide is easily oxidized in air to nitrogen dioxide (NO_2), which dissolves in water to form nitric acid (HNO_3) and nitrous acid (HNO_2):

$$2\,NO + O_2 \longrightarrow 2\,NO_2$$

$$H_2O + 2\,NO_2 \longrightarrow HNO_3 + HNO_2$$

Nitric acid is readily soluble in rain, clouds, or ground moisture and thus increases nitrate concentration in soil.

Another major source of nitrogen replenishment in soil is dead organisms and animal wastes. Even in the absence of legumes, this can be an adequate source of nitrogen.

Like nitrogen, phosphorus must be in mineral or inorganic form before it can be used by plants. Unlike nitrogen, phosphorus comes totally from the mineral content of the soil. Orthophosphoric acid (H_3PO_4) loses hydrogen ions to form the dihydrogen and monohydrogen ions ($H_2PO_4^-$ and HPO_4^{2-}), which are the dominant phosphate ions in soils of normal pH (Fig. 13–4). Because of the great concentration of electric charge associated with the trivalent phosphate ion, phosphates tend to be held to positive centers in the soil structure and are not as easily leached by groundwater as are nitrate ions. The nitrate ion has only one negative charge; nitrate salts are generally soluble in water.

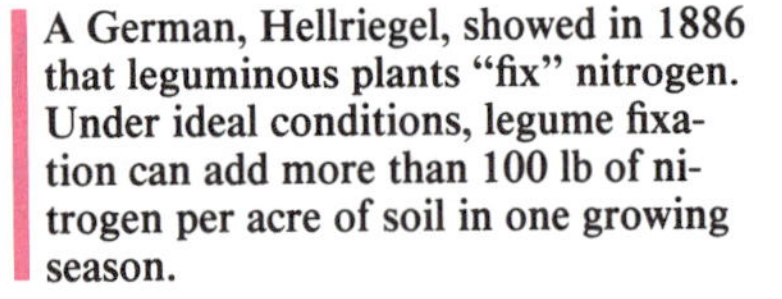

A German, Hellriegel, showed in 1886 that leguminous plants "fix" nitrogen. Under ideal conditions, legume fixation can add more than 100 lb of nitrogen per acre of soil in one growing season.

Potassium is absorbed as the free ion, $K^+(aq)$.

Potassium is a key element in the enzymatic control of the interchange of sugars, starches, and cellulose. Although potassium is the seventh most

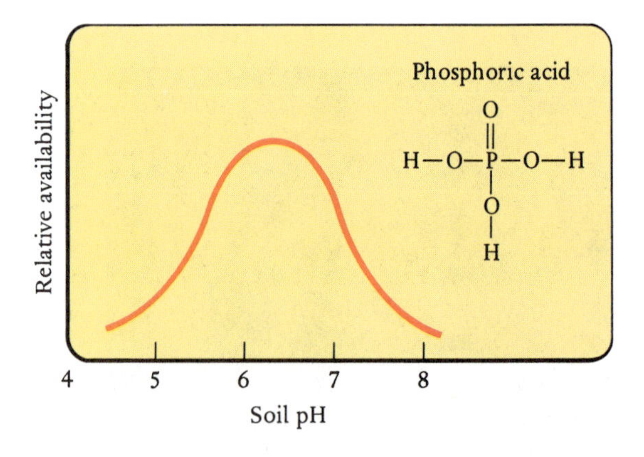

Figure 13–4 The availability of phosphate in the soil is a function of pH. The dominant species present for phosphoric acid at pH 5 to pH 8 are $H_2PO_4^-$ and HPO_4^{2-}. At very low pH values, the phosphorus is in the form of the acid H_3PO_4 (all three protons on the acid structure). At a very high pH, all three protons are removed, and the phosphorus is in the form of the PO_4^{3-} ion. Low soil temperatures in temperate regions significantly reduce phosphorus uptake by plants.

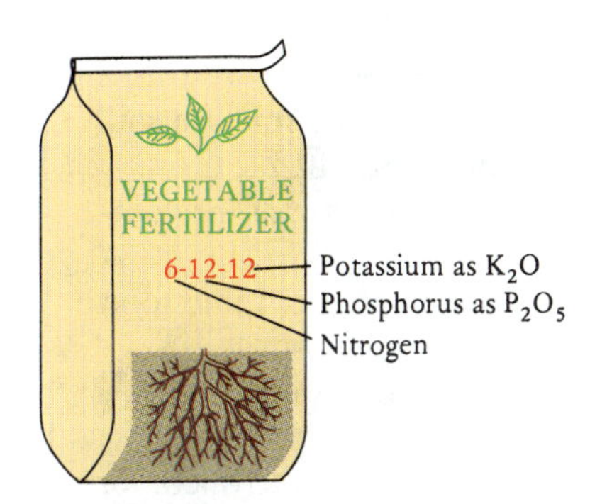

Figure 13–5 Fertilizer analysis numbers refer to the percentage by weight of N (nitrogen), P_2O_5 (for phosphate), and K_2O (for potassium). Following the lead of Liebig, Samuel William Johnson, an American student of Liebig and the author of *How Crops Grow,* burned plants and analyzed their ashes. He expressed the nutrient concentrations in the oxide form present in the ashes as P_2O_5, K_2O, and so on, a practice that has continued to this day.

abundant element in the Earth's crust, soil used heavily in crop production can be depleted in this important metabolic element, especially if it is regularly fertilized with nitrate with no regard to potassium content (Fig. 13–5). Some fungus plants in the soil produce chemicals that cause bound potassium to be released into a soluble form that can be taken in through the plant root system in excessive amounts or simply leached out by the flow of soil water.

Secondary Nutrients

Calcium and magnesium are available as Ca^{2+} and Mg^{2+} ions in small amounts as well as in complex ions and crystalline formations. These abundant elements are bound tightly enough so they are not readily leached yet loosely enough to be available to plants. When in the soil as sulfate (SO_4^{2-}), sulfur is readily available to plants.

Chlorophyll requires nitrogen and magnesium from the soil. Magnesium deficiencies, like nitrogen deficiencies, cause *chlorosis,* a condition of low chlorophyll content.

Micronutrients

Only very small amounts of micronutrients are required by plants; therefore, unless extensive cropping or other factors deplete the soil of these nutrients, sufficient quantities are usually available.

Iron is also an essential component of the catalyst involved in the formation of chlorophyll, the green plant pigment. When the soil is iron deficient or when too much lime, $Ca(OH)_2$, is present in the soil, iron availability decreases. This condition is usually indicated by plant leaves that are light or yellowish in color. Often a gardener or lawn worker will apply phosphate and lime to adjust soil acidity, only to see green plants turn yellow. What happens in such cases is that both phosphate and the hydroxide from the lime tie up the iron and make iron unavailable to the plants:

Lime, as CaO or $Ca(OH)_2$, is the number-five commercial chemical (see listing inside front cover).

$$Fe^{3+} + 2\,\underset{\text{Phosphate}}{PO_4^{3-}} \longrightarrow \underset{\text{Tightly bound complex}}{Fe(PO_4)_2^{3-}}$$

$$Fe^{3+} + 3\,OH^- \longrightarrow \underset{\text{Insoluble hydroxide}}{Fe(OH)_3}$$

Boron is absolutely necessary in trace amounts, but there is a relatively narrow concentration range above which boron is toxic to most plants.

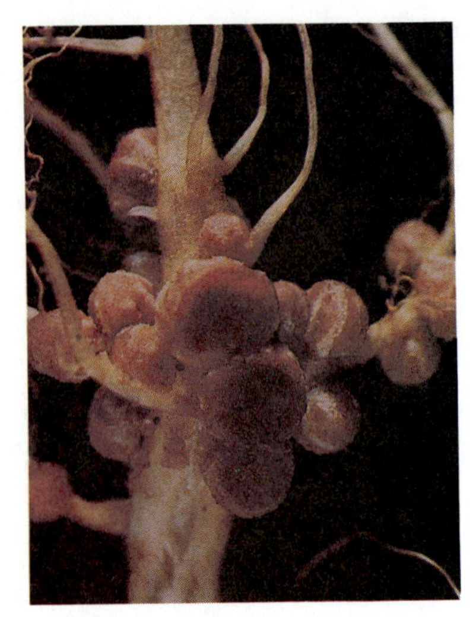

Atmospheric nitrogen is fixed in root nodules on leguminous plants such as the soybean pictured here. Rather than requiring fertilization for nitrogen, such plants can be grown to enrich the soil with nitrogen compounds and produce a valuable crop at the same time. (Metcalfe, Williams, and Castka and Walter O. Scott)

SELF-TEST 13A

1. Rank the following types of soils from those with the smallest soil particles to those with the largest: silts, sandy soils, loams, clays.
2. Acidic soils are described as __________ because of the common taste of aqueous acids.
3. Carbon dioxide causes soils to be ( ) acidic ( ) basic. Limestone ($CaCO_3$) causes soils to be ( ) acidic ( ) basic.
4. The two factors that determine the percolation of a soil are __________ and __________.
5. Which is more acidic, a monovalent ion like Na^+ or a trivalent ion like Fe^{3+}? __________
6. A well-decomposed, dark-colored plant residue that is relatively resistant to further decomposition is known as __________.
7. The primary elemental plant nutrients necessary in the soil for healthy plant growth are __________, __________, and __________.
8. The secondary elemental plant nutrients are __________, __________, and __________.
9. Nitrogen fixation involves breaking a nitrogen-nitrogen triple bond and combining nitrogen with another __________.
10. Some micronutrients can poison plants: True ( ) False ( )

FERTILIZERS SUPPLEMENT NATURAL SOILS

Vast amounts of acres have been torched each year in the Amazon basin. Current efforts have curbed this deforestation but have not halted it.

Chinese farmers added calcined bones to their soil 2000 years ago.

Crop yield explosions: (1) U.S. corn—25 bushels per acre in 1800, 110 bushels per acre in the 1980s; (2) English wheat—below 10 bushels per acre from A.D. 800 to 1600, above 75 bushels per acre in the 1980s. (3) Rice in Japan, Korea, and Taiwan—fourfold increase in the last 40 years.

Primitive people raised crops on a cultivated plot until the land lost its fertility; then they moved to a virgin piece of ground. In many cases, the slash-burn-cultivate cycle was no more than a year in length, and few found a piece of ground anywhere that could support successful cropping for more than five years without fertilization. Farming villages, developed in ancient times and prevalent throughout the Middle Ages, demanded innovation in fertilization, because they had to use the same land for many years. With the use of legumes in crop rotations, manures, dead fish, or almost any organic matter available, the land was kept in production.

An estimated 4 billion acres are used worldwide in the cultivation of crops for food, less than 0.8 acre per person. This acreage would likely be sufficient if modern chemical fertilization were employed on all of it. If about

As chemical fertilizers became widely available during the early part of the 20th century, farmers were eager to use them because of the explosive increase in crop yields obtained. Tractor-powered spreading equipment can now easily fertilize a 40-ft path across the field in a side-dressing to the growing plant, a general application to the soil surface or subsurface. (*The World of Chemistry*, Program 2, "Color")

THE WORLD OF CHEMISTRY

Agriculture

When any covalent bond forms, energy is released. In turn, the same amount of energy is needed to break a bond.

Consider nitrogen. There is nitrogen in all living things. Muscles, hair, and DNA all contain nitrogen bonded to other elements. But 80% of the atmosphere is nitrogen molecules held together by strong triple bonds. How do living things get the form of nitrogen they need? Lightning helps. The electrical flash in the sky has enough energy to break apart nitrogen molecules, which then react with oxygen in the air, eventually forming nitric acid. The natural acid dissolves in rain and falls to earth as a dilute solution. There it is absorbed and metabolized by plants.

Some plants, however, convert molecular nitrogen in a different way. Soybeans and other legumes, like peas and peanuts, host a unique bacterium in their roots. It is this bacterium that converts the nitrogen molecule into a nitrogen compound, ammonia, which the plant can then use to make amino acids.

Exactly how the bacterium works is the subject of vigorous research. Don Keister, of the U.S. Department of Agriculture, claims, "This is one of the very unique enzymes in all of nature, because it is the only solution that nature has evolved for biologically reducing nitrogen."

The soybean and the bacterium have a symbiotic relationship. The plant houses and feeds the bacterium and, in turn, it receives the nitrogen it needs. But not all plants can host these nitrogen fixers. They have to rely on rain and natural fertilizers, as well as expensive manufactured fertilizers, like ammonium nitrate.

Keister goes on to point out, "We are currently using something like 300 million barrels of oil per year in the United States alone to produce nitrogen fertilizers. We forget sometimes, that we're going to need to double the food supply over the next 20 years." He further asks the unanswered questions, "Where is that energy going to come from? Where is the fertilizer going to come from?"

For feeding the world, there are two basic options: We can either produce more fertilizer at greater cost and some risk to the environment, or we can create new varieties of nitrogen-fixing plants. Both options are being pursued worldwide.

The World of Chemistry (Program 8) "Chemical Bonds."

$40 were spent on fertilizer for each cultivated acre, world crop production would increase by 50%, the equivalent of having 2 billion more acres under cultivation. However, the cost to produce this additional food would approach a prohibitive $160 billion, and the environmental impact of such a large dispersion of fertilizer chemicals would likely be massive. For example, at the present time, aquifer contamination with nitrates due to corn crop fertilization renders well water from these natural underground basins unfit for drinking in large areas of the U.S. corn belt.

Peruvian guano, a natural source of nitrates was first imported into the United States in 1824.

Figure 13-6 At normal temperatures ammonia can be maintained as a liquid under tank pressure. On release from pressure the ammonia turns to a gas and is readily dissolved in the moisture of the soil if the soil pH is slightly acid, the ammonia is immediately converted to the ammonium ion and enters into the natural nitrogen pathways of the soil (Fig. 13-3). (Grant Heilman)

The nonmineral nutrient elements, carbon, hydrogen, and oxygen, are readily available in a continuing supply to plants from the air and water. The primary nutrients, nitrogen, phosphorus, and potassium, are easily depleted and must be resupplied through natural or chemical fertilization. The cheapest source of nitrogen is the air, but it must be combined with relatively expensive hydrogen, obtained from petroleum, to form ammonia. The process is named the **Haber process** after Fritz Haber, the chemist who learned to control this reaction for the production of industrial qualtities of ammonia.

$$N_2 + 3\,H_2 \longrightarrow 2\,NH_3$$

The ammonia can then be applied directly to the soil (Fig. 13-6) or converted into numerous agrichemicals (Fig. 13-7). Phosphorus is readily available in the form of phosphate rock, which can be transformed into the needed fertilizers (Fig. 13-8). World deposits of phosphate rock are limited, and costs for supplying phosphorus fertilizers will increase as deposits are depleted. Potassium in the form of potash (K_2CO_3) exists in enormous quantities throughout the world. Also, potassium chloride, KCl, is readily available. Mineral resources are abundant for supplying the secondary nutrients, calcium, magnesium, and sulfur (Table 13-2).

TABLE 13-2 Some Chemical Sources for Plant Nutrients

Element	Source Compound(s)
Nonmineral Nutrients	
C	CO_2 (carbon dioxide)
H	H_2O (water)
O	H_2O (water)
Primary Nutrients (Fig. 13-5)	
N	NH_3 (ammonia), NH_4NO_3 (ammonium nitrate), H_2NCONH_2 (urea)
P	$Ca(H_2PO_4)_2$ (calcium dihydrogen phosphate)
K	KCl (potassium chloride)
Secondary Nutrients	
Ca	$Ca(OH)_2$ (calcium hydroxide [slaked lime]), $CaCO_3$ (calcium carbonate [limestone]), $CaSO_4$ (calcium sulfate [gypsum])
Mg	$MgCO_3$ (magnesium carbonate), $MgSO_4$ (magnesium sulfate [epsom salts])
S	Elemental sulfur, metallic sulfates
Micronutrients	
B	$Na_2B_4O_7 \cdot 10H_2O$ (borax)
Cl	KCl (potassium chloride)
Cu	$CuSO_4 \cdot 5H_2O$ (copper sulfate pentahydrate)
Fe	$FeSO_4$ (iron(II) sulfate, iron chelates)
Mn	$MnSO_4$ (manganese(II) sulfate, manganese chelates)
Mo	$(NH_4)_2\,MoO_4$ (ammonium molybdate)
Na	NaCl (sodium chloride)
V	V_2O_5, VO_2 (vanadium oxides)
Zn	$ZnSO_4$ (zinc sulfate, zinc chelates)

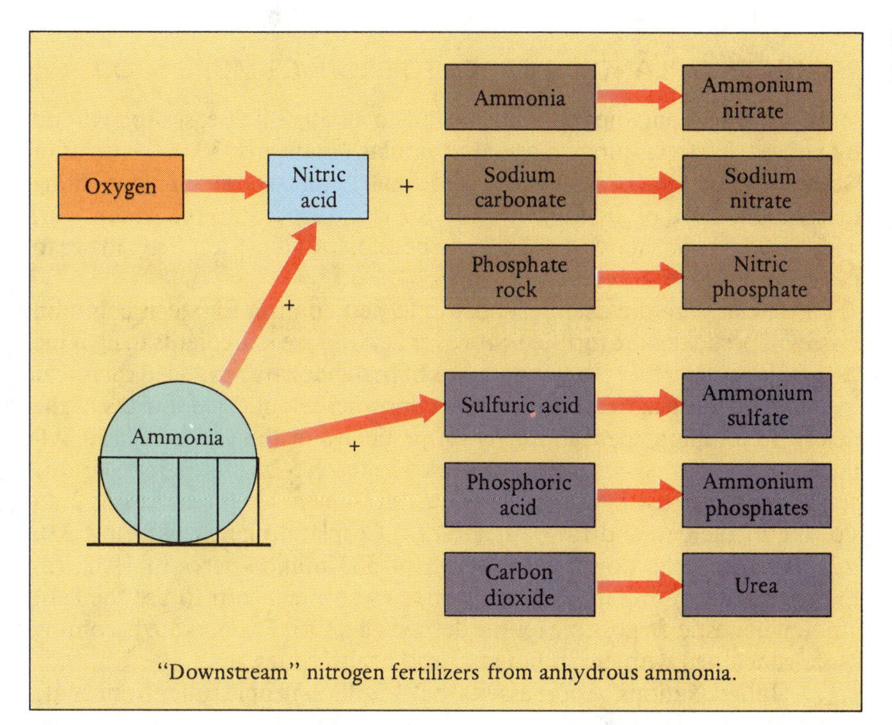

Figure 13–7 Nitrogen fertilizers, produced from anhydrous ammonia.

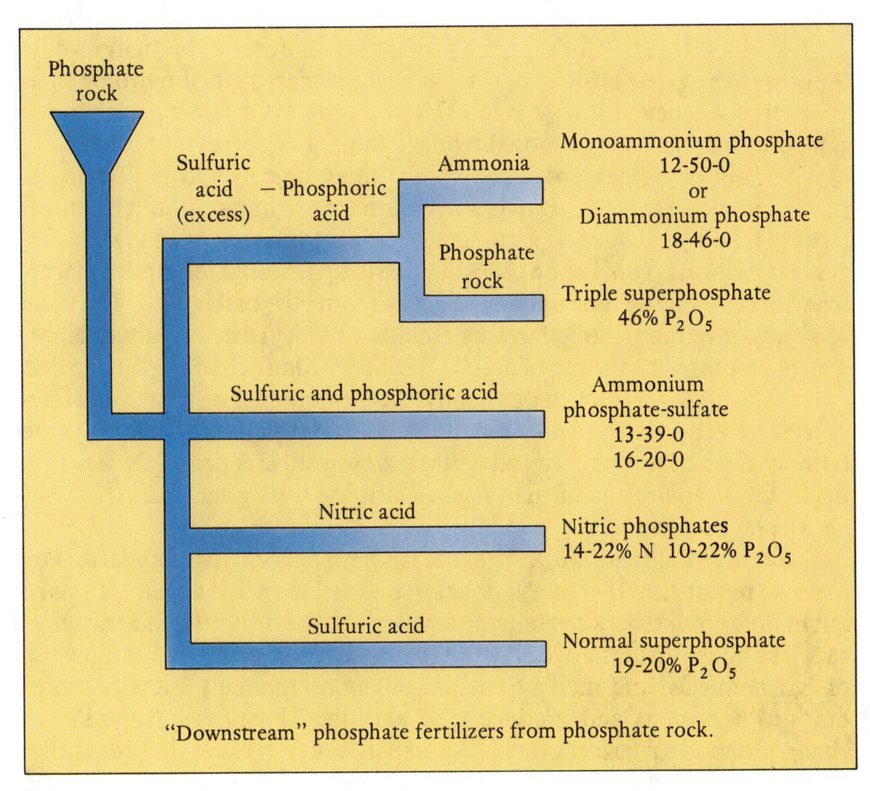

Figure 13–8 Fertilizers produced from phosphate rock.

PROTECTING PLANTS IN ORDER TO PRODUCE MORE FOOD

The natural enemies of plants include over 80,000 diseases brought on by viruses, bacteria, fungi, algae, and similar organisms: 30,000 species of weeds; 3,000 species of nematodes; and about 10,000 species of plant-eating insects. One third of the food crops in the world are lost to pests each year, with the loss going above 40% in some developing countries. Crop losses to pests amount to $20 billion per year.

Pesticides are the chemical answer to pest control. Eighteen common classes of pesticides are fortified with over 2600 active ingredients to fight the battle with pests. Over 5 billion pounds of pesticides are produced each year in the world. The agricultural market consumes about three fourths of the nearly $6.0 billion spent annually on pesticides in the United States. Although the dollar cost is up and expected to reach $7.5 billion in 1995, the actual poundage use began a decline in 1987. There are three reasons for a slowing in the demand for pesticides: (1) Cropland planted is below 330 million U.S. acres, down from a high of 383 million acres in 1982. (2) Farming is becoming more cost effective as farmers learn to use the least amount of pesticide possible for the desired effect. (3) Farmers are becoming more concerned with environmental and health issues.

A United Nations' report asserts that 1 million people suffer from acute pesticide poisoning each year, with at least 20,000 fatalities. On a national level complacency in the use of agrichemicals is turning into concern for the environment as well as the individual. In a 1987 study by *U.S. News and World Report,* 51 food samples in each of three cities were tested with 87 tests for pesticides. Forty-six of the samples showed no traces of pesticides while the other five did test positive but below legal limits set by the Environmental Protection Agency (EPA). Three of these five failed a new negligible-risk standard that has been proposed by the National Academy of Sciences.

Unchecked insects consume on average up to one third of the grower's efforts, and in extreme cases all of the food produced. (*The World of Chemistry,* Program 24, "Genetic Code")

Without pesticides, crop production, on average, would be 20% lower than at present. **Organic farming,** a new name for farming without chemical fertilizers and pesticides, is gaining a significant number of followers. Organic farming uses only about 40% of the energy required for modern farming with synthetic chemicals and produces about 90% of the yield. The costs of energy saved in organic farming are offset by the costs of human labor, which is required by the use of natural fertilizers. Many claims are made that organic farming produces a better product for human consumption. However, there is no real evidence that these claims are generally true. Organic farming does have one clear advantage, however; it is definitely less of a threat to the environment than regular farming is if agrichemicals are not very carefully controlled.

Alternative agriculture, a term that has been particularly popularized by a special report (1989) of the National Research Council, is an effort to stake out middle ground between organic farming and the heavy use of agrichemicals. The goals of alternative agriculture are to improve profits, limit the use of agrichemicals, and increase the use of environmentally friendly procedures to fight pests and produce food and fiber. Examples of significant changes in modern farming include (a) expanding crop rotations because the same pests do not afflict every crop; (b) using multiple crops in alternate plantings within a given planting field; (c) using as much natural fertilizer as possible before using agrichemicals; (d) increasing the use of biological pest controls; (e) employing renewed efforts at soil and water conservation; and (f) having a diversification in livestock as well as field crops on the same farm.

Insecticides

Before World War II, the list of insecticides included only a few compounds of arsenic, petroleum oils, nicotine, pyrethrum, rotenone, sulfur, hydrogen cyanide gas, and cryolite. DDT, the first of the chlorinated organic insecticides, was originally prepared in 1873, but it was not until the beginning of World War II that it was recognized as an insecticide.

The use of synthetic insecticides increased enormously on a worldwide basis after World War II. As a result, insecticides such as DDT have found their way into our environment. There is a great variety of pesticides, and their use frequently leads to severe damage to other forms of animal life, such as fish and birds. The toxic reactions and peculiar biological side effects of many of the pesticides were often not thoroughly studied or understood prior to their widespread use.

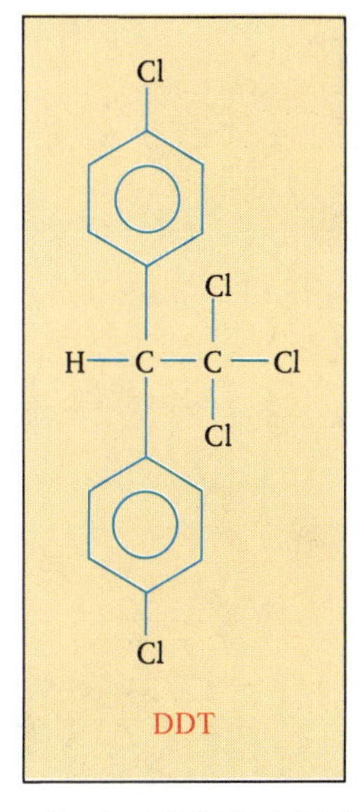

Figure 13–9 DDT [1,1-bis(*p*-chlorophenyl)-2,2,2-trichloroethane].

A case in point is DDT (Fig. 13–9). This insecticide, which has not been shown to be toxic to humans in doses as high as those received by factory workers involved in its manufacture (400 times the average exposure), does have peculiar biological consequences. The structure of DDT is such that it is not metabolized (broken down) very rapidly by animals; instead it is deposited and stored in the fatty tissues. The biological half-life of DDT is about eight years; that is, it takes about eight years for an animal to metabolize half of the amount it assimilates. If ingestion continues at a steady rate, DDT builds up within the animal over time.

Half-life is the time required for half of the substance to disappear.

For many animals this buildup of DDT is not a problem, but for some predators, such as the eagle and osprey, that feed on other animals and fish, the consequences are disastrous. The DDT in the fish eaten by a bird is

Aerial application of an insecticide to a field crop. This once widely used method has been largely discarded for local farming operations because of the lack of control over the spread of the expensive insecticide. (*The World of Chemistry,* Program 25, "Chemistry and the Environment")

The massive use of DDT for insect control nearly wiped out several species of birds owing to the interference of this insecticide in the formation of egg shells. (*The World of Chemistry,* Program 25, "Chemistry and the Environment")

concentrated in the bird's fatty tissue; the bird then attempts to metabolize the insecticide by altering its normal metabolic pattern. This alteration involves the use of compounds that normally regulate calcium metabolism in the bird and are vital to the female's ability to lay eggs with thick shells. When these compounds are diverted to their new use, they are chemically modified and are no longer available for the egg-making process. As a consequence, the eggs the bird does lay are easily damaged, and the survival rate of the species decreases drastically. This process has led to the nearly complete extinction of eagles and ospreys in some parts of the United States.

The buildup of DDT in natural waters is not an irreversible process; the EPA reported a 90% reduction of DDT in Lake Michigan fish by 1978 as a result of a ban on the use of the insecticide.

Dieldrin and aldrin were banned by the courts from their major uses (such as on corn) in 1974.

DDT and other insecticides such as **dieldrin** and **heptachlor** are referred to as **persistent pesticides.*** Other substances with biodegradable structures are now substituted as much as possible; the compound **chlordan** is an example of just such a substitution. Heptachlor, because of its persistent nature, and chlordan, because of its short-term toxic effects on test animals, were banned for most garden and home use in December 1975.

Many insecticides are much more toxic to humans than is DDT. These include organic materials based on arsenic compounds, as well as organic phosphates, such as **parathion** and **malathion,** (see Figure 10–11.) Called anticholinesterase poisons (Chapter 10), these compounds are readily hydrolyzed to less toxic substances that are not residual poisons.

The goal of the insecticide quest: a selectively toxic chemical that is quickly biodegradable.

The choice of solutions to the problems of pesticides is not an easy one. By using insecticides, we introduce them into our environment and our water supplies. If we fail to use them, we must tolerate malaria, plague, sleeping sickness, and the consumption of a large part of our food supply by insects. Continuing research in the development of more effective and safer insecticides is intense and new products are introduced each year. For example, Du Pont is touting Asana® as the pyrethroid insecticide of the 1990s. The active ingredient in this new product is a chemical derivative of pyrethrum, one of the oldest natural insecticides, which was originally extracted from chrysanthemum plants. The search goes on!

* The use of DDT was banned in the United States in 1973, although it is still in use in some other parts of the world. Over 1.8 billion kg of DDT have been produced and used.

Herbicides

Herbicides kill plants. They may be **selective** and kill only a particular group of plants, such as the broad-leaved plants or the grasses, or they may be **nonselective,** making the ground barren of all plant life.

Because there are so many varieties and such large quantities of herbicides produced, (about 300,000 tons yearly in the United States), they are considered a separate category from pesticides.

Nonselective herbicides usually interfere with photosynthesis and thereby starve the plant to death. On application, the plant quickly loses its green color, withers, and dies. Selective herbicides act like a hormone, a very selective biochemical catalyst that controls a particular chemical change in a particular type of organism at a particular stage in its development. Most selective herbicides in use today are growth hormones; they cause cells to swell, so that leaves become too thick for chemicals to be transported through them and roots become too thick to absorb needed water and nutrients.

The traditional method for the control of weeds in agriculture was **tillage.** Only in the early 1900s was it recognized that some fertilizers were also weed killers. For example, when calcium cyanamide (CaNCN) was used as a source of nitrogen, it was found to retard the growth of weeds. Arsenites, arsenates, sulfates, sulfuric acid, chlorates, and borates have also found use as weed killers. A typical product still in commercial use contains 40% sodium chlorate ($NaClO_3$), 50% sodium metaborate ($NaBO_2$), and 10% inert filler. These herbicides are nonselective and must be used with considerable care to protect the desired plants.

Tillage is land prepared for agricultural use by plowing, harrowing, etc.

Nitrophenol was used in 1935 as the first selective, organic herbicide (Fig. 13-10a). It was also in the 1930s that work began on the auxins, or hormone-type weed killers. The most widely used herbicide today, 2,4-D (Fig. 13-10b), came out of this work. The corresponding trichloro-com-

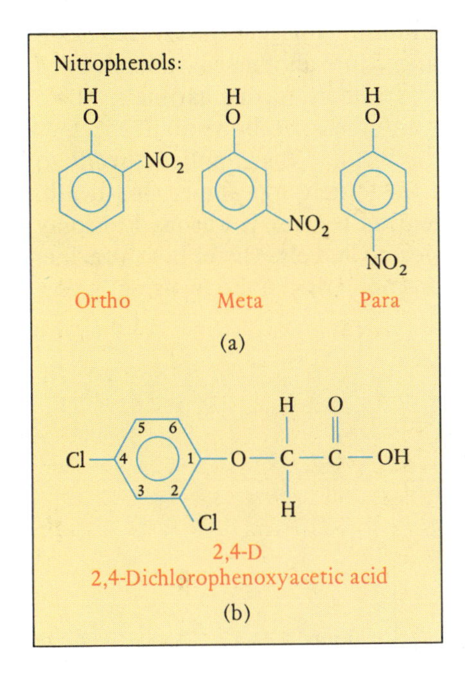

Figure 13-10 Organic herbicides. (a) The first organic herbicides used were the three nitrophenols: *ortho, meta,* and *para.* (b) A widely used organic herbicide today is 2,4-dichlorophenoxyacetic acid (2,4-D).

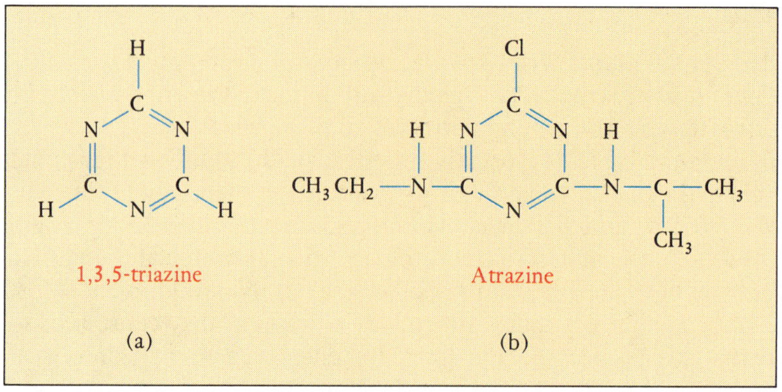

Figure 13–11 (a) 1,3,5-Triazine; (b) Atrazine.

pound (common name: 2,4,5-T) has also been shown to be highly effective. The only difference between it and 2,4-D is the additional chlorine atom on the benzene ring in the fifth position. Agent Orange, widely used as a defoliant during the Vietnam War, is a mixture of these two compounds. The second compound, 2,4,5-T, has been banned by the EPA because of a number of health problems associated with its use. It is probable that most of the problems associated with Agent Orange were caused by the presence of an impurity, dioxin, described in Chapter 10. It will be interesting to see whether 2,4,5-T, which is now commercially produced free of dioxin, will be reestablished as a herbicide. Both of these compounds result in an abnormally high level of RNA in the cells of the affected plants, causing the plants to grow themselves to death.

No-till farming is the control of weeds by herbicides without cultivation.

Several different triazines have been effective as herbicides, the most famous one being atrazine (Fig. 13–11). Atrazine is widely used in no-till corn production or for weed control in minimum tillage. Atrazine is a poison to any green plant if it is not quickly changed into another compound. Corn and certain other crops have the ability to render atrazine harmless, which weeds cannot do. Hence, the weeds die and the corn shows no ill effect.

Paraquat is also used as a contact herbicide. When applied directly to susceptible plants, it quickly causes a frostbitten appearance and death. Paraquat has received considerable attention because it was used to spray illegal poppy and marijuana fields in Mexico and elsewhere. Like atrazine, paraquat has a nitrogen atom in each aromatic ring of the two-ring system (Fig. 13–12).

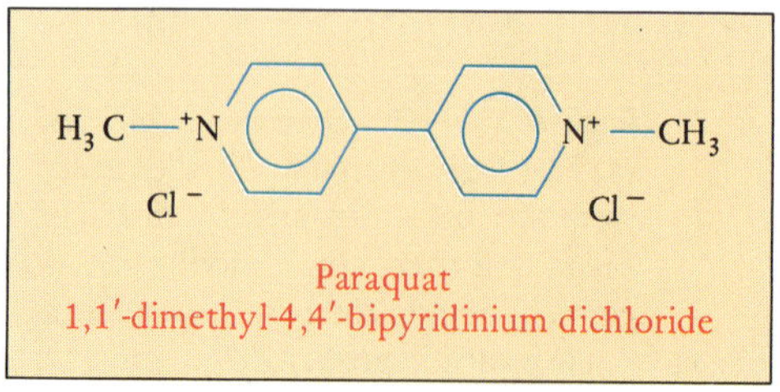

Figure 13–12 Paraquat.

The amount of energy saved by herbicides used in no-till farming is enormous. The saving of topsoil is also considerable, because the cover from the previous crop holds the soil against wind and water runoff. However, agriculturists who use herbicides are highly dependent on agricultural research institutions for the selection of herbicides that will do the desired job without harmful side effects. Such selections depend on considerable research, much of which is carried out on a trial-and-error basis on test plots. A procedure that is recommended today may be outdated by the next growing season.

GENETIC ENGINEERING FOR PEST AND DISEASE CONTROL

Armed with the ability to insert genes into organisms (see Chapter 9), it follows that we might be able to introduce genes into food plants and animals in order to fight pests, control diseases, and improve the quality of the food being produced. Also, it should be possible to introduce genes into organisms to cause that organism to produce a particular food or chemical that we

THE WORLD OF CHEMISTRY

Pheromones

One important area in insect research is the development of pheromones, the sex attractants used by female insects to entice males. They offer a safe, nontoxic way to lure insects into traps and avoid the use of pesticides. Dr. Meyer Schwartz makes such synthetic insect pheromones in his laboratory.

An important part of Dr. Schwartz's work is confirming that he has made exact copies of the natural pheromones. He uses infrared spectroscopy to verify the structure of the molecules he's tried to duplicate in the lab.

Once an accurate copy is made, it's tried out to see if it works. A miniature wind tunnel lets Agriculture Department scientists see just how tantalizing their creation is. The pheromone is placed at one end of the tunnel, and a love-struck male gypsy moth flies against the wind to get it. It's bad enough that he's going to all this trouble for a synthetic chemical, but the Agriculture Department scientists have another trick up their sleeve. They can control the speed of a striped conveyor belt on the wind tunnel floor. The distracted moth sees the belt moving by and thinks he's flying full tilt toward a female. By carefully adjusting the belt speed, the researchers can stop the moth's forward progress. The speed required to stop the moth gives them a good measure of how strong a sex lure their pheromone is. When he finally is allowed to reach the end, all he finds are synthetic pheromones in a cold steel cage.

The World of Chemistry (Program 10) "Signals From Within."

A male moth is attracted to a synthetic pheromone. By placing the bait in a wind tunnel with a moving floor, the moth can be quantitatively measured in its efforts to reach the "female." Thus, the efficacy of the chemical is measured. (*The World of Chemistry,* Program 10, "Signals from Within")

After the laboratory work of gene modification, the test plants must be evaluated in greenhouse and field tests to see if a sought-for characteristic is obtained. (*The World of Chemistry,* Programs 26, 2, and 24, "Futures," "Color," and "Genetic Code")

desire or to provide a living system in which the effects of a particular gene can be studied. Specific progress has been made in the following areas: growing plants that produce their own insecticide, making plants resistant to viral and bacterial infections, matching the chemistry of a plant with a protecting herbicide, growing animals such as hogs and cows and plants such as corn that produce a better food product, and the introduction of human genes into test animals to produce "models" of human diseases for further study. Since 1983, more than 50 species of fruits, vegetables, and grains have been genetically altered.

A transgenic animal has a genetic code that has been altered under human control.

A number of "transgenic animals" have been produced including goats, rabbits, and mice; these animals are used as drug "pharms" because their new genes cause them to produce marketable quantities of desirable pharmaceuticals. Sometimes the new animals are less than ideal. Transgenic pigs did produce leaner meat but they had arthritis, lethargy, and a low sex drive. A group of transgenic beef calves did better in general health but it is as yet unclear if the quality of beef will be improved.

Natural breeding methods require up to ten years to produce plants suitable for field testing that show resistance to a particular virus. The genetic engineer has accomplished this in just one year! DNA segments from the tobacco mosaic virus have been introduced into the genetic code of tomato plants (Fig. 13–13). The implants caused the tomato plants to be strongly resistant to attack by this virus. Natural breeding can use only a selected gene pool characteristic of the species involved. The genetic engineer can move genes from species to species, which multiplies enormously the genes that are available for experimentation. The possibilities for life forms appear open ended.

BHT is a natural hormone that causes cows to give more milk. This hormone is readily produced when the proper gene is introduced into selected bacteria. A considerable health controversy has arisen relative to the safety of this milk even though the hormone is identical to that produced in higher animals. Two other questions that are moral or ethical are: Do we

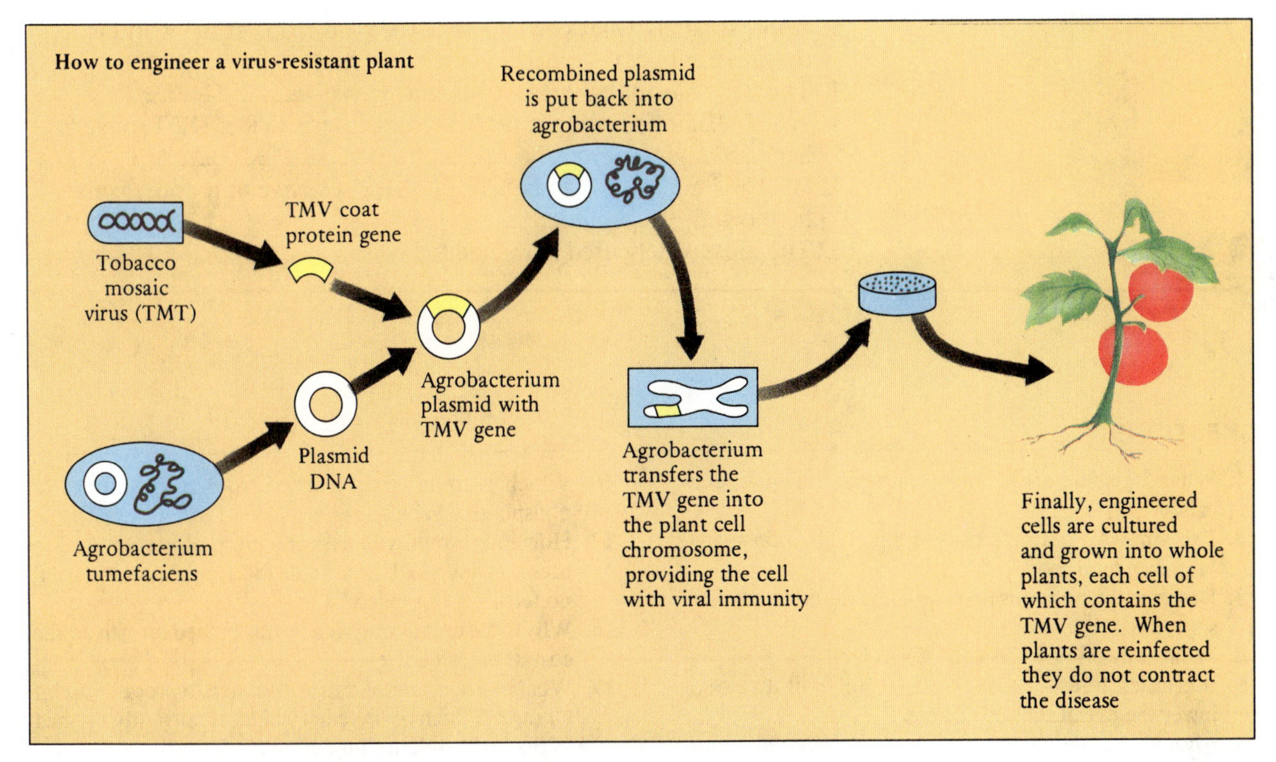

Figure 13–13 Steps necessary to insert a gene from tobacco mosaic virus into a tomato plant in order to make the tomato plant resistant to a particular viral disease.

have the right to produce higher life forms that are dramatically different from the naturally selected ones in nature? Should we risk producing new traits that could be multiplied in nature and make hardier pests? Science alone cannot answer such questions to the satisfaction of society as a whole.

SELF-TEST 13B

1. Most virgin soils can support crop production for a decade or more before fertilization is needed: True ( ) False ( )
2. What do the numbers 6-12-8 on a fertilizer mean? The 6 is the percentage of ______________; the 12 is the percentage of ______________; and the 8 is the percentage of ______________ in the fertilizer.
3. Would the nitrogen in ammonia be considered "fixed" nitrogen? ______________
4. Pure ammonia under ordinary conditions is a ( ) solid, ( ) liquid, ( ) gas.
5. The chemical formula for potash is ______________.

6. Approximately what percentage of the food crops of the world is lost to pests each year? ____________
7. The first chlorinated organic insecticide was ____________.
8. Which of the following is not a persistent insecticide: DDT, dieldrin, heptachlor, or chlordan? ____________
9. Which is more likely to be a hormone, a selective or a nonselective herbicide? ____________
10. The most widely used herbicide today is ____________.

QUESTIONS

1. What are some of the major categories of agrichemicals?
2. Explain why animals can't properly be considered "fundamental food."
3. Describe the horizons that would be found in a typical soil.
4. What causes soil to be sour? Sweet?
5. If crushed limestone is spread on soil, will it raise or lower the pH of the soil? Explain.
6. About how many pounds of water are required to produce one pound of food?
7. Guano was used as a fertilizer in colonial America. If you do not know the meaning of this word, look it up in the dictionary. What nutrients does guano add to the soil?
8. What are the three nonmineral nutrients obtained from water and air required for normal plant growth?
9. The two principal elements in the soil are first oxygen and then silicon. How does this compare to the composition of the crust of the earth? The earth as a whole? The universe? Refer back to Chapter 2.
10. Which groups of elements are first leached from soils, the Group IA and IIA metals or the Group IIIA and transition metals? What is the effect of this selective leaching on soil pH?
11. What are two important roles of humus in the soil? Do leaves turned into the soil to produce humus raise or lower the soil pH?
12. What are the three primary mineral plant nutrients that are considered first in fertilizer formulations?
13. Which is more likely to be a problem in farming, a soil shortage of N, P, and K, or a shortage of Ca, Mg, and S? Give a reason for your answer.
14. What does the term *nitrogen fixation* mean? Give two natural ways that this process occurs.
15. Explain the numbers 6-12-6 as might be found on a fertilizer bag.
16. Which is more easily leached from soils: nitrates or phosphates? Why?
17. How long would you expect a piece of new ground to present good yields of a single food crop such as corn if no fertilizer is applied?
18. Why is the cost of nitrogen fertilizers tied directly to the cost of petroleum?
19. What two herbicides were formulated to produce Agent Orange? Which of these herbicides is currently banned in the United States for agricultural use?
20. Write an essay giving the benefits and possible harms in using pesticides in agriculture. What conclusions can you draw on this controversial issue?
21. Organic farming saves energy in one area but loses it in another. Explain.
22. Draw and explain a block diagram showing the relationship between traditional agriculture, organic farming, and alternative agriculture.
23. How is it possible that your great-grandfather might have been happy with 25 bushels of corn per acre, whereas farmers today who cannot average over 100 bushels per acre cannot adequately compete in the corn market?
24. In the period after World War II, most farmers fertilized "enough to be sure." Farmers today are likely to have the soil analyzed and have a fertilizer formulated on prescription. What is the cause of this change? Can you see how this change in farming practice might have an effect on water pollution?
25. Investigate a no-till farming operation. What herbicides were used? How is energy saved and, at the same time, how is additional energy required? What is the effect of no-till farming on the conservation of topsoil?
26. Give examples of persistent pesticides.
27. What are transgenic animals? Give some examples.
28. Trace the rise and fall of the use of DDT in agriculture.

Debate whether it has been more good than bad for the human race.

29. Debate the proposition that herbicides such as paraquat should be sprayed from airplanes to destroy crops grown to produce illegal drugs.

30. Contact the U.S. Department of Agriculture through the Soil Conservation Service in your area. Find out if there is documentation of a micronutrient problem in the agriculture in your state. Define the problem if one is found, and outline a chemical solution.

The Dessert **by Cezanne depicts a bowl of fruit—nature's nutritional start for new life.** (Philadelphia Museum of Art: The Mr. and Mrs. Carroll S. Tyson Collection)

CHAPTER

14 Nutrition: The Basis of Healthy Living

Humans, at the apex of animal intelligence, have demonstrated a remarkable tendency toward excesses and deficiencies in meeting their nutritional needs. Nutrition, on the back burner of intellectual disciplines for a number of years, is now at the center of considerable attention at levels ranging from national governments to individual concerns. However, there is still a significant gap between what our society knows about good nutrition and what it practices.

1. What are the basic nutrients needed for good health?
2. What is the role of diet and exercise in control of excessive caloric intake?
3. What are the functions of carbohydrates, fats, and proteins in our bodies?
4. Why are saturated fats a health problem?
5. How can one control hypoglycemia, hyperglycemia, and diabetes mellitus?
6. What minerals are necessary in our diet? Is sodium a problem?
7. Are we in danger of not getting enough vitamins?
8. What are the purposes for adding chemicals to our processed foods?
9. Will artificial sweeteners replace refined sugar in our diets, should they?

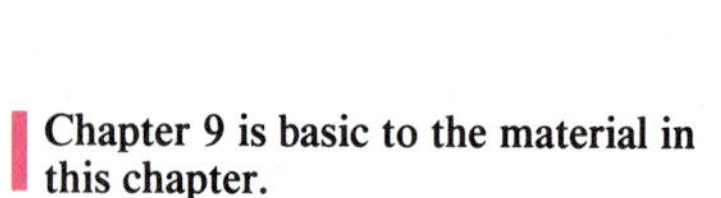

Chapter 9 is basic to the material in this chapter.

Nutrition is the science that deals with diet and health. The old saying "we are what we eat" is true in the sense that we are continually replacing parts of our bodies and that the material to make these replacements comes from our food. The skin that covers us now is not the same skin that covered us seven years ago. The fat beneath our skin is not the same fat that was there just a year ago. Our oldest red blood cells are 120 days old. The entire linings of our digestive tracts are renewed every three days. Many chemical reactions are required to replace these tissues, and all of these reactions are supplied ultimately by what we eat. Nutrition, then, is concerned with the chemical requirements of the body—the nutrients. The six classes of nutrients are carbohydrates, fats, proteins, vitamins, minerals, and water.

GENERAL NUTRITIONAL NEEDS

Why do we need carbohydrates, fats, proteins, vitamins, minerals, and water to sustain life? Basically, we need these compounds in the nutrients to compose and supply the tissues, organs, and systems of the body. The major elements in the human body, on a weight percentage basis, are oxygen (65%), carbon (18%), hydrogen (10%), and nitrogen (3%). Practically all of these elements are in the form of water or organic compounds. Other elements present (and required) are calcium (2%), phosphorus (1%), potassium (0.35%), sulfur (0.25%), sodium (0.15%), chlorine (0.15%), and magnesium (0.05%). The total thus far equals over 99.9% of the total body weight. The rest of the body is composed of trace amounts of other elements. Either a lack of proper nutrients or an excess of either proper or improper nutrients can produce **malnutrition.**

Vegetables are a good source of complex starch. (*The World of Chemistry,* Program 10, "Signals from Within")

There is general agreement in the health community that a good balanced diet should contain no more than 30% fat, along with about 12% protein and 58% carbohydrate (38% starch, 10% naturally occurring sugars, and 10% sucrose). Such a diet is significantly lower in fat and sucrose and higher in starches than the average diet of U.S. citizens of today, which is 42% fat, 12% protein, and 46% carbohydrate (20% starch, 6% naturally occurring sugars, and 20% sucrose) (Fig. 14–1). Table 14–1 presents ten basic rules that are widely supported by nutritionists.

Caloric Needs

Heat is needed by humans to maintain body temperature at about 37 °C (98.6 °F under the tongue) and to energize endothermic chemical reactions.

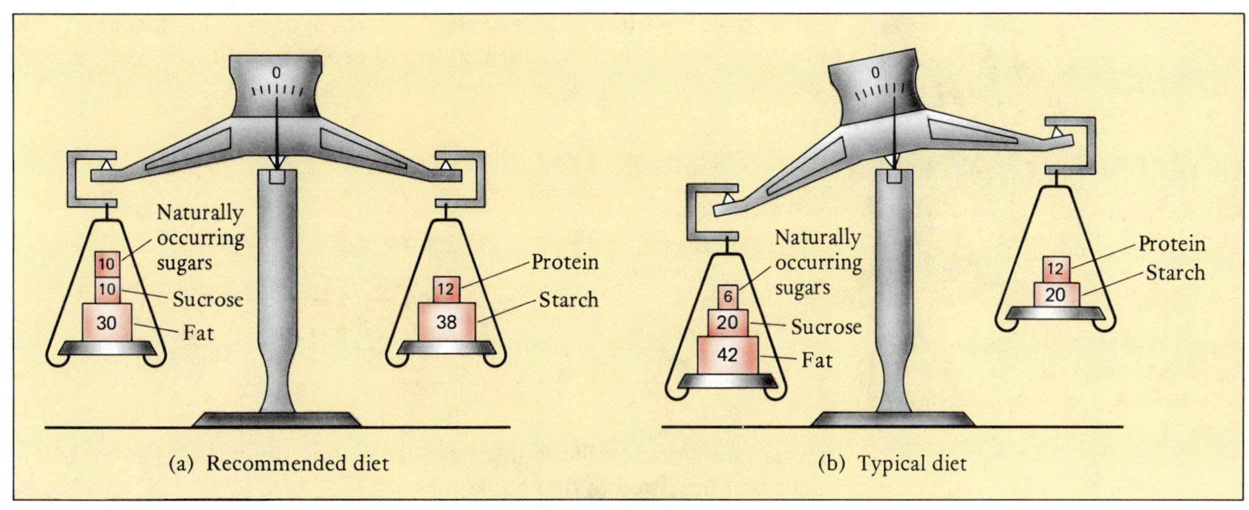

Figure 14–1 Successful nutrition means eating a balanced diet. Scientific studies recommend the percentages shown in (a). However, the average citizen of the United States consumes the percentages shown in (b), which are low in naturally occurring sugars and complex starches and high in fats and refined sugar.

TABLE 14–1 General Dietary Recommendations

1. Eat a variety of foods.
2. Maintain ideal body weight.
3. Avoid too much saturated fat and cholesterol.
4. Eat foods with adequate starch and fiber.
5. Avoid excess sugar.
6. Avoid excess sodium.
7. Drink alcoholic beverages in moderation.
8. Fluoride should be available in water supplies or in other sources for prevention of tooth decay.
9. Calcium intake should be increased for adolescent girls and adult women.
10. Iron intake should be improved, especially for children, adolescents, and women of child-bearing age, by eating lean red meats, fish, certain beans, iron-rich cereals, and whole-grain products. Improved iron intake is of special concern for low-income families.

Dietary guidelines for Americans.

The principal source of this heat is the oxidation of some fats and carbohydrates. The oxidation of proteins and various other exothermic reactions provides the rest of the heat for the body.

Some fats, such as those made from linoleic acid, are required for chemical reactions in the body other than oxidations.

Fats in the diet produce more than twice as much energy per gram as do carbohydrates and proteins (Table 14–2). The values from this table can be used to calculate the caloric value of a food, if the composition of the food is known and if the complete quantity of each food is oxidized. For example, if

A food, or dietary, calorie is a kilocalorie (kcal).

A pole vaulter needs high strength, speed, and energy to soar. (*The World of Chemistry*, Program 14, "Molecules in Action")

a steak is 49% water, 15% protein, 0% carbohydrate, 36% fat, and 0.7% minerals, then 3.5 oz (about 100 g) would produce about 384 kcal, or 384 food Cal.

Nutrient	Weight kcal/g	Total
Water	49 g × 0 kcal/g =	0 kcal
Protein	15 g × 4 kcal/g =	60
Carbohydrate	0 g × 4 kcal/g =	0
Fat	36 g × 9 kcal/g =	324
Minerals	0.7 g × 0 kcal/g =	0
	Total	384 kcal

Calorie values of most foods are calculated by this method, and these are the values that are listed in diet books.

TABLE 14–2 Calorie Data for Fats, Carbohydrates, and Proteins

Foodstuff	kcal/g	RDA	Actually Consumed in U.S. Daily Diet	kcal Produced by Daily Intake	Percentage of Daily Calorie Output
Fat	9	—	100–150 g	900–1350	30–50
Carbohydrate	4	—	300–400 g	1200–1600	35–45
Protein	4	46–56 g (10 oz meat)	80–120 g	320–480	10–15

A 150-lb person who skis at 10 mph requires 600 kcal/hr. (Gordon Wiltsie)

TABLE 14–3 Approximate Energy Expenditure by a 150-lb Person in Various Activities

Activity	Energy (kcal/hr)	Activity	Energy (kcal/hr)
Bicycling, 5.5 mph	210	Roller skating	350
13 mph	660	Running, 10 mph	900
Bowling	270	Skiing, 10 mph	600
Domestic work	180	Square dancing	350
Driving an automobile	120	Squash and handball	600
Eating	150	Standing	140
Football, touch	530	Swimming, 0.25 mph	300
tackle	720	Tennis	420
Gardening	220	Volleyball	350
Golf, walking	250	Walking, 2.5 mph	210
Lawn mowing (power mower)	250	3.75 mph	300
Lying down or sleeping	80	Wood chopping or sawing	400

Energy spent for normal maintenance activities of the body is the **basal metabolic rate (BMR).** These maintenance activities include the beating of the heart, breathing, maintenance of life in each cell, maintenance of body temperature, and the sending of nerve impulses from the brain to direct these automatic activities. Energy for these activities must be supplied before energy can be taken for digesting food, running, walking, talking, or other activities (Table 14–3). BMR, usually expressed as kilocalories per hour, is defined as the energy spent by a body at rest after a 12-hr fast. To get a rough estimate of your BMR (kcal/day), multiply your weight (in pounds) by 10.

Basal metabolic rate is energy required to do nothing willfully.

INDIVIDUAL NUTRIENTS—WHY WE NEED THEM IN A BALANCED AMOUNT

Proteins

Of the 20 amino acids identified in human protein, 8 are considered essential for adults, meaning that the adult body cannot synthesize them and therefore must obtain them from ingested food. Infants also require arginine because they cannot make it fast enough to have a supply for both protein synthesis and urea synthesis. The lack of an essential amino acid in one meal is not supplied by an excess of the amino acid in another because excess amino acids are not stored very long except in functioning proteins. If proteins are eaten at only one meal per day, the liver must store a full day's supply from that one meal.

Functions in the Human Body

Humans must have proteins to provide the structural compounds for repairing and maintaining muscles and most organs. Proteins are part (the apoenzyme) of the some 80,000 known enzymes. Some hormones, transport molecules (such as hemoglobin and transferrin), antibodies, and fibrinogin (for blood clotting) contain proteins.

Hormone (Greek *hormaein,* "to set in motion, spur on"): a chemical substance, produced by the body, that has a specific effect on the activity of a certain organ.

Daily Needs

Proteins are nearly the only source of nitrogen in the diet. An adult male has about 10 kg of protein, about 300 g of which is replaced daily. Part of the 300 g is recycled, and part comes from intake. Various studies indicate that, on the average, 25 to 38 g of high-quality protein (as in meat, chicken eggs, and cow's milk) or 32 to 42 g of lower quality proteins (as in corn and wheat) are required in the daily diets of healthy adult humans in order to maintain nitrogen equilibrium in the body. The average daily intake has remained near 100 g of protein per person since 1910, although there was a small drop in protein intake during the Depression of the 1930s. Methionine is the essential amino acid required in the greatest amount (2 g of the total of 7.1 g of all of the essential amino acids). Protein is lost in urine (as urea, a by-product of protein metabolism), fecal material, sweat, hair and nail cuttings, and sloughed skin.

Meats such as beef and poultry are the primary sources of protein in the diets of most human beings. (*The World of Chemistry,* Programs 5 and 13, "A Matter of State" and "The Driving Forces")

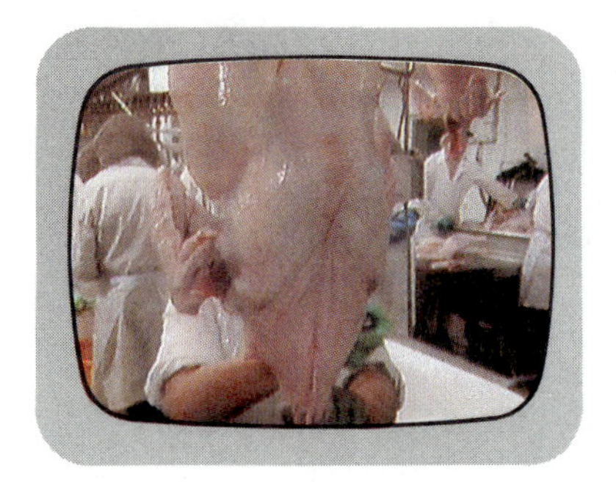

Fats, Functions in the Human Body

Properties and structures of fats and fatty acids are given in Chapter 9.

Structures of glycerol and triglycerides are in Chapter 9.

Fats are essential structural parts of cell membranes. They are the most concentrated source of food energy in our diets; they furnish 9000 cal/g when oxidized for energy, compared with glucose, which furnishes about 3800 cal/g. Stored fat is a potential energy source for the body. Fats insulate thermally, pad the body, and are packing material for various organs. Fatty, or adipose, tissue is composed mainly of specialized cells, each featuring a relatively large globule of triglycerides.

Solid animal fat. (*The World of Chemistry,* Program 9, "Molecular Architecture")

Daily Consumption of Fats

The daily consumption of fat in the United States has risen continuously from about 125 g per person in 1910 to about 155 g per person today. The fat in today's diet is about 40% saturated, 40% monounsaturated, and 20% polyunsaturated. The ratios of saturated and unsaturated fatty acids in common fats and oils are listed in Table 14–4. Note that coconut oil is very high in saturated fatty acids although coconut oil is a vegetable oil. Since coconut oil is cheap, it is used in some foods. Those people who tend toward artery blockage from atherosclerotic plaque should avoid foods that have coconut oil.

A Problem Associated with Eating Saturated Fats

Hydrogenation of oils to solid fats is discussed in Chapter 9.

Conjugated structures have alternating double and single bonds in the carbon chain or ring.

Ingestion of hydrogenated polyunsaturated fats (oils) causes biochemical problems (Fig. 14–2). Hydrogenation is used to convert oils into solid fats to make margarine, cooking fats, and similar products. However, hydrogenation of vegetable oils (liquid fats) decreases some of the double bonds, forms unnatural *trans*-fatty acids from natural *cis*-fatty acids, and moves double bonds around to form conjugated structures. The *trans*-fatty acids are not readily metabolized in the human system because the *trans*-fatty acids are "straight" molecular structures and pack together into solids like the saturated acids. Both stack similar to sticks of wood. By contrast, the *cis*-fatty acids are bent like broken but attached sticks of wood and do not pack into solid masses easily. Another problem with *trans*-fatty acids is their effect on cholesterol. *Cis*-fatty acids participate but *trans*-fatty acids do not participate in the formation of cholesterol esters, the principal storage form of cholesterol. By the inactivity of *trans*-fatty acids, cholesterol is free to

TABLE 14-4 Ratios of Saturated and Unsaturated Fatty Acids from Common Fats and Oils*

	Percentage of Total Fatty Acids by Weight		
Oil or Fat	*Saturated*	*Monounsaturated*	*Polyunsaturated*
Coconut oil†	93	6	1
Corn oil	14	29	57
Cottonseed oil	26	22	52
Lard	44	46	10
Olive oil	15	73	12
Palm oil	57	36	7
Peanut oil†	21	49	30
Safflower oil	10	14	76
Soybean oil	14	24	62
Sunflower oil	11	19	70
Canola oil	6	58	36

* Recall that *saturated* means a full complement of hydrogen (only C—C single bonds); *monounsaturated* means one C=C double bond per fatty acid molecule; *polyunsaturated* means two or more C=C double bonds per molecule of fatty acid. The chief unsaturated fatty acid is linoleic acid.

† Although derived from vegetable rather than animal fats, both coconut oil and peanut oil have been associated recently with hardening of the arteries when combined with a high cholesterol intake.

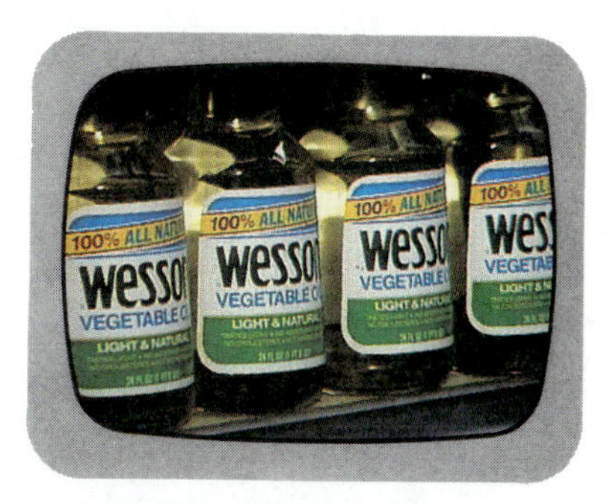

Liquid vegetable oil. (*The World of Chemistry,* Program 9, "Molecular Architecture")

The number of adipose (fat) cells is fixed by adulthood. The more adipose cells one has, the harder it is to lose weight.

roam through the bloodstream and become entangled in a blockage in an artery, **atherosclerosis.** Both cholesterol and straight-structured fatty acids (saturated and *trans*) are implicated as the ingredients for forming the solid blockages in an artery, particularly damaging to the small arteries serving the heart. The site of blockage is often a damaged or rough lining of the artery caused by heredity, chemicals (smoking is implicated), or physical injury.

Problems related to cholesterol, triglycerides, atherosclerotic plaque, and heart attack are discussed further in Chapter 15.

Cholesterol is necessary in making cell membranes stronger and more flexible.

Figure 14-2 Saturated fatty acids (a) and *trans*-unsaturated fatty acids (top structure in b) are linear and tend to pack into solid masses. *Cis*-unsaturated fatty acids (c and bottom structure in b) are bent, do not pack as well as straight structures, and tend to be liquid at normal temperatures. (*The World of Chemistry,* Program 9, "Molecular Architecture")

(a)

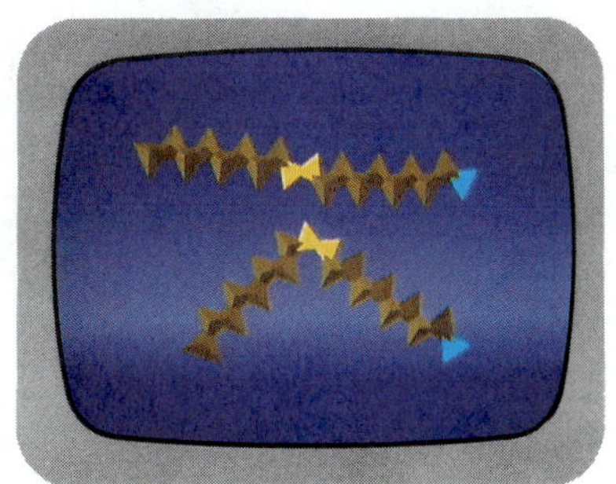

(b)

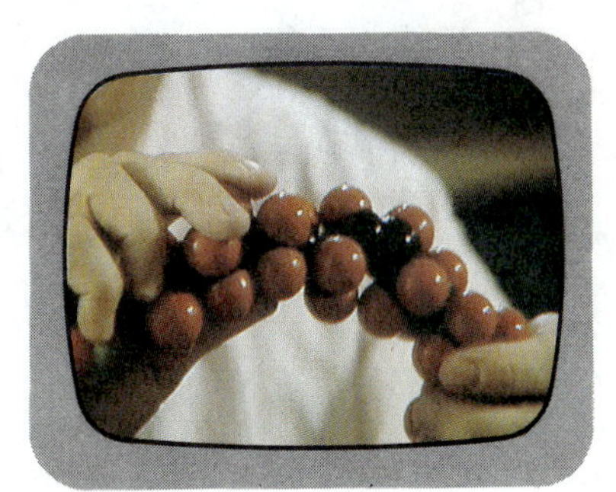

(c)

Bread, the staff of life.

Fake Fats

Another solution to problems with fats is to use "fake fats," substances that give the fat or oil taste and consistency but are not fats or oils. Some of the fake fats either now on the market or about ready to be sold commercially are Simplesse, Olestra, emulsified starch, and emulsified protein. Simplesse was developed by the makers of NutraSweet, G. D. Searles Co., and is a butter substitute with only 15% of the real fat calories in butter. An ounce of cheese made from Simplesse instead of butterfat drops from 82 Cal to 36. Simplesse is made from egg white or milk proteins. It feels creamy on the tongue. However, Simplesse is not suitable for cooking because heat makes it tough. Olestra, developed by Proctor and Gamble Co., is a sucrose polyester made from sugar and fatty acids. Olestra contributes no calories because it is indigestible, and it can be used in cooking. Emulsified starch is now used in Hellmann's light mayonnaise and salad dressing. It is not used for cooking but can be used in ice cream and yogurt. Emulsified protein (Unilever) is an emulsion of gelatin and water that cuts margarine calories in half. It can be used for baking and light frying.

Carbohydrates

Properties and structures of carbohydrates are given in Chapter 9.

Carbohydrates in foods include digestible simple sugars (glucose, fructose, galactose), disaccharides (sucrose, maltose, lactose), and polysaccharides (amylose, amylopectin, glycogen). Indigestible carbohydrates consumed include cellulose, hemicellulose, lignin, plant gums, sulfated polysaccharides, carrageenan, and cutin.

Functions in the Human Body

The only beneficial function of digestible carbohydrates is to provide energy at the rate of approximately 4 kcal/g of glucose oxidized. Excess digestible carbohydrates are stored first as glycogen, principally in the liver; further excesses are converted into fats and stored as such. The indigestible carbohydrates serve as roughage in the diet, along with bran and fruit pulp.

Daily Needs

A daily caloric intake of 2000 kcal would require the ingestion of about 500 g of glucose (or its equivalent). Fats and proteins are also oxidized for energy, so less digestible carbohydrate is required. Daily consumption of digestible carbohydrate has declined from 500 g per person in 1910 to 380 g

One source of roughage in the diet is fruit pulp. Eating whole oranges provides delectable juice, vitamin C, and needed roughage. (James W. Morgenthaler.)

per person in the 1980s. The decline in total amount of carbohydrates includes a rise in the amounts of refined sugars: 150 g in 1910 to 200 g in the 1980s.

Control of Blood Sugar

Serious health problems can be encountered when the glucose level is either too low or too high in the blood. After about 10 hr without food, the normal person will have from 80 to 120 mg of glucose per 100 mL of blood. Persons with low blood sugar, a condition known as **hypoglycemia,** must carefully control the amounts of carbohydrates in their diets in order to maintain normal blood sugar levels. A low glucose concentration may lead to sluggish or dizzy feelings and possibly fainting. Eating meals with a very high concentration of carbohydrates can also cause problems for persons with hypoglycemia as the body overcorrects with excess insulin, which quickly drives the glucose concentration below the normal level.

Hyperglycemia is the condition of elevated blood sugar concentrations. Above a concentration of 160 mg of glucose per 100 mL of blood, the kidneys begin to excrete glucose in the urine. If such high blood sugar concentrations are chronic without medical or dietary control, the individual has **diabetes mellitus** and, without treatment, is likely to have symptoms of thirst, frequent urination, weakness, low resistance to infection, slowness to heal, and, in later stages, blindness and coma. About 5% of Americans have diabetes.

There are two types of diabetes, which have been termed Type I (insulin-dependent, formerly juvenile-onset) and Type II (non–insulin-dependent, formerly maturity-onset). Type I diabetics, about 10% of those who have diabetes, do not produce enough insulin at the *islets of Langerhans* in the pancreas. The needed insulin, which is necessary for the glucose to move from the blood to the cells, has to be obtained by daily injections, as this compound is destroyed in the digestive tract. Type II diabetes is usually found in older, obese people. These people generally produce plenty of insulin, but the insulin receptors on their cells do not respond properly to the insulin and consequently do not move the glucose from the blood into the cells. Type II diabetics can generally control the disease by diet and oral medication.

Diabetes:
Type I: insulin deficient
Type II: plenty of insulin but the insulin receptors on the cells are inactive

Insulin, a protein, is hydrolyzed (digested) in the gastrointestinal tract if taken orally.

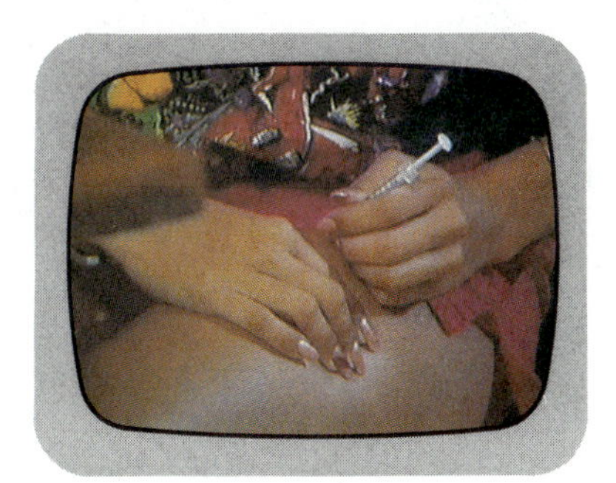

Some substances such as insulin have to be injected to save life (shown here). Other substances are injected daily by millions for a short-term "high" and sooner or later the result is the destruction of the human system. (*The World of Chemistry,* Program 24, "Genetic Code")

TABLE 14–5 Refined Sugar Added to Some Commercially Processed Foods*

Food	Sugar (%)
Cherry Jello	82.6
Coffeemate	65.4
Shake'N Bake, barbecue style	50.9
Wishbone Russian dressing	30.2
Heinz ketchup	28.9
Sealtest chocolate ice cream	21.4
Libby's peaches (in heavy syrup)	17.9
Skippy peanut butter	9.2
Coca Cola	8.8

* According to *Consumer Reports,* "Too Much Sugar" (March 1978) pp. 136–142; percents by weight.

The yearly sugar consumption in the United States in 1750 was 2 lb per person; today it is 110 to 135 lb per person. Sixty percent of the sugar comes from sugar cane; the other 40% comes from sugar beets. The sugar content of some commercially processed foods is given in Table 14–5.

Some problems with refined sugar are due to the removal of required nutrients during the refining process and the dumping of too much refined sugar into the bloodstream too quickly. The production of white sugar (almost pure sucrose) removes all other nutrients, such as B vitamins, manganese, and chromium, which generally coexist in natural foods with sucrose in the appropriate amounts for proper metabolism in the human body. Therefore, a large intake of refined sugar means that the B vitamins and certain minerals must be obtained from another food source. Brown sugar supplies more minerals than white sugar because brown sugar is darkened with molasses, the residue from sugar cane that is rich in essential minerals.

SELF-TEST 14A

1. The six classes of nutrients are ____________, ____________, ____________, ____________, ____________, and ____________.
2. The nutrient that produces the most heat per gram is ____________.
3. A food calorie is the same value as ____________ kilocalorie(s).
4. The amount of heat required to operate the body at rest is ____________, which in kilocalories is about ten times your ____________.
5. Which nutrient is generally recommended to be decreased in the American diet? ____________
6. Proteins compose the ____________ part of enzymes.
7. Proteins are nearly the only source of ____________ in the diet.

8. Most of the lipids in the diet are ____________.
9. The one essential fatty acid is ____________.
10. Which type of fatty acids forms esters with cholesterol?
____
11. Give the primary functions for digestible and indigestible carbohydrates in the diet.
Digestible: ____________________________
Indigestible: ____________________________
12. What disease is associated with low blood sugar? ____________
13. ( ) *Trans-* or ( ) *cis-* fatty acids have straight molecular structures and stack together to make a solid mass.

Minerals

As nutrients, minerals are substances that are needed for good health and contain elements other than C, H, O, and N. On vitamin and mineral supplement labels and elsewhere, nutrient elements are called minerals, the two terms being used interchangeably in nutrition. Most of the elements needed for nutrition are obtained from soluble inorganic salts either in foods or in food supplements. Magnesium is an exception in that it is obtained primarily from organic chlorophyll.

Carbon, hydrogen, oxygen, and nitrogen are supplied by organic fats, carbohydrates, and proteins.

There are 46 different minerals in the human body, 21 of which are *known* to be essential. The seven **macrominerals** make up about 4% of body weight. They are calcium, phosphorus, magnesium, sodium, potassium, chloride, and sulfur. The macrominerals are necessary in building bones, maintaining body fluids, maintaining proper pH in body tissues, transmitting nerve impulses, maintaining cell membrane structures, and facilitating enzyme action. The 14 trace minerals known to be essential for mammals are iron, copper, zinc, manganese, iodine, fluoride, chromium, selenium, molybdenum, cobalt, nickel, vanadium, silicon, and arsenic. Definitive studies for the essential nature of nickel, vanadium, silicon, and arsenic have not been made for humans as yet.

The nutrient minerals have varied functions, including as components of enzymes, as structural components (calcium and phosphorus in bones and teeth), in electrolyte balance in body fluids, and as transport vehicles (iron in hemoglobin transports oxygen; iron and cobalt transport electrons in electron transport cycles). Not only does the human body need minerals for its functions, but the minerals must be maintained in balanced amounts, with no deficiencies and no excesses. Many of the body's minerals are excreted daily in the feces, urine, and sweat and must therefore be replenished. For most of the elements, the amount excreted each day is very nearly the amount ingested.

One way to ensure ingestion of an ample supply of each mineral nutrient, particularly the trace nutrients, is to eat a variety of whole foodstuffs grown in different places. Mineral supplements are also available.

Calcium slows down the heartbeat by increasing electrical resistance across nerve membranes. The movement of potassium and sodium ions across the membrane is constrained, and the nerve impulse rate is thus decreased. Calcium is metabolized in the body by a hormone synthesized

The role of calcium in regulating the heartbeat is discussed in Chapter 15.

from calciferol (vitamin D). The calciferol also brings about synthesis of a substance called calcium-binding protein (CBP), which carries calcium through the small intestine wall. Fat slows down the transfer, and lactose speeds up calcium absorption.

We could not support ourselves nor eat without calcium. The hard part of bone is a calcium compound, and the structure of teeth is mostly hydroxy-apatite, a calcium compound.

The structure of apatite is given and the role of fluoride in tooth decay is discussed in Chapter 16.

Excess calcium may lead to the formation of kidney stones, but the body has a protein, **calmodulin,** that collects excess calcium and then binds to a number of enzymes to mediate their activity. By the use of calmodulin, the body monitors the amount of calcium in the bloodstream. A possible benefit of excess calcium is that it will make a person taller.

A deficiency in calcium can occur in postmenopausal women, who produce less estrogen than premenopausal women. The estrogen suppresses bone dissolution. Further bone dissolution can be suppressed by long-term medication with estrogen, but this has produced toxic side effects in some women and therefore seems unwise. Taking calcium supplements may help.

The principal function of *iodine* in the human body is the proper operation of the thyroid glands located at the base of the neck. Thyroid hormones, collectively known as **thyroxine,** go into every cell and regulate the rate at which the cell uses oxygen. Thyroxine thus regulates the BMR.

Iodine is absolutely necessary to the production of thyroxine. If there is a deficiency of iodine, the thyroid glands sometimes swell to as large as a person's head; this swelling is called a goiter. In 1960 it was estimated that 7% of the world's population (200 million) had goiters. Treatment with iodized salt (0.1% KI) with the hormone thyroxine decreases the size of or even eliminates small goiters. Large goiters may require surgery. Because an excess amount of iodine also causes goiters, balance is the key to health.

The lack of either vitamin B_9 or B_{12} can cause pernicious anemia.

Anemia can be caused by a deficiency of *iron,* but it can also be due to heredity, an improper level of vitamin B_6, lack of folic acid (vitamin B_9), and lack of vitamin B_{12} (pernicious anemia). Iron-deficient anemia is not necessarily fatal; a person with only 20% of the normal amount of hemoglobin still has the energy and strength to walk.

In the case of some nutrient elements, good health depends on the element being present in the proper amount and in the *proper ratio* to one or more other elements. An example of an important ratio is the potassium/sodium ratio (K/Na ratio), which has to be within certain limits to facilitate the transmission of electrical signals between nerve cells.

Foods high in potassium (mg K/3.5 oz of food):

Raw prunes	940
Raw raisins	763
Banana (1)	740
Turkey breast	411
Boiled potato	404
Orange (1)	300
Apple (1)	165
Cottage cheese	81

Typical values of the **K/Na ratio** are greater than 1. Some K/Na ratios for specific tissues follow: muscle, 4; liver, 2.5; heart, 1.8; brain, 1.7; and kidney, 1.0. For individual cells, potassium ions (K^+) concentrate inside the cell, whereas sodium ions (Na^+) concentrate outside in the fluid that bathes the cell. Natural, unprocessed food has high K/Na weight ratios. Fresh, leafy vegetables average a K/Na ratio of 35. Fresh, nonleafy vegetables and fruits average a ratio of 360, with extreme values of 3 for beets and 840 for bananas. K/Na ratios in meats range from 2 to 12. Thus, when such foods are eaten, the body has K/Na ratios greater than 1. However, problems occur with processed and cooked foods. Potassium and sodium compounds are quite soluble in water. During processing (and cooking, if foods are boiled), both potassium and sodium compounds are dissolved by water and discarded. The sodium is replenished by "salting" of the food (addition of sodium chloride). Potassium is usually not added to the food. One solution to the problem is to eat unprocessed, natural food, which "naturally" has the proper K/Na ratios. Another solution is to "salt" food with a commercial product that contains both potassium and sodium, such as Morton's Lite Salt.® In summary, do not add much NaCl, if any, to food, and eat fresh vegetables and fruits high in potassium.

Normal daily urinary excretion of *sodium* is in the range of 1.4 to 7.8 g for adults. If excess sodium is not eliminated, water is retained, which may lead to edema (swollen legs and ankles). Various clinical studies have shown that increased levels of sodium raise the blood pressure of some individuals but have no effect on the blood pressure of others. The high-salt diets of 70 g NaCl per day in certain areas of Japan have traditionally produced an unusually high frequency of heart attacks. Sodium levels in the bloodstream are regulated by **aldosterone,** which is secreted from the adrenal gland. Aldosterone works in the kidney to reabsorb sodium from the urine. The secretion of aldosterone is controlled by receptors that measure salt concentration in the blood. If the blood sodium concentration is too high, less aldosterone is excreted and less sodium is reabsorbed from the urine.

The word *salary* derives from the Latin *sal,* for "salt." Roman soldiers were given an allowance for salt.

Vitamins

A vitamin is an organic constituent of food that is consumed in relatively small amounts (less than 0.1 g/kg of body weight per day) and is essential to the maintenance of life. Vitamins are not synthesized in the cells of human beings; they are synthesized by plants, our principal natural source of them. Of the some million organic compounds eaten in a normal diet, only about 100 are of proper size and stability to be absorbed from the digestive tract into the bloodstream without digestion or breakdown. Vitamins are included in this group of compounds.

There are thirteen known essential vitamins (Table 14–6).

Two sources of vitamins: green leafy vegetables and laboratory-synthesized tablets.

The structures of vitamins divide them into two classes: oil-soluble and water-soluble. The oil-soluble vitamins—A, D, E, F, and K—tend to be stored in the fatty tissues of the body (especially the liver). The structures of oil-soluble vitamins have nonpolar hydrocarbon chains and rings that are compatible with nonpolar oil and fat. For good health and nutrition, it is important to store enough oil-soluble vitamins, but not too much.

The water-soluble vitamins tend to pass through the body and are not stored readily. Water-soluble vitamins are the B group (called vitamin B complex) and C. The structures of these vitamins have polar hydroxy (—OH) and carboxyl (—COOH) groups, which are attracted to polar water. Fewer problems are caused by excessive intake of water-soluble than of oil-soluble vitamins.

Vitamin D is the most toxic of all of the vitamins; avoid excesses.

Table 14–6 lists the vitamins, their USRDAs, some food sources, and their deficiency effects. Since vitamins are synthesized by plants, a good natural source is plants that are not overcooked.

Contrary to popular belief, carrots provide the **provitamin** β-carotene, not retinol (vitamin A). The body converts β-carotene into retinol during the transfer of the provitamin through the intestinal wall. Night blindness is

TABLE 14-6 Vitamin Summary Chart

Name	USRDA*	Deficiency Effect	Sources
Water-soluble			
Thiamin (B_1)	1.5 mg	Beriberi	Seeds, pork, whole-wheat bread
Riboflavin (B_2)	1.7 mg	Cheilosis (shark skin)	Organ meats, yeast, wheat germ
Niacin (B_3) (nicotinic acid)	2 mg	Pellagra	Meat, yeast, legumes
Pantothenic acid (B_5)	10 mg	Neuromotor disturbance	Yeast, liver, eggs
Vitamin B_6 (pyridoxine)	2 mg	Skin lesions, anemia	Liver, nuts, wheat germ
Biotin (B_7)	0.3 mg	Dermatitis	Liver, yeast, grains
Folic acid (B_9) (folacin)	0.4 mg	Anemia, gastrointestinal changes	Green leafy vegetables, liver
Vitamin B_{12} (cobalamin)	6 μg	Pernicious anemia	Intestinal bacteria, organ meats
Ascorbic acid (vitamin C)	60 mg	Scurvy	Fruits, vegetables
Oil-soluble			
Vitamin A (retinol)	5000 IU	"Night blindness" (xerophthalmia)	Liver, fruits, vegetables
Vitamin D (calciferol)	400 IU	Rickets	Fish liver oil
Vitamin E (tocopherol)	30 IU	Lack of hemoglobin in blood cells (hemolysis)	Plant oils
Vitamin K (phylloquinone)	†	Blood loss	Green leafy vegetables

* USRDA values are for adults and children over 4 years old. (USRDA = United States Recommended Dietary Allowances, from the Food and Drug Administration.)

† There is no USRDA value for vitamin K, but an estimated need is 0.1 mg per day.

Eat polar bear liver sparingly. Thirty grams contain 450,000 IU of retinol; continued ingestion causes peeling of the skin from head to foot.

Vitamin E is the only vitamin destroyed by the freezing of food.

Is vitamin E the fountain of youth?

A free radical has one or more unpaired valence electrons. An example is the methyl group, $CH_3\cdot$.

prevented by regeneration of rhodopsin (visual purple) from retinol. **Vitamin A** aids in the prevention of infection by barring bacteria from entering and passing through cell membranes. The vitamin performs its sentinel duty by producing and maintaining mucus-secreting cells. Bacteria stick to the mucus and are thus trapped.

The function of **vitamin E** as an antioxidant has been well established. Vitamin E is particularly effective in preventing the oxidation of polyunsaturated fatty acids, which readily form peroxides. Perhaps this is why vitamin E is always found distributed among fats in nature. The fatty acid peroxides are particularly damaging because they can lead to runaway oxidation in the cells. Vitamin E protects the integrity of cell membranes, which contain considerable fat. In addition, vitamin E helps maintain the integrity of the circulatory and central nervous systems, and it is involved in the proper functioning of the kidneys, lungs, liver, and genital structures. Vitamin E also detoxifies poisonous materials absorbed into the body.

According to some theories that view aging as the cumulative effects of the action of free radicals running wild, vitamin E, with its antioxidant properties, is considered a good candidate as an agent to inhibit aging or at least to help avoid premature aging.

The B group of vitamins (the **B-complex vitamins**) work together, primarily as coenzymes in biochemical reactions leading to growth and to energy production. Their place of action is in the mitochondria of the cells. Being water-soluble, the B vitamins are easily eliminated during the processing and cooking of food. The effectiveness of vitamins B_3 and B_6 is diminished in the presence of light, especially if the food is hot.

Pyridoxine (B_6), considered the "master vitamin," is involved in 60 known enzymatic reactions, mostly in the metabolism and synthesis of proteins.

Vitamin C is involved in the destruction of invading bacteria, in the synthesis and activity of interferon, which prevents the entry of viruses into cells, and in decreasing the ill effects of toxic substances, including drugs and pollutants. The question of whether vitamin C decreases the incidence of the common cold has been studied for many years. Results of the studies show an average decrease of about 30% in illness (particularly upper respiratory infection) as a result of ingestion of vitamin C supplements. Not as well publicized or studied, vitamin A in large doses also decreases colds and the effects of colds. In avoiding or breaking colds, some persons respond better to vitamin A than to vitamin C, others respond better to vitamin C than to vitamin A, and still others respond to neither. In any case, for either vitamin to be effective, it must be taken preferably before but no later than at the early onset of a cold. It is recommended that the vitamins not be taken in combination, since this seems to prolong the cold symptoms.

English sailors are called Limeys because the British admiralty ordered a daily ration of lime juice (vitamin C) to prevent scurvy.

The major deficiency diseases of today: anemia and osteoporosis.

Total Parenteral Nutrition

Parenteral means bypassing the intestine.

Are our present knowledge and understanding of human nutrition sufficient to maintain excellent human health by supplying the essential nutrients in digested form directly to the blood stream? The astronauts rejected any such approach to their nutritional needs in space and opted for "real food," along with the associated waste disposal problems. The truth is that we are very close to being able to do just that, even though no one claims to know an absolutely *complete* list of essential fats, vitamins, amino acids, or minerals.

In 1970, Judy Taylor, a young Canadian mother, completely lost the use of her small intestine because of a blood clot and consequent gangrene. Her life was saved through total parenteral nutrition (TPN), in which all of the known chemicals needed to sustain life were transfused into her body each evening through one of the major veins. Judy Taylor was able to live a relatively normal life! However, the bag of chemicals each day turned out to be incomplete. In 1977 Judy Taylor's blood sugar tested in the diabetic range even though her age and body weight would not suggest this disease. The attending physician and nutritionist, K. N. Jeejeebhoy, recalled an animal study wherein there was a relationship between a trace amount of chromium and blood sugar levels. When chromium was added to Judy Taylor's daily transfusion, her blood sugar levels returned to normal, and, as a consequence, chromium was added to the essential mineral list of humans.

What does TPN have to say about diet? Although our understanding of nutrition is extensive and is able to save lives that otherwise would be lost, it is hard to argue at this point equality for artificial diets relative to a balance of natural foods.

Whole grain wheat, oats, and barley supply fiber and vitamins, particularly the B-vitamins, to the diet.

FOOD ADDITIVES

Many chemicals with little or no nutritive value are added to food for a variety of reasons (Fig. 14–3). The chemicals are added during the processing and preparation of food for the purpose of preserving the food from oxidation, microbes, and the effects of metals. Food additives add and enhance flavor. They color the food, control pH, prevent caking, stabilize, thicken, emulsify, sweeten, leaven, and tenderize among other effects.

The GRAS List

The GRAS list is a noble effort—but it is not foolproof.

The Food and Drug Administration lists about 600 chemical substances **"generally recognized as safe" (GRAS)** for their intended use. A small portion of this list is given in Table 14–7. It must be emphasized that an additive on the GRAS list is safe *only if it used in the amounts and in the foods specified.* The GRAS list was published in several installments in 1959 and 1960. It was compiled from the results of a questionnaire asking experts in nutrition, toxicology, and related fields to give their opinions about the safety of various materials used in foods. Since its publication, few substances have been added to the GRAS list, and some, such as the cyclamates, carbon black, safrole, and Red Dye No. 2, have been removed.

There are more than 2500 known food additives, and many more chemicals than those that appear on the GRAS list are approved (or at least, not banned) for use as food additives by the FDA.

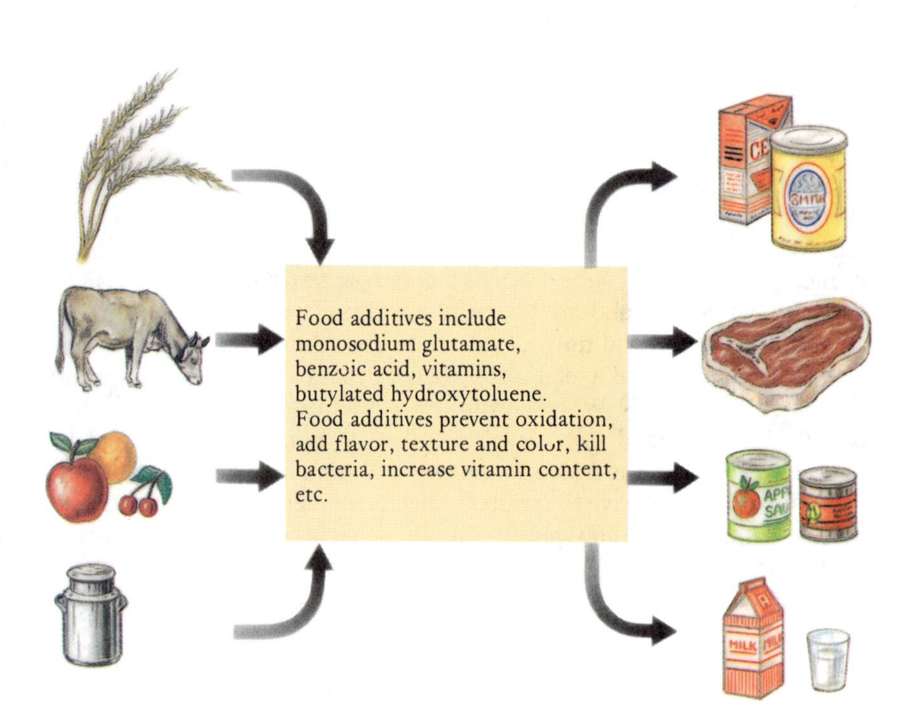

Food additives include monosodium glutamate, benzoic acid, vitamins, butylated hydroxytoluene. Food additives prevent oxidation, add flavor, texture and color, kill bacteria, increase vitamin content, etc.
MILK

Figure 14–3 Between the harvested and the consumer-ready food, one often finds a large variety of food additives.

TABLE 14-7 A Partial List of Food Additives Generally Recognized as Safe*

Anticaking Agents
- Calcium silicate
- Iron ammonium citrate

Acids, Alkalies, and Buffers
- Acetic acid
- Calcium lactate
- Citric acid
- Lactic acid
- Phosphates, Ca^{2+}, Na^{+}
- Potassium acid tartrate
- Sorbic acid
- Tartaric acid

Surface Active Agents (Emulsifying Agents)
- Glycerides: mono- and diglycerides of fatty acids
- Sorbitan monostearate

Polyhydric Alcohols
- Glycerol
- Mannitol
- Propylene glycol
- Sorbitol

Preservatives
- Benzoic acid
  - Sodium benzoate
- Propionic acid
  - Propionates, Ca^{2+}, Na^{+}
- Sorbic acid
  - Sorbates, Ca^{2+}, K^{+}, Na^{+}
- Sulfites, Na^{+}, K^{+}

Antioxidants
- Ascorbic acid
  - Ascorbates, Ca^{2+}, Na^{+}
- Butylated hydroxyanisole (BHA)
- Butylated hydroxytoluene (BHT)
- Lecithin
- Sulfur dioxide and sulfites

Flavor Enhancers
- Monosodium glutamate (MSG)
- 5′-Nucleotides
- Maltol

Sweeteners
- Aspartame
- Mannitol
- Saccharin
- Sorbitol

Sequestrants
- Citric acid
- EDTA, Ca^{2+}, Na^{+}
- Pyrophosphate, Na^{+}
- Sorbitol
- Tartaric acid
  - NaK (tartrate)

Stabilizers and Thickeners
- Agar-agar
- Algins
- Carrageenin

Flavorings
- Amyl butyrate (pearlike)
- Bornyl acetate (piney, camphor)
- Carvone (spearmint)
- Cinnamaldehyde (cinnamon)
- Citral (lemon)
- Ethyl cinnamate (spicy)
- Ethyl formate (rum)
- Ethyl vanillin (vanilla)
- Geranyl acetate (geranium)
- Ginger oil (ginger)
- Menthol (peppermint)
- Methyl anthranilate (grape)
- Methyl salicylate (wintergreen)
- Orange oil (orange)
- Peppermint oil (peppermint)
- Wintergreen oil (wintergreen) (methyl salicylate)

* For precise and authoritative information on levels of use permitted in specific applications, consult the regulations of the U.S. Food and Drug Administration and the Meat Inspection Division of the U.S. Department of Agriculture.

Preservation of Foods

Foods generally lose their usefulness and appeal a short time after harvest. Bacterial decomposition and oxidation are the prime reasons steps must be taken to lengthen the time that a foodstuff remains edible. Any process that prevents the growth of microorganisms or retards oxidation is generally an effective preservation process. Perhaps the oldest technique is the drying of grains, fruits, fish, and meat. Water is necessary for the growth and metabolism of microorganisms, and it is also important in oxidation. Dryness thus thwarts both the oxidation of food and the microorganisms that feed on it.

Dry foods tend to be stable.

Chemicals may also be added as preservatives. Salted meat and fruit preserved in a concentrated sugar solution are protected from microorganisms. The abundance of sodium chloride or sucrose in the immediate

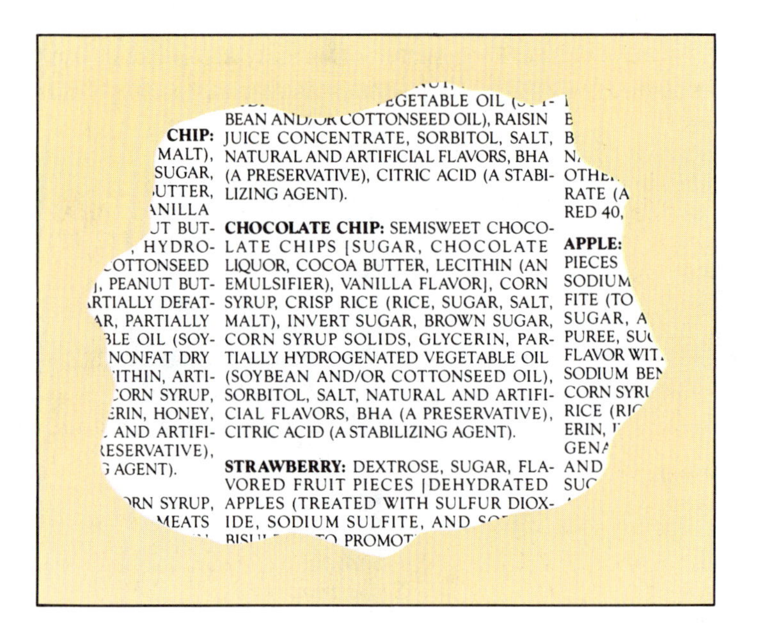

JUICE CONCENTRATE, SORBITOL, SALT,
NATURAL AND ARTIFICIAL FLAVORS, BHA
(A PRESERVATIVE), CITRIC ACID (A STABI-
LIZING AGENT).
CHOCOLATE CHIP: SEMISWEET CHOCO-
LATE CHIPS [SUGAR, CHOCOLATE
LIQUOR, COCOA BUTTER, LECITHIN (AN
EMULSIFIER), VANILLA FLAVOR], CORN
SYRUP, CRISP RICE (RICE, SUGAR, SALT,
MALT), INVERT SUGAR, BROWN SUGAR,
CORN SYRUP SOLIDS, GLYCERIN, PAR-
TIALLY HYDROGENATED VEGETABLE OIL
(SOYBEAN AND/OR COTTONSEED OIL),
SORBITOL, SALT, NATURAL AND ARTIFI-
CIAL FLAVORS, BHA (A PRESERVATIVE),
CITRIC ACID (A STABILIZING AGENT).
STRAWBERRY: DEXTROSE, SUGAR, FLA-
VORED FRUIT PIECES [DEHYDRATED
APPLES (TREATED WITH SULFUR DIOX-
IDE, SODIUM SULFITE, AND

Food additives and other contents of a popular granola bar. Note sorbitol, BHA, and citric acid, which are discussed in this chapter.

A hypertonic solution is more concentrated than solutions in its immediate environment.

Osmosis is the flow of water from a more dilute solution through a membrane into a more concentrated solution.

environment of the microorganisms forms a **hypertonic** condition in which water flows by **osmosis** from the microorganism to its environment. Salt and sucrose have the same effect on microorganisms as does dryness; both dehydrate them.

The canning process for preserving food, developed around 1810, involves first heating the food to kill all bacteria and then sealing it in bottles or cans to prevent access of other microorganisms and oxygen. Some canned meat has been successfully preserved for over a century. Newer techniques for the preservation of food include vacuum freezing, pasteurization, cold storage, irradiation, and chemical preservation.

Antimicrobial Preservatives

A preservative must interfere with microbes but be harmless to the human system—a delicate balance.

Food spoilage caused by microorganisms is a result of the excretion of toxins. A preservative is effective if it prevents multiplication of the microbes during the shelf life of the product. Sterilization by heat or radiation, or

Cooling food slows down the rate of oxidation and enzyme action and thus is one way to preserve food longer. (*The World of Chemistry,* Program 17, "The Precious Envelope")

THE WORLD OF CHEMISTRY

Food Spoilage

Food is shipped across America and around the world. As a natural consequence, the rate of food spoilage is a vital factor for our global economy. As a matter of fact, the kinds of reactions that cause food to spoil are not very different from those the chemists study in the laboratory.

How do scientists study food spoilage and preservation?—We have asked Dr. Theodore Labuza, a food chemist.

The way I always like to talk about food, it's the study of messy chemistry. And the reason I say that is that, in a food which has so many different organic compounds and inorganic compounds together, there are lots and lots of different reactions that could have caused spoilage. What we can do, however, is narrow them down to several classes. For example, when you bite into an apple, you see it start to brown. That's an enzyme reaction. There are reactions that make food go rancid. Potato chips, for example, if they sit around for a long time, the fat goes rancid.

Modern techniques of food preparation and refrigeration have greatly reduced spoilage. Still, the shopper might like further proof that reactions have been retarded, that food is as fresh as possible.

One of the more interesting things that we're doing in our laboratory here, which is an application of chemical kinetics, is the study of little devices that can be used to monitor time-temperature when a food goes through a distribution cycle. Foods deteriorate at a faster rate when the temperature goes higher, and at a slower rate when the temperature goes lower. If you had a device that could be placed on a package that would essentially integrate the time-temperature exposure and show a color change that could be related to the loss of quality of the food, then, by simply picking up a package and looking at a device that may be on the package, you could tell how much more shelf life is left. Then you'd know whether to consume it.

The World of Chemistry (Program 13) "The Driving Forces."

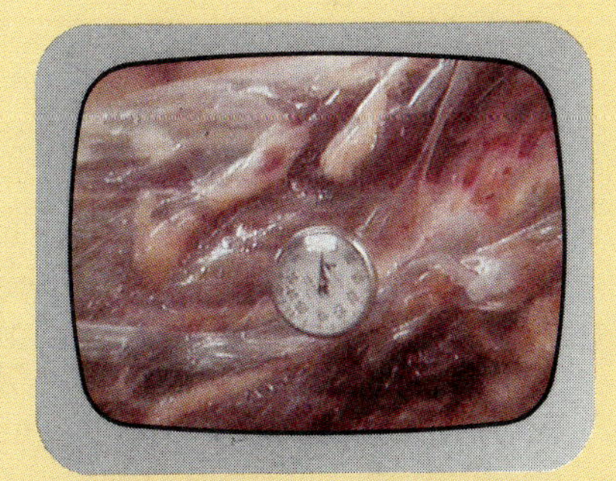

A time-temperature monitor of food. (*The World of Chemistry,* Program 13, "The Driving Forces")

inactivation by freezing, is often undesirable, since it impairs the quality of the food. Chemical agents seldom achieve sterile conditions but can preserve foods for considerable lengths of time.

Antimicrobial preservatives are widely used in a large variety of foods. For example, in the United States sodium benzoate is permitted in nonalcoholic beverages and in some fruit juices, fountain syrups, margarines, pickles, relishes, olives, salads, pie fillings, jams, jellies, and preserves. Sodium propionate is legal in bread, chocolate products, cheese, pie crust, and fillings. Depending on the food, the weight of the preservative permitted ranges up to a maximum of 0.1% for sodium benzoate and 0.3% for sodium propionate.

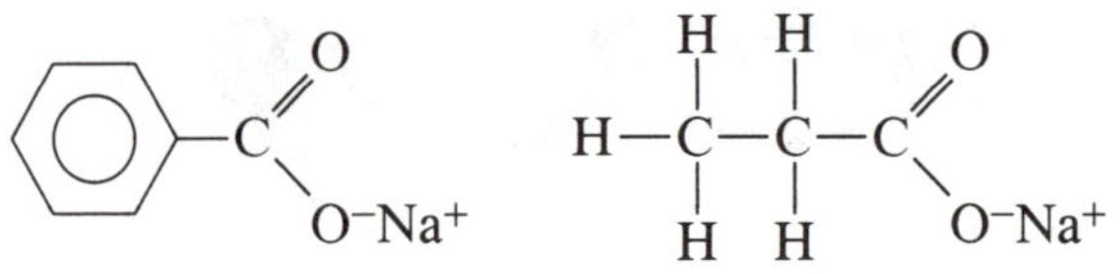

Sodium benzoate

Sodium propionate

Atmospheric Oxidation

Microbial activity results in oxidative decay of food, but it is not the only means of oxidizing food. The direct action of oxygen in the air, **atmospheric oxidation,** is the chief cause of the destruction of fats and fatty portions of food. Foods kept wrapped, cold, and dry are relatively free of air oxidation. An antioxidant added to the food can also hinder oxidation. Antioxidants most commonly used in edible products contain various combinations of butylated hydroxyanisole (BHA) and butylated hydroxytoluene (BHT).

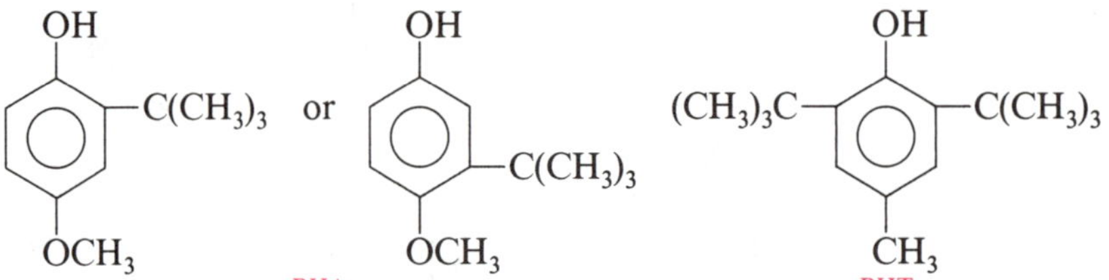

BHA

BHT

To prevent the oxidation of fats, the antioxidant can donate the hydrogen atom (H·) in the —OH group to reactive species; this effectively stops the reaction between fats and oxygen. If antioxidants are not present, the oxidation of fats leads to a complex mixture of volatile aldehydes, ketones, and acids, which cause a rancid odor and taste.

Sequestrants

Metals get into food from the soil and from machinery during harvesting and processing. Copper, iron, and nickel, as well as their ions, catalyze the oxidation of fats. However, molecules of citric acid bond with the metal ions, thereby rendering them ineffective as catalysts. With the competitor metal ions tied up, antioxidants such as BHA and BHT can accomplish their task much more effectively.

To sequester means "to withdraw from use." The sequestering ability of EDTA accounts for its use in treating heavy-metal poisoning (Chapter 10).

Citric acid belongs to a class of food additives known as **sequestrants.** For the most part sequestrants react with trace metals in foods, tying them up in complexes so the metals will not catalyze the decomposition or oxidation of food. Sequestrants such as sodium and calcium salts of EDTA (ethylenediaminetetraacetic acid) are permitted in beverages, cooked crab meat, salad dressing, shortening, lard, soup, cheese, vegetable oils, pudding mixes, vinegar, confections, margarine, and other foods. The amounts range from 0.0025% to 0.15%. The structural formula of EDTA bonded to a metal ion is shown in Figure 14–4.

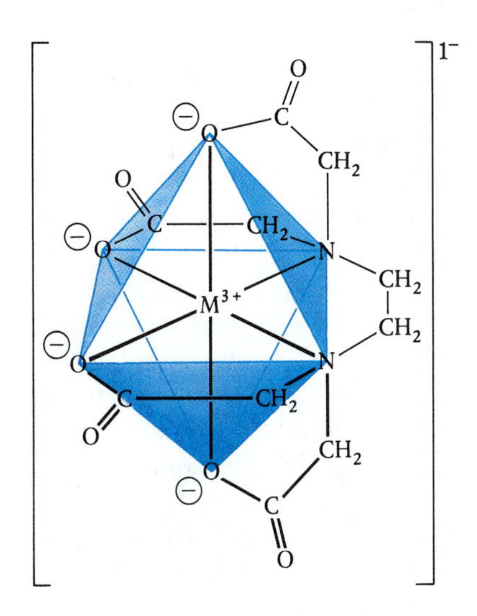

Figure 14–4 The structural formula for the metal chelate of ethylenediaminetetraacetic acid (EDTA).

Flavor in Foods

Flavors result from a complex mixture of volatile chemicals. Since we have only four tastes—sweet, sour, salty, and bitter—much of the sensation of taste in food is smell (Fig. 14–5). For example, the flavor of coffee is determined largely by its aroma, which in turn is due to a very complex

Some 1700 natural and synthetic substances are used to flavor foods, making flavors the largest category of food additives.

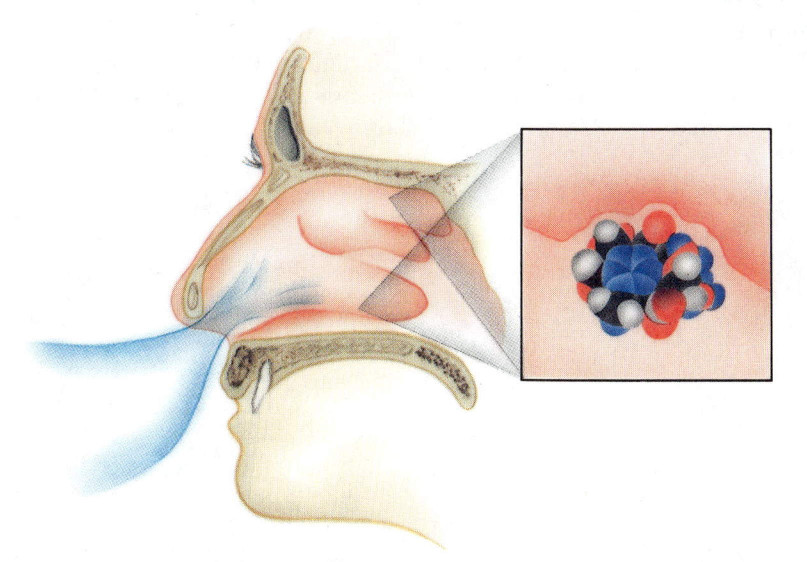

Figure 14–5 A stereochemical interpretation of the sensation of smell. The substance fits a cavity in the back of the oral cavity. If the atoms are properly spaced, they sensitize nerve endings that transmit impulses to the brain. The brain identifies these sensations as a particular smell. A complete explanation of smell is certainly more involved than the simple idea presented here.

mixture of over 500 compounds, mostly volatile oils. These compounds are of undetermined toxicity.

Most flavor additives originally came from plants. The plants were crushed and the compound extracted with various solvents such as ethanol or carbon tetrachloride. Sometimes a single compound was extracted; more often the residue contained a mixture of several compounds. By repeated efforts, relatively pure oils were obtained. Oils of wintergreen, peppermint, orange, lemon, and ginger, among others, are still obtained in this way. These oils, alone or in combination, are then added to foods to produce the desired flavor. Gradually, analyses of the oils and flavor components of plants have revealed the active compounds responsible for the flavors. Today synthetic preparations of the same flavors actively compete with natural extracts.

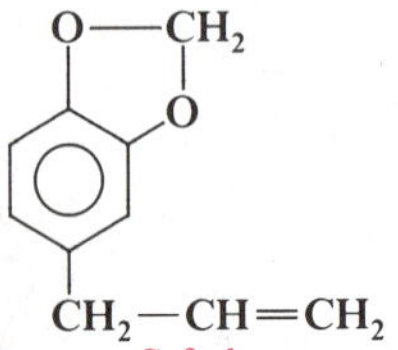

Safrole

The FDA has banned some of the naturally occurring flavoring agents that used to be used, including safrole, the primary root beer flavor, found in the root of the sassafras tree.

Flavor Enhancers

In some people MSG causes the so-called Chinese restaurant syndrome, an unpleasant reaction characterized by headaches and sweating that usually occurs after an MSG-rich Chinese meal. Tomatoes and strawberries affect some individuals in the same way.

MSG is a natural constituent of many foods, such as tomatoes and mushrooms.

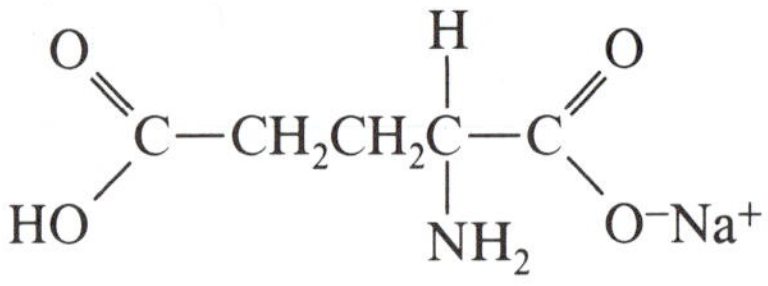

Flavor enhancers have little or no taste of their own but amplify the flavors of other substances. They exert synergistic and potentiation effects. **Synergism** is the cooperative action of discrete agents such that the total effect is greater than the sum of the effects of each used alone. **Potentiators** do not have a particular effect themselves but exaggerate the effects of other chemicals. The 5′-nucleotides, for example, have no taste but enhance the flavor of meat and the effectiveness of salt. Potentiators were first used in meat and fish but now are also used to intensify flavors or cover unwanted flavors in vegetables, bread, cakes, fruits, nuts, and beverages. Three commonly used flavor enhancers are **monosodium glutamate (MSG), 5′-nucleotides** (similar to inosinic acid), and **maltol.**

Monosodium glutamate

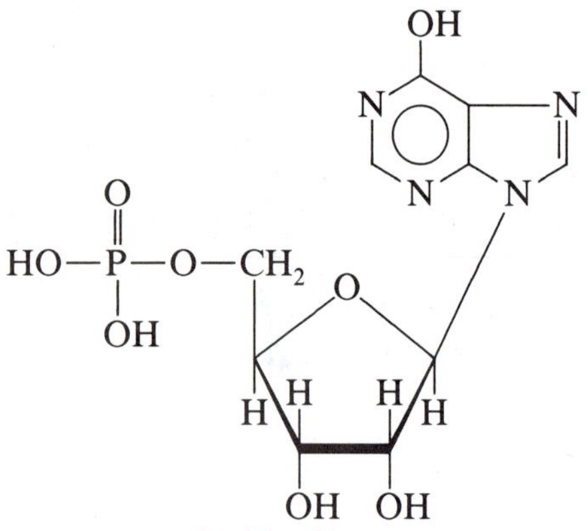

Inosinic acid
(a 5′-nucleotide)

When MSG is injected in very high doses under the skin of 10-day-old mice, it causes brain damage. When these laboratory results were reported, considerable discussion ensued concerning the merits of MSG. National investigative councils have suggested that MSG be removed from baby foods because infants do not seem to appreciate enhanced flavor. However, in the absence of hard evidence that MSG is harmful in the amounts used in regular food, no recommendations were made relative to its use.

Sweeteners

Sweetness is characteristic of a wide range of compounds, many of which are completely unrelated to sugars. Lead acetate $[Pb(CH_3COO)_2]$ is sweet but poisonous. A number of **artificial sweeteners** are allowed in foods. Artificial sweeteners are important in special diets such as those of diabetics and for those who wish to control their caloric intake while still enjoying sweet-tasting foods.

Saccharin

The first common artificial sweetener was saccharin, which is 300 times sweeter than ordinary sugar (sucrose). When ingested, saccharin passes through the body unchanged and consequently has no caloric value. Saccharin has a somewhat bitter aftertaste which is offset in commercial products such as Sweet'n Low by the addition of small amounts of naturally-occurring sweeteners. Such products do have a small caloric value because of the natural sweeteners added.

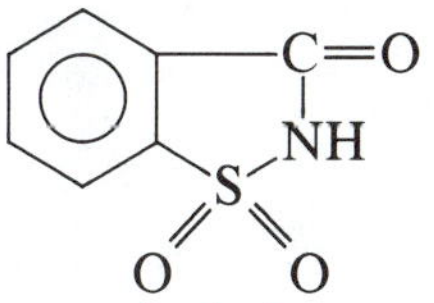

Saccharin is a synthetic chemical.

Saccharin

High doses of saccharin have been shown to cause cancer in mice, and commercial products containing saccharin are required by law to have a warning label: "This product contains saccharin which has been determined to cause cancer in laboratory animals."

Aspartame

Aspartame has replaced saccharin as the principal artificial sweetener and, sold under the trade name NutraSweet, accounts for 75% of the $1 billion worldwide artificial sweetener market. In a recent year, NutraSweet accounted for over 20% of the $909 million pretax earnings of Monsanto Company. Aspartame is about 200 times sweeter than table sugar and is used in over 3000 products.

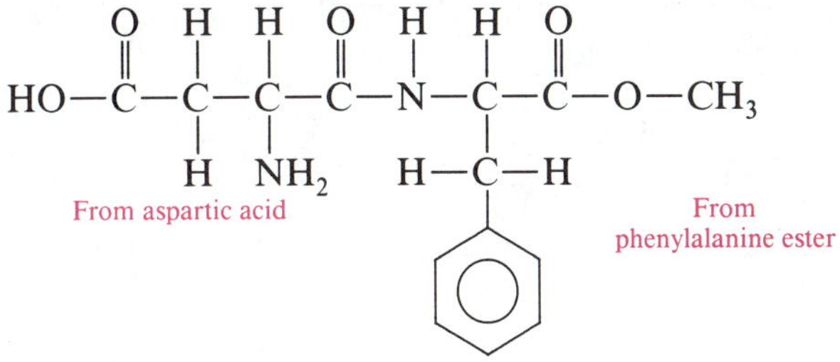

From aspartic acid

From phenylalanine ester

Aspartame

Aspartame is digested, and the caloric value is approximately equal to that of proteins. However, since much smaller amounts of aspartame are needed for sweetness in comparison with table sugar, many fewer calories are consumed in the sweetened food. Aspartame is not stable at cooking temperatures, limiting its general use as a sugar substitute. Saccharin sweeteners can be substituted for large fractions of the sugar called for in recipes for cooked foods. Aspartame does not have a bitter aftertaste.

Discovered Sweetness Through Molecular Design

The sweetness of saccharin and aspartame was discovered after these compounds had been prepared. The terms *accident* and *serendipity* can be and have been associated with the discovery and commercialization of the sweet taste of these compounds. In recent years, theoretical models of the interaction of molecules of sweet compounds and the biological receptors on the tongue have proven useful in predicting new compounds that should be sweet when synthesized. Early it was noted that if chlorine atoms replaced hydroxyl atoms on the sucrose molecule, the derivative was sweeter than the sucrose. In fact, a product that is predicted to become a great commercial success is **sucralose,** in which three hydroxyl groups are replaced by chlorine atoms. Sucralose is 650 times sweeter than sucrose.

Compare the structure for sucrose with those for simple sugars and polysaccharides given in Chapter 9.

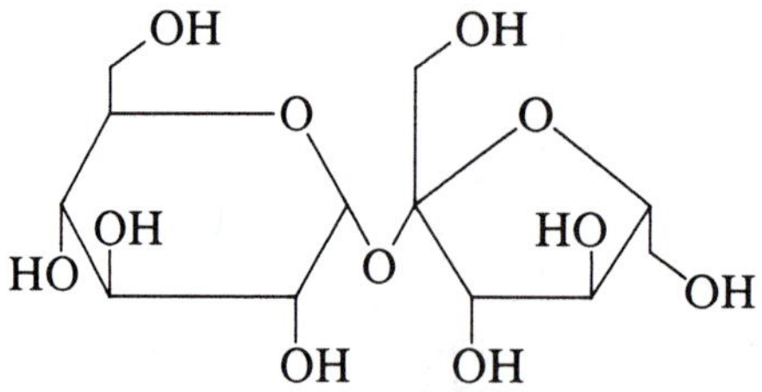

Sucrose

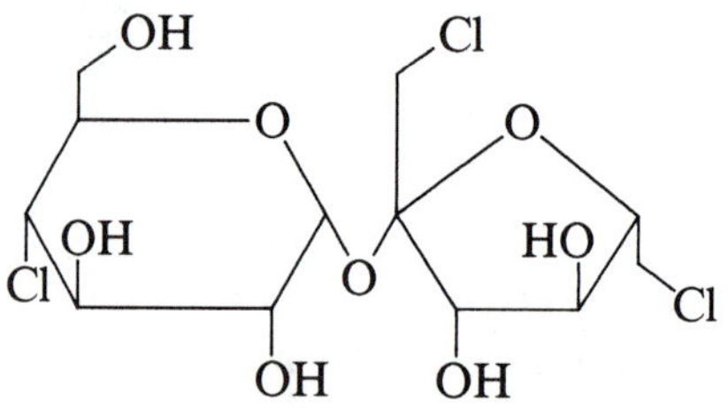

Sucralose

The peptide bond described in Chapter 9 links two amino acid units together in the protein structure.

Notice the peptide bond in the formula for aspartame given above. Following the commercial discovery of the sweetness of aspartame, there has been intensive research into the sweetness of a large number of peptide compounds from both theoretical and trial-and-error approaches. **Alitame,** a dipeptide sweetener, was prepared first in 1979 and has a sweetness potency over 2000 times that of table sugar! It has a sucrose-like sweetness with no aftertaste. Also, alitame is more stable than aspartame in cooked food applications. If alitame receives approval by the FDA as expected, it will have exciting commercial applications. One problem with this and other high-potency sweeteners, that will likely be discovered, is that sweetness control in the food may be difficult when such small amounts of the chemical are required in the food sample.

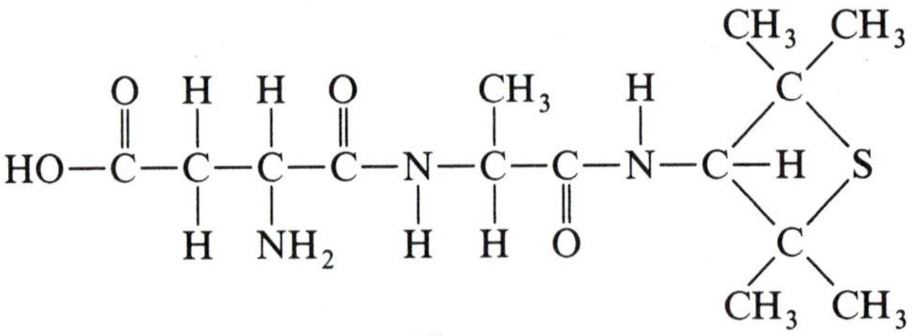

Alitame

Food and Esthetic Appeal

Food Colors

There are about 30 chemical substances used to color food. All are under investigation by the FDA, and some may be prohibited as the investigations progress. About half of the food colors are laboratory synthesized, and half are extracted from natural materials. Most food colors are large organic molecules with several double bonds and aromatic rings. The electrons of these conjugated structures can absorb certain wavelengths of light and pass the rest; the wavelengths passed give the substances their characteristic colors. Beta-carotene, an orange-red substance in a variety of plants that gives carrots their characteristic color, has a conjugated system of electrons and is used as a food color. Beta-carotene is a precursor (provitamin) of vitamin A.

Colored organic substances often are conjugated molecules, having alternating double and single bonds in the carbon chain or ring.

Because one of the food colors, Yellow No. 5, causes allergic reactions (mainly rashes and sniffles) in an estimated 50,000 to 90,000 Americans, the FDA has required manufacturers to list Yellow No. 5 on the labels of any food products containing it.

pH Control in Foods

Weak organic acids are added to such foods as cheese, beverages, and dressings to give a mild acidic taste. They often mask undesirable aftertastes. Weak acids and acid salts, such as tartaric acid and potassium acid tartrate, react with bicarbonate to form CO_2 in the baking process.

Some acid additives control the pH of food during the various stages of processing as well as in the finished product. In addition to single substances, there are several combinations of substances that adjust and then maintain a desired pH; these mixtures are called **buffers.** An example of one type of buffer is potassium acid tartrate, $KHC_4H_4O_6$.

Buffer solutions resist change in acidity and basicity; pH remains constant.

Adjustment of fruit juice pH is allowed by the FDA. If the pH of the fruit is too high, it is permissible to add acid (called an **acidulant**). Citric acid and lactic acid are the most common acidulants used, since they are believed to impart good flavor, but phosphoric, tartaric, and malic acids are also used. These acids are often added at the end of the cooking time to prevent extensive hydrolysis of the sugar. In the making of jelly they are sometimes mixed with the hot product immediately after pouring. To raise the pH of a fruit that is unusually acid, buffer salts such as sodium citrate or sodium potassium tartrate are used.

Small amounts of certain acids are allowed to be added to some foods.

The versatile acidulants also function as preservatives to prevent the growth of microorganisms, as synergists and antioxidants to prevent rancidity and browning, as viscosity modifiers in dough, and as melting-point modifiers in such food products as cheese spreads and hard candy.

Anticaking Agents

Anticaking agents are added to hygroscopic foods—in amounts of 1% or less—to prevent caking in humid weather. Table salt (sodium chloride) is particularly subject to caking unless an anticaking agent is present. The additive (magnesium silicate, for example) incorporates water into its structure as water of hydration and does not appear wet as sodium chloride does

Hygroscopic substances absorb moisture from the air.

when it absorbs water physically on the surface of its crystals. As a result, the anticaking agent keeps the surface of sodium chloride crystals dry and prevents crystal surfaces from codissolving and joining together.

Stabilizers and Thickeners

Stabilizers and thickeners are types of emulsifying agents.

Stabilizers and thickeners improve the texture and blends of foods. The action of carrageenan (a polymer from edible seaweed) is shown in Figure 14–6. Most of this group of food additives are polysaccharides (Chapter 9), which have numerous hydroxyl groups as a part of their structure. The hydroxyl groups form hydrogen bonds with water to prevent the segregation of water from the less polar fats in the food and to provide a more even blend of the water and oils throughout the food. Stabilizers and thickeners are particularly effective in icings, frozen desserts, salad dressings, whipped cream, confections, and cheeses.

Surface Active Agents

Cholic acid

Surface active agents are similar to stabilizers, thickeners, and detergents in their chemical action. They cause two or more normally incompatible (nonpolar and polar) chemicals to disperse in each other. If the chemicals are liquids, the surface active agent is called an **emulsifier.** If the surface active

Figure 14–6 The action of carrageenan to stabilize an emulsion of water and oil in salad dressing. An active part of carrageenan in a polysaccharide, a portion of which is shown here. The carrageenan hydrogen-bonds to the water, which keeps it dispersed. The oil, not being very cohesive, disperses throughout the structure of the polysaccharide. Gelatin (a protein) undergoes similar action in absorbing and distributing water to prevent the formation of ice crystals in ice cream.

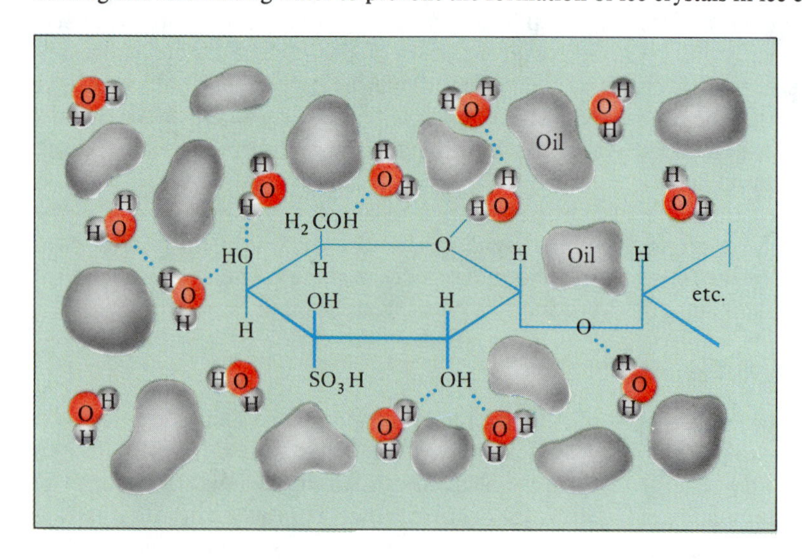

agent has a sufficient supply of hydroxyl groups, as does cholic acid, the groups form hydrogen bonds to water. Cholic acid and its associated group of water molecules are distributed throughout dried egg yolk in a manner quite similar to that of carrageenan and water in salad dressing.

Some surface active agents have both hydroxyl groups and a relatively long nonpolar hydrocarbon end. Examples are diglycerides of fatty acids, polysorbate 80, and sorbitan monostearate. The hydroxyl groups on one end of the molecule are anchored by hydrogen bonds in the water, and the nonpolar end is held by the nonpolar oils or other substances in the food. This provides tiny islands of water held to oil. These islands are distributed evenly throughout the food.

Hydrogen bonding plays a major role in stabilizers, thickeners, surface active agents, and humectants.

Polyhydric Alcohols

Polyhydric alcohols are allowed in foods as humectants, sweetness controllers, dietary agents, and softening agents. Their chemical action is based on their multiplicity of hydroxyl groups that hydrogen-bond to water. They thus hold water in food, soften it, and keep it from drying out. Tobacco is also kept moist by the addition of polyhydric alcohols such as glycerol. An added feature of polyhydric alcohols is their sweetness. The two polyhydric alcohols mentioned earlier for their sweetness are mannitol and sorbitol. The structures of these alcohols are strikingly similar to the structure of glucose (Chapter 9), and all three have a sweet taste.

D-Sorbitol	D-Mannitol
CH_2OH	CH_2OH
H—C—OH	HO—C—H
HO—C—H	HO—C—H
H—C—OH	H—C—OH
H—C—OH	H—C—OH
CH_2OH	CH_2OH

SELF-TEST 14B

1. Our mineral needs can be obtained by eating a ____________ of ____________ foodstuffs grown in ____________.
2. Minerals involved in transmitting a nerve impulse are ____________ and ____________.
3. For proper balance, the K/Na weight ratio in the body should be slightly ( ) greater than, ( ) less than 1.
4. Calmodulin, a protein, collects excess calcium from the bloodstream and helps prevent ____________ stones.
5. Goiter is caused by a deficiency of ____________.
6. Vitamins are synthesized by cells of the body: True ( ) False ( )
7. Vitamins A, D, E, F, and K are ____________ -soluble, whereas vitamins B-complex and C are ____________ -soluble.
8. The relationship between β-carotene and vitamin A is that β-carotene is a ____________ of vitamin A.
9. Vitamin E is effective as a(n) ____________, particularly in preventing the deterioration of polyunsaturated fatty acids.
10. Because it is involved in so many biochemical reactions, vitamin ____________ is considered the master vitamin.
11. B-complex vitamins function generally as ____________ involved in the process of ____________.
12. Flavor enhancers exert a(n) ____________ or ____________ effect on the flavors of foods.

13. Name two of the oldest means for preserving foods. ____________ and ____________
14. Antimicrobial preservatives make foods sterile: True ( ) False ( )
15. Antioxidants are ( ) more ( ) less easily oxidized than the food into which they are placed.
16. Citric acid is an example of a(n) ____________. Such compounds tie up metal in stable complexes.
17. A flavor in food can usually be traced to a single compound: True ( ) False ( )
18. Monosodium glutamate is a(n) ____________ ____________.
19. GRAS is an acronym for ____________ ____________ ____________ ____________.
20. Which has the sweetest taste per unit of weight in taste tests: table sugar, aspartame, saccharin, or alitame? ____________

QUESTIONS

1. Name the six classes of nutrients needed by the human body in order to maintain good health.
2. Based on current generally accepted standards, what should be the maximum percentage of fat in your diet?
3. What nutrient should be about 60% of your diet?
4. In your opinion, what is the reason Americans tend to eat too much fat in their diets?
5. What two metals should be given special attention in planning for an adequate diet for the typical family? Give the importance of each.
6. Which nutrient produces the greatest number of kilocalories per gram when digested and metabolized in the human body?
7. What nutrient is the primary source of nitrogen for the body?
8. Is it acceptable to eat enough protein in one day to last for three days? Why or why not?
9. For the average adult male, what percentage of the protein is replaced in the body each day?
10. What are two sources of dietary fiber, and why is fiber important in the diet?
11. What problem is caused by the hydrogenation of polyunsaturated fats (oils) used in foods?
12. What is the name of the essential fatty acid? What vitamin is designated as this fatty acid?
13. Look at the labels in your food store to see what fats are most often used in prepared foods. Are you able to decide what fats are "good" and what fats are "bad"?
14. Give two examples of dietary components in which balance is especially important.
15. What is lost in the purification of cane sugar, sucrose?
16. There are 46 different mineral elements that are known to be in your body. Would you argue that all 46 of these are essential? Are not necessarily essential? Why?
17. What activities go on within your body while a measurement of your basal metabolic rate (BMR) is being made?
18. What are the functions of fat, protein, and carbohydrates in the body?
19. Based on their solubilities, what are the two classes of vitamins? An excess of which class of vitamins causes fewer problems?
20. Why is vitamin B_6 called the "master vitamin"?
21. Why should one eat the whole fruit rather than, for example, drinking the juice and throwing the rest of the fruit away?
22. What is the assumption if one is to believe that all nutritional needs can be met through total parenteral nutrition?
23. Should salt tablets be taken after the body is acclimated to the heat? Why or why not?
24. Why may a calcium deficiency occur in some women after menopause?
25. Does vitamin C decrease the symptoms of the common cold? Does vitamin A?
26. How does salt preserve food?
27. a. Distinguish between refined sugar and complex carbohydrates with respect to how they are assimilated by the body.
    b. What problems may arise from consuming too much (1) refined sugar? (2) refined grains (white flour)?
28. A label on a brand of breakfast pastries lists the following additives: dextrose, glycerin, citric acid, potassium

sorbate, vitamin C, sodium iron pyrophosphate, and BHA. What is the purpose of each substance?

29. What is a common flavor enhancer? How do flavor enhancers work?

30. Choose a label from a food item, and try to identify the purpose of each additive.

31. What are the pros and cons of eating "natural" foods, as opposed to foods containing chemical additives?

32. What is the GRAS list?

33. What foods have you eaten during the past week that did not have chemicals added or applied to them?

The Physician **by Franz Christophe Janneck depicts the art and science of medicine in the 18th century.** (Courtesy Fisher Scientific)

CHAPTER

15 Chemistry and Medicine

Chemistry plays an important role in improving mental and physical health. Heart disease and cancer account for about 60% of the deaths in the United States. New drugs are being developed to combat these diseases. Other current health topics include steroids, drugs and the brain, and drugs used in the treatment of AIDS.

1. What are atherosclerosis and hypertension, and how are they related to heart disease?
2. Does aspirin cut the risk of heart disease?
3. What is cancer chemotherapy, and what cancers have increased survival rates because of it?
4. What are hormones?
5. What are neurotransmitters?
6. What are natural opiates?
7. What drugs are currently being abused?
8. What happens when the AIDS virus invades the body?

The average life expectancy for men in the United States has risen from 53.6 years in 1920 to 72.1 years in 1990, an increase of 34.5%. During this same period, the life expectancy for women has risen from 54.6 years to 79.0 years, an increase of 44.7%. What roles have chemistry and chemical technology played in raising life expectancy and in improving the quality of life? The primary focus has been on the discovery of new and improved medicines and therapeutic drugs for treating both the mind and the rest of the body.

Within the past decade, remarkable progress has been made in understanding how chemical reactions control and regulate biological processes. As knowledge about chemistry of biological processes and the chemical mechanism of drug action improves, drug design will receive increased emphasis over the earlier trial-and-error methods used for screening chemicals for drug use. In this chapter the emphasis is on illustrating the role chemistry plays in improving mental and physical health.

MEDICINES

Americans spend over $34 billion per year on medicines. The top ten prescription drugs, based on the number of prescriptions written (Table 15–1), show an interesting cross-section of medicinal uses. The generic name for a drug is the drug's generally accepted chemical name. The trade or brand name is the name used by the drug manufacturer. For example, the antibiotic amoxicillin (number one drug) is sold under brand names such as

TABLE 15–1 Ten Most Prescribed Drugs in the United States in 1990*

Trade Name	Generic Name	Use
1. Amoxil	Amoxicillin trihydrate	Antibiotic
2. Lanoxin	Digoxin	Heart disease
3. Zantac	Ranitidine hydrochloride	Antiulcer agent
4. Premarin	Mixture of estrogens	Menopausal symptoms
5. Xanax	Alprazolam	Tranquilizer
6. Dyazide	Triamterene/hydrochlorothiazide	Blood pressure (diuretic/antihypertensive)
7. Cardizem	Diltiazem hydrochloride	Heart disease (calcium blocker)
8. Synthroid	Levothyroxine sodium	Thyroid hormone
9. Ceclor	Cefaclor	Antibiotic
10. Seldane	Terfenadine	Antihistamine

* Data from *American Druggist,* Vol. 199, No. 2, p. 56, 1991.

Amoxil, Amoxidall, Amoxibiotic, Infectomycin, Moxaline, Utimox, and Wymox. If the generic name is used, the prescription is often cheaper, particularly if the drug is not protected by patents and can be manufactured and marketed competitively by several companies.

LEADING CAUSES OF DEATH

In 1900 five of the ten leading causes of death in the United States were infectious diseases. By 1987, only pneumonia and influenza remained in the top ten, causing 3.3% of the deaths (Fig. 15-1). This dramatic decrease can be attributed to the successful development of a variety of **antibiotics.** However, the lack of a cure for AIDS and the experimental nature of drugs currently being used to treat AIDS patients are likely to result in an increase in the number of deaths due to infectious diseases during the next decade.

HEART DISEASE

Heart disease is the number-one killer of Americans, claiming about 760,000 lives per year, or 36% of all deaths. However, a variety of surgical techniques along with many new drugs are being used to decrease the death rate and improve the quality of life for persons suffering from heart disease. Since 1970, the number of deaths due to heart disease has dropped from 369 to 312 per 100,000 population. Medical observers attribute progress against

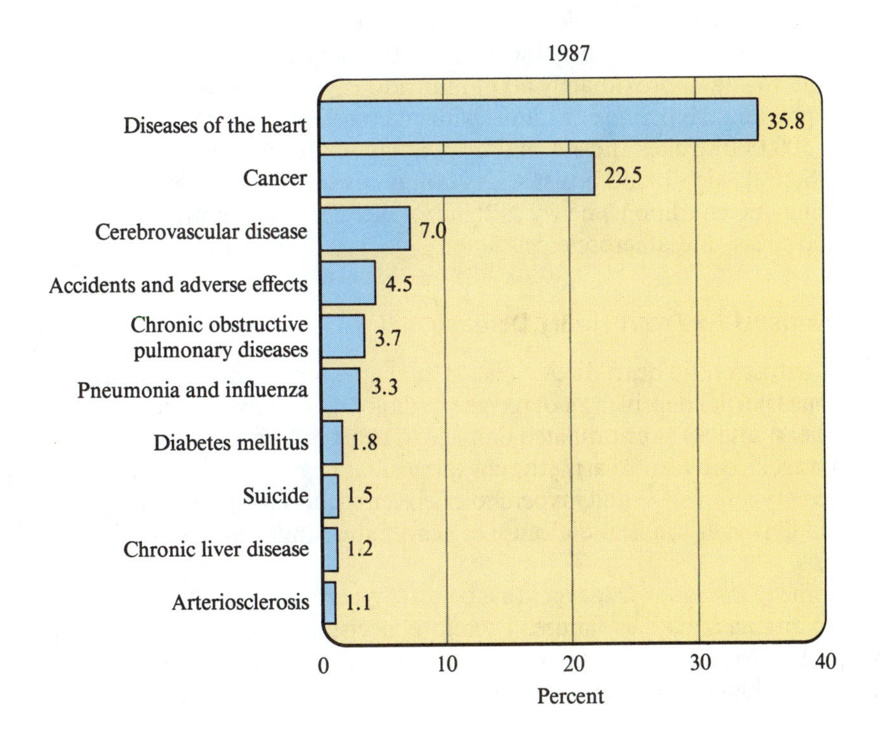

Figure 15-1 The ten leading causes of death as a percentage of all deaths. (Source: National Center for Health Statistics, Division of Vital Statistics, National Vital Statistics System)

heart disease to reduction of **atherosclerosis** (plaque build-up on artery walls), the most common cause. Risk factors that contribute to cardiovascular diseases include smoking, eating too much saturated fat and cholesterol, and stress. These factors together with genetic influences may lead to **hypertension** (high blood pressure) and **hypercholesterolemia** (high concentrations of blood cholesterol).

Heart disease is an assortment of diseases, and death rates remain high despite the progress in recent years. Of about 1.5 million persons who experience heart attacks in the United States each year, approximately one third die. About 98% of all heart attack victims have atherosclerosis.

Atherosclerosis

Review the discussion of cholesterol in Chapter 9.

Atherosclerosis is the build-up of fatty deposits called **plaque** on the inner walls of arteries. Cholesterol is a major component of atherosclerotic plaque. Many scientists believe that a high level of cholesterol in the blood, along with high blood levels of triglycerides, contributes to the build-up of this plaque. The plaque build-up reduces the flow of blood in general and through the heart in particular. If a coronary artery is blocked by plaque, a heart attack occurs as a result of a reduction in the flow of blood carrying oxygen to the heart muscle. In the vast majority of cases the reduction in the blood supply can be traced to the formation of a blood clot in the plaque of a coronary artery. If prolonged, such an attack can cause part of the heart muscle to die.

Hypertension

Hypertension, or high blood pressure, is a common medical problem in the United States, as shown by the number six position of the drug, dyazide, in Table 15–1. Approximately 60 million adults, or one in four, have high blood pressure. Hypertension, along with atherosclerosis, is the major cause of the 500,000 strokes and 1.5 million heart attacks each year. Factors that contribute to high blood pressure are family history of hypertension, age, race, diabetic conditions, heavy salt and/or alcohol consumption, obesity, chronic stress, and atherosclerosis.

Drugs Used to Treat Heart Disease

Atherosclerotic heart disease results in angina (chest pain on exertion), ischemia (partial deprivation of oxygenated blood), and myocardial infarction (heart attack). Accumulated damage to heart muscle by ischemia and heart attacks can lead to arrhythmias (abnormal heart rhythms) and heart failure. Hypertension and hypercholesterolemia accelerate atherosclerosis, and hypertension is also a cause of heart failure independent of atherosclerosis.

Some of the same drug types that control hypertension are also effective against angina and heart failure. Therefore, decisions about which antihypertensive drugs to prescribe must be weighed against their effects on arrhythmias, blood cholesterol, and heart failure.

Treatment of hypertension begins with loss of excess weight and restrictions on dietary sodium ions (Na^+). If these measures do not lower blood pressure sufficiently, drugs are required. The first choice is a **diuretic** such as dyazide which stimulates the production of urine and excretion of Na^+ Because diuretics can also cause excess excretion of K^+, which is important in nerve conduction and muscle contraction, patients whose hearts are subject to arrhythmia must receive either a potassium supplement or a diuretic that does not cause K^+ excretion.

Clinical studies have shown that increased levels of Na^+ raise the blood pressure in some individuals. See Chapter 14.

The ultimate method of prevention of most forms of heart disease is prevention of the build-up of atherosclerotic plaque in the arteries. In 1954 John W. Gofman of the University of California at Berkeley discovered different categories of lipoproteins in the blood. About 65% of the cholesterol in the blood is carried by low density lipoproteins (LDL), whereas only 25% of the cholesterol in the blood is carried by high density lipoproteins (HDL). In 1968 John A. Glomset of the University of Washington showed that HDLs are effective in removing cholesterol from arterial walls and transporting it to the liver, where it is metabolized. This discovery opened up the possibility that already-formed atherosclerotic plaque might be dissolved by using a normal blood component. Scientists have also been attempting to develop drugs that either raise the level of HDLs or lower the lever of LDLs for patients who cannot control these lipoproteins by proper diet.

Cholesterol-lowering drugs include niacin (nicotinic acid), lovastatin, and cholestyramine. Niacin and lovastatin lower cholesterol levels in the blood by interfering with cholesterol synthesis in the liver. Cholestyramine resin lowers blood cholesterol by lowering the concentration of bile acids in the intestines, and this causes the liver to convert more cholesterol into bile acids. Of the three drugs, lovastatin is the most recent drug on the market and is more effective for treating patients with very high blood cholesterol and/or LDL levels. Early indications are that lovastatin causes a drop in LDL levels of 19% to 39%.

Vasodilator drugs are used to relieve the pain from angina attacks. Vasodilators dilate veins, thus reducing the blood pressure against which the heart must work. Angina occurs because of insufficient oxygen delivery to the heart muscle. Angina attacks are brought on by exercise or anxiety, which increase the work of the heart and thus its oxygen demand. Symptoms include a crushing sensation in the chest, abdominal pain, and/or pain radiating from the chest to the left arm, the throat and jaw, and sometimes the right arm. Treatment of an attack centers on easing the work of the heart and thus its oxygen demand. Drugs used to treat angina include vasodilators such as nitroglycerin, amyl nitrate, and digoxin (number two drug in Table 15–1), as well as various beta blockers and calcium channel blockers.

Beta Blockers and Calcium Channel Blockers

In 1948, Raymond P. Ahlquist of the Medical College of Georgia discovered that heart muscle contains receptors for epinephrine and norepinephrine. He called these **beta receptors.** Stimulation of these heart muscle receptors results in an increase in the number of heart beats. In 1967 Alonzo M. Lands, a pharmacologist in Rensselaer, New York, discovered two dif-

Epinephrine and norepinephrine are neurotransmitters. See Figure 15–8.

ferent beta receptors, $beta_1$ and $beta_2$. $Beta_1$ sites are located primarily in the heart but are also found in the kidneys. $Beta_2$ receptors are involved in the relaxation of the peripheral blood vessels and the bronchial tubes.

With the knowledge about beta receptors gained by Ahlquist, chemists began to explore the action of chemicals that would compete with epinephrine and norepinephrine at the beta receptor sites. If these sites could be blocked, the heart rate would decrease. For a heart already overworked from the build-up of plaque in the arteries supplying heart tissue with blood (and oxygen), this might produce enough relaxation to allow recovery from an impending attack. In addition, these drugs might be able to relieve high blood pressure and migraine headaches.

The first drugs of this type, called **beta blockers** because of their action of blocking beta receptor sites, came into use in the late 1950s and early 1960s but were later withdrawn because of undesirable side effects. In 1967 the beta blocker propranolol (trade name Inderal) was first prescribed. Its first use was for cardiac arrhythmias, but now it is also used in the treatment of angina, hypertension, and migraine headaches. Propranolol has high lipid solubility, so it passes through the blood-brain barrier and builds up in the central nervous system, where it is more slowly metabolized. This build-up of the chemical in the central nervous system can cause side effects of fatigue, lethargy, depression, and confusion. In spite of these possible side effects, propranolol is a widely prescribed drug, and it ranked 30th in prescribed drug sales in 1990.

A second beta blocker, metoprolol (trade name Lopressor), was introduced in the United States in 1978 for the treatment of hypertension. Because metoprolol is selective in its beta-blocking effects, blocking only $beta_1$ sites and not the $beta_2$ sites of the peripheral blood vessels or the bronchial tube, it is safe for asthma sufferers and for patients with severe blood vessel disorders. Metoprolol is not as soluble in lipids as propranolol and does not accumulate in the central nervous system. It is also more slowly metabolized by the liver, and this allows more widely spaced doses, usually twice a day. Metoprolol ranked 23rd in prescribed drug sales in 1990. A third beta blocker, atenolol (trade name Tenormin), was 12th in prescribed drug sales in 1990.

Other heart disease drugs are the **calcium channel blockers.** Research has found that calcium ions move into the heart muscle by means of gated channels in the phospholipid membrane surrounding the heart muscle. In the muscle cells the calcium ions cause an interaction between the parallel protein filaments myosin and actin, and this interaction causes the cell to contract (Fig. 15–2). In addition, the double positive charge on the calcium ion neutralizes some of the negative charge of the muscle cell, also causing the muscle to contract. Movement of the calcium ions out of the cell restores the negative charge, and the cell relaxes. Blocking the flow of calcium ions into the cell causes the muscles to relax. When the smooth muscles in the walls of the coronary arteries are relaxed, these arteries expand and increase the supply of blood to the heart. Some calcium blockers also decrease the force of contraction, which lowers the heart work rate and thus decreases the oxygen requirements of the heart.

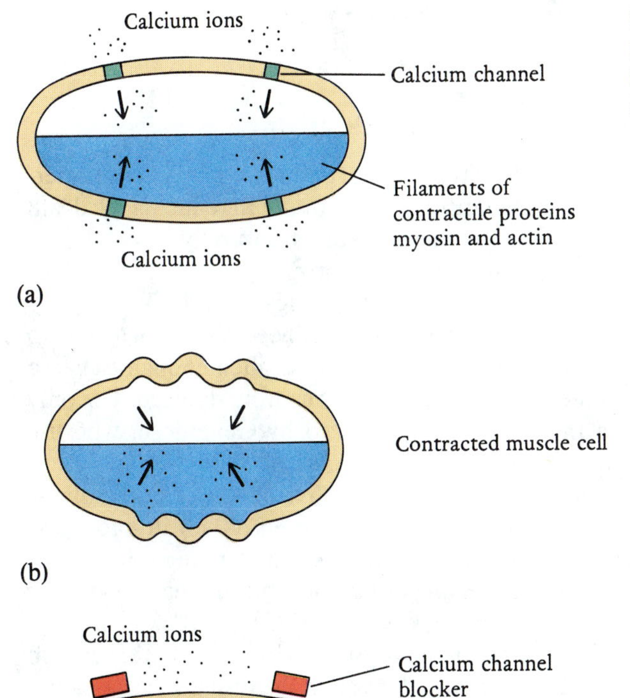

Figure 15–2 (a) Calcium ions flowing into a muscle cell. (b) Contracted muscle cell caused by the presence of calcium ions in contractile protein fiber bundles. (c) Relaxed muscle cell with calcium channels blocked by drug molecules.

Calcium blockers, unlike beta blockers, can prevent spasms of the coronary arteries. These spasms cause a blockage of blood flow to the heart and the intense pain of the angina attack. The causes of these spasms are poorly understood, but the calcium blockers do dilate the arteries and lessen the possibility of an angina attack.

After 19 years of use in Europe, verapamil (trade names Isoptin and Calan) became available in the United States and was first used to treat angina in 1981. In the same year a second calcium channel blocker, nifedipine (trade name Procardia), was introduced. Nifedipine is a powerful dilator of coronary arteries that has an immediate effect on patients suffering from angina. Although calcium blockers have fewer side effects than beta blockers, they may cause headaches, dizziness, flushing of the skin, and lightheadedness. FDA-approved uses for calcium blockers now include angina, hypertension, and arrhythmias. Diltiazem (trade name Cardizem) has fewer side effects than the other calcium blockers and as a result is the top-ranked calcium blocker—seventh among prescribed drugs in 1990 (Table 15–1).

Emergency Treatment of Heart Attack Victims

New "clot-dissolving" drugs, given to heart-attack victims in the emergency room (or even in the ambulance), show promise in reducing the death rate. **Plasmin,** which helps dissolve the blood clot naturally by catalyzing chain breaking in the blood-clotting protein fibrin, is generated from plasminogen by enzymes called **kinases.** Drug companies have developed drugs such as **urokinase, streptokinase,** and **tissue-plasminogen activator (TPA)** to treat heart attacks. Intravenous injection of TPA is often enough to halt a heart attack within minutes and to save heart tissue from damage. The only TPA now approved in this country is produced by Genentech, who use biotechnology to produce TPA from hamster ovary cells.

Controversy has erupted over both the relative costs of clot-dissolving drugs and their effectiveness. Costs per treatment are $200 for streptokinase and $2000 for TPA or urokinase. A 1991 study of 46,000 patients in Europe and the United States indicates that all clot-dissolving drugs have about the same effectiveness, reducing heart attack deaths by 25% to 30%. Aspirin has also been used to treat heart-attack patients. Studies have shown that aspirin can reduce heart-attack deaths by 23% if given at the onset of chest pain.

One problem in reducing deaths by heart attack is the delay in seeking treatment—more than 350,000 heart attack victims die every year before reaching the hospital, even though an average of 3 hours elapse between the initial appearance of symptoms and death.

Does Aspirin Cut the Risk of a Heart Attack?

As mentioned above, aspirin is used to treat heart-attack patients. Studies have shown that aspirin can also cut the risk of a heart attack by almost half even for those who have no overt signs of cardiovascular disease, apparently by working to reduce the incidence of clot formation. The results of a six-year study covering 22,000 male physicians aged 40 to 84 with no history of heart disease or stroke showed that taking a 325-mg tablet of aspirin every other day can reduce the risk of an initial heart attack by 47%. Aspirin works by blocking the manufacture of prostaglandins, which are instrumental in the formation of blood clots. There are potentially serious side effects, however. Persons susceptible to ulcers or bleeding should not take aspirin frequently.

Preventive Measures for Heart Disease

Ultimately, the most effective measures are those that depend on an individual's own initiatives. Preventive measures taken by individuals are as important as having drugs for treating heart disease. Persons who exercise, do not smoke, have a diet low in fat and sodium, and avoid consuming more than 2 ounces of 100-proof whiskey, 8 ounces of wine, or 24 ounces of beer per day are less likely to develop heart disease.

ANTICANCER DRUGS

Cancer is not one but perhaps 100 different diseases, caused by a number of factors. The sites of attack for major types of cancer are shown in Figure 15–3. A cancer begins when a cell in the body starts to multiply without restraint and produces descendants that invade tissues in the vicinity. It seems reasonable, then, that some drugs might be able either to stop this undesirable spreading of cancer cells or to prevent cancer from happening at all.

Cancers are treated by (1) surgical removal of whole areas affected by them, as well as the cancerous growths themselves, (2) irradiation to kill the cancer cells, and (3) chemicals to kill the cancer cells **(chemotherapy).** These treatment methods have resulted in dramatic improvements in the rates of survival of patients with certain cancers. A group of cancer patients can be considered cured if, after their treatment, they die at about the same rate as the general population. Another way of judging success in cancer therapy is by the number of patients who survive for five years after the treatment. Estimates of current survival rates are shown in Figure 15–4. In the 1930s less than one cancer patient in five was alive at least five years after treatment; in the 1940s it was one in four; in the 1960s it was one in three; and today it is about one in two.

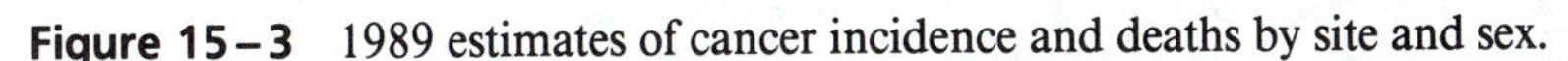

Figure 15–3 1989 estimates of cancer incidence and deaths by site and sex.

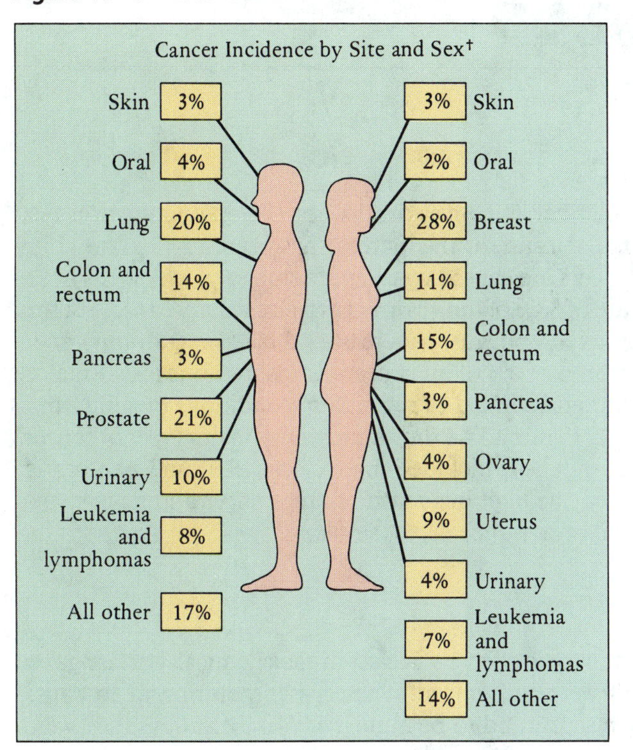

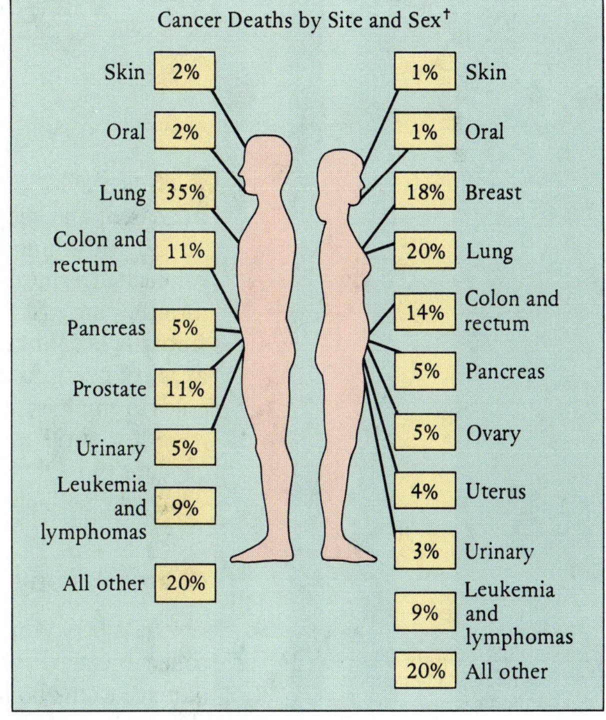

† Excluding nonmelanoma skin cancer and carcinoma in situ.

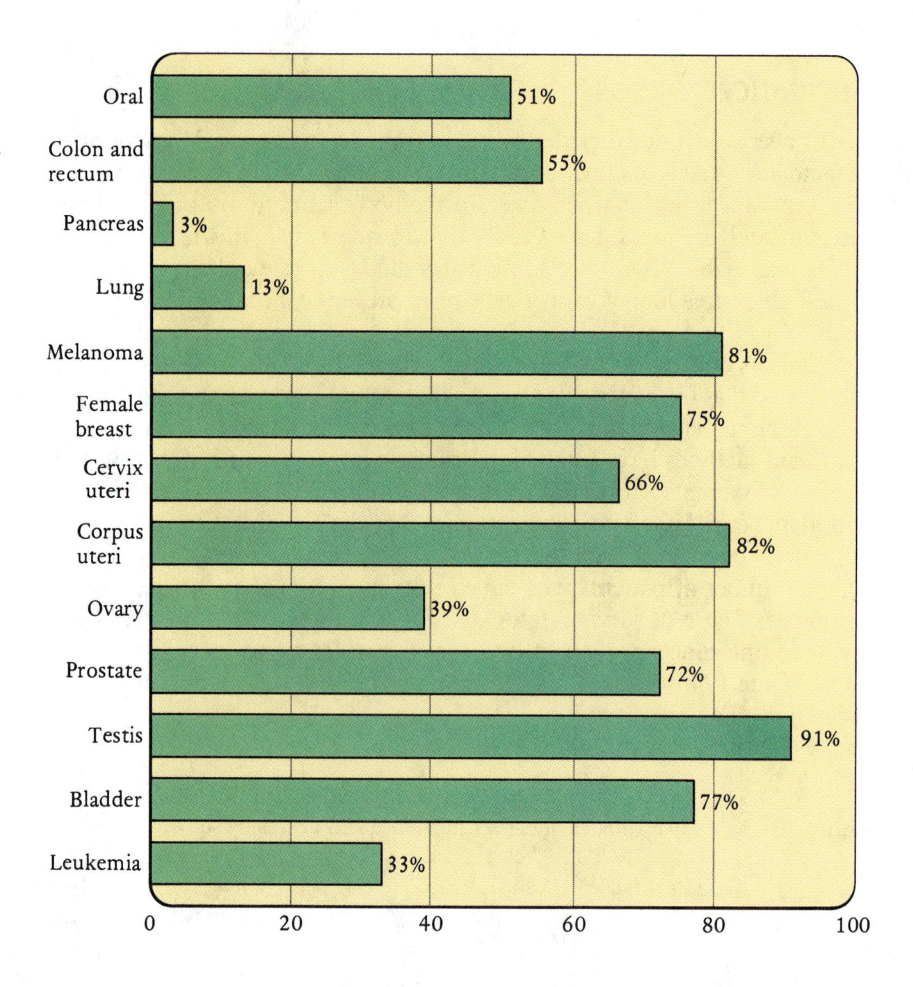

Figure 15–4 Five-year cancer survival rates for selected cancers. The rates have been adjusted for normal life expectancy, and the figure is based on cases diagnosed from 1980 to 1985. (Source: National Center for Health Statistics)

Since 1930 the number of cancer deaths per 100,000 population has increased, and the major cause of the increase has been lung cancer (Fig. 15–5). Most lung cancers are caused by cigarette smoking, which has been implicated as the cause of 83% of lung cancers as well as many cancers of the mouth, pharynx, larynx, esophagus, pancreas, and bladder. Not only does smoking account for about 30% of all cancer deaths, but it is also a major cause of heart disease and is linked to respiratory conditions ranging from colds to emphysema. As Figure 15–4 shows, only a few major types of cancer have low survival rates. These include lung cancer (13%) and pancreatic cancer (3%). Obviously, the high incidence of lung cancer and its low survival rate highlight the importance of not smoking.

Chemotherapy

The name *mustard gas* comes from its mustardlike odor; mustard gas, however, is not a gas but a high-boiling liquid that was dispersed as a mist of tiny droplets.

In World War I the toxic effects of a class of the chemical-warfare gases called mustard gases were recognized. These gases were found to cause damage to the bone marrow and to be mutagenic. In these ways they were acting like X rays, which are also toxic to cells and cause mutations.

$$Cl{-}CH_2CH_2{-}S{-}CH_2CH_2{-}Cl$$

Mustard gas

Beginning around 1935, other mustards of the nitrogen family were synthesized. They, too, caused mutations in some laboratory animals. In addition, they caused cancers in some animals.

$$R{-}N(R'{-}Cl)(R''{-}Cl)$$

Nitrogen mustard (general formula)

$$CH_3CH_2{-}N(CH_2CH_2{-}Cl)_2$$

A nitrogen mustard

After World War II the secrecy surrounding the mutagenic nature of these chemicals was lifted, and it occurred to cancer researchers that cancers might be treated with chemicals that selectively destroy unwanted cells.

Chemical warfare agents are discussed in Chapter 10. See Table 10–7.

One of the most widely used anticancer drugs is cyclophosphamide, a compound that contains the nitrogen mustard group (shown in color).

Cyclophosphamide: ring of $CH_2{-}NH$, CH_2, $CH_2{-}O$ bonded to $P(=O)$, with $P{-}N(CH_2{-}CH_2{-}Cl)_2$ (nitrogen mustard group shown in color)

Cyclophosphamide

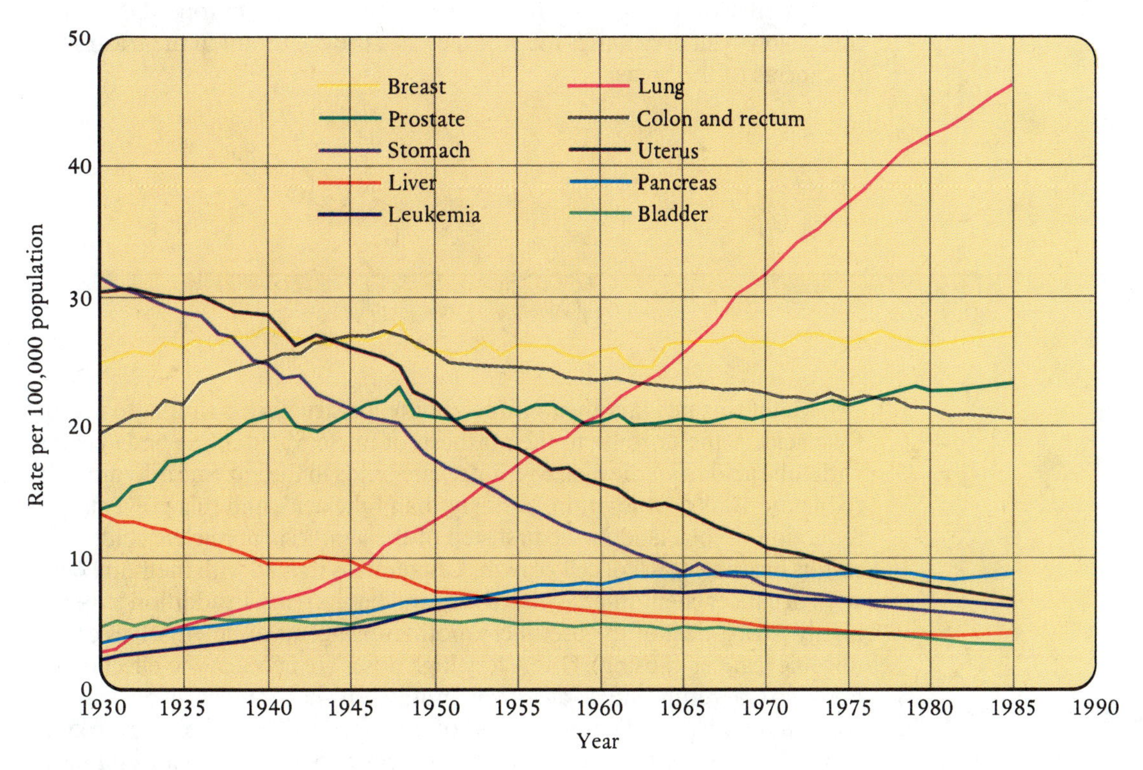

Figure 15–5 U.S. cancer death rates by site, 1930 to 1985. The rate for the population is standardized for age based on the 1970 U.S. population. Rates are for both sexes combined except for breast and uterus (female population only) and prostate (male population only). (Source: National Center for Health Statistics and Bureau of the Census, United States)

Compounds such as cyclophosphamide belong to the alkylating class of anticancer drugs. **Alkylating agents** are reactive organic compounds that transfer alkyl groups in chemical reactions. Their effectiveness as anticancer agents is due to the transfer of alkyl groups to the nitrogen bases in DNA, particularly guanine. The presence of the alkyl group in the guanine molecule blocks base pairing and prevents DNA replication, which stops cell division. Although alkylating agents attack both normal cells and cancer cells, the effect is greater for rapidly dividing cancer cells.

The discovery of anticancer properties of cisplatin was discussed in Chapter 1.

Another widely used anticancer drug, **cisplatin,** blocks DNA replication by a similar mechanism.

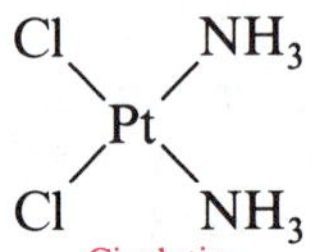

Cl NH$_3$ Pt Cl NH$_3$

Cisplatin

Inside the cell, the Cl^- ions are displaced, and the Pt(NH$_3$)$_2$ unit binds to the nitrogen sites on the guanine bases in DNA. As with alkylating agents, the blocking action prevents base pairing and DNA replication.

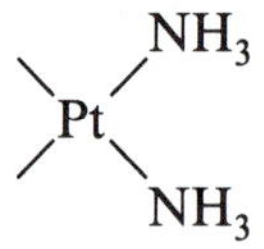

Antimetabolites, another group of chemotherapeutic agents, interfere with DNA synthesis. One of these chemicals, 5-fluorouracil, gets involved in the synthesis of a nucleotide, which inhibits the formation of a thymine-containing nucleotide necessary for DNA synthesis. Lack of proper DNA synthesis slows cell division. 5-Fluorouracil has proved useful in the treatment of cancers of the breast.

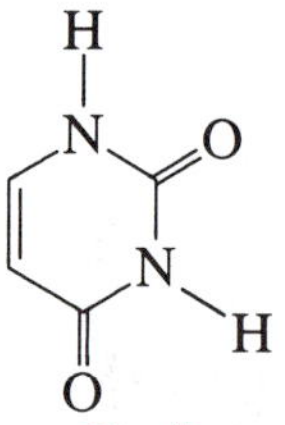

Uracil

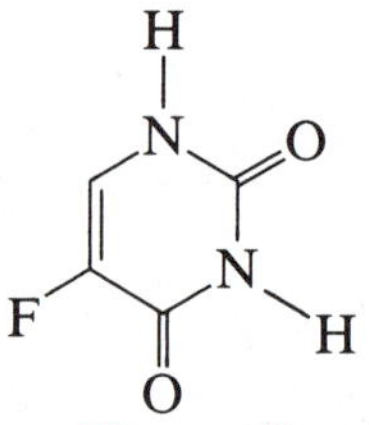

5-fluorouracil

Methotrexate is another antimetabolite. Its structure is similar to that of folic acid. A major route to the synthesis of nucleic acid in the body begins with folic acid from the diet. Methotrexate binds to the enzyme dihydrofolic reductase 10,000 times more strongly than folic acid, and this prevents the reduction of folic acid in the first step of the synthesis of nucleic acid. This results in a slowdown of cell growth. Leukemia is treated with methotrexate.

All cancer chemotherapy is tedious and has its risks. In addition to being highly toxic, most of the useful chemotherapy agents are themselves carcinogenic (cancer-causing). Often very high doses are necessary to effect treatment. As a result, single-agent chemotherapy has largely given way to combination chemotherapy because of the success of additive, or even synergistic, effects when two or more anticancer drugs are used. For example, a

Synergism **is the working together of two things to produce an effect greater than the sum of the individual effects.**

combination that is used in the treatment of several cancers is cisplatin and cyclophosphamide. Because of their synergistic action, lower doses of each compound can be used when they are given together than if they were each given alone, and this reduces the harmful side effects of chemotherapy.

Childhood cancers respond most favorably to chemotherapy. Most children with leukemia who are treated with chemotherapy drugs enter a period of relapse-free survival. In terms of the definition given earlier, they are cured. During the early 1950s about 1900 children under the age of 5 years died of cancer per year in the United States. Today the number is fewer than 700 per year. A few other cancers, such as Hodgkin's disease, have shown similar increases in survival rates due to chemotherapy. Not all cancers have been so treatable, however, because cancer is so many different diseases.

SELF-TEST 15A

1. The accepted chemical name for a drug is called its ______________ name.
2. The most widely prescribed drug in the United States is used as a(n) ______________.
3. The two ingredients of atherosclerotic plaque are ______________ and ______________.
4. When beta sites in the heart are stimulated by epinephrine, the heart beats ( ) faster or ( ) slower.
5. A drug that blocks calcium ions from flowing into heart muscles would have the effect of ( ) exciting or ( ) relaxing the heart.
6. Cholesterol is carried in the blood by both HDLs and LDLs. Which carries the greater percentage of cholesterol? ______________
7. Most anticancer drugs can also cause cancer: True ( ) False ( )
8. The nitrogen mustards act on cancer cells by blocking ______________ replication.
9. Chemicals that interfere with DNA synthesis are called ______________.
10. Which cancer has shown the greatest response to chemotherapeutic drugs? ______________
11. ______________, a "clot-dissolving" drug produced by biotechnology, is given to heart attack victims to halt heart attacks and reduce heart damage.
12. Aspirin is effective in treating heart-attack patients: True ( ) False ( )
13. The leading cause of death in the United States is ______________.

ANTIBACTERIAL DRUGS

Modern chemotherapy began with the work of German chemist Paul Ehrlich (1854–1915; 1908 Nobel Prize recipient). In 1904 he realized that infectious diseases could be conquered if toxic chemicals could be found that

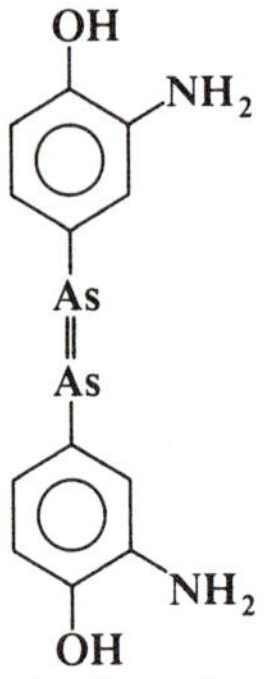

Arsphenamine

attacked parasitic organisms within the body to a greater extent than they did host cells. After observing that certain dyes used to stain bacteria for microscopic examination also killed the bacteria, he prepared arsenic analogs of dyes and tested them for treatment of African sleeping sickness and syphilis. One of these drugs, arsphenamine, was for many years the most widely used agent for the treatment of syphilis. Ehrlich also introduced the term *receptor* in 1907 and proposed concepts of receptor binding, bioactivation, the therapeutic index, and drug resistance that are still valid in principle.

After experimenting with several drugs, Gerhard Domagk, a pathologist in the I. G. Farbenindustrie Laboratories in Germany, found in 1935 that Prontosil, a dye, was effective against bacterial infection in mice. Actually, Domagk discovered that Prontosil did not have antibacterial activity in vitro (outside the living body) but is metabolized in vivo (inside the body) to **sulfanilamide,** the actual antibacterial agent.

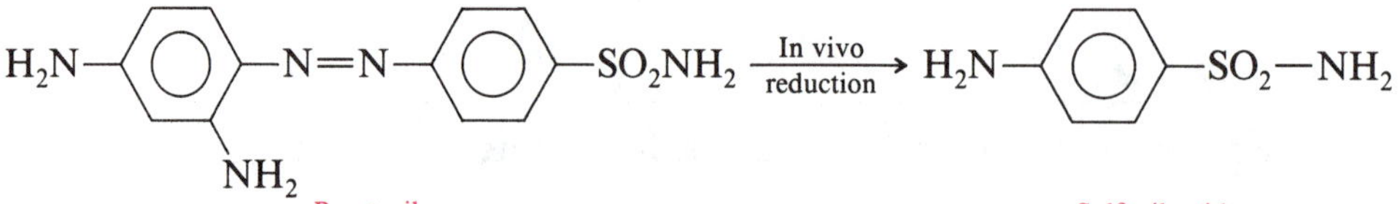

Prontosil —In vivo reduction→ Sulfanilamide

Sulfa Drugs

Sulfanilamide destroys bacteria that cause pneumonia, diphtheria, and gonorrhea.

Many drugs in use today are not active themselves but are metabolized to the active agent. Drugs of this type are often referred to as precursors.

Sulfanilamide, the first of a class of drugs referred to as sulfa drugs, is effective against streptococci, staphylococci, pneumococci, gonococci, meningococci, and dysentery bacteria. However, sulfanilamide has harmful side effects and, when used in high doses, can cause kidney damage. The success of sulfanilamide led to the synthesis and testing of over 5000 derivatives in an attempt to find more effective drugs with fewer side effects. All of these compounds are referred to as sulfonamides, or **sulfa drugs,** and have the following general formula:

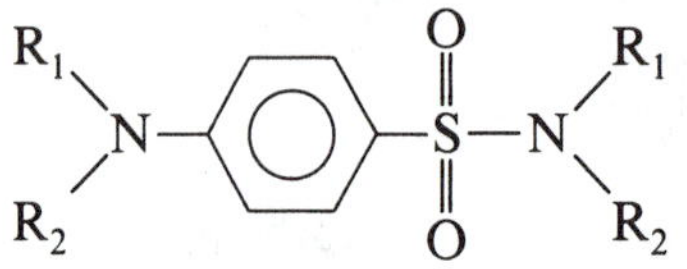

(R_1 and R_2 may be hydrogen or an organic group.)

This attempt to find better drugs by synthesizing and testing derivatives of a proven drug provides an excellent example of the strategy often used by the pharmaceutical industry in the search for better drugs. Structure-activity correlations based on the testing results of sulfonamides are shown in Figure 15-6. Only about 30 of the more than 5000 known sulfonamides actually reached the point of being used as drugs.

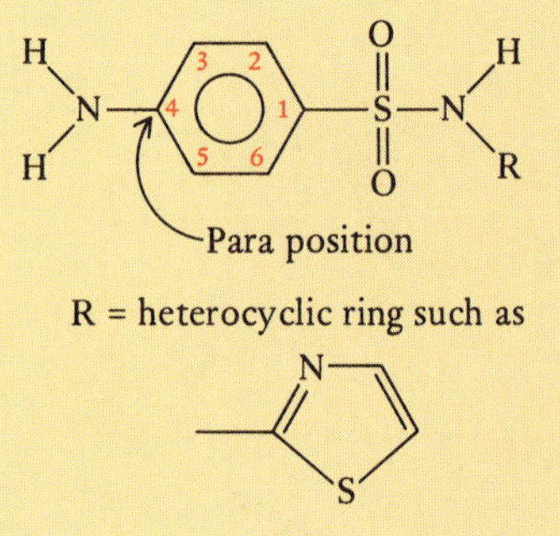

Para position

R = heterocyclic ring such as

Figure 15-6 5000 compounds showed the highest activity for compounds that fit the following structural limitations: (1) the amino groups must be in the *para* position to the sulfonamide group and be unsubstituted; (2) the benzene ring can be substituted in positions 1 and 4 only; (3) the sulfonamide amine nitrogen can be monosubstituted only and heteroaromatic substituents increase the activity.

Sulfa drugs inhibit bacteria by preventing the synthesis of folic acid, a vitamin essential to their growth. The drugs' ability to do this lies in their structural similarity to *para(p)*-aminobenzoic acid, a key ingredient in the folic acid synthesis in bacteria.

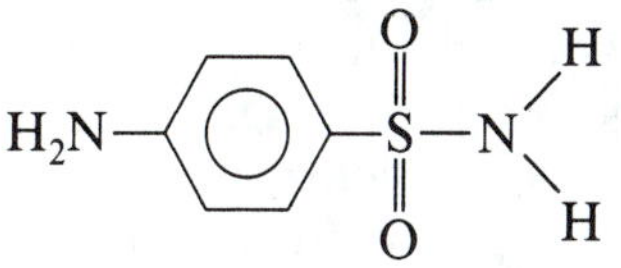

Sulfanilamide
(a typical sulfa drug)

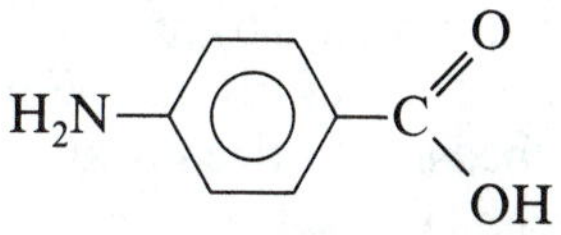

p-aminobenzoic acid

The close structural similarity of sulfanilamide and *p*-aminobenzoic acid permits sulfanilamide instead of *p*-aminobenzoic acid to be incorporated into the enzymatic reaction sequence. By bonding tightly, sulfanilamide shuts off the production of the essential folic acid, and the bacteria die of vitamin deficiency. Because humans and other higher animals obtain folic acid from their diet, *p*-aminobenzoic acid is not necessary for folic acid synthesis, and sulfa drugs do not interfere with normal cell growth.

Recall that methotrexate is an effective anticancer agent because it takes the place of folic acid on an enzyme that promotes cell growth.

Sulfa drugs were the miracle drugs during the late 1930s and early 1940s. During World War II thousands of lives were saved by the effective use of sulfa drugs to prevent infection in wounds. However, in the 1940s another group of drugs, the antibiotics, were found to be more effective antibacterial drugs.

Antibiotics

An antibiotic is a substance produced by a microorganism which inhibits the growth of other organisms. Its job generally is to aid the white blood cells by stopping bacteria from multiplying. When a person falls victim to or is killed by a bacterial disease, it means that the invading bacteria multiplied faster than the white blood cells could devour them and that the bacterial toxins increased more rapidly than the **antibodies** could neutralize them. The action of the white blood cells and antibodies plus an antibiotic is generally enough to repulse an attack of disease germs. **Penicillins** were the first class of antibiotics to be discovered.

An antibody is a specific protein produced to protect the organism from harmful invading molecules.

The Penicillins

Penicillin was discovered in 1928 by Alexander Fleming, a bacteriologist at the University of London, who was working with cultures of *Staphylococcus aureus,* a germ that causes boils and some other types of infections. In order to examine cultures with a microscope, Fleming had to remove the covers of the culture plates for a while. One day as he started work he noticed that one culture was contaminated by a blue-green mold. For some distance around the mold growth, the bacterial colonies were being destroyed. Upon further investigation, Fleming found that the broth in which this mold had grown also had an inhibitory or lethal effect on many pathogenic (disease-causing) bacteria. The mold was later identified as *Penicillium notatum* (the spores sprout and branch out in pencil shapes; hence the name). Although

Another *Penicillium* strain that proved to be an excellent source of the new antibiotic was discovered on a moldy cantaloupe in a Peoria, Illinois, market.

Fleming showed that the mold contained an active antibacterial agent which he called **penicillin,** he was not able to purify the active substance. In 1940 Howard Florey and Ernst Chain of Oxford University succeeded in isolating a product from the mold, called **penicillin G,** and by 1943 penicillin G was available for clinical use. By the end of World War II, penicillin G was saving many lives threatened by pneumonia, bone infections, gonorrhea, gangrene, and other infectious diseases. For their work, Fleming, Florey, and Chain received the Nobel Prize for medicine and physiology in 1945.

Penicillin G had to be injected because it was destroyed by stomach acid when taken orally. Bacterial strains also became resistant to penicillin G. This led to the testing of other penicillins such as amoxicillin, which was the number one prescribed drug in 1990. It can be taken orally and is effective against a wide variety of bacteria.

All penicillins kill growing bacteria by preventing them from making their normal cell walls. Because animal cells do not have cell walls, bacteria can be destroyed without damaging animal cells.

Tetracyclines

Antibiotics other than penicillin work in a variety of ways. Many interfere with the making or functioning of DNA in bacteria. The tetracyclines —streptomycin and erythromycin—prevent bacteria from making proteins from their DNA.

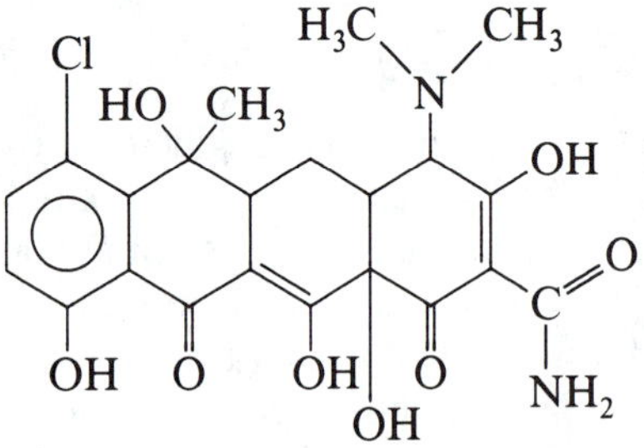

Cl, HO, CH_3, H_3C, CH_3, N, OH, C, O, NH_2, OH, O, OH, O, OH

Chlortetracycline
(aureomycin)

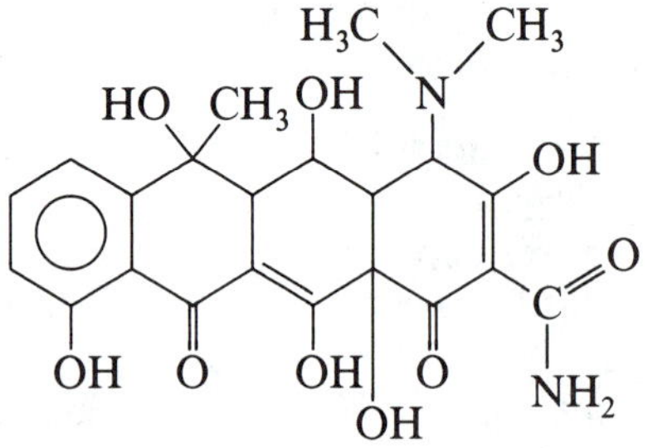

HO, CH_3, OH, H_3C, CH_3, N, OH, C, O, NH_2, OH, O, OH, O, OH

Oxytetracycline
(terramycin)

Cephalosporins

Cephalosporins, another class of antibiotics, were discovered in the 1950s. They are widely used because of their broad antimicrobial activity and low risk of serious adverse effects. In 1990 cefaclor (trade name Ceclor) was the top-ranked cephalosporin in sales, ranking ninth in total prescriptions in the United States.

Problems with Antibiotics

Although antibiotics have saved millions of lives by controlling infectious diseases, they also have some disadvantages. Repeated use of broad-spectrum antibiotics such as tetracyclines kills beneficial and harmless bacteria in the digestive tract and causes diarrhea. Some people are allergic to

THE WORLD OF CHEMISTRY

Folk Medicine

There's a long history of folk medicine based on plants, especially in the Near and Far East. Now the question is: which of these things really works and how much of it is just a rumor? First the chemist checks out the plant rumored to be good and analyzes its extracts to see if there is any activity. One such substance is called fredericamycin. Fredericamycin is an antibiotic of some interest as a possible antitumor compound. It comes from a soil organism, a bacterium, found in the soil in Frederick, Maryland.

Dr. Kathlyn Parker, of Brown University, and other researchers like her, first analyze the naturally occurring substance in the lab. If they find an active component, they look further. As Dr. Parker says:

Then, if you are really interested in drug development, you have to isolate the active component, purify it, determine its structure, and then it gets handed over to the pharmaceutical people who decide how to package it. For some pharmaceuticals, what you really need is to be able to make a large amount of stuff really cheap. One solution is that you would develop methods so that it was so cheap to make something that you could distribute it to people in a way that they could afford it.

The World of Chemistry (Program 21) "Carbon."

Kathlyn Parker (*The World of Chemistry*, Program 21, "Carbon")

antibiotics, especially penicillin, and suffer allergic responses occasionally severe enough to cause death. Hypersensitivity to penicillin is the most common side effect. Adverse reactions to penicillin may occur immediately or within 20 minutes. More delayed reactions are not as severe, and the risk of hypersensitivity reactions increase with prolonged therapy.

Another problem is the development of genetic resistance. As antibiotics are used more and more for both people and livestock, resistant bacteria result. Strains of malaria, typhoid fever, gonorrhea, and tuberculosis have emerged that are resistant to various penicillins, streptomycin, and a number of tetracyclines. Certain bacteria can gain the ability to produce enzymes that inactivate the antibiotic. Other bacteria can become resistant to antibiotics by preventing the antibacterial agent from entering the bacterial cell.

ALLERGIES AND ANTIHISTAMINES

About one person in ten suffers from some form of allergy; more than 16 million Americans suffer from hay fever. When persons are exposed to airborne substances to which they are allergic, special cells in their nose and breathing passages release histamine. Histamine accounts for most of the symptoms of hay fever, bronchial asthma, and other allergies.

Histamine causes runny noses, red eyes, and other hay fever symptoms.

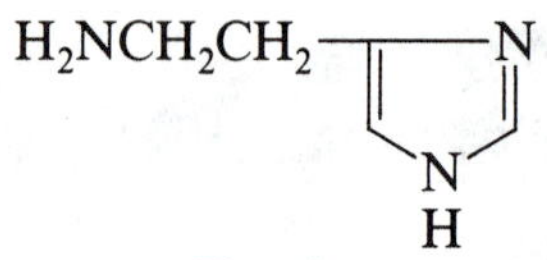

$H_2NCH_2CH_2$–(imidazole ring, N, N–H)

Histamine

Epinephrine (adrenalin), steroids, and antihistamines are effective drugs in treating allergies. The first two are particularly effective in treating bronchial asthma; the **antihistamines,** introduced commercially in the United States in 1945, are the most widely used drugs for treating allergies. More than 50 antihistamines are offered commercially in the United States, and one of these, terfenadine, ranked tenth in prescribed drug sales in 1990.

HORMONES

Hormones are substances produced by glands that serve as chemical messengers to regulate biological processes. They are chemically diverse but are primarily proteins or steroids. Hormones are synthesized by specific glands—hypothalamus, pituitary, thymus, thyroid, parathyroid, pancreas, adrenals, and gonads—and are secreted directly into the blood.

Insulin

Review the discussion of diabetes in Chapter 14.

Diabetes mellitus, a disease that affects several million people, arises from a deficiency of the hormone insulin. Insulin is a polypeptide hormone that promotes the entry of glucose, some other sugars, and amino acids into muscle and fat cells. Diabetes is characterized by an elevated level of glucose in the blood and in the urine because the deficiency of insulin prevents sufficient transfer of glucose to muscle and fat cells.

Steroid Hormones

The structure of steroids and the importance of cholesterol was discussed in Chapter 9. The sex hormones are structurally related to cholesterol. One female sex hormone, **progesterone,** differs only slightly in structure from an important male hormone, **testosterone.**

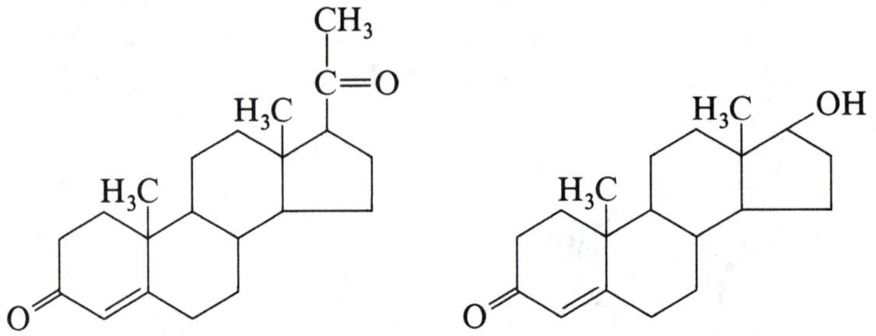

CH_3, C=O, H_3C, H_3C, O

Progesterone

H_3C, OH, H_3C, O

Testosterone

Other female hormones are estradiol and estrone, together called **estrogens.** The estrogens differ from the steroids discussed earlier in that they contain an aromatic ring, shown by the circle inside the first six-membered ring.

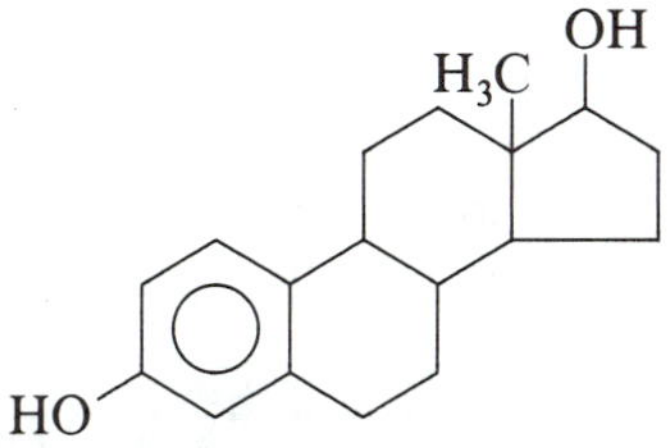

Estradiol

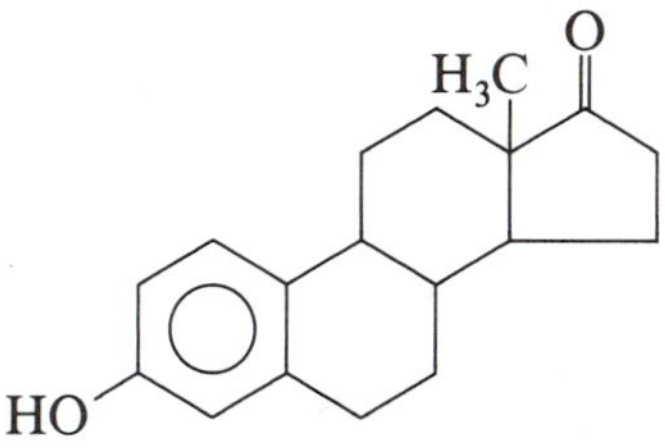

Estrone

The estrogens and progesterone are produced by the ovaries. Estrogens are important to the development of the egg in the ovary, whereas progesterone causes changes in the wall of the uterus and after pregnancy prevents release of a new egg from the ovary (ovulation). Birth control drugs use derivatives of estrogens and progesterone to simulate the hormonal processes resulting from pregnancy and thereby prevent ovulation.

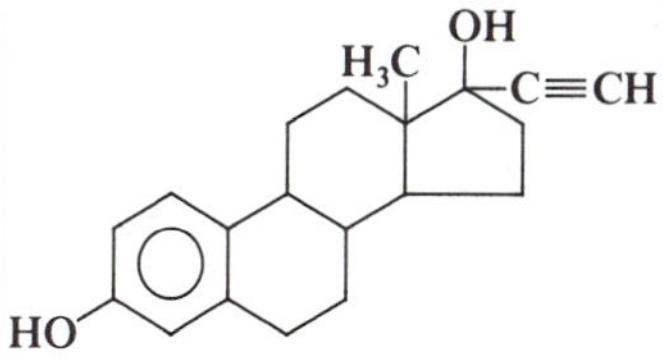

Ethynyl estradiol

Birth Control Pills

One of the most revolutionary medical developments of the 1950s was the worldwide introduction and use of "the pill." Now there are two types of oral contraceptives, "the pill" and "the minipill."

The pill contains small amounts of synthetic analogs of estrogens and progesterone. The common ones are mestranol, a synthetic estrogen derivative, and norethindrone, a synthetic progesterone derivative. The estrogen derivative regulates the menstrual cycle, and the progesterone derivative establishes a state of false pregnancy resulting in the prevention of ovulation.

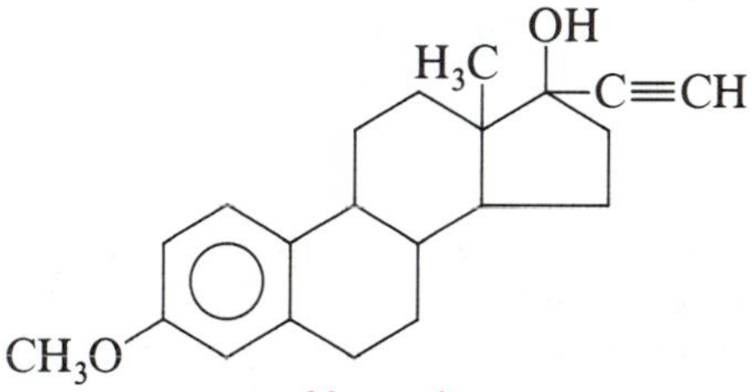

Mestranol
(synthetic estrogen)

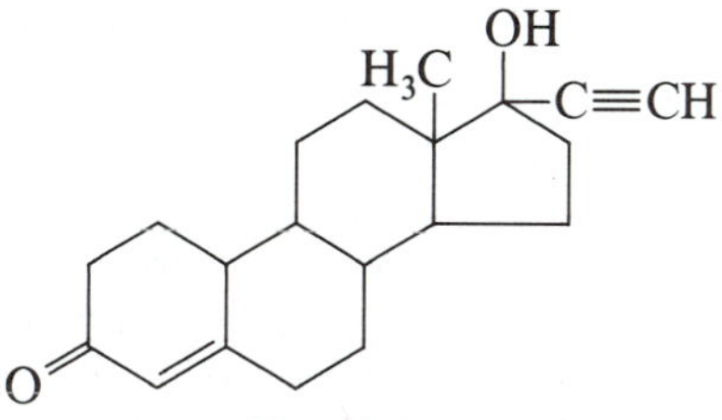

Norethindrone
(synthetic progesterone)

Early versions of the pill contained larger doses of the estrogen and progesterone derivatives. However, since the 1960s, researchers have succeeded in reducing the steroid content of the pill from about 200 mg to as little as 2.6 mg. In addition, "biphasic" and "triphasic" products were developed to reflect more accurately a woman's natural hormonal changes. These products are formulated to provide the lowest effective dosages of both progesterone and estrogen on different days of the menstrual cycle. An example of a biphasic oral contraceptive is Ortho-Novum 10/11, which provides 0.5 mg of norethindrone for 10 days, then 1.0 mg of norethindrone for 11 days; 35 μg of ethynyl estradiol is administered consistently for 21 days. Triphasic oral contraceptives change the amount of norethindrone three times instead of twice.

1 mg = 1×10^{-3} g
1 μg = 1×10^{-6} g or 1×10^{-3} mg

All oral contraceptives are prescription drugs and should be taken only after a checkup by a doctor. There are risks associated with the use of the pill. The risk of blood clots, the most common of the serious side effects, increases with age and with heavy smoking (more than 15 cigarettes per day). The chance of a fatal heart attack is about 1 in 10,000 in women between the ages of 30 and 39 who use oral contraceptives and smoke, compared with about 1 in 50,000 in users who do not smoke and 1 in 100,000 in nonusers who do not smoke.

Women are warned not to use oral contraceptives if they have or have had a heart attack or stroke, blood clots in the legs or lungs, angina pectoris, known or suspected cancer of the breast or sex organs, or unusual vaginal bleeding. In addition, they should not use the pill if they are pregnant or suspect they are pregnant.

The "minipill," developed in the 1970s, contains much smaller amounts of the synthetic progesterones (0.1 – 0.2 mg) and no synthetic estrogen. The minipill was introduced in the hope that its users would experience fewer side effects than users of the pill because the estrogen component of the pill is regarded as the cause of many of the serious side effects. However, there is not sufficient information available to support this concept. Although the mechanism of minipills is not fully understood, these contraceptives are thought to stop conception by preventing release of the egg, by keeping the sperm from reaching the egg, and by making the uterus unreceptive to any fertilized egg that reaches it.

A new birth control option for women is an implant, called Norplant, which prevents pregnancy for up to five years. The method involves implanting in a woman's upper arm six silicone rubber capsules, each the size of a 1 ½-inch matchstick. The capsules contain levonorgestrel, which has been used in birth control pills for many years. The porosity of the silicone rubber allows slow release of the hormone over a five-year period.

Levonorgestrel is the levo isomer of norgestrel, which has a structure very similar to that of norethindrone.

Steroid Drugs in Sports

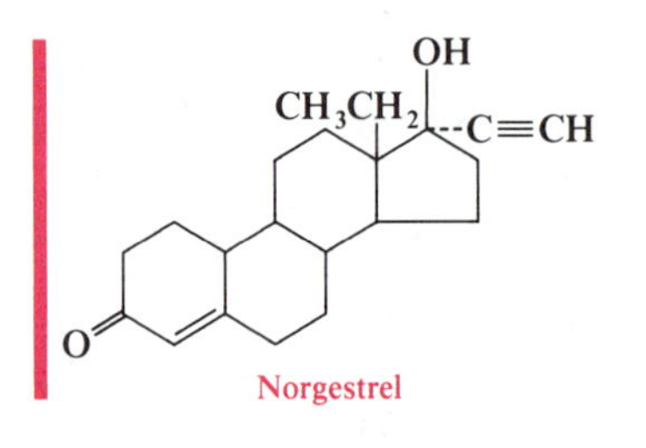

Norgestrel

The steroid testosterone is responsible for the muscle building that boys experience at puberty, in addition to the development of adult male sexual characteristics. Synthetic steroids have been developed in part to separate the masculinizing (androgenic) effects and muscle-building (anabolic) effects of testosterone. These steroids have been prescribed by physicians to correct hormonal imbalances or to prevent the withering of muscle in persons who are recovering from surgery or starvation.

Healthy athletes discovered that synthetic steroids appeared to have an anabolic effect on them as well. Initially these anabolic steroids were used by weight lifters and by athletes in track-and-field events like the shot-put and hammer throw. Later, some inconclusive evidence surfaced suggesting that anabolic steroids increased endurance, and this caused weight lifters, runners, swimmers, and cyclists to begin using them.

Such sports organizations as the National Collegiate Athletic Association and the International Olympic Committee have banned the use of anabolic steroids and other drugs by athletes. Although few human studies have been carried out on the use of anabolic steroids by healthy individuals, a number of harmful side effects have been identified.

The side effects of anabolic steroid use include acne, baldness, and changes in sexual desire. Some men experience enlargement of the breasts. Accompanying these noticeable changes are testicular atrophy and decreased sperm production. This is caused by an imbalance among the testes, pituitary, and hypothalamus due to the increased concentration of these male sex hormones in the bloodstream. High levels of male sex hormones cause the hypothalamus to signal the pituitary gland to lower production of two other hormones, luteinizing hormone and follicle-stimulating hormone, which stimulate sperm production in the testes. Although these changes appear to be reversible, additional testing is needed. In women the use of anabolic steroids produces facial hair, male-pattern baldness, deepening of the voice, and changes in the menstrual cycle. Most of these changes are not reversible.

In addition to these problems, oral-dose anabolic steroids are toxic to the liver. Testosterone taken orally is not very effective because most of it is rapidly metabolized by the liver before it reaches the bloodstream. However, several of the common anabolic steroids are active when taken orally, in part because of an alkyl group in addition to the hydroxyl group at the carbon-17 position of the steroid nucleus. This alkyl structure slows metabolism in the liver and thus allows more of the dose to reach the bloodstream, but it also increases liver toxicity. Some liver cancer has been reported in anabolic steroid users.

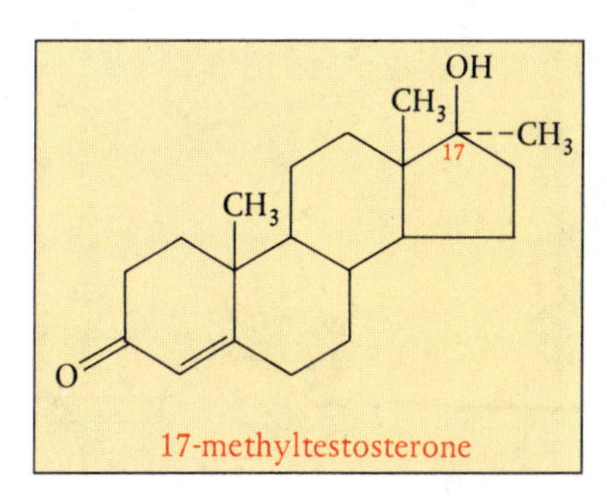

17-methyltestosterone

A small percentage of athletes continue to use anabolic steroids in spite of the publicity about harmful side effects and the ban on their use. David L. Black, director of the Athletic Drug Testing Laboratory at Vanderbilt University, reports that 1% to 2% of athletes in international competitions now test positive for steroids; in college athletics, 5% test positive; and in professional football, 7% to 8% test positive. Although there is no firm estimate of the number of people who take anabolic steroids, the percentages above together with the results of a 1988 survey of teenaged boys that indicated 7% used anabolic steroids suggest at least a million users in the United States, and the concern is that the demand is growing.

SELF-TEST 15B

1. Sulfa drugs inhibit bacteria by preventing the synthesis of ____________, a vitamin essential to their growth.
2. Penicillins kill bacteria by preventing them from making ____________, which are not found in animal cells.
3. ____________ is the common allergy mediator that accounts for most of the symptoms of allergies.
4. The male sex hormone is called ____________.
5. Two female sex hormones are ____________ and ____________.
6. Anabolic as used in the term *anabolic steroid* means ____________.
7. Three undesirable side effects of anabolic steroids are ____________, ____________, and ____________.
8. Diabetes mellitus arises from a deficiency in the hormone ____________.

BRAIN CHEMISTRY

Everything we do is controlled by nerve signals racing between our brain and various parts of our nervous system. Estimates of the number of nerve cells, or **neurons,** in the human brain range from 10 billion (10^{10}) to 1 trillion (10^{12}). The brain along with the spinal cord make up the central nervous system (CNS) (Fig. 10–10). Other parts of the nervous system are the peripheral nervous system and the autonomic nervous system. The brain receives, processes, and acts upon information originating within the central nervous system or brought to it by the peripheral and autonomic nervous systems. The brain controls both our voluntary actions such as walking, talking, eating, and the involuntary functions of our body such as regulation of heartbeat, gland secretions, and the smooth-muscle action of blood vessels. Research in brain chemistry has made rapid progress in the last decade, and most of the material discussed here was not known 30 years ago. However, much is still unknown about the brain because of the complexity and variety of actions controlled by the brain.

The brain functions through its billions of neurons and millions of neuron networks. Neurons are like tiny circuits that pass electrical impulses (messages) along a network of other neurons that result in a specific action (Fig. 15–7). The electrical impulse is transmitted in less than a thousandth of a second along the axon of one neuron across a small gap (synapse) to a dendrite on an adjacent neuron. The electrical impulse is based on changes in the concentration of Na^+ and K^+ inside and outside the neuron. A typical neuron may have 10,000 or more dendrites, and each dendrite can receive the input of numerous neurons.

The importance of the K/Na ratio was discussed in Chapter 14. The K/Na ratio for normal brain cells is 1.7.

The important thing to keep in mind is that these neuron networks are not physically connected. Chemicals known as **neurotransmitters** are released, which cross the synapse and bind to the dendrites of the adjacent neuron and trigger the flow of Na^+ and K^+ that allows the elecrical impulse to travel to the adjacent neuron.

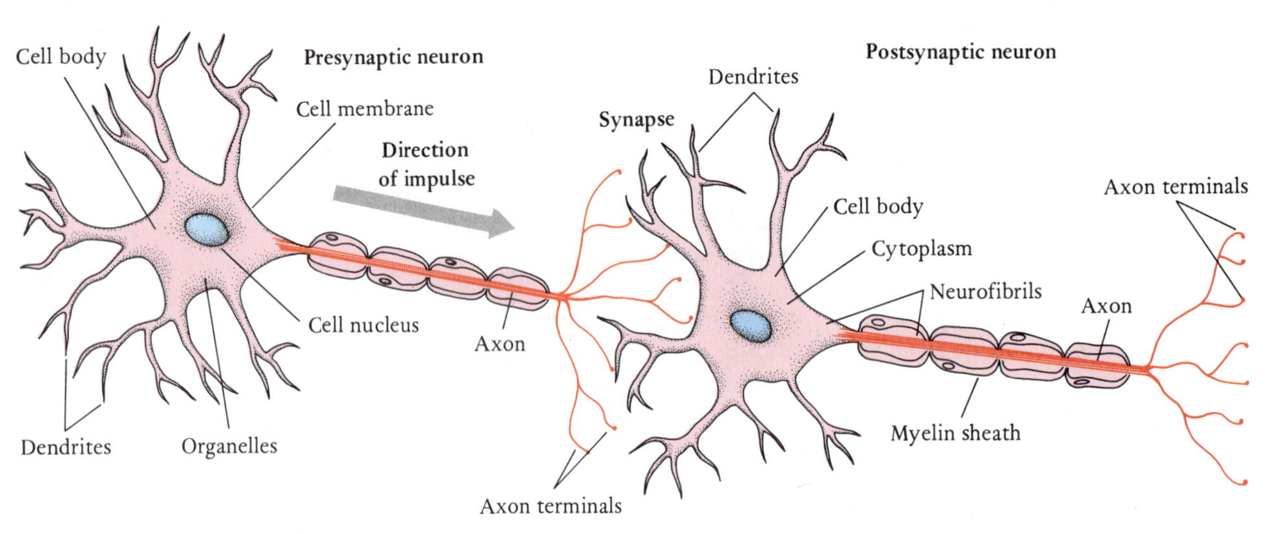

Figure 15–7 Two associated neurons. An electrical impulse is transmitted to another neuron across the synapse by a neurotransmitter. Neurotransmitter synthesis occurs in the axon terminals.

Neurotransmitters

Different neurons use different transmitters. Some of the principal neurotransmitters are shown in Figure 15–8. Some excite the receiving neuron to send the electrical impulse on to another neuron, and others halt the electrical impulse.

Most drugs that affect either the brain or the nervous system interact with the neurotransmitters or with their binding sites. The emphasis here is on those neurotransmitters that are affected by drugs or that have been identified with certain diseases such as Parkinson's disease.

Acetylcholine, an important neurotransmitter, was discussed in Chapter 10 in connection with neurotoxins. Several research groups have found that patients with Alzheimer's disease have significantly lower concentrations of the enzyme choline acetyltransferase in their brains. This enzyme is needed in the synthesis of acetylcholine. However, attempts to treat the disease by increasing the concentration of acetylcholine in the brain have not been successful.

Norepinephrine is found in several different parts of the brain. These include (a) the cerebellum and cerebral cortex, which affect the fine coordi-

Figure 15–8 Some of the known neurotransmitters.

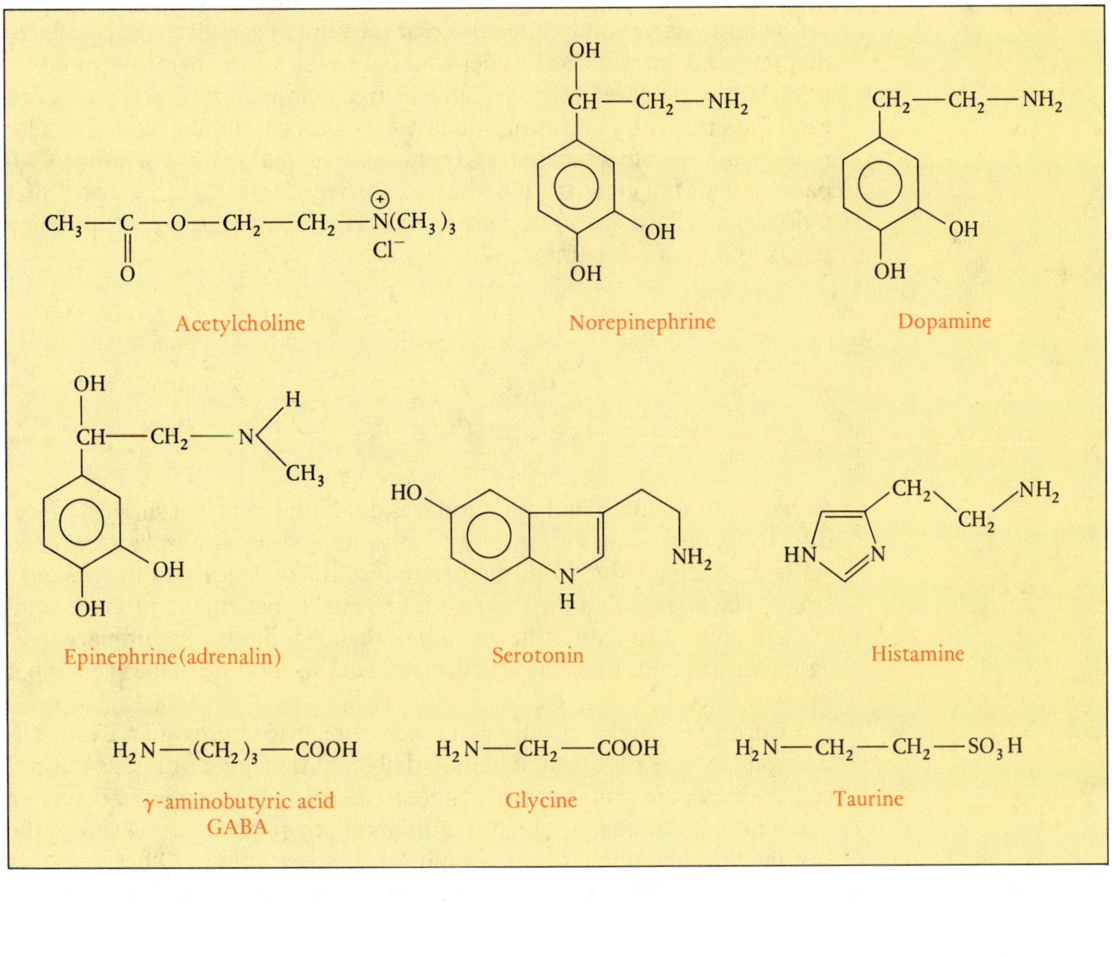

$CH_3-C(=O)-O-CH_2-CH_2-\overset{\oplus}{N}(CH_3)_3\ Cl^-$

Acetylcholine

Norepinephrine

Dopamine

Epinephrine (adrenalin)

Serotonin

Histamine

$H_2N-(CH_2)_3-COOH$

γ-aminobutyric acid
GABA

H_2N-CH_2-COOH

Glycine

$H_2N-CH_2-CH_2-SO_3H$

Taurine

nation of body movement and balance, alertness, and emotion; and (b) the hypothalamus, which controls hunger, thirst, temperature regulation, blood pressure, reproduction, and behavior. In terms of the latter, norepinephrine affects mood, dreaming, and the sense of satisfaction. Mood swings are caused by changes in levels of norepinephrine. An excess causes elation and a deficiency causes depression. The difference between normal persons and manic-depressive mental patients can be measured in the excess or deficiency and the degree of difference between the two extremes. Because most drugs are able to control only one problem—either excess or deficiency—there are limitations to the use of drugs in the treatment of this condition. Lithium salts seem to have the ability to control both mania and depression and to control the mood swings. Lithium carbonate has been used for 30 years in the treatment of manic-depressives. The reason for its action is likely tied to the electrical balance because it seems to regulate both an excess and a deficiency of norepinephrine.

Epinephrine, or adrenalin, is both a neurotransmitter and a hormone released by the medulla of the adrenal gland. This is the reason that a sudden discharge of adrenalin from the adrenal gland produces effects similar to those caused by stimulation of the autonomic nervous system. These include increased blood pressure, dilation of blood vessels, widening of the pupils, and erection of the hair.

Dopamine is found in several areas of the brain and is involved with the action and integration of fine muscular movement as well as the control of memory and emotion. An understanding of the brain chemistry of dopamine led to the development of an effective treatment for Parkinson's disease. Patients with Parkinson's disease experience trembling and muscular rigidity, among other symptoms, because of a deficiency of dopamine. Dopamine does not cross the blood-brain barrier, but research indicated that L-dopa is an effective drug because it crosses the blood-brain barrier and then reacts to produce dopamine.

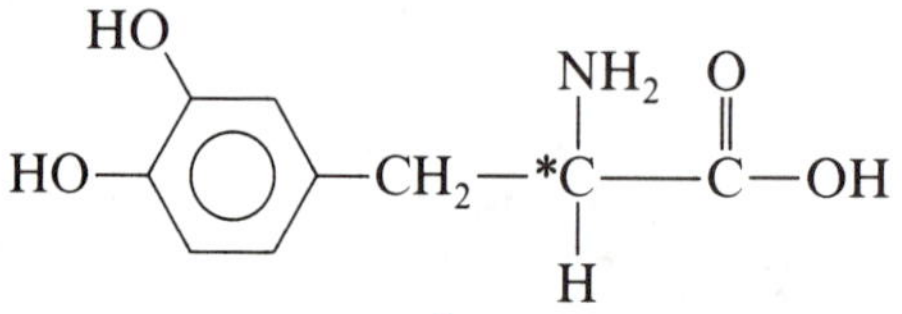

$$(HO)_2C_6H_3{-}CH_2{-}{}^{*}CH(NH_2){-}C(=O){-}OH$$

(*Asymmetric carbon)

L-dopa

Because the adrenal gland also produces dopamine, a recent surgical procedure for patients with Parkinson's disease has involved transplanting dopamine-producing tissue from the patient's adrenal glands into the patient's brain. The hope is that the transplant will enable patients to produce sufficient dopamine to reduce the severity of their symptoms. Preliminary indications are that the procedure is effective, but the duration of the effect is not known.

Current research indicates that tissue from dead fetuses offers extraordinary possibilities for the treatment of diabetes, Alzheimer's disease, Parkinson's disease, and Huntington's chorea. Scientists have reported reversing the effects of Parkinson's disease in monkeys by implanting cells from the substantia nigra area of the midbrain of monkey fetuses. Other scientists have observed a reduction in chemically induced symptoms of Huntington's chorea, a fatal genetic brain disorder, in rats implanted with fetal nerve tissue. A moratorium has been placed on the use of fetal tissue in any federally funded research projects while the ethical dilemmas of its use are discussed.

Schizophrenia has been related to an excess of dopamine, and drugs that block dopamine receptor sites are used to treat this condition.

Serotonin controls sensory perception, the onset of sleep, and body temperature and may affect mood.

Drugs and the Brain

Drugs that affect the brain or nervous system generally work by inhibiting or promoting the production of neurotransmitters or by blocking the receptor sites occupied by neurotransmitters. Analgesics are painkillers. They include both narcotic and non-narcotic drugs.

Strong Analgesics

The **opiates** are analgesics with a strong narcotic action, producing sedation and even loss of consciousness. Opium, obtained from the unripened seed pods of opium poppies, contains at least 20 different compounds called **alkaloids** (organic nitrogenous bases described in Chapter 7). Some of the more important opiates are given in Figure 15–9. About 10% of crude opium is **morphine,** which is primarily responsible for the effects of opium.

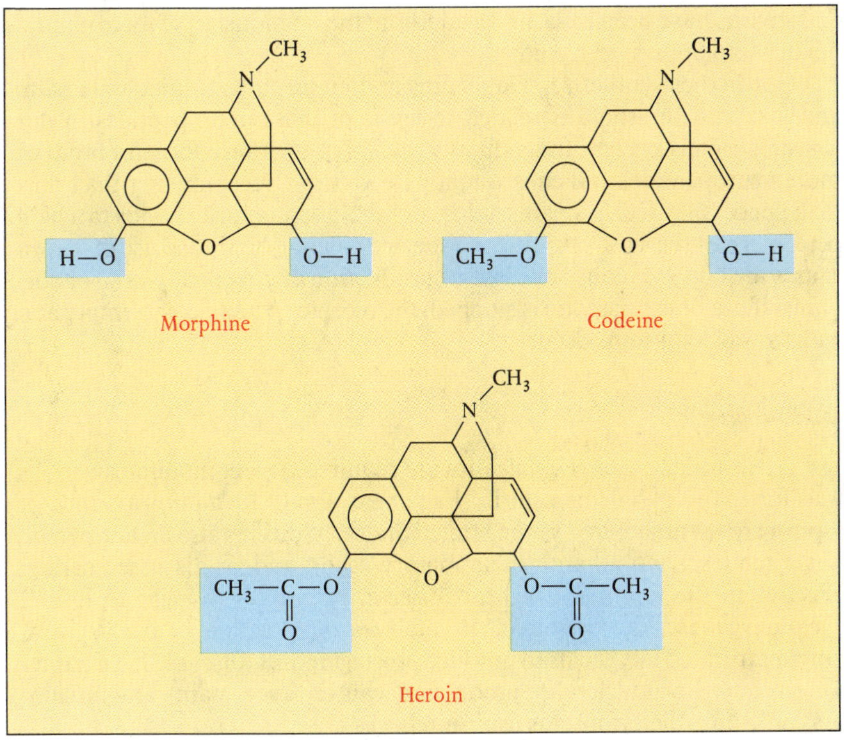

CH_3
N
H—O
O
O—H
Morphine

CH_3
N
CH_3—O
O
O—H
Codeine

CH_3
N
CH_3—C(=O)—O
O
O—C(=O)—CH_3
Heroin

Figure 15–9 Some of the more important opiates.

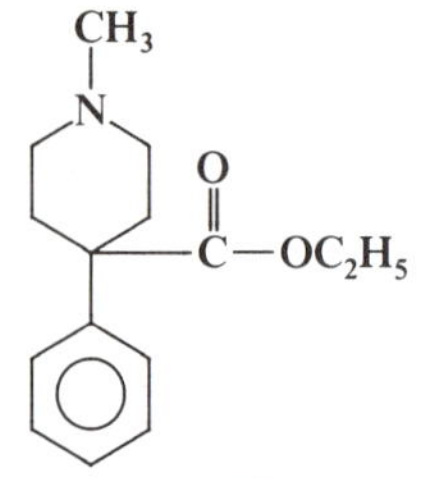

Meperidine (Demerol)

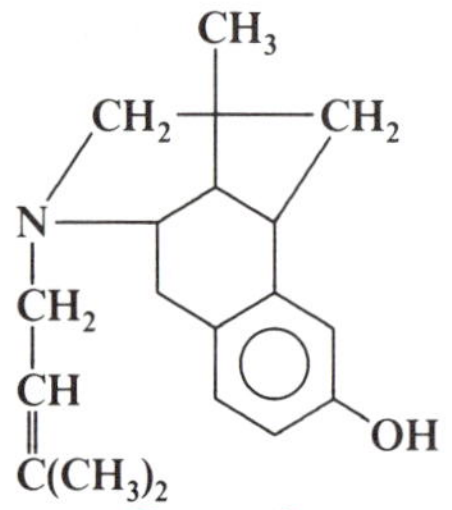

Pentazocine (Talwin)

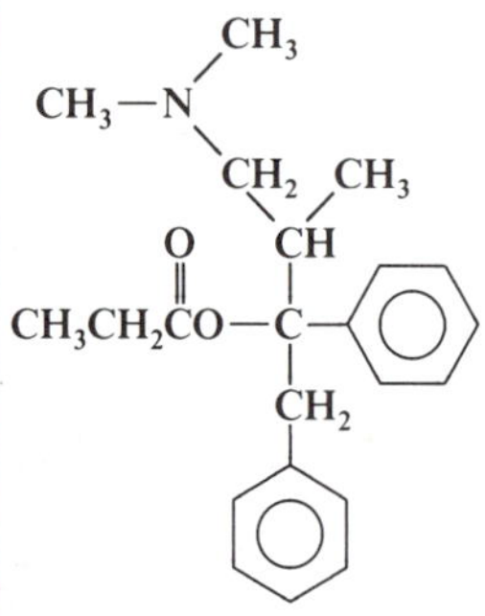

Propoxyphene (Darvon)

Two derivatives of morphine are of interest. **Codeine,** a methyl ether of morphine, is one of the alkaloids found in opium and is used in cough syrup and for relief of moderate pain. Codeine is less addictive than morphine, but its analgesic activity is only about one-fifth that of morphine. Another derivative of morphine is **heroin,** the diacetate ester of morphine, which does not occur in nature but can be synthesized from morphine. Heroin is much more addictive than morphine and for that reason has no legal use in the United States.

One of the most effective substitutes for morphine is **meperidine,** first reported in 1931 and now sold as Demerol. It is less addictive than morphine. Two other relatively strong pain relievers used today are **pentazocine** (Talwin) and **propoxyphene** (Darvon).

Considerable progress has been made in understanding the drug action of opiates. For many years scientists speculated about the action of opiates in the brain and the possible relationship to the human response to pain. Solomon Snyder and coworkers at Johns Hopkins University discovered in 1973 that the brain and spinal cord contain specific bonding or receptor sites that the opiate molecules fit as a key fits into a lock. This enhanced the search for opiate-like neurotransmitters. In 1975, John Hughes and Hans Kosterlitz, of the University of Aberdeen, Scotland, isolated two peptides with opiate activity from pig brains. They decided to call these peptides **enkephalins** (from the Greek *en kephale,* meaning "within the head"); specifically, the two pentapeptides they isolated are known as methionine-enkephalin and leucine-enkephalin (Fig. 15-10). A year later, Roger Guillemin and coworkers at the Salk Institute isolated a longer peptide, called **beta-endorphin,** from extracts of the pig hypothalamus. Beta-endorphin is 50 times more potent than morphine. Since this early work other enkephalins and endorphins have been isolated. In addition, the relationship of these natural opiates to pain has been studied.

Natural opiates include enkephalins (pentapeptides) and endorphins (peptides with about 30 amino acids).

Our bodies synthesize enkephalins and endorphins to moderate pain, and our pain threshold is related to levels of these neuropeptides in our central nervous system. Individuals with a high tolerance for pain produce more neuropeptides and consequently tie up more receptor sites than normal; hence, they feel less pain. A dose of heroin temporarily bonds to a high percentage of the sites, resulting in little or no pain. Continued use of heroin causes the body to reduce or cease its production of enkephalins and endorphins. If use of the narcotic is stopped, the receptor sites become empty and withdrawal symptoms occur.

Mild Analgesics

When milder general analgesics are required, few compounds work as well for many people as **aspirin.** Each year about 40 million pounds of aspirin are manufactured in the United States. Aspirin is also an antipyretic (fever reducer) and an anti-inflammatory agent, and, as discussed earlier, effective in the treatment of heart disease. Aspirin is thought to inhibit cyclo-oxygenase, the enzyme that catalyzes the reaction of oxygen with polyunsaturated fatty acids to produce prostaglandins (discussed in Chapter 9). Excessive prostaglandin production causes fever, pain, and inflammation—just the symptoms aspirin relieves.

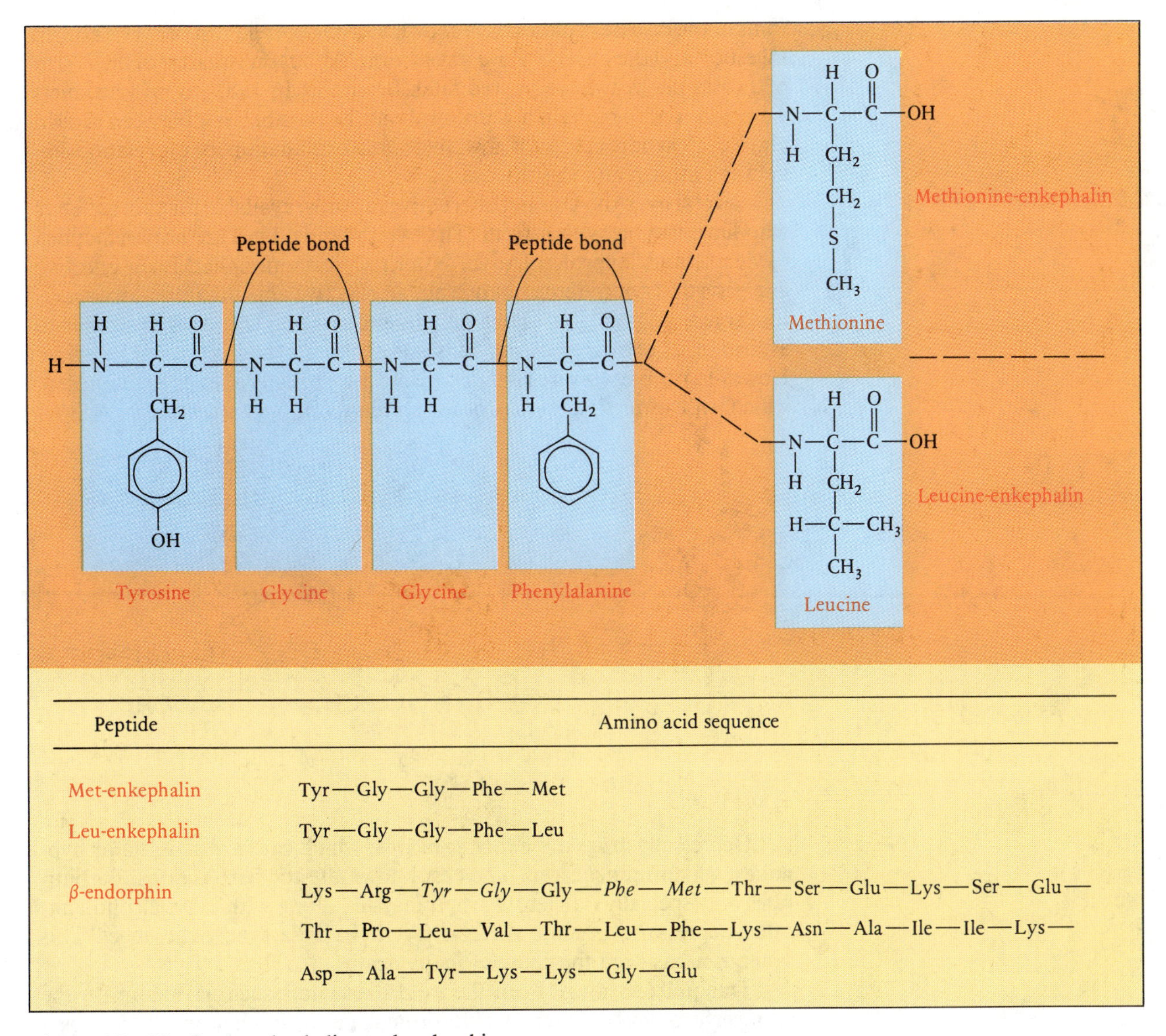

Peptide	Amino acid sequence
Met-enkephalin	Tyr—Gly—Gly—Phe—Met
Leu-enkephalin	Tyr—Gly—Gly—Phe—Leu
β-endorphin	Lys—Arg—*Tyr*—*Gly*—Gly—*Phe*—*Met*—Thr—Ser—Glu—Lys—Ser—Glu—Thr—Pro—Leu—Val—Thr—Leu—Phe—Lys—Asn—Ala—Ile—Ile—Lys—Asp—Ala—Tyr—Lys—Lys—Gly—Glu

Figure 15–10 Some enkephalins and endorphins.

A danger presented by aspirin is stomach bleeding, caused when an undissolved aspirin tablet lies on the stomach wall. As the aspirin molecules pass through the fatty layer of the mucosa, they appear to injure the cells, causing small hemorrhages. The blood loss for most individuals taking two 5-grain tablets is between 0.5 mL and 2 mL. However, some persons are more susceptible than others. Early aspirin tablets were not particularly fast dissolving, which aggravated this problem greatly. Today's aspirin tablets are formulated to disintegrate quickly.

Many aspirin tablets contain starch to hasten their disintegration in the stomach.

A greater potential danger of aspirin is its possible link to **Reye's syndrome,** a brain disease that also causes fatty degeneration in organs such as the liver. Reye's syndrome can occur in children recovering from the flu or chicken pox. Vomiting, lethargy, confusion, and irritability are the symptoms of the disease. Studies have shown a strong correlation between aspirin ingestion and the onset of Reye's syndrome. About one quarter of the 200 to 600 cases per year have proved fatal. Beginning in 1982, aspirin products were required to contain a warning about the possible link between aspirin and Reye's syndrome. As of now, there is no explanation for the relationship between aspirin and this disease.

Several over-the-counter alternatives are now available for pain sufferers who have trouble with aspirin. The two principal ones are **acetaminophen** (Tylenol) and **ibuprofen** (Advil, Nuprin). Acetaminophen is an effective analgesic and antipyretic, but it is not an effective anti-inflammatory agent. Ibuprofen, originally available only by prescription (Motrin), is similar to aspirin in its effectiveness but causes less stomach and intestinal bleeding. However, people who are allergic to aspirin are usually allergic to ibuprofen, which isn't surprising in view of the similarity in their structures.

Acetylsalicylic acid (aspirin)

Acetaminophen (Tylenol)

Ibuprofen (Advil, Nuprin)

Depressants

Depressant drugs are either **sedatives,** which cause relaxation, or **hypnotics,** which induce sleep. The best-known are the **barbiturates.** Barbiturates are especially dangerous when ingested along with ethyl alcohol, another depressant, because the two together give a synergistic effect. This synergism has been the cause of many deaths.

Tranquilizers range from the mild diazepams such as Valium to the stronger promazines such as chlorpromazine, which is used to treat psychotic disorders.

Stimulants

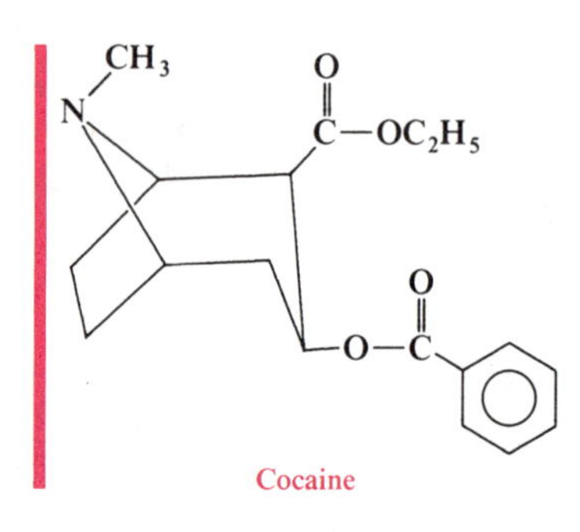

Cocaine

Stimulants increase the concentration of neurotransmitters such as norepinephrine and serotonin in the brain. **Cocaine,** derived from the leaves of the coca plant of South America, prevents norepinephrine uptake after it is released by neurons. This increases the concentration of norepinephrine that is available to neurons and causes uncontrolled firing of electrical impulses to adjacent neurons. The problems of cocaine addiction are described later in this chapter.

Amphetamines are derivatives of phenylethylamine (Fig. 15-11). Note the similarity of the structure of amphetamine and methamphetamine to norepinephrine. This similarity helps to explain the stimulant activity of

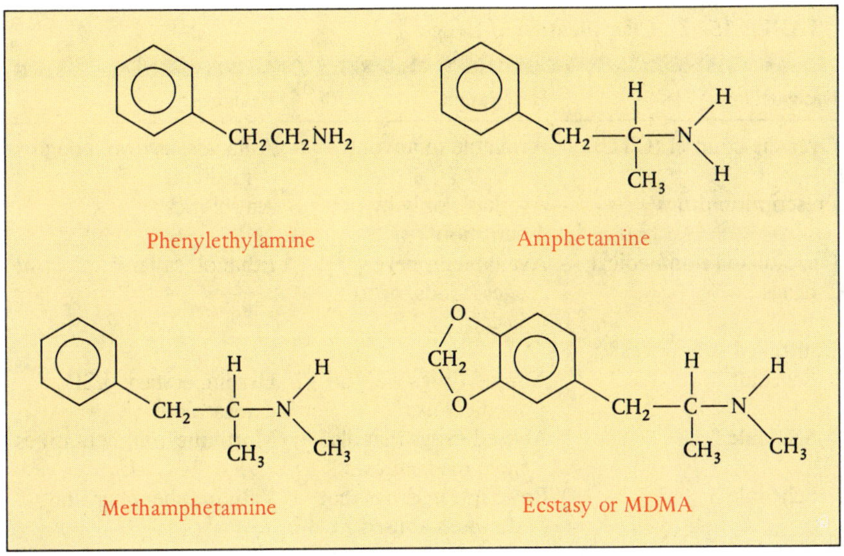

Phenylethylamine $CH_2CH_2NH_2$ — Amphetamine — Methamphetamine — Ecstasy or MDMA

Figure 15–11 Amphetamines are derivatives of phenylethylamine.

amphetamines. They have been used for years as a mood elevator and stimulant by persons who must stay awake. They are still used as an appetite suppressant. The two optical isomers of amphetamine have different activity. The right-handed isomer, dextro-amphetamine (Dexedrine) is four times more active than the equal mixture of the two isomers, sold under the trade name Benzidrene. However, amphetamines are abused drugs (next section).

Natural stimulants include the alkaloids **caffeine** and **nicotine** (see Fig. 7–5).

Drug Abuse

Although the list of drugs of abuse is wide ranging, a central theme is their relationship to brain chemistry and the effect of the drug on neurotransmitters or their receptor sites. Many of the drugs mentioned above are widely abused. These include amphetamines, narcotics, stimulants, and ethyl alcohol. Before discussing these, it is important to have an understanding of the extent of legal control of abused drugs. Table 15–2 lists the various classifications. Note that examples of controlled substances include drugs that are narcotics, hallucinogens, depressants, and stimulants.

Problems with drug abuse and drug addiction are as old as civilization itself. However, a new dimension has been added in the 1980s as a result of three developments, "crack," "ice," and "designer drugs."

Crack

Crack is a purified form of cocaine obtained by heating a mixture of cocaine and sodium bicarbonate for 15 min. The reaction is an acid–base reaction since the base, sodium bicarbonate, is neutralizing cocaine hydro-

TABLE 15-2 Classification of Drugs

Designation	Description	Examples
Over-the-counter (OTC)	Available to anyone	Antacids, aspirin, cough medicines
Prescription drugs	Available only by prescription	Antibiotics
Unregulated nonmedical drugs	Available in beverages, foods, or tobacco	Ethanol, caffeine, nicotine
Controlled substances*		
Schedule 1	Abused drugs with no medical use	Heroin, ecstasy, LSD, mescaline
Schedule 2	Abused drugs that also have medical uses	Morphine, amphetamines
Schedule 3	Prescription drugs that are often abused	Valium, phenobarbital

* The sale, distribution, and possession of drugs classified as controlled substances are controlled by the Drug Enforcement Administration of the U.S. Department of Justice.

chloride, the normal form of cocaine. Because the reaction releases carbon dioxide gas, the term "crack" came about as a result of the crackling sound of the heated mixture during the release of carbon dioxide.

More than 20 million Americans have used cocaine, and about 4 million are using or abusing it now. The appearance of crack has caused an increase in the number of cocaine addicts because crack is much more addictive than cocaine. The reason for its addictiveness is that crack is more potent than cocaine and is smoked rather than sniffed, giving the user a much quicker, more intense high. Because the high lasts less than 10 min, users have a tendency to use crack repeatedly over a short period, and many users become addicted after only one try. The problem is enhanced by the ready availability and low cost of crack ($10-15 per fix).

Ice

The newest drug abuse problem involves a smokable form of methamphetamine, nicknamed "ice." Ice is regarded as the number one drug problem in Hawaii and will likely become a serious national problem in the next few years.

Designer Drugs

Designer drugs are chemical substances that are structurally similar to legal drugs. Because of their action, they are potential drugs of abuse. All the designer drugs that have been discovered so far are either narcotics or hallucinogens. For example, fentanyl (Fig. 15-12) is a powerful narcotic marketed under the trade name Sublimaze. Fentanyl is about 150 times more potent than morphine and just as addictive, but very short-acting. Fentanyl

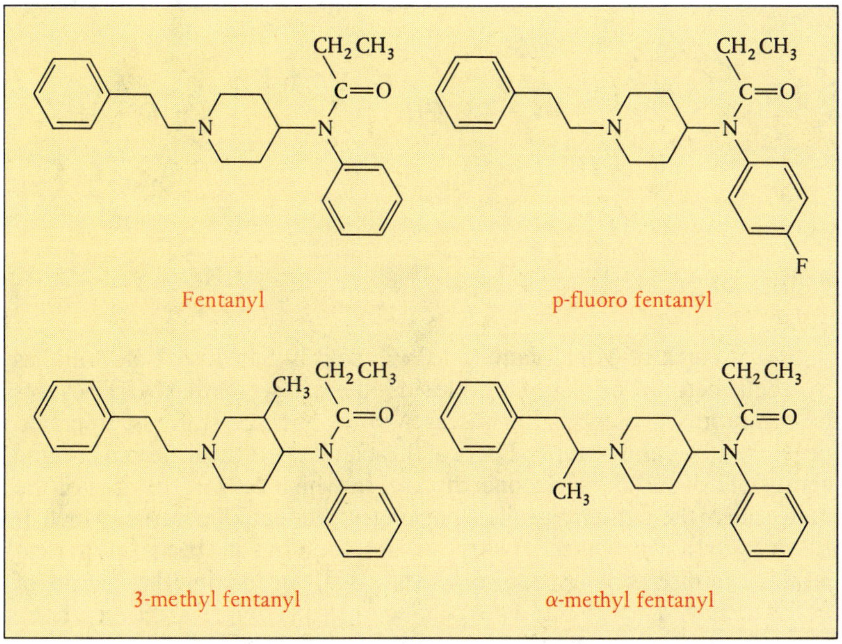

Fentanyl

p-fluoro fentanyl

3-methyl fentanyl

α-methyl fentanyl

Figure 15–12 Fentanyl and several of its derivatives.

is used in up to 70% of all surgical procedures in the United States. The derivatives of the fentanyl molecule are also potent narcotics. These drugs were called designer drugs when they first appeared on the streets because they had obviously been designed by some unscrupulous chemists for consumption by drug addicts. These fentanyl derivatives were every bit as potent as heroin, but because they were not listed on the U.S. Drug Enforcement Administration (DEA) list of controlled substances, they could be sold legally. Until a compound is recognized as being abused and is classified as a dangerous drug—a process known as *scheduling*—no laws apply to it.

In the past few years several fentanyl derivatives (Fig. 15–12) have appeared in California. Samples ranged from pure white powder, sold as China White, to a brown material. First came α-methyl fentanyl, then *p*-fluoro fentanyl, then α-methyl acetyl fentanyl, and in early 1984 3-methyl fentanyl, a compound 3000 times more potent than morphine. Because of this potency and because heroin addicts can use the fentanyl derivatives interchangeably with heroin, fentanyl derivatives, mostly 3-methyl fentanyl, have been responsible for over 100 overdose deaths in California.

Another group of designer drugs is derived from meperidine (Demerol). One of these is MPPP, which is short for 1-methyl-4-phenyl-4-propionoxypiperidine. This compound was first synthesized in 1947, never used commercially, and never scheduled as a controlled substance. MPPP is about three times more potent than morphine and 25 times more potent than meperidine. It is structurally so close to meperidine that one has to look closely at the structures to see the difference (hint: look at the ester linkage). If the synthesis of MPPP is carried out at too high a temperature or at too low a pH, the product is MPTP, 1-methyl-4-phenyl-1,2,3,6-tetrahydropyridine.

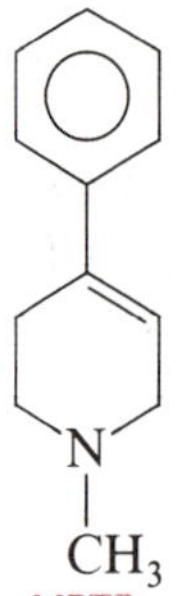

MPTP 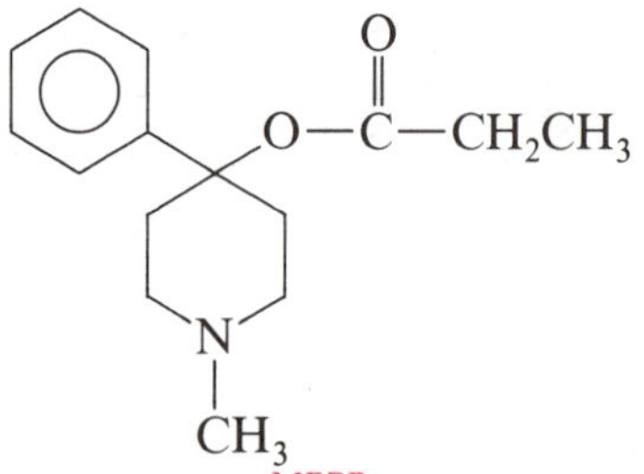 MPPP 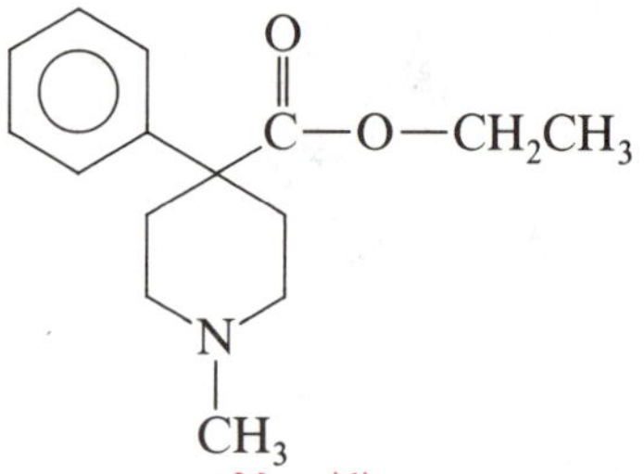 Meperidine

In 1982 a batch of MPTP-tainted MPPP, sold in San Jose, California, as "synthetic heroin," produced terrible side effects. It seems that MPTP causes the symptoms of Parkinson's disease, which include stiffness, impaired speech, rigidity, and tremors. Users of this batch of synthetic heroin became victims of advanced Parkinson's disease, in which cells in the area of the brain called the substantia nigra no longer produce dopamine, which is necessary for normal muscle control. A substance that had been used to treat Parkinson's disease, L-dopa, also proved useful in treating the victims of MPTP toxicity. L-Dopa could not be used to effect complete recovery, however, because it also causes hallucinations and exaggerated movements.

The link of MPTP to Parkinson's disease symptoms has stimulated research in this area. In 1987, over 150 papers on MPTP were published, and the study of MPTP and its effect on dopamine was central to this research. The selective destruction of dopamine neurons by MPTP has been established, and the extensive research in this area is likely to aid in a better understanding of the cause of Parkinson's disease.

The control of designer drugs has been easier since passage of the Comprehensive Crime Control Act of 1984, which gave the DEA emergency scheduling authority. Now any drug can be designated a controlled substance within 30 days. This scheduling lasts for one year while additional data are gathered to determine final scheduling authority. All the designer drugs described here have been placed in Schedule 1, which precludes their use for any legal purpose.

Hallucinogens

Hallucinogens are chemicals that cause vivid illusions, fantasies, and hallucinations. The most common hallucinogens obtained from natural sources are mescaline, which comes from the fruit of the peyote cactus, and lysergic acid diethylamide (LSD), which is made from lysergic acid derived from either the morning glory or ergot, a fungus that grows on wheat, rye, and other grasses. Marijuana is a mild hallucinogen and sedative made from the flowering tops, seeds, leaves, and stems of the female hemp plant, *Cannabis sativa.* The structures of the common hallucinogens are shown in Figure 15–13.

Although marijuana is not physically addicting and thus gets less attention than addicting drugs such as crack and cocaine, persons can become psychologically dependent on the drug. An estimated 6 million Americans between the ages of 18 and 25 smoke marijuana every day, and another 16

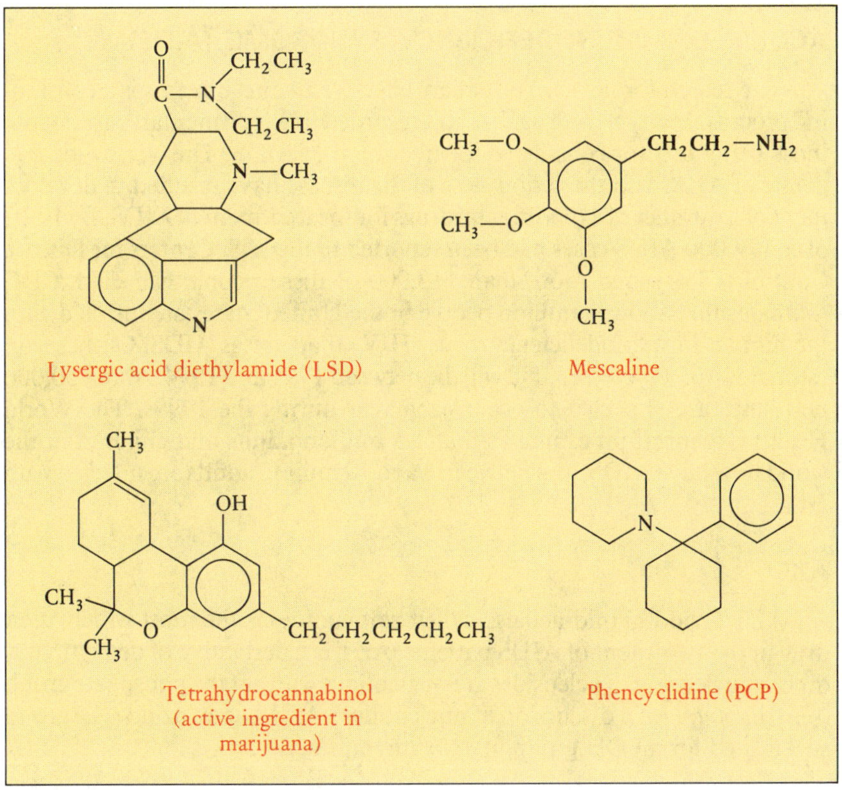

Lysergic acid diethylamide (LSD)

Mescaline

Tetrahydrocannabinol (active ingredient in marijuana)

Phencyclidine (PCP)

Figure 15–13 Some hallucinogens.

million use it occasionally. Although these users may regard marijuana as a "safe" drug, recent studies indicate otherwise. Besides the dangers associated with psychological dependence, the smoke from marijuana and its active ingredient, tetrahydrocannabinol (THC), cause damage to the lungs, impede brain function, and hamper the immune system.

A hallucinogenic designer drug that appeared in the late 1970s is Ecstasy, also called XTC, Adam, or MDMA (Fig. 15–11). Before its appearance on the streets as a designer drug, Ecstasy was used by some psychiatrists who found it useful in the treatment of schizophrenia, depression, and anxiety. By the early 1980s, this drug was being widely used as a recreational substance and was sold openly in bars, often accompanied by advertisements in windows. On July 1, 1985, the drug was placed in the Schedule 1 classification, which prevented its legal use by anyone. Nevertheless, there is evidence that the drug continues to be popular on college campuses. Results of surveys of college students at several campuses in 1987 indicated that 17% to 30% of students had tried the drug at least once.

MDMA works by increasing the release of serotonin. Although the neurotoxicity is unknown, tests with rats show the MDMA destroys serotonin nerve terminals and results in long-term depletions of brain serotonin levels. Because the Drug Enforcement Agency and the Food and Drug Administration reaffirmed MDMA's status as a Schedule 1 drug in 1988, studies of human neurotoxicity are currently not possible.

ACQUIRED IMMUNODEFICIENCY SYNDROME (AIDS)

The search for new drugs that are effective against diseases is a continuing process. Many new drugs are still regarded as experimental, particularly those being used or considered for treatment of AIDS. The worldwide epidemic of AIDS and the seriousness of the disease have resulted in development of a number of experimental drugs for the treatment of AIDS. By 1991, over 179,000 AIDS cases had been reported to the U.S. Centers for Disease Control (CDC), and more than 113,000 of these people had died. CDC estimates that about 1 million people in the United States are infected with the human immunodeficiency virus (HIV) that causes AIDS. Of these, an estimated 165,000 to 215,000 will die between 1991 and 1994. About 50,000 new cases are expected to occur each year during the 1990s. The World Health Organization estimates that 1.5 million adults and children in the world now have AIDS, and between 8 and 10 million adults are infected with HIV.

AZT

AZT (azidothymidine, also called zidovudine) is the most widely used drug in the treatment of AIDS patients. AZT is a derivative of deoxythymidine, a nucleoside. Nucleosides are nucleotides without the phosphate group (see Chapter 9 for a discussion of nucleotides). AZT has an azido (N_3) group substituted for an OH group in deoxythymidine.

In 1987, the FDA approved the use of AZT for the treatment of AIDS.

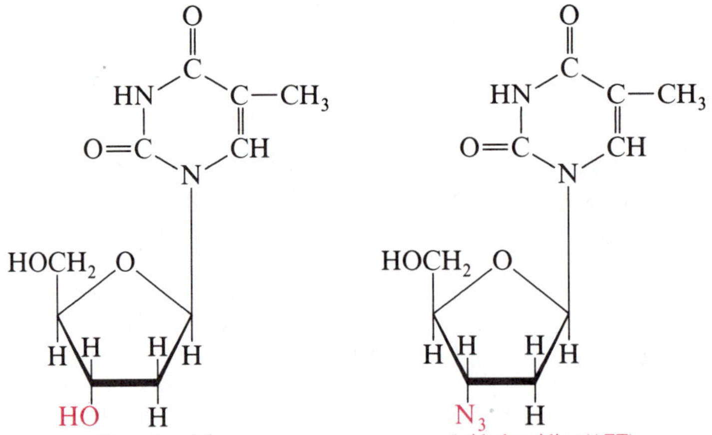

O, C, HN, C—CH_3, O=C, CH, N, $HOCH_2$, O, H, H, H, H, HO, H

Deoxythymidine
(a nucleoside)

O, C, HN, C—CH_3, O=C, CH, N, $HOCH_2$, O, H, H, H, H, N_3, H

Azidothymidine (AZT)

AIDS is caused by a **retrovirus,** a virus with an outer double layer of lipid material that acts as an envelope for several types of proteins, an enzyme called reverse transcriptase, and RNA. The term *retrovirus* is used because the virus enzyme carries out RNA-directed synthesis of DNA rather than the usual DNA-directed synthesis of RNA (see Fig. 9–21).

T cells are white blood cells that play a crucial role in controlling the immune response of the body.

The AIDS retrovirus penetrates the T cell, a key cell in the immune system of the body. Once the retrovirus is inside the T cell, the reverse transcriptase of the AIDS virus translates the RNA code of the virus into the T cell's double-stranded DNA, directing the T cell to synthesize more AIDS

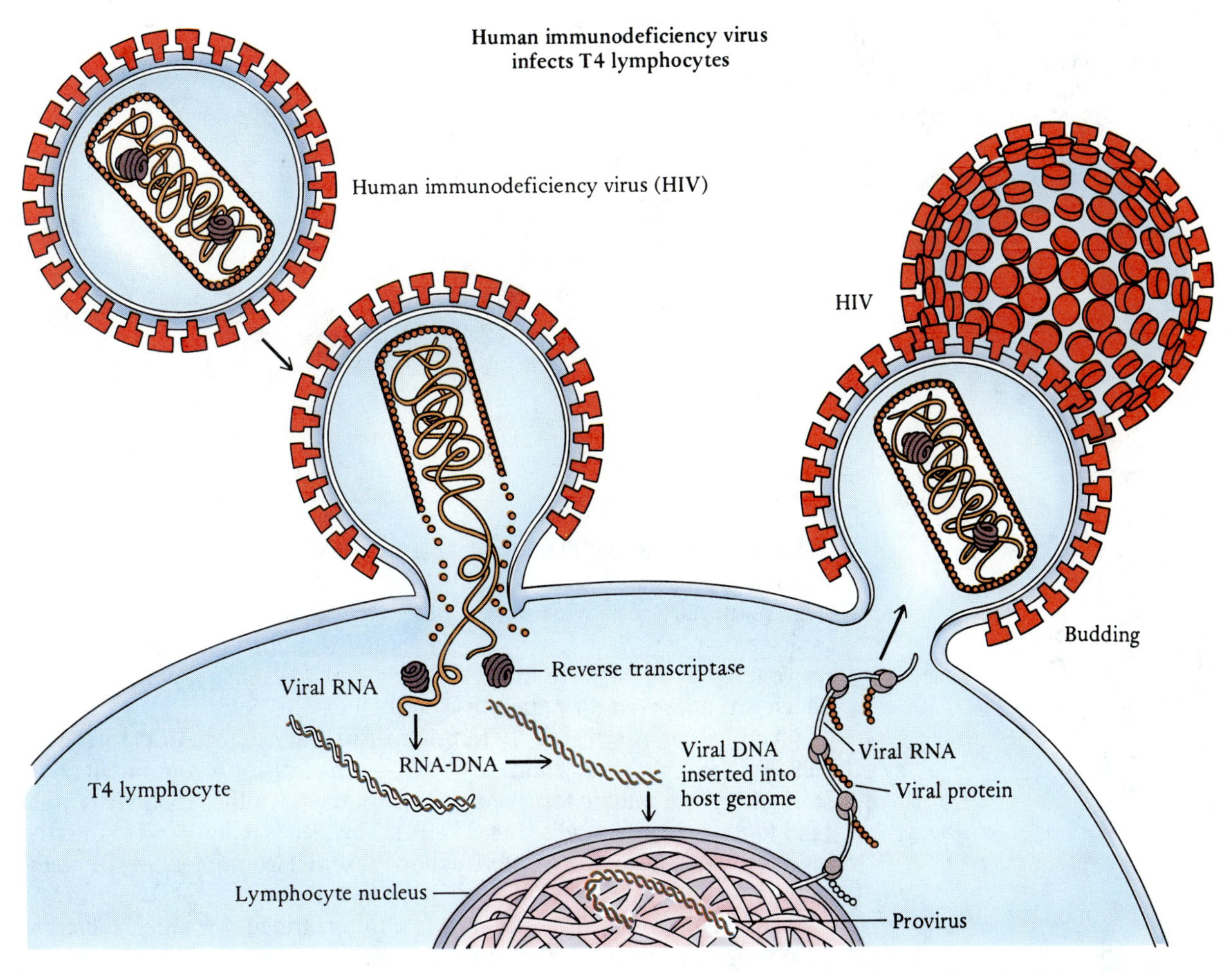

Figure 15–14 Schematic diagram of the AIDS virus attacking T cells.

viruses. Eventually the T cell swells and dies, releasing more AIDS viruses to attack other T cells (Fig. 15–14).

AZT apparently works by being accepted by reverse transcriptase in place of thymidine. After AZT has become a part of the DNA chain, its structure prevents additional nucleosides from being added onto the DNA chain.

Azidothymidine was first synthesized in the 1960s by Jerome Horwitz, who was looking for a compound that would stop cancer cells from multiplying. He reasoned that the incorporation of a "fake nucleoside" into a DNA chain would prevent additional nucleosides from being added and thus cause cell division to stop. The idea didn't work because tumor cells recognized that AZT was not thymidine and didn't incorporate AZT into DNA chains. However, it appears that the enzyme in the AIDS virus is fooled by the fake nucleoside.

Although AZT has been shown effective in retarding the progression of AIDS, it is not a cure. In addition, there are a number of problems with its use. AZT has toxic side effects that many patients cannot tolerate; the cost of

Figure 15–15 Structures of some nucleoside derivatives that are being tested for treatment of AIDS. Note the similarity of their structures.

2′,3′-dideoxyinosine (DDI)

2′,3′-dideoxycytidine (DDC)

2′,3′-dideoxyadenosine (DDA)

the drug is about \$10,000 per year; and the AIDS virus can become resistant to AZT.

Preliminary tests with several other dideoxynucleosides (Fig. 15–15) indicate that they have fewer side effects and with further testing they may prove to be more effective than AZT. The most promising of these is ddI, which was approved for patient use by the FDA in 1991.

Another approach is to try to prevent the entry of the AIDS virus into cells. For example, the protein CD4, which is made using recombinant DNA technology, is a soluble form of the receptor to which the AIDS virus must bind in order to infect cells (Fig. 15–14). The idea is to inject CD4 into the bloodstream to tie up the AIDS virus and prevent it from spreading throughout the body.

Although these drugs offer hope for the treatment of AIDS, there has been little progress in developing a cure for it. A cure would require eliminating from the body the virus that causes AIDS. The problems of developing a cure are associated with the latency period before AIDS develops and the genetic variability of the virus. However, the use of GP120, a glycoprotein that has a molecular weight of 120,000, shows promise. A small segment of the protein, CD4, was described above as having therapeutic value. David Ho of the University of California at Los Angeles School of Medicine has found that antibodies made by rabbits injected with a segment of GP120, referred to as C21E, blocked infection of human cells by several strains of HIV.

Computer-generated ball-and-stick model of AZT. (Courtesy of Phillips Petroleum)

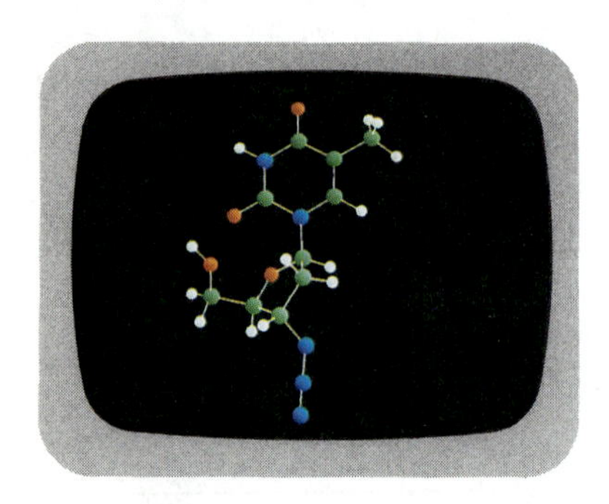

SELF-TEST 15C

1. Neurons signal the production of chemicals known as ______________, which cross the synapse and bind to dendrites of adjacent neurons.
2. Patients with Parkinson's disease have a deficiency of the neurotransmitter ______________.
3. ______________ is a derivative of morphine that is not found in nature.

4. ______________ are pentapeptides found in the brain which are referred to as natural opiates.
5. Crack is a purified form of ______________.
6. Fentanyl derivatives that have been sold on the streets tend to act like what other drug of abuse? ______________
7. AZT is used to treat ______________.
8. The difference between nucleotides and nucleosides is a ______________ group.
9. The AIDS retrovirus penetrates ______________, which are white blood cells that play a crucial role in controlling the immune response of the body.
10. An illegal drug that may have pronounced psychological or addictive effects and differs little in a molecular sense from a legal drug is often called a ______________.
11. Mood swings are caused by changes in the levels of the neurotransmitter ______________.
12. Studies have shown that the brains of Alzheimer's disease patients have lower concentrations of the enzyme needed to synthesize the neurotransmitter ______________.
13. Ice, the nickname for a smokable form of ______________, is the newest drug abuse problem.

QUESTIONS

1. What is atherosclerotic plaque? What are its two major ingredients?
2. Explain how a beta blocker can lower blood pressure.
3. What effect does the calcium ion have on the heart?
4. What do HDL and LDL stand for? Which one should be at high levels and which one should be at low levels to reduce danger of heart attack? Explain.
5. What happens when beta receptor sites in heart muscles are stimulated?
6. Why is aspirin thought to cut the risk of a heart attack?
7. How do the "clot-dissolving" drugs given to heart attack victims work?
8. Nitrogen mustards are alkylating agents that interfere with DNA replication. Explain.
9. If the chemotherapeutic agents kill living cells, why do they have a preferential effect on cancer cells?
10. What is one of the dangers of chemotherapy using an alkylating agent or an antimetabolite?
11. Describe the action of antihistamines.
12. Describe how the sulfa drug sulfanilamide works.
13. How was penicillin discovered?
14. What is Reye's syndrome? What common drug is associated with Reye's syndrome?
15. Anabolic steroids that can be taken orally are unlike testosterone with respect to what structural feature?
16. How do the estrogens differ structurally from the other female hormone progesterone?
17. What designer drug is toxic to the part of the brain called the substantia nigra? The effects of this drug are similar to what disease?
18. The Drug Enforcement Administration has three different classifications of controlled substances. Explain the differences among these and give an example of each.
19. Look up the drugs scheduled by the Drug Enforcement Administration as Schedule 1 drugs. How many are there?
20. What are enkephalins and endorphins? How do they differ?
21. Indicate which of the following are narcotics and which are stimulants: cocaine, morphine, heroin, amphetamines, ecstasy.
22. Explain the function of acetylcholine, norepinephrine, dopamine, and serotonin in the brain.
23. What neurotransmitter affects mood changes in an individual? Explain how this information could be used in treating mental illness.
24. What is a retrovirus? How does the AIDS retrovirus attack the immune system of the body?
25. Why is AZT effective in the treatment of AIDS?
26. What are hallucinogens? Give three examples.

Andy Warhol's *The Four Marilyns* is a good example of how chemistry can enhance natural beauty.

CHAPTER

16

Consumer Chemistry—Our Money for Chemical Mixtures

There are a large number of consumer products that cleanse, beautify, and protect from the Sun's harmful rays. These products are the result of chemical discovery and applied technology to accomplish a specific goal. How wisely we spend our money on these products can be affected by our understanding of what they do.

1. What are cosmetics?
2. What is the structure of skin and hair?
3. How can the shape and color of hair be changed?
4. How do skin products work?
5. Are sunscreens effective?
6. How do deodorants work?
7. How do cleansing agents do their job?

We spend our money for food, clothing, housing, beauty and health products, transportation, and entertainment. To reap the full benefit of the things we purchase, we should know something of the types and the availability of the raw materials used, the physical and chemical modifications made in them, the properties that give the desired effects, and the precautions to be taken in their use. For example, it is common to find products with the same formulation from the same raw materials that differ considerably in cost as a result of differences in packaging, appearance, and advertising. Whether it is the selection of a nonpoisoning material for the water pipes in your home, a cleansing medium for costly apparel, or the preservation of a work of art, your knowledge of chemistry can help you make better choices.

A considerable number of consumer products have been discussed in the previous chapters. In this chapter we give attention to items that command consumer attention in the areas of cosmetics and cleansing agents.

The authors believe that you, using the approach presented here, can apply it to other products of interest to you; and, with a minimum of library research, base your consumer spending on chemical knowledge and understanding rather than on advertising hype. We live at a time when it is more and more necessary to **"read and understand the label."**

COSMETICS

The use of chemical preparations, which are applied to the skin to cleanse, beautify, disinfect, or alter appearance or smell, is older than recorded history. Such preparations are known as **cosmetics.** There is at best a fuzzy distinction between cosmetics and drugs. Traditionally drugs alter body functions, but cosmetics do not. However, antiperspirants are considered cosmetics but do stop the secretions of the sweat glands. The distinction is becoming even more difficult with the introduction of creams to stop the aging process in wrinkling skin (Retin A) and to promote the growth of hair (minoxidil). Perhaps the best distinction is the level of governmental control in the introduction of new products. Drugs require elaborate safety testing prior to receiving approval by the U.S. Food and Drug Administration, but cosmetics do not.

Cosmetics are being expanded from their traditional use of adornment for esthetic reasons to scientific applications for skin care, protection from ultraviolet light, conditioning hair, and beautification with increasingly milder ingredients. One brand of cosmetics advertises the molecular formula of its humectant (sodium hyaluronate) right on the product label for a $50 tube of skin cream — an indication of the movement from an art to a science. Although there is considerable debate over the level of scientific understanding in cosmetics, there is no doubt about public appetite for cosmetics and the profit potential in the manufacture and sale of them. Approximately $30 billion is currently being spent each year for cosmetics in the United States.

Annual retail sales for cosmetics in Western Europe are approximately $30 billion; in Japan, $15 billion.

Skin, Hair, Nails, and Teeth — A Chemical View

The skin, hair, and nails are protein structures. Skin (Figs. 16–1 and 16–2), like other organs of the body, is not composed of uniform tissue and has several functions made possible by its structure; they are protection, sensation, excretion, and body temperature control. The exterior of the epidermis is called the stratum corneum, or **corneal layer,** and is where most cosmetic preparations for the skin act. The corneal layer is composed principally of dead cells with a moisture content of about 10% and a pH of about 4. The principal protein of the corneal layer is **keratin,** which is composed of about 20 different amino acids. Its structure renders it insoluble in, but slightly permeable to, water. Dry skin is uncomfortable, but an excessively moist skin is a good host for fungus organisms. An oily secretion, sebum, is

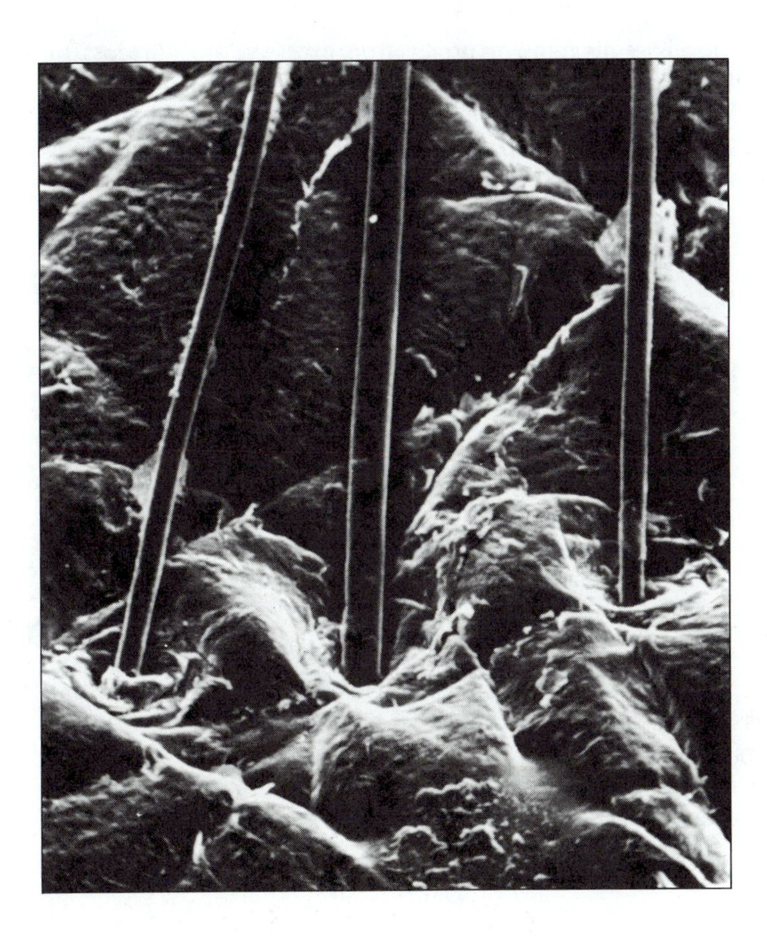

Figure 16–1 Replica of the surface of human forearm skin, showing three hairs emerging from the skin (×225). (Courtesy of E. Bernstein and C. B. Jones: *Science,* Vol. 166, pp. 252–253, 1969. © 1969 by the American Association for the Advancement of Science)

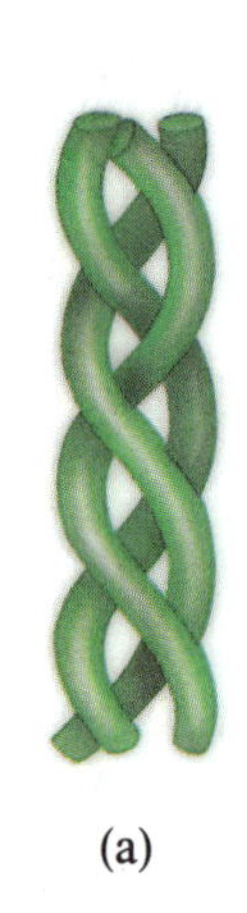

(a)

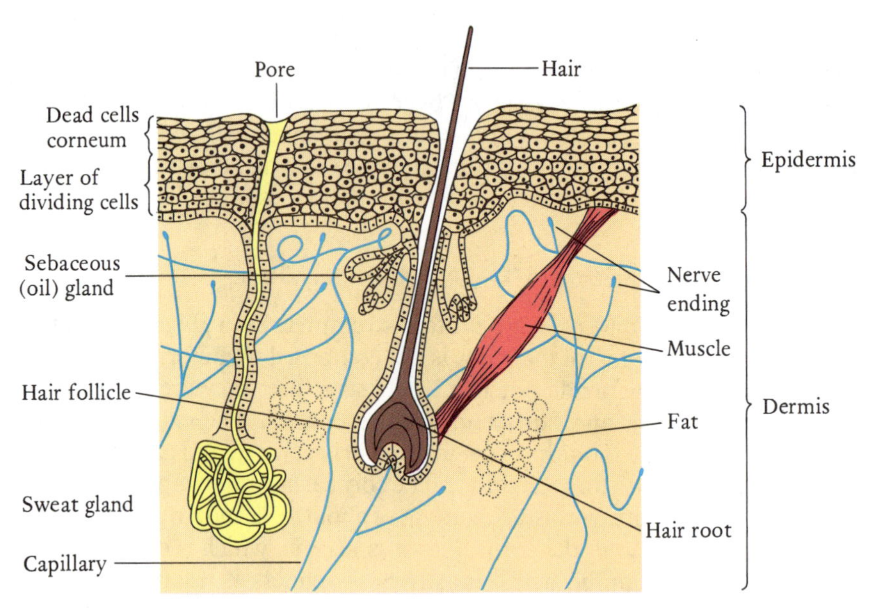

Figure 16–2 Cross section of the skin.

(b)

Graphic of hair structure. (a) The structure of hair begins with the spiral proteins, which are covered in strands (b) as they are bound together. This allows for hair to be fine or coarse in texture.

produced by the sebaceous glands to protect from excessive moisture loss. In order to control the moisture content of the corneal layer so that it does not dry out and slough off too quickly, moisturizers may be added to the skin.

Hair is also composed principally of keratin (Fig. 16–3). An important difference between hair keratin and other proteins is its high content of the amino acid cystine (Fig. 16–4). About 16% to 18% of hair protein is cystine, but only 2.3% to 3.8% of the keratin in corneal cells is cystine. This amino acid plays an important role in the structure of hair.

The toughness of both skin and hair is due to the bridges between different protein chains, such as hydrogen bonds and —S—S— linking bonds, called disulfide bonds.

$$
\begin{array}{ccccc}
| & & & & | \\
O{=}C & & & & N{-}H \\
| & & & & | \\
H{-}C{-}CH_2{-}S & {-} & S{-}CH_2 & {-} & C{-}H \\
| & & \uparrow & & | \\
H{-}N & & & & C{=}O \\
| & & & & |
\end{array}
$$

Disulfide bonds (cross-links)

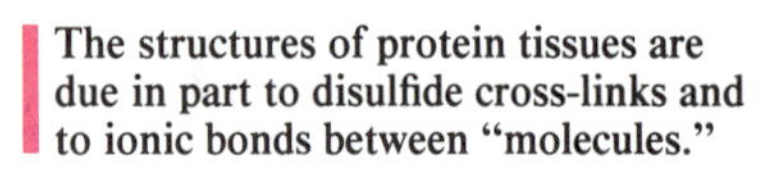
The structures of protein tissues are due in part to disulfide cross-links and to ionic bonds between "molecules."

Another type of bridge between two protein chains, which is important in keratin as well as in all proteins, is the ionic bond. Consider the interaction between a lysine $—NH_2$ group and a carboxylic group —COOH of glutamic acid on a neighboring protein chain. At pH 4.1, protons are added to the $—NH_2$ groups and removed from the —COOH groups, resulting in $—NH_3^+$ and $—COO^-$ groups on adjacent chains. If the two charged groups approach closely, an ionic bond is formed.

$$\text{H}\overset{|}{\underset{|}{\text{C}}}\text{CH}_2\text{CH}_2\text{CH}_2\text{CH}_2\text{CH}_2\text{NH}_2 + \text{HOOCCH}_2\text{CH}_2\overset{|}{\underset{|}{\text{C}}}\text{H} \xrightarrow{\text{at pH 4.1}}$$

Lysine — Glutamic acid

$$\text{H}\overset{|}{\underset{|}{\text{C}}}\text{CH}_2\text{CH}_2\text{CH}_2\text{CH}_2\text{CH}_2\text{NH}_3^{+}\,{}^{-}\text{OOCCH}_2\text{CH}_2\overset{|}{\underset{|}{\text{C}}}\text{H}$$

Ionic bond

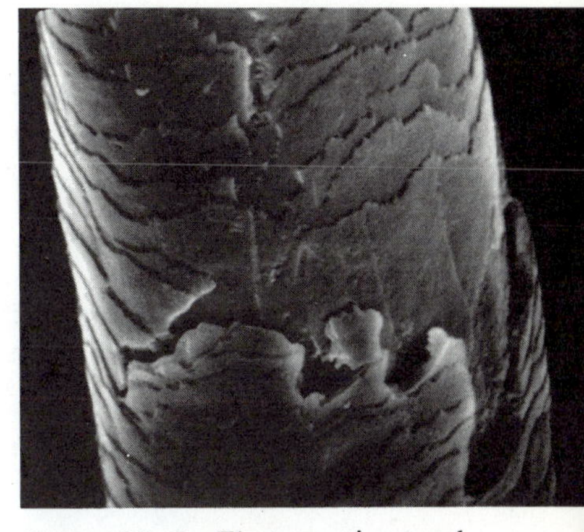

Figure 16–3 Electron micrograph of human hair. Note the layers of keratinized cells.

As the pH rises above 4, keratin swells and becomes soft as these cross-links are broken. This is an important aspect of hair chemistry because the pH of most shampoos and even water is above 4. Finger and toe nails are composed of hard keratin, a very dense type of this protein. These epidermal cells grow from epithelial cells lying under the white crescent at the growing end of the nail. Like hair, the nail tissue beyond the growing cells is dead.

HAIR PRODUCTS

Changing the Shape of Hair

When hair is wet, it can be stretched to one and a half times its dry length because water (approximately pH 7) weakens some of the ionic bonds and causes swelling of the keratin. Imagine the disulfide cross-links remaining between two protein chains in hair as in Figure 16–5. Winding the hair on rollers causes tension to develop at the cross-links (b). In "cold" waving, these cross-links are broken by a reducing agent (c), relaxing the tension. Then, an oxidizing agent regenerates the cross-links, (d) and the hair holds

Refer to Chapter 9 for the structure of proteins.

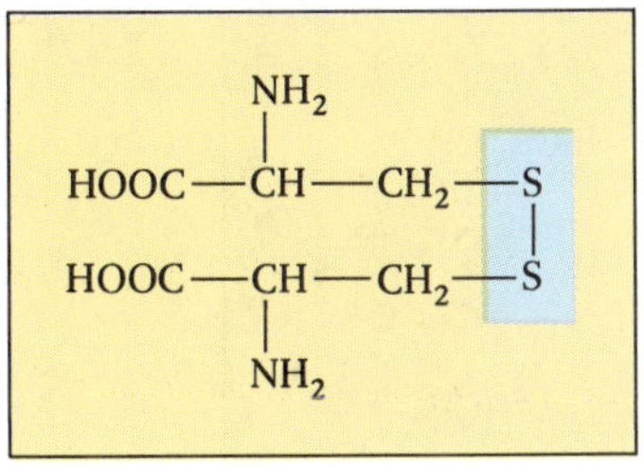

$$\begin{array}{l} \quad\quad\quad\ \text{NH}_2 \\ \quad\quad\quad\ \ | \\ \text{HOOC—CH—CH}_2\text{—S} \\ \quad\quad\quad\quad\quad\quad\quad\ \ | \\ \text{HOOC—CH—CH}_2\text{—S} \\ \quad\quad\quad\ \ | \\ \quad\quad\quad\ \text{NH}_2 \end{array}$$

Figure 16–4 Molecular bonding in cystine. Note the disulfide bond shown in blue. See Chapter 9 for a discussion of this amino acid.

Figure 16–5 A schematic diagram of a permanent wave.

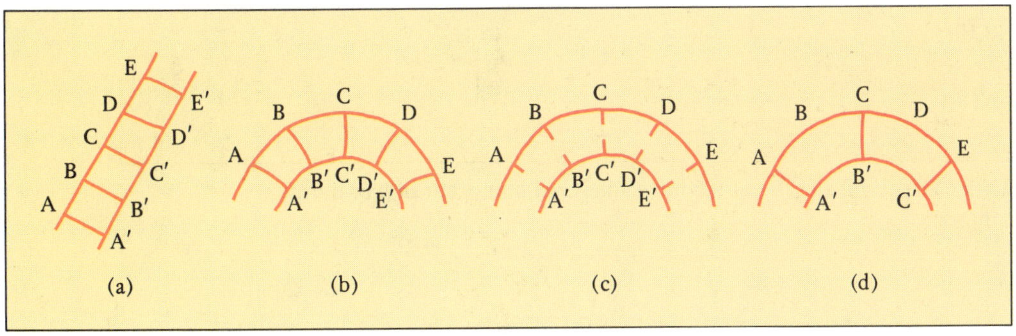

(a) A B C D E / A′ B′ C′ D′ E′ — (b) A B C D E / A′ B′ C′ D′ E′ — (c) A B C D E / A′ B′ C′ D′ E′ — (d) A B C D E / A′ B′ C′

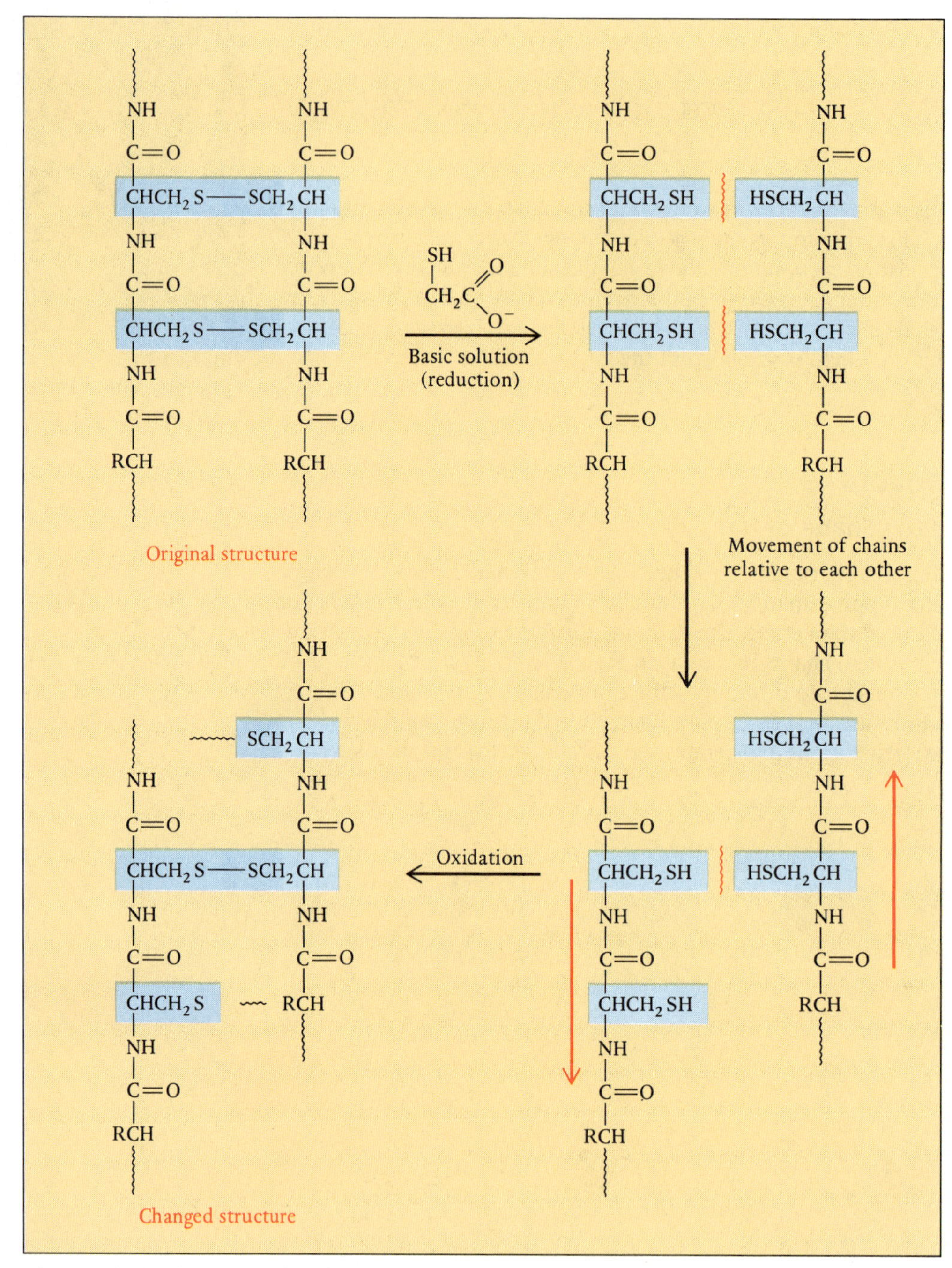

Figure 16–6 Structural changes at the molecular level that occur in hair during a permanent wave.

the shape of the roller. The chemical reactions in simplified form are shown in Figure 16–6.

The most commonly used reducing agent is thioglycolic acid (Fig. 16–7). The common oxidizing agents used include hydrogen peroxide, perborates ($NaBO_2 \cdot H_2O_2 \cdot 3H_2O$), and sodium or potassium bromate ($KBrO_3$). A typical neutralizer solution contains one or more of the oxidizing agents dissolved in water. The presence of water and a strong base in the oxidizing solution also helps to break and re-form hydrogen bonds between adjacent protein molecules. However, too-frequent use of strong base causes hair to become brittle and "lifeless."

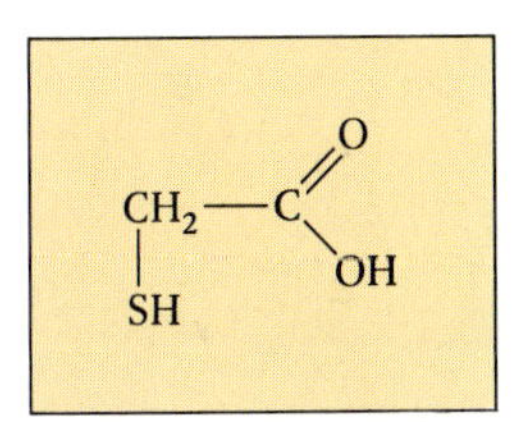

Figure 16–7 Thioglycolic acid. Note the relationship of the molecular structure to acetic acid and the basic structure in amino acids.

Hair can be straightened by the same solutions. It is simply "neutralized" (or oxidized) while straight (no rolling up).

Various additives are present in both the oxidizing and the reducing solutions in order to control pH, odor, and color, and for general ease of application. A typical waving lotion contains 5.7% thioglycolic acid, 2.0% ammonia, and 92.3% water.

Coloring and Bleaching Hair

Melanin: brown-black; iron pigment: red.

Hair contains two pigments: brown-black melanin and an iron-containing red pigment. The relative amounts of each actually determine the color of the hair. In deep black hair melanin predominates and in light-blond, the iron pigment predominates. The depth of the color depends upon the size of the pigment granules.

Formulations for dyeing hair vary from temporary coloring (removable by shampoo), which is usually achieved by means of a water-soluble dye that acts on the surface of the hair, to semipermanent dyes, which penetrate the hair fibers to a great extent (Fig. 16–8). The more permanent dyes often consist of cobalt or chromium complexes of dyes dissolved in an organic solvent and are generally "oxidation" dyes. They penetrate the hair and then

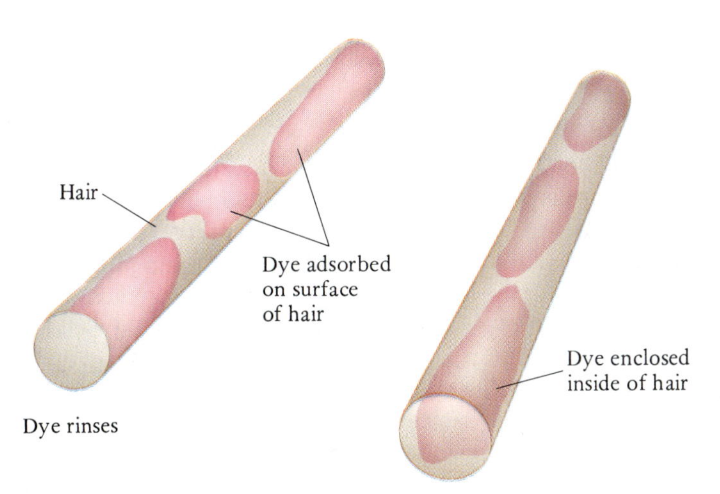

Figure 16–8 Methods of dyeing hair.

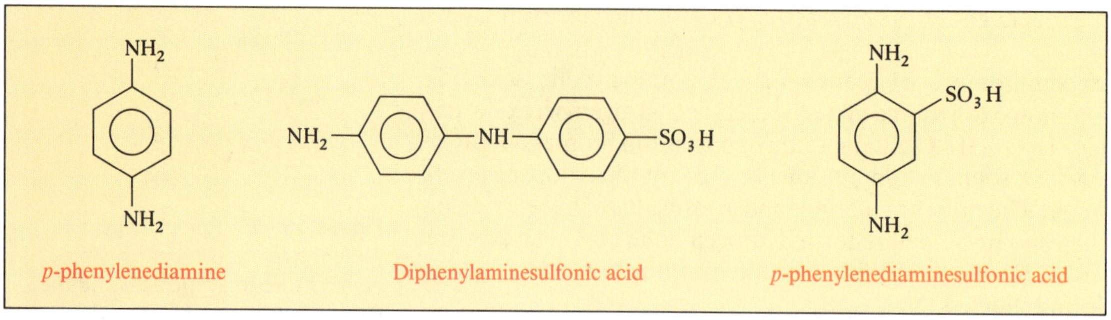

Figure 16–9 Hair dyes.

Some hair dyes are suspected of being carcinogenic.

are oxidized to give a colored product that either is permanently attached to the hair by chemical bonds or is much less soluble in shampoo water than the reactant molecule. Permanent hair dyes generally are derivatives of phenylenediamine. Phenylenediamine dyes hair black. A blond dye can be formulated with *p*-aminodiphenylaminesulfonic acid or *p*-phenylenediaminesulfonic acid. The structures of these three dye molecules are given in Figure 16–9.

Just about any shade of hair color can be prepared by varying the modifying group on certain basic dye structures.

The active compounds are applied in an aqueous soap or detergent solution containing ammonia to make the solution basic. The dye material is then oxidized by hydrogen peroxide to develop the desired color. The amine groups in the dye molecules are oxidized to nitro compounds.

$$\underset{\text{Amine}}{-NH_2} + 3\,H_2O_2 \xrightarrow{\text{Oxidation}} \underset{\text{Nitro compound}}{-NO_2} + 4\,H_2O$$

Hair can be bleached by a more concentrated solution of hydrogen peroxide, which destroys the hair pigments by oxidation. The solutions are made basic with ammonia to enhance the oxidizing power of the peroxide. Parts of the chemical process are given in Figure 16–10. This drastic treatment of hair does more than just change the color. It may destroy sufficient structure to render the hair brittle and coarse.

Hair Control—Sprays, Conditioners, and Mousses

There are several dangers in breathing the vapor of hair sprays, such as possible harm from chemicals acting on delicate lung tissue and the danger of asphyxiation by the plastic coating the lining of the lungs.

Wet hair can be shaped better than dry hair because some of the ionic bonds maintaining the protein shape are broken as water hydrates the ionic centers and thereby isolates the ionic charges from each other. Hair sprays are solutions of resins in a volatile solvent. When the solvent evaporates, the hair is coated with a film of sufficient strength to hold the hair in place (Fig. 16–11). A suitable resin is a copolymer of vinylpyrrolidone and vinyl acetate (Fig. 16–12). In addition to the resin, solvent, and propellant (the liquid that gasifies when the valve is opened), a hair spray is likely to have a plasticizer to enhance elasticity and a silicone oil to impart a sheen to the hair.

An after-shampoo conditioner attempts to manage hair without the film. Added to the water–alcohol dispersing medium, the conditioner contains emollients, oils, waxes, resins, and proteins that adhere to and penetrate into the hair to produce a more pliable and elastic fiber that is not as likely to

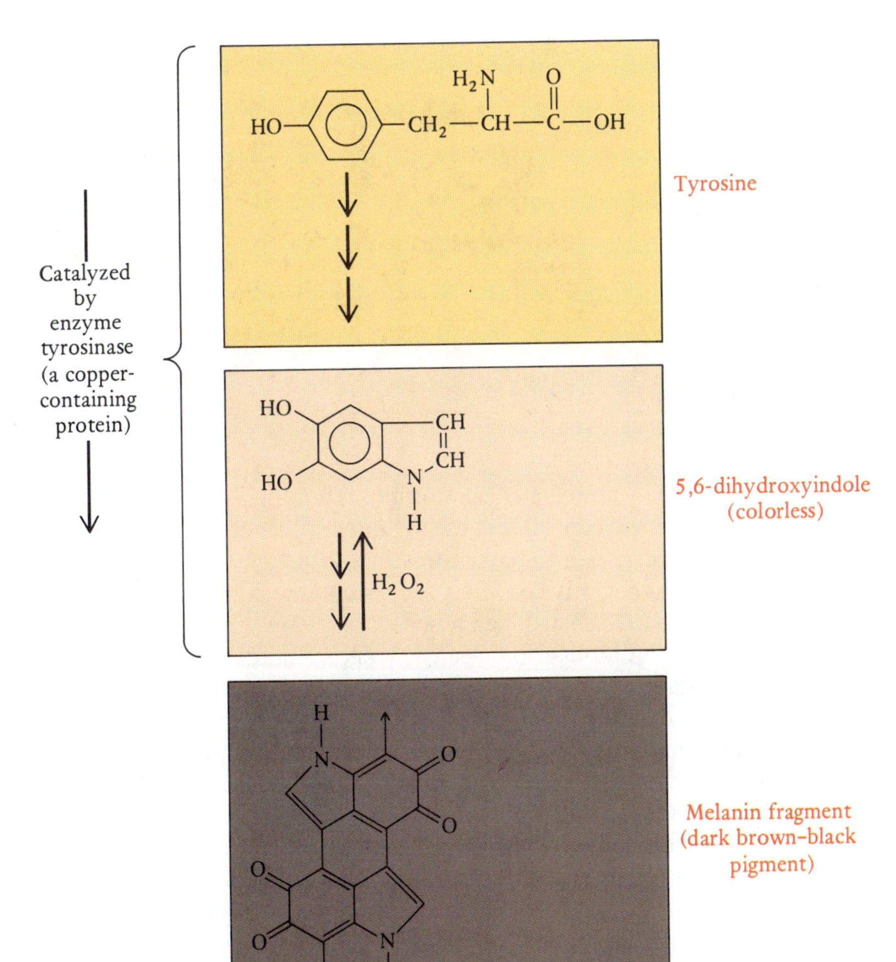

Figure 16–10 Bleaching of the hair by hydrogen peroxide. There are several chemical intermediates between the amino acid, tyrosine, and the hair pigment, melanin, which is partly protein. Hydrogen peroxide oxidizes melanin back to colorless compounds, which are stable in the absence of the enzyme tyrosinase (found only in the hair roots). Melanin is a high-molecular-weight polymeric material of unknown structure. The structure shown here is only a segment of the total structure.

Figure 16–11 Film of hair spray. Hair spray was allowed to dry on the white surface and was then pulled up to reveal film.

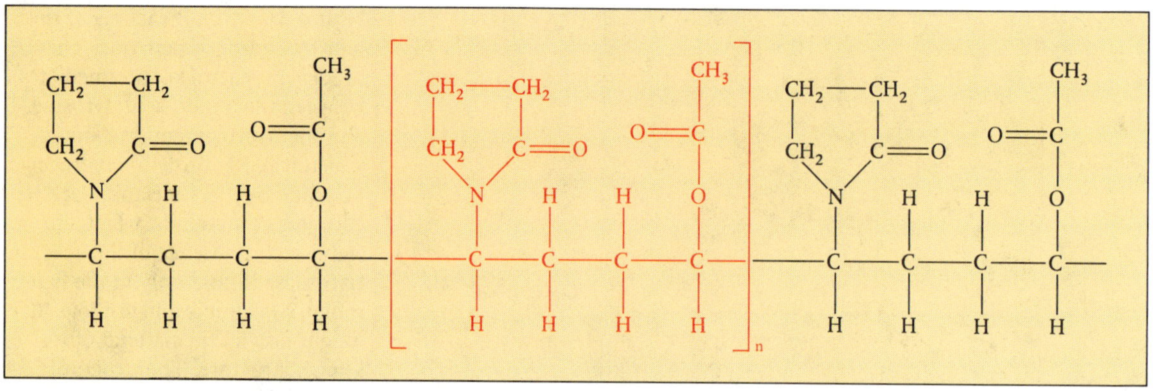

Figure 16–12 A copolymer of vinylpyrrolidone and vinyl acetate is a resin used in hair sprays. Recalling the formula for acetic acid, see if you can identify the vinyl acetate unit in the polymeric chain.

dry and be affected by atmospheric conditions. Holding the correct amount of moisture is the key to control because too much water causes the hair to be limp and too little causes it to be fly-away.

The mousse has recently become a very popular technique for the application of chemicals to the hair. The French word means "froth, foam, lather, or whipped cream." A mousse foam can deliver any haircare chemical with a delightful advantage. It isn't messy, which is very important in coloring and curling hair, and it gives pinpoint accuracy with no overspray or runoff. Although the technique is as old as whipped cream from a bottle pressurized with carbon dioxide, the mousse in haircare products has only become popular since 1983.

Because the permanent-waving process depends upon breaking and reforming cystine bonds present in hair keratin, it appears possible to actually add protein structures to the hair by covalent bonding. Products are now on the market that make this claim; in addition to chemically bonding protein bulk to the hair, the process is thought to restore deteriorated hair by adding protein bridges between broken fragments within a strand.

Removing Hair—Depilatories

The purpose of a depilatory is to remove hair chemically. Because skin is sensitive to the same kind of chemicals that attack hair, such preparations should be used with caution and, even then, some damage to the skin is almost unavoidable. Because of this, depilatories should be used only weekly and should never be used on skin that is infected or when a rash is present. Depilatories should not be used with a deodorant with astringent action. If the sweat pores are closed by the deodorant, the caustic chemicals are retained and can do considerable harm.

The chemicals used as depilatories include sodium sulfide, calcium sulfide, strontium sulfide (water-soluble sulfides), and calcium thioglycolate [$Ca(HSCH_2COO)_2$], the calcium salt of the compound used to break S—S bonds between protein chains in permanent waving. These active chemicals are added to a cream base.

The water-soluble sulfides are all strong bases in water, as indicated by the hydrolysis of the sulfide ion because of its high affinity for protons.

$$\underset{\text{Sulfide}}{S^{2-}} + H_2O \longrightarrow HS^- + \underset{\text{Hydroxide}}{OH^-}$$

A dilute solution of sodium sulfide may have a pH of 13 (strongly basic). This basic solution breaks some peptide bonds in the protein chain, and the result is a mixture of peptides and amino acids that can be washed away in a detergent solution.

Hair is also removed by electrical cauterization, commonly called "electrolysis." The hair follicle is destroyed by a high-voltage electric spark.

SELF-TEST 16A

1. Drugs and cosmetics are regulated in the United States by the ______________.
2. The principal protein making up the corneal layer is ______________. This same protein is also the principal component of hair: True ( ) False ( )
3. About 18% of the protein on hair is made up of the amino acid ______________, which contains disulfide linkages.
4. At approximately what pH does the protein of hair and skin soften? ______________
5. What kinds of chemical bonds are broken and remade when hair undergoes a permanent wave? ______________
6. What two natural color pigments are found in human hair? ______________ and ______________
7. A depilatory has a pH (greater than 7), (less than 7).
8. When a sulfide dissolves in water, the pH of the solution (a) increases (becomes basic), (b) decreases (becomes acidic), (c) doesn't change.
9. A common oxidizer used in hair dyes to develop colors is (a) oxygen, (b) chlorine, (c) hydrogen peroxide, (d) ozone.

SKIN PRODUCTS

Creams and Lotions

To remain healthy, the moisture content of skin must stay near 10%. If it is higher, microorganisms grow too easily; if lower, the corneal layer flakes off. Washing skin removes fats that help retain the right amount of moisture. If dry skin is treated with a fatty substance after washing, it is protected until enough natural fats have been regenerated.

An **emollient** is a skin softener. Lanolin, an excellent emollient, is a complex mixture of esters from hydrated wool fat. The esters are derived from 33 different alcohols of high molecular weight and 37 fatty acids. Cholesterol is a common alcohol in lanolin and is found both free and in the esters. Cholesterol has hydrophilic properties due to the hydroxyl groups (—OH) and causes the fat mixtures to hydrogen-bond with water. Any preparation that holds moisture in the skin is also termed a **moisturizer** (Fig. 16–13). Emollients also include other natural and synthetic esters, ethers, fatty alcohols, hydrocarbons, and silicones.

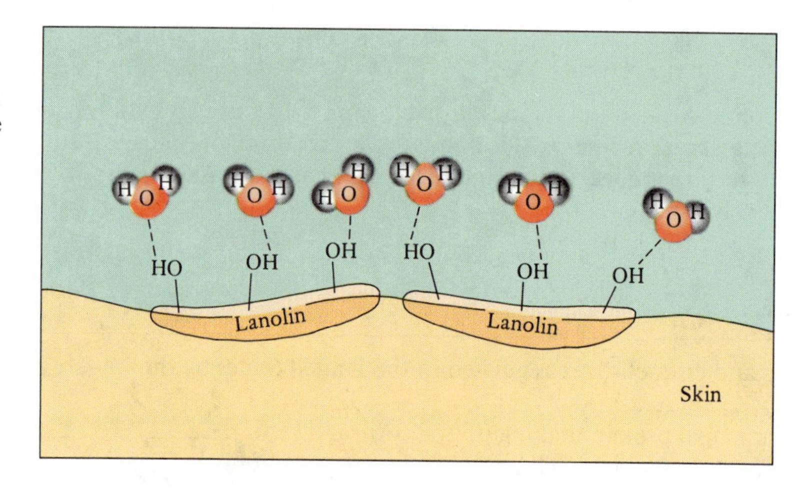

Figure 16–13 The hydroxyl groups of lanolin form hydrogen bonds with water and keep the skin moist. The fat parts of the molecule are "soluble" in the protein and fat layers of the skin.

You may wish to review the concept of colloidal sizes of particles in Chapter 2.

Creams are generally emulsions of either an oil-in-water type or a water-in-oil type. An **emulsion** is simply a colloidal suspension of one liquid in another. The oil-in-water emulsion has tiny droplets of an oily or waxy substance dispersed throughout a water medium; homogenized milk is an example. The water-in-oil emulsion has tiny droplets of a water solution dispersed throughout an oil; examples are natural petroleum and melted butter. An oil-in-water emulsion can be washed off the skin surface with tap water, whereas a water-in-oil emulsion gives skin a greasy, water-repellent surface when rubbed under running water. The product is a **lotion** if the oil or water content is increased to provide fluidity.

Creams offer an excellent application base for other cosmetic preparations or medical applications as in the cases of cream deodorants and antifungal foot products. The cream is the vehicle for chemical delivery into the skin structure.

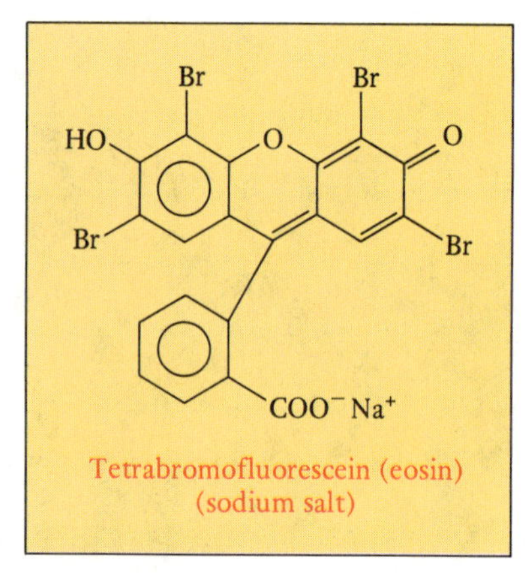

Figure 16–14 Tetrabromofluorescein, a purple dye used in lipstick.

Lipstick

The skin on our lips is covered by a thin corneal layer that is free of fat and consequently dries out easily. A normal moisture content is maintained from the mouth. In addition to being a beauty aid, lipstick, with or without color, can be helpful under harsh conditions that tend to dry lip tissue.

Lipstick consists of a solution or suspension of coloring agents in a mixture of high-molecular-weight hydrocarbons or their derivatives, or both. Consistency of the mixture over a wide temperature range is all important in a product that must have even application, holding power, and a resistance to running of the coloring matter at skin temperature. Lipstick is perfumed and flavored to give it a pleasant odor and taste. The color usually comes from a dye, or "lake," from the eosin group of dyes. A **lake** is a precipitate of a metal ion (Fe^{3+}, Ni^{2+}, Co^{3+}) with an organic dye. The metal ion modifies the natural color of the dye and usually produces a more intense color; the metal also keeps the dye from dissolving in the oil medium, thus preventing the color from running. Two commonly used dyes are dibromofluorescein (yellow-red) and tetrabromofluorescein (purple) [Fig. 16–14]. The ingredients in a typical formulation of lipstick are given in Table 16–1.

A lake is a coloring agent made up of an organic dye adhering to an inorganic substance called a mordant. Some lakes are also approved as food colors.

TABLE 16-1 A Typical Lipstick Formulation

Ingredient	Purpose	Percentage
Dye	Furnishes color	4-8
Castor oil, paraffins, or fats	Dissolves dye	50
Lanolin	Emollient	25
Carnauba and/or beeswax	Makes stick stiff by raising the melting point	18
Perfume	Gives pleasant odor	Small amount
Flavor	Gives pleasant taste	Small amount

Note: Carnauba wax and beeswax are mixtures of high-molecular-weight esters.

Face and Body Powder

Powder is used to give the skin a smooth appearance and a dry feel. Face powders often contain dyes to impart color or shading to the skin. The principal ingredient in body powder is talc ($Mg_3(OH)_2Si_4O_{10}$), a natural mineral able to absorb both water and oil. The absorptive properties of talc are due to the electronegative oxygen atoms, which can hydrogen-bond to water, and to the extensive amount of surface area resulting from the fineness of the ground powder. A binder, such as zinc stearate, is necessary to increase adherence to the skin. Zinc oxide or another astringent is added to shrink tissue and reduce fluid flow in oily skins. A general formulation for body powder is given in Table 16-2.

Powders are less effective for the application of medicines but preferred when dryness is desired, as in foot powders.

Eye Makeup

Eye makeup consists of emollients, solvents, preservatives, and colors. Mixtures of fats, oils, petrolatum, lanolin, beeswax, and paraffin can be blended to give the desired consistency and melting point. Eyebrow pencils can be colored black with lampblack (carbon), brown with a mixture of iron oxide and lampblack, or a variety of colors with other dyes.

Petrolatum, or petroleum jelly, is a semisolid mixture of hydrocarbons (saturated, $C_{16}H_{34}$ to $C_{32}H_{66}$; and unsaturated, $C_{16}H_{32}$; etc.; melting point, 34-54°C.

The molding and sticky qualities of mascara are obtained by increasing the amounts of soap and waxes in the mixture. Chromic oxide imparts a dark green color to mascara while ultramarine (a sodium and aluminum silicate admixed with sodium sulfide) gives a blue color. Mascara may be made

TABLE 16-2 General Formula for Body Powder

Ingredient	Purpose	Percentage
Talc	Absorbent	56
Precipitated chalk ($CaCO_3$)	Absorbent	10
Zinc oxide	Astringent	20
Zinc stearate	Binder	6
Dye	Color	Trace
Perfume	Odor	Trace

water-soluble or water-resistant depending on the emollients and solvents used.

A typical eye shadow base is composed of 60% petroleum jelly, 6% lanolin, 10% fats and waxes (beeswax, spermaceti oil, and cocoa butter are commonly used), and the balance zinc oxide (white) plus tinting or coloring dyes. Cocoa butter is composed of triglycerides of stearic, palmitic, and lauric acids.

Glycerides (see Chapter 9) are triesters of glycerine and fatty acids. Refer to the section on soap making later in this chapter.

PROTECTION FROM THE SUN

Sunbather. (*The World of Chemistry,* Program 10, "Signals from Within")

Ultraviolet light darkens skin as the skin responds to the light by increasing the concentration of the natural pigment **melanin.** The melanin absorbs some of the ultraviolet photons and changes the energy to heat, thus protecting the molecular structure of the skin from damage by these photons. However, even with melanin's protection, ultraviolet light in the short-wavelength region causes a general degradation of the skin and, in extreme cases, skin cancer. The problem is more acute for fair-skinned people, whose skin has smaller amounts of melanin.

Since chemicals selectively absorb particular wavelengths of light, it is possible to screen the skin from the harmful radiation by placing an absorbing chemical in a lotion to be applied to the skin. The chemical must function in a manner similar to melanin. The first popular **sunscreens** contained *p*-aminobenzoic acid and some of its derivatives as the active ingredients. The absorption spectrum for *p*-aminobenzoic acid (PABA) is shown in Figure 16–15, showing that it absorbs an appreciable amount of the radiation in the dangerous portion of the spectrum.

Sunbathers refer to *p*-aminobenzoic acid and its close derivatives as PABA.

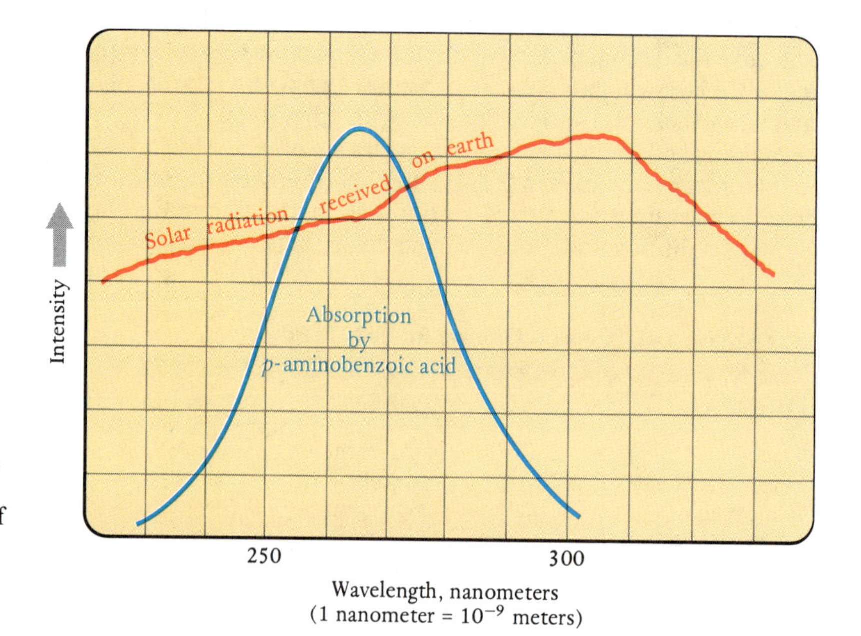

Figure 16–15 Absorption spectrum of *p*-aminobenzoic acid and its relationship to solar ultraviolet radiation received on Earth. Maximum absorption occurs at 265 nm, although it absorbs at other wavelengths as shown. The maximum of the deep-burning ultraviolet radiation received on Earth is about 308 nm.

Coppertone assembly line. (*The World of Chemistry*, Program 10, "Signals from Within")

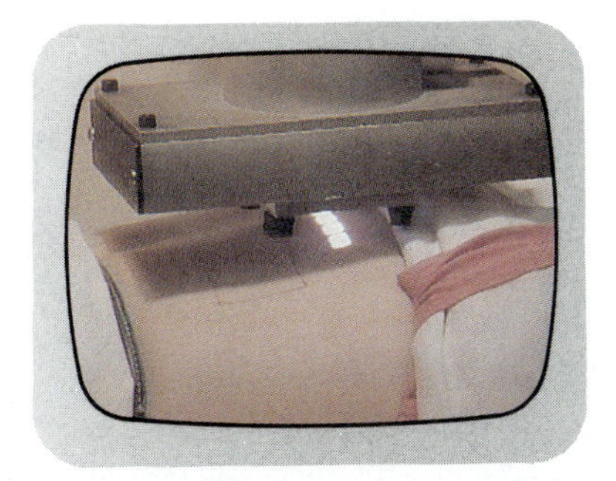

Sunscreen testing. (*The World of Chemistry*, Program 10, "Signals from Within")

There is a growing concern relative to possible toxic effects with the *p*-aminobenzoic acid compounds. In Europe, the market is moving toward cinnamates, derivatives of cinnamic acid, and in the United States the compounds on the rise in this market are the benzophenones.

Cinnamic acid   Benzophenone

The sun protection factor (SPF) for a sunscreen gives the ratio of the protection of the screen to that in natural skin. An SPF of 4, then, would provide four times the skin's natural sunburn protection. Although numbers above 30 are advertised, the FDA has questioned whether numbers above 15 are realistic.

A recent application of sunscreening is the introduction to the market of eyedrops that claim protection from ultraviolet light for several hours.

Some suntan products contain a dye, which dyes the skin. Local anesthetics such as benzocaine are added to some suntan preparations to overcome the pain associated with overexposure to the sun.

DEODORANTS

The 2 million sweat glands on the surface of the body are primarily used to regulate body temperature via the cooling effect produced by the evaporation of the water they secrete. This evaporation of water leaves solid constituents, mostly sodium chloride, as well as smaller amounts of proteins and other organic compounds. Body odor results largely from amines and hydrolysis products of fatty oils (fatty acids, acrolein, etc.) emitted from the body and from bacterial growth within the residue from sweat glands. Sweating is both normal and necessary for the proper functioning of the

Body odor is promoted by bacterial action.

human body; sweat itself is quite odorless, but the bacterial decomposition products are not.

An astringent closes the pores, thus stopping the flow of perspiration.

There are three kinds of deodorants: those that directly "dry up" perspiration by acting as astringents, those that have an odor to mask the odor of sweat products, and those that remove odorous compounds by chemical reaction. Among those that have astringent action are hydrated aluminum sulfate, hydrated aluminum chloride ($AlCl_3 \cdot 6H_2O$), aluminum chlorohydrate [actually aluminum hydroxychloride, $Al_2(OH)_5Cl \cdot 2H_2O$ or $Al(OH)_2Cl$ or $Al_6(OH)_{15}Cl_3$], and alcohols. Compounds that act as deodorizing agents include zinc peroxide, essential oils and perfumes, and a variety of mild antiseptics to stop the bacterial action. Zinc peroxide removes odorous compounds by oxidizing the smelly amines and fatty acid compounds. The essential oils and perfumes absorb or otherwise mask the odors.

SELF-TEST 16B

1. Another name for a skin-softener is a(n) ____________.
2. Any product that holds moisture in the skin is called a(n) ____________.
3. Cholesterol, a naturally occurring alcohol in lanolin, can bond to water molecules by what type of bonding? ____________
4. What gives lipstick its color? ____________
5. What is the purpose of zinc oxide in a body powder? ____________
6. What is the purpose of lampblack found in some mascaras? ____________
7. Ultraviolet light darkens skin by stimulating production of the pigment called ____________.
8. The ingredient called PABA, found in many sunscreen formulations, functions as ____________.
9. The sun protection factor (SPF) is found on many sunscreens. Which would be better to screen out ultraviolet radiation, SPF = 1 or SPF = 4? ____________
10. Many deodorants contain salts of what metal? ____________. These compounds act as ____________.

CLEANSING AGENTS

Surfactants stabilize suspensions of nonpolar materials in polar solvents or vice versa. Over 6 billion pounds are produced in the United States per year. Examples include soaps, detergents, wetting agents, and foaming agents.

Dirt can be defined as matter in the wrong place. Tomato soup may be tasty food, but on your shirt it is dirt. There are over 1200 commercial cleansing, or surface-active, compounds **(surfactants)** capable of removing dirt with results similar to that shown in Figure 16–16. The classic surfactant, soap, dates back in recorded history to the Sumerians in 2500 B.C., in what is now Iraq and Iran. Soap has always been made from the reaction of a fat with an alkali. The Greek physician Galen referred to this recipe and stated further that soap removed dirt from the body as well as serving as a medicament. What is now new is that soap can be made in a very pure state

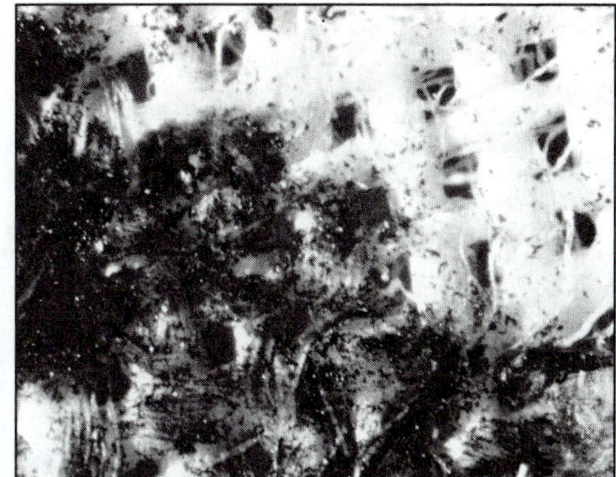

Figure 16–16 Photomicrograph of clean cotton cloth *(left)* and soiled cotton cloth *(right)*. The proper application of surface-active agents should return the soiled cloth to its original state.

and many other compounds, both natural and synthetic, have been found to be excellent surfactants and are commercially employed in a variety of applications.

Soap

The reaction of fats and oils in strongly basic solutions is a hydrolysis reaction that produces glycerol and salts of the fatty acid. Such hydrolysis reactions are called **saponification** reactions: the sodium or potassium salts of the fatty acids formed are **soaps.** Pioneers prepared their soap by boiling animal fat with an alkaline solution obtained from the ashes of hardwood. The resulting soap could be "salted out" by adding sodium chloride, making use of the fact that soap is less soluble in a salt solution than in water. Inventory of cleaning materials in a typical modern household might include half a dozen or more formulated products designed to be the most suitable for a specific job, whether cleaning clothes, floors, or the family car.

Review Chapter 9 for the molecular structure of fats and oils.

Principal fats and oils for soap making: tallow from beef and mutton, coconut oil, palm oil, olive oil, bone grease, and cottonseed oil.

$$\begin{array}{l} CH_3(CH_2)_{16}COO-CH_2 \\ \qquad\qquad\qquad\quad | \\ CH_3(CH_2)_{16}COO-CH \\ \qquad\qquad\qquad\quad | \\ CH_3(CH_2)_{16}COO-CH_2 \end{array} + 3\,NaOH \longrightarrow 3\,CH_3(CH_2)_{16}COO^-Na^+ + \begin{array}{l} HO-CH_2 \\ \qquad\; | \\ HO-CH \\ \qquad\; | \\ HO-CH_2 \end{array}$$

Tristearin (glyceryl tristearate) — Sodium stearate (a soap) — Glycerol

The cleansing action of soap can be explained in terms of its molecular structure. Substances that are water soluble can be readily removed from the skin or a surface by simply washing with an excess of water. To remove a sticky sugar syrup from one's hands, the sugar is dissolved in water and rinsed away. Many times the material to be removed is oily, and water merely runs over the surface of the oil. Because the skin has natural oils, even substances such as ordinary dirt that are not oily themselves can adhere to the skin quite strongly. The cohesive forces (forces between molecules tending to hold them together) within the water layer are too large to allow the oil

Floating soaps float because of trapped air.

Toilet soaps are generally pure soap to which dyes and perfumes are added.

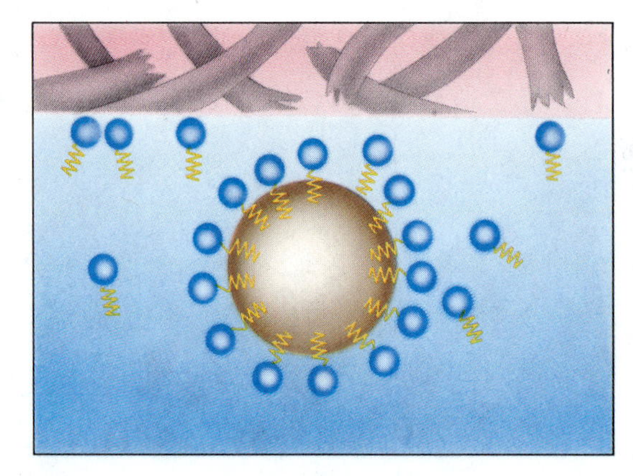

(a)

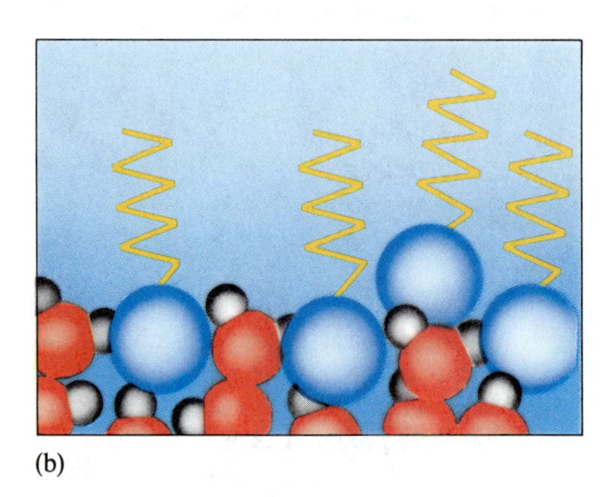

(b)

Figure 16–17 The cleaning action of soap. (a) Soap molecules in water interact strongly with the water through electrical interaction at the salt end of the molecule. The hydrocarbon end of the soap is "pulled" along into solution by the water-salt interaction. (b) The soap molecule, with its oil-soluble and water-soluble ends, becomes oriented at an oil-water interface such that the hydrocarbon chain is in the oil (with molecules that are electrically similar, nonpolar) and the salt group ($—COO^-Na^+$) is in the water. When greasy dirt is broken up in soapy water, a process that is aided by mechanical agitation, the oily particles are surrounded and insulated from each other by the soap molecules.

Soap, water, and oil together form an emulsion, with the soap acting as the emulsifying agent.

and water to intermingle (Fig. 16–17). When present in an oil-water system, soap molecules such as sodium stearate move to the interface between the

$$CH_3CH_2CH_2CH_2CH_2CH_2CH_2CH_2CH_2CH_2CH_2CH_2CH_2CH_2CH_2CH_2CH_2C\begin{matrix}{\diagup\!\!\!\!\diagup O} \\ \diagdown O^-Na^+\end{matrix}$$

two liquids. The hydrocarbon chain, which is a nonpolar organic structure, mixes readily with the nonpolar grease molecules, whereas the highly polar $—COO^-Na^+$ group enters the water layer because the polar groups become hydrated (Fig. 16–17b). The soap molecules then tend to lie across the oil-water interface. The grease is then broken up into small droplets by agitation, each droplet surrounded by hydrated soap molecules. The surrounded oil droplets cannot come together again since the exterior of each droplet is covered with $—COO^-Na^+$ groups that interact strongly with the surrounding water. If enough soap and water are available, the oil is swept away, leaving a clean and water-wet surface.

Stearates: sodium—hard soap; potassium—soft soap; ammonium—liquid soap.

Pure soap is a mildly basic material because it is the salt (salts are ionic) of a strong base and a weak acid (Stearate ions + $H_2O \rightarrow$ Hydrogen stearate + OH^-).

Shampoos

Shampoos are generally more complex than a simple soap solution, with a number of ingredients to satisfy different requirements for maintaining clean and healthy-looking hair. Condensation products like that obtained from diethanolamine and lauric acid are often used because good surfactant properties are obtained without the alkaline properties characteristic of soap.

$$\underset{\text{Diethanolamine}}{HN(CH_2CH_2OH)_2} + \underset{\text{Lauric acid}}{CH_3(CH_2)_{10}COOH} \longrightarrow \underset{\text{An amide detergent}}{CH_3(CH_2)_{10}\overset{\overset{O}{\|}}{C}-N(CH_2CH_2OH)_2} + H_2O$$

Some shampoos contain anionic detergents. Sodium lauryl sulfate is an example of an anionic detergent.

$$\underset{\text{Sodium lauryl sulfate}}{CH_3(CH_2)_{11}OSO_3^-Na^+}$$

The hair is more manageable and has a better sheen if all the shampoo is removed. An anionic detergent can be removed by using a **rinse** containing a dilute solution of a cationic detergent, which electrically attracts the anions and facilitates their removal. Caution should be taken with the cationic rinse because of the possible irritation to the eyes.

Shampoos also contain compounds to prevent the calcium or magnesium ions in hard water from forming a precipitate. Ethylenediaminetetraacetic acid (EDTA, see Fig. 14–4), a metal complexing agent, added to a shampoo, ties up the calcium, magnesium, and iron and avoids the sticky precipitate.

Calcium, magnesium, and iron soaps are insoluble in water, forming a sticky precipitate, a common problem when using soap in "hard water." Soap and rain water—clean hair; soap and hard water—a sticky mess.

Lanolin and mineral oil (or their substitutes) are often added to shampoos to replace the natural oils in the scalp, thus preventing it from drying out and scaling. The presence of oil additives and stabilizers gives the shampoo a pearlescent appearance.

Clear shampoos are preferred in the United States, 70% of the market; Europeans buy 80% pearlescent products.

Synthetic Detergents

There is a synthetic detergent for almost every type of cleaning problem.

A typical laundry detergent might contain surfactants, builders, ion exchangers, alkalies, bleaches, fabric softeners, anticorrosion materials, anti-redeposition materials, enzymes, optical brighteners, fragrances, dyes, and fillers.

Synthetic detergents ("syndets") are derived from organic molecules designed to have the same cleansing action but less reaction than soaps with the cations found in hard water. As a consequence, synthetic detergents are more effective in hard water than soap. Soap leaves undesirable precipitates that have no cleansing action and tend to stick to laundry.

There are many different synthetic detergents on the market. The molecular structure of a detergent molecule consists of a long oil-soluble (hydrophobic) group and a water-soluble (hydrophilic) group. The hydrophilic groups include sulfate ($-OSO_3^-$), sulfonate ($-SO_3^-$), hydroxyl ($-OH$), ammonium ($-NH_3^+$), and phosphate [$-OPO(OH)_2$] groups.

$$\underset{\text{Sulfate group}}{R-O-\overset{\overset{O}{\uparrow}}{\underset{\underset{O}{\downarrow}}{S}}-O^-} \qquad \underset{\text{Sulfonate group}}{R-\overset{\overset{O}{\uparrow}}{\underset{\underset{O}{\downarrow}}{S}}-O^-}$$

Cationic (positively charged) detergents are almost all quaternary ammonium halides with the general formula:

$$R_1-\overset{\overset{R_2}{|}}{\underset{\underset{R_4}{|}}{N^+}}-R_3X^-$$

where one of the R groups is a long hydrocarbon chain and another frequently includes an —OH group. X^- represents a halogen ion such as chloride. In these the water-soluble portion is positively charged; so they are sometimes called invert soaps (in soaps the water-soluble portion is negatively charged).

Cationic detergents act as disinfectants.

Cationic detergents frequently exhibit pronounced bactericidal qualities. Cationic detergents are incompatible with anionic detergents. When they are brought together, a high-molecular-weight insoluble salt precipitates out, and this has none of the desired detergent properties of either detergent.

In commercial production, approximately 62% of the surfactants are anionic, 29% nonionic, and 9% cationic, with nonionics increasing their share faster.

Some detergents are nonionic. They have a polar, but not an ionic, grouping attached to a large organic grouping of low polarity. Consider the following formula that has a long hydrocarbon chain attached through an ester group to a carbon chain containing multiple ether links.

$$\underbrace{CH_3(CH_2)_{11}}_{\text{Carbon chain}}\underbrace{COO}_{\text{Ester group (—COO—)}}(CH_2)_2O(CH_2CH_2O)_2CH_2CH_2OH \quad \text{Ether links (—O—)}$$

Hydrophilic substances have an attraction for water, usually through dipole–dipole attraction or hydrogen bonding. Cotton, for example, develops a moist feeling in humid air.

The carbon chain is oil-soluble and the rest of the molecule is hydrophilic, the properties needed for the molecule to be a detergent.

The nonionic detergents have several advantages over ionic detergents. Since nonionics contain no ionic groups, they cannot form salts with calcium, magnesium, and iron ions and consequently are unaffected by hard water. For the same reason, nonionic detergents do not react with acids and may be used even in relatively strong acid solutions.

In general, the nonionic detergents foam less than ionic surface-active agents, a property that is desirable where nonfoaming detergents are required, as in dishwashing. Nonionics tend to be viscous liquids with melting points below room temperature. Although scarcely used in 1970, liquid detergents hold one third or more of the detergent market in the early 1990s.

In spite of the major impact of synthetics on the detergent industry, soap is still the number one surfactant, holding approximately 39% of the market.

Whiter Whites and Bleaching Agents

Bleaching agents are compounds used to remove color from textiles. Most commercial bleaches are oxidizing agents such as sodium hypochlorite. Optical brighteners are quite different because they act by converting a portion of the incoming invisible ultraviolet light into visible blue or blue-green light, which is emitted. Together or separately, these two classes of compounds find their way into commercial laundry and cleaning preparations because they seem to make clothes cleaner.

In earlier times textiles were bleached by exposure to sunlight and air. In 1786, the French chemist Berthollet introduced bleaching with chlorine, and subsequently this process was carried out with sodium hypochlorite, an oxidizing agent prepared by passing chlorine into aqueous sodium hydroxide:

$$2\,Na^+ + 2\,OH^- + Cl_2 \longrightarrow \underbrace{Na^+ + OCl^-}_{\text{Sodium hypochlorite}} + Na^+ + Cl^- + H_2O$$

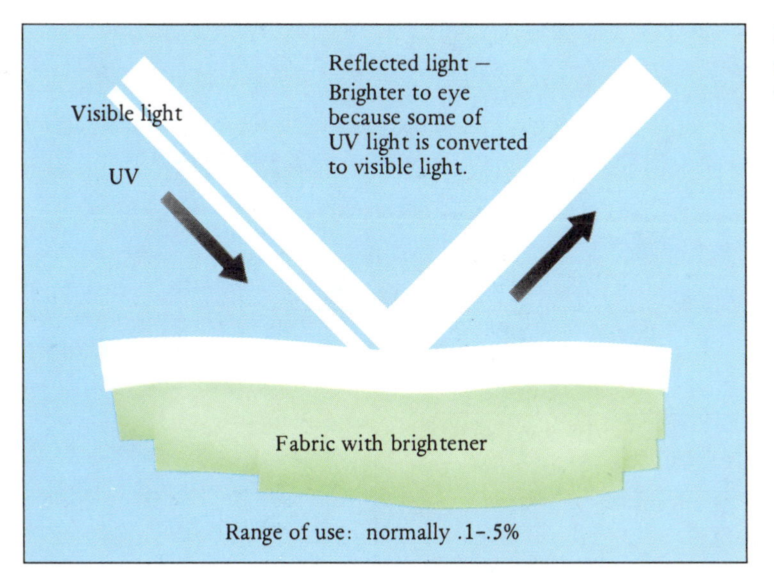
Reflected light –
Brighter to eye
because some of
UV light is converted
to visible light.
Visible light
UV
Fabric with brightener
Range of use: normally .1-.5%

Figure 16–18 Optical brighteners absorb unseen ultraviolet light and emit the energy as visible light.

Shortly thereafter, hydrogen peroxide was introduced as a textile bleach. Later, a number of other oxidizing agents based on chlorine were developed and introduced.

One way to decolorize materials is to remove or immobilize those electrons in the material activated by visible light.

$$\text{Colored (or stained) material} \xrightarrow{-\text{electrons}} \text{White material}$$

The hypochlorite ion, because it is an oxidizing agent, is capable of removing electrons from many colored materials. In this process, the hypochlorite ion is reduced to chloride and hydroxide ions.

$$ClO^- + H_2O + 2\,e^- \longrightarrow Cl^- + 2\,OH^-$$

Optical brighteners are fluorescent compounds. A fluorescent material absorbs light of a shorter wavelength and emits light of a longer wavelength (Fig. 16–18). When optical brighteners are incorporated into textiles or paper, they make the material appear brighter and whiter. An example of such a brightener is represented by the formula in Figure 16–19 along with its absorption and emission spectra in the ultraviolet and visible regions, respectively.

Spot and Stain Removers

To a large extent, stain removal procedures are based on solubility patterns or chemical reactions. Many stains, such as those due to chocolate or other fatty foods, can be removed by treatment with the typical dry-cleaning solvents such as tetrachloroethylene, $Cl_2C{=}CCl_2$.

Stain removers for the more resistant stains are almost always based on a chemical reaction between the stain and the essential ingredients of the stain

Many stains can be removed by an appropriate solvent or chemical reagent.

(a)

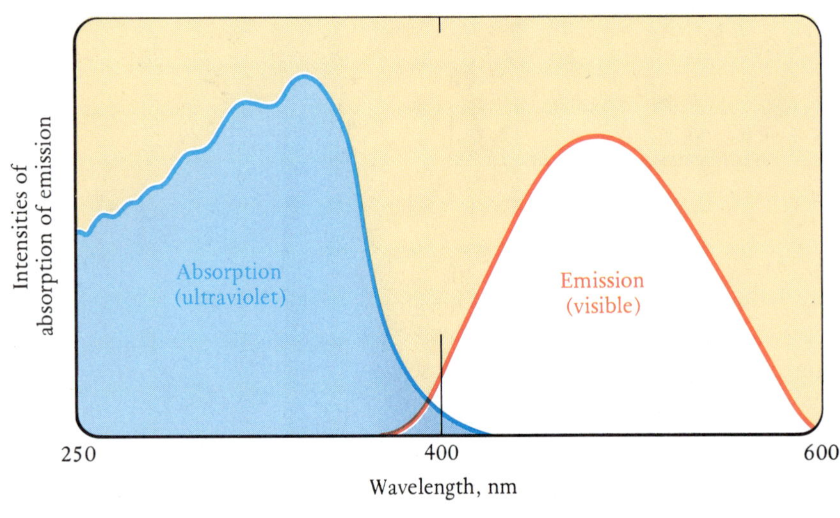

(b)

Figure 16–19 (a) Optical brighteners are composed of complex organic molecules containing numerous double bonds between the atoms. (b) A typical absorption and emission spectral pattern for an optical brightener. The energy of the ultraviolet is "rendered visible" as it is shifted to visible wavelengths. Compare this to special glasses that allow you to see better under night-darkness conditions.

remover. A typical example is an iodine stain remover, which is simply a concentrated solution of sodium thiosulfate. The stain removal reaction is:

$$\underset{\text{Iodine}}{I_2} + 2\,Na_2S_2O_3 \longrightarrow \underset{\text{(soluble in water and colorless)}}{2\,NaI + Na_2S_4O_6}$$

Other examples are given in Table 16–3.

Toothpaste

The structure of tooth enamel is essentially that of a stone composed of calcium carbonate and calcium hydroxy phosphate (apatite) (Fig. 16–20). Such structures are readily attacked by acid. Because the decay of some food particles produces acids and because bacteria convert plaque, a deposit of dextrins, to acids, it is important to keep teeth clean and free from prolonged contact with these acids if the hard, stonelike enamel is to be preserved.

The two essential ingredients in toothpaste are a detergent and an abrasive. The abrasive serves to cut into the surface deposits, and the detergent assists in suspending the particles in a water medium to be carried away in the rinse. Abrasives commonly used in toothpaste formulations in-

TABLE 16–3 Some Common Stains and Stain Removers*

Stain	Stain Remover
Coffee	Sodium hypochlorite
Lipstick	Isopropyl alcohol, isoamyl acetate, Cellosolve ($HOCH_2CH_2OCH_2CH_3$)
Rust and ink	Oxalic acid, methyl alcohol, water
Airplane cement	50/50 amyl acetate and toluene or acetone
Asphalt	Carbon disulfide
Blood	Cold water, hydrogen peroxide
Berry, fruit	Hydrogen peroxide
Grass	Sodium hypochlorite or alcohol
Nail polish	Acetone
Mustard	Sodium hypochlorite or alcohol
Antiperspirants	Ammonium hydroxide
Perspiration	Ammonium hydroxide, hydrogen peroxide
Scorch	Hydrogen peroxide
Soft drinks	Sodium hypochlorite
Tobacco	Sodium hypochlorite

* Before any of these stain removers are used on clothing, the possibility of damage should be checked on a portion of the cloth that ordinarily is hidden. Some stain removers are toxic, such as methanol and carbon disulfide.

clude hydrated silica (a form of sand, $SiO_2 \cdot nH_2O$); hydrated alumina ($Al_2O_3 \cdot nH_2O$); and calcium carbonate ($CaCO_3$). It is difficult to select an abrasive that is hard enough to cut the surface contamination yet not so hard as to cut the tooth enamel. The choice of detergent is easier; any good detergent such as sodium lauryl sulfate will do quite well.

Because the necessary ingredients in toothpaste are not very palatable, it is not surprising to see the inclusion of flavors, sweeteners, thickeners, and colors to appeal to our senses.

One addition to the toothpaste mixture has caused a dramatic decrease in the amount of tooth decay in our population; it is the addition of stannous

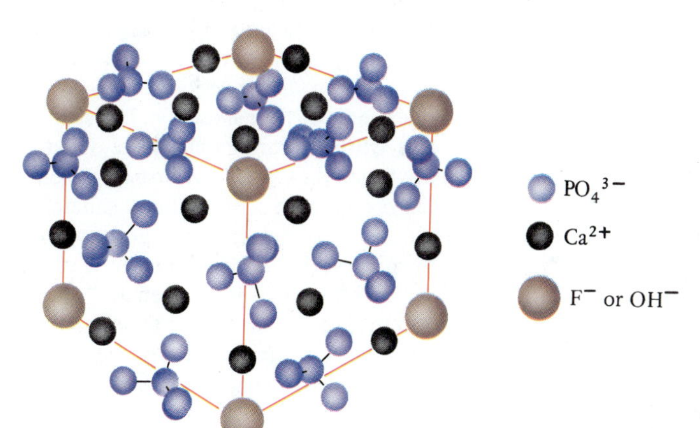

Figure 16–20 Structure of apatite and fluoroapatite. The grey circles denote Ca^{2+} ions; purple, phosphorus; red, oxygen; and brown, OH^- or F^- groups.

"No cavities!" (*The World of Chemistry,* Program 3, "Measurement: The Foundation of Chemistry")

fluoride (SnF_2) to provide a low level of fluoride ion concentration in the brushing medium. Some of the fluoride ions actually replace the hydroxide ions in the hydroxyapatite structure [$Ca_{10}(PO_4)_6(OH)_2$], to form fluoroapatite [$Ca_{10}(PO_4)_6F_2$]. The fluoride ion forms a stronger ionic bond because of its high concentration of negative charge in the crystalline structure, and, as a result, the fluoroapatite is harder and less subject to acid attack than the hydroxyapatite. Hence, there is less tooth decay. The fluoride ion is also introduced into essentially all of the public water supplies for this same purpose.

Most teeth are now lost as a result of gum disease, which results from the lack of proper massage, irritating deposits below the gum line, bacterial infection, and poor nutrition. More attention is being given to toothpastes containing disinfectants such as peroxides in addition to the other ingredients.

SELF-TEST 16C

1. Another name for surface-active compounds is ____________.
2. The hydrolysis of a fat or oil to make glycerol and fatty acids is called ____________.
3. The salt of a fatty acid is called a(n) ____________.
4. Soap can be made to float by adding ____________.
5. What ingredient is often added to shampoos to replace natural oils lost while washing hair? ____________
6. A laundry bleach ( ) oxidizes ( ) reduces many forms of dirt found on soiled clothes.
7. What would be best for removing a mustard stain from a piece of clothing, sodium hypochlorite or cold water? ____________

QUESTIONS

1. Look at the cosmetics you own or use and list one or more ingredients found in each. Try to figure out what each of the ingredients listed does. Note that the ingredients are listed in order of decreasing percentage.
2. Explain why your skin "prunes" when you stay in some waters too long. Why would your skin prune more easily in some water than in others?
3. Why does your hair often become unmanageable on humid days?
4. Explain in general terms how hair is waved or straightened by the application of chemicals. What other means can be used to wave or straighten hair? How are these means alike in terms of what is going on in the hair structure?
5. Name one danger of coloring hair with chemical dyes.
6. Explain how a depilatory works.
7. Name three skin or hair care products that might contain lanolin. What is lanolin and what is its source? What is the function of lanolin in cosmetics?
8. A sunscreen has a SPF of 6. What does this mean?
9. What is a danger of too much ultraviolet radiation striking the skin for long periods of time?
10. Look at the labels of two or three deodorants and decide whether they (a) act as astringents, (b) act as a deodorizing agent by masking, or (c) act to remove odorous compounds by a mild oxidizing action.
11. Look at the ingredients of your shampoo. Does it contain a soap or a synthetic surfactant? Write down its name. (Usually water is listed as the first ingredient.) Does your shampoo contain sodium lauryl sulfate?
12. Look at ten laundry detergents in a local grocery store. Write down the names and beside each indicate which contain a bleach, a whitener, and a softener.
13. While you are in the grocery store, look at several stain removers and list the active ingredients of each.
14. What is the purpose of fluoride in toothpaste? Find two toothpastes that do not contain fluoride.

The International System of Units (SI)

A coherent system of units known as the Système International (SI system), bearing the authority of the International Bureau of Weights and Measures, has been in effect since 1960 and is gaining increasing acceptance among scientists. It is an extension of the metric system that began in 1790, with each physical quantity assigned a unique SI unit. An essential feature of both the older metric system and the newer SI is a series of prefixes that indicate a power of 10 multiple or submultiple of the unit (see Table 2–1).

The SI system was introduced in Chapter 2.

Seven fundamental units are required to describe what is now known about the universe (Table A–1). Figures A–1 through A–5 have been selected to give the student a "feel" for some of the important units used in this text.

Other necessary units are derived from these seven. For example, volume is defined in terms of cubic length (cubic centimeters [cm^3]).

UNITS OF LENGTH

The international spelling of the unit of length is metre.

The standard unit of length, the **meter,** was originally meant to be 1 ten-millionth of the distance along a meridian from the North Pole to the equator. However, the lack of precise geographical information necessitated a better definition. For a number of years the meter was defined as the distance between two etched lines on a platinum-iridium bar kept at 0°C (32°F) in the International Bureau of Weights and Measures at Sèvres, France. However, an inability to measure this distance as accurately as desired and its inaccessibility to most scientists prompted a recent redefinition of the meter as being 1,650,763.73 times the wavelength of a certain wavelength of light given off in the emission spectrum of krypton-86.

Figure A–1 The length of the King's foot, while serving vanity well, was short lived as a standard for the measurement of length if there is a desire for accuracy and reproducibility. (*The World of Chemistry,* Program 3)

TABLE A–1 Fundamental Units of Measure

Physical Quantity	Name of Unit	Symbol
1. Length	meter	m
2. Mass	kilogram	kg
3. Time	second	s
4. Thermodynamic temperature	kelvin	K
5. Luminous intensity	candela	cd
6. Electric current	ampere	A
7. Amount of a substance	mole	mol

Figure A–2 One inch is equal to 2.54 centimeters (cm). (Charles Steele)

UNITS OF MASS

The primary unit of mass is the **kilogram** (1000 g). This unit is the mass of a platinum-iridium alloy sample deposited at the International Bureau of Weights and Measures. One pound contains a mass of 453.6 g (a five-cent nickel coin contains about 5 g).

Conveniently enough, the same prefixes defined in the discussion of length are used in units of mass, as well as in other units of measure.

UNITS OF VOLUME

The SI unit of volume is the **cubic meter** (m^3). However, the volume capacity used most frequently in chemistry is the liter, which is defined as 1 cubic decimeter (1 dm^3). Since a decimeter is equal to 10 cm, the cubic decimeter is equal to 10 cm^3 or 1000 cubic centimeters (cc). One cc, then, is equal to one milliliter (the thousandth part of a liter). The mL (or cc) is a common unit that is often used in the measurement of medicinal and laboratory quantities. There are then 1000 liters (L) in a kiloliter or cubic meter.

Figure A–3 (a) A sample is weighed on a laboratory balance. (b) Dr. Hoffmann at an analytical balance. (*The World of Chemistry,* Program 3)

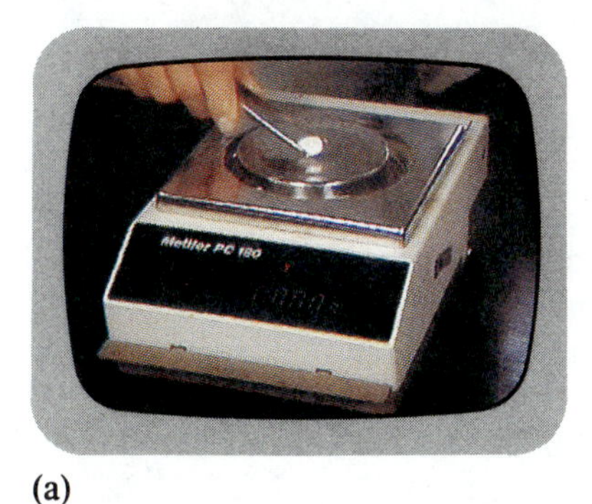

(a)

(b)

(a)

(b)

Figure A–4 (a) The volume of a liter is slightly more than that of a quart. One liter equals 1.06 quarts. (b) Volumetric equipment is designed to do a specific job to a defined level of precision in the easiest way; hence, such equipment takes many forms. (Charles D. Winters)

UNITS OF ENERGY

The SI unit for energy is the **joule** (J), which is defined as the work performed by a force of one newton acting through a distance of one meter. A newton is defined as that force which produces an acceleration of one meter per second when applied to a mass of one kilogram. Conversion units for energy are:

$$1 \text{ calorie} = 4.184 \text{ J}$$

$$1 \text{ kilowatt-hour} = 3.5 \times 10^6 \text{ J}$$

OTHER SI UNITS

Other SI units are listed below.

Time	second (sec)
Temperature	Kelvin (K)
Electric current	ampere (A) = 1 coulomb per second
Amount of molecular substance	mole = 6.023×10^{23} molecules
Pressure	pascal (Pa) = 1 newton per square meter
Power	watt (W) = 1 joule per second
Electric charge	coulomb (C) = 6.24196×10^{18} electron charges = 1.036086×10^{-5} faraday

TABLE A–2 Conversion Factors*

Length	1 inch (in)	= 2.54 centimeters (cm)
	1 yard (yd)	= 0.914 meter (m)
	1 mile (mi)	= 1.609 kilometers (km)
	1 meter (m)	= 3.28 feet (ft)
	1 kilometer (km)	= 0.622 mile (mi)
Volume	1 ounce (oz)	= 29.57 milliliters (mL)
	1 liter (L)	= 1.06 quarts (qt)
	1 gallon (gal)	= 3.78 liters (L)
Mass (weight)†	1 ounce (oz)	= 28.35 grams (g)
	1 pound (lb)	= 453.6 grams (g)
	1 ton (tn)	= 907.2 kilograms (kg)
	1 kilogram (kg)	= 2.20 pounds (lb)

* Common English units are used.

† Mass is a measure of the amount of matter, whereas weight is a measure of the attraction of the earth for an object at the earth's surface. The mass of a sample of matter is constant, but its weight varies with position and velocity. For example, the space traveler, having lost no mass, becomes weightless in Earth orbit. Although mass and weight are basically different in meaning, they are often used interchangeably in the environment of Earth's surface.

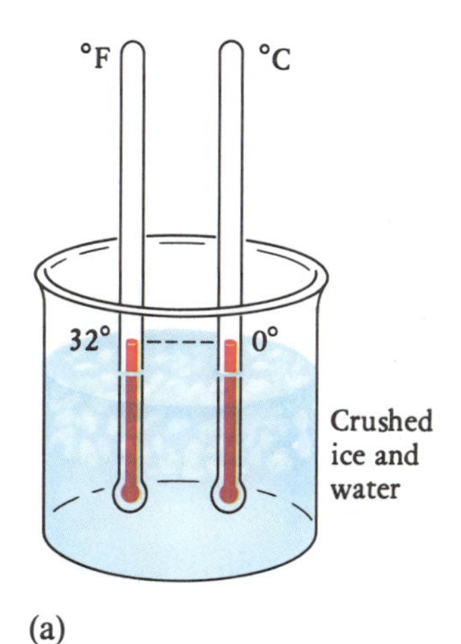

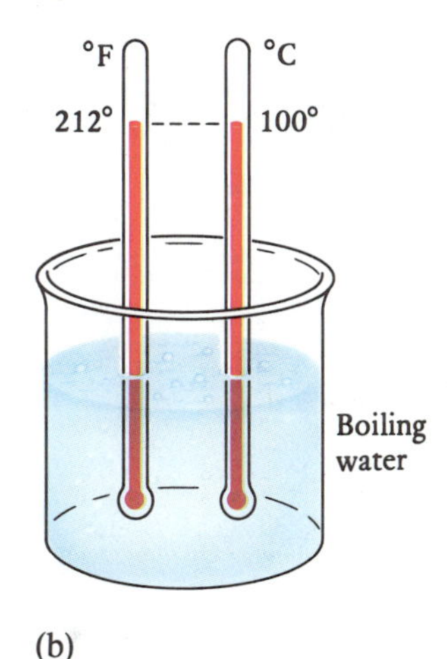

Figure A–5 Temperature scales are defined in terms of the expansion of common materials such as mercury and in terms of fixed reference points such as the changes of state of water (a and b) and other common materials. It is only a matter of preference and convenience whether one scale or another is used.

Conversion of measurements from one system to the other is a common problem. Some commonly used English–SI equivalents (conversion factors) are given in Table A–2.

Using the SI system requires no higher level of thinking than using the English system, and problem solving is usually easier in the newer system if one is familiar with the units. Because the prefixes are defined in terms of the decimal system, the conversion from one metric length to another involves only shifting the decimal point. Mental calculations are quickly accomplished.

How many centimeters are in a meter? Think: Because a centimeter is the one-hundredth part of a meter, there would be 100 cm in a meter.

Consider the following three examples:

EXAMPLE 1

1. English: How many feet are in 0.5 mile?

$$? \text{ feet} = 0.5 \cancel{\text{mile}} \times 5280 \text{ ft}/\cancel{\text{mile}} = 2640 \text{ ft}$$

2. SI: How many meters are in 2 km?
   The prefix kilo- means 1000 times, so

$$? \text{ m} = 2 \cancel{\text{km}} \times 1000 \text{ m}/\cancel{\text{km}} = 2000 \text{ m}$$

It is hardly worth the trouble to write anything down in this solution. One just thinks 2000 meters as one thinks one dollar for ten dimes.

EXAMPLE 2

1. English: How many ounces are in 1.50 gal?
   You might remember there are 32 oz/qt and 4 qt/gal. Your solution then would be:

$$? \text{ ounces} = 1.50 \,\cancel{\text{gal}} \times 4 \,\cancel{\text{qt}}/\cancel{\text{gal}} \times 32 \text{ oz}/\cancel{\text{qt}} = 192 \text{ oz}$$

2. SI: How many milligrams are in a coin that weighs 5 g?
   The prefix milli- means one-thousandth of, so there are 1000 mg in a gram. Five grams then would be 5000 mg.

EXAMPLE 3

A typical piece of white bread contains 70 dietary calories (Cal). One dietary calorie is equal to 1000 small (scientific) calories (cal). How many small calories are in a typical piece of white bread? 70,000 small calories.
Note: You can see why the dietitians like the larger unit; with it, they can use smaller numbers in their notations.

If you wish to make "Perfect Brownies" (Table A–3), there is really little choice between the SI and English systems; the ease depends on the measuring instrument at hand and your familarity with the units to be measured.

Further information on SI units can be obtained from "SI Metric Units —An Introduction," by H. F. R. Adams, McGraw-Hill Ryerson Ltd., Toronto, 1974.

TABLE A–3 Recipe for Perfect Brownies in English and Metric Measurements

Ingredient	English	Metric (SI)
Unsweetened chocolate squares	2 oz	60 g
Butter or margarine	$\frac{1}{2}$ c	120 mL
Sugar	1 c	240 mL
Eggs	2	2
Vanilla	1 tsp	5 mL
Sifted enriched flour	$\frac{1}{2}$ c	120 mL
Chopped walnuts	$\frac{1}{2}$ c	120 mL
Oven	325°F	163°C
Pan	8 × 8 × 2 in.	20 × 20 × 5 cm

APPENDIX

B Calculations with Chemical Equations

The basis for calculations with chemical equations was presented in Chapter 2. Problems of a more complex nature and a systematic approach to their solution are presented in the following examples. Finally, a list of exercise problems is given for further study.

EXAMPLE 1

BALANCED EQUATIONS EXPRESS NUMBER RATIOS FOR PARTICLES

In the reaction of hydrogen with oxygen to form water, how many molecules of hydrogen are required to combine with 19 oxygen molecules?

Solution A chemical equation can be written for a reaction only if the reactants and products are identified and the respective formulas determined. In this problem, the formulas are known and the unbalanced equation is:

$$H_2 + O_2 \longrightarrow H_2O$$

It is evident that one molecule of oxygen contains enough oxygen for two water molecules and the equation, as written, does not account for what happens to the second oxygen atom. As it is, the equation is in conflict with the conservation of atoms in chemical changes. This conflict is easily corrected by balancing the equation:

$$2\ H_2 + O_2 \longrightarrow 2\ H_2O$$

Now all atoms are accounted for in the equation and it is obvious that two hydrogen molecules are required for each oxygen molecule. In other words, two hydrogen molecules are equivalent to one oxygen molecule in their usage. This can be expressed as follows:

2 hydrogen molecules {are equivalent to} 1 oxygen molecule

or

$$2\ H_2 \text{ molecules} \approx O_2 \text{ molecule}$$

or

$$\frac{2\ H_2 \text{ molecules}}{O_2 \text{ molecule}}$$

which can be read as two hydrogen molecules per one oxygen molecule.

Using the factor-label approach, the solution is readily achieved.

$$?\ H_2 \text{ molecules} = 19\ O_2 \text{ molecules}$$

$$?\ H_2 \text{ molecules} = 19\ \cancel{O_2 \text{ molecules}} \times \frac{2\ H_2 \text{ molecules}}{\cancel{O_2 \text{ molecule}}}$$

$$= 38\ H_2 \text{ molecules}$$

Note: The reader is likely to say at this point that the method is cumbersome and that he can quickly see the answer to be 38 H_2 molecules without "the method." However, problems to follow are made much easier if a systematic method is used.

EXAMPLE 2

LABORATORY MOLE RATIOS IDENTICAL WITH PARTICLE NUMBER RATIOS

How many moles of hydrogen molecules must be burned in oxygen (the reaction of Example 1) to produce 15 moles of water molecules (about a glassful)?

Solution The balanced equation

$$2\,H_2 + O_2 \longrightarrow 2\,H_2O$$

tells us that two molecules of hydrogen produce two molecules of water, or

$$\text{2 molecules hydrogen} \approx \text{2 molecules water}$$

and therefore

$$\text{1 molecule hydrogen} \approx \text{1 molecule water}$$

It is obvious then that the number of water molecules produced is equal to the number of hydrogen molecules consumed regardless of the actual number involved. Therefore,

$$6.02 \times 10^{23} \text{ molecules of hydrogen} \approx 6.02 \times 10^{23} \text{ molecules of water}$$

Because 6.02×10^{23} is a number called the mole, it follows that 1 mole of hydrogen molecules produce 1 mole of water molecules. The general conclusion, then, is the following: the ratio of particles in the balanced equation is the same as the ratio of moles in the laboratory. The solution to the problem logically follows:

$$\left.\begin{matrix}\text{? moles of hydrogen}\\ \text{molecules}\end{matrix}\right\} = \left\{\begin{matrix}15\ \cancel{\text{moles}}\\ \cancel{\text{water molecules}}\end{matrix}\right\} \times \frac{\text{1 mole hydrogen molecules}}{1\ \cancel{\text{mole water molecules}}}$$

$$= \text{15 moles hydrogen molecules}$$

Note: Again the solution to the problem looks simple enough without resorting to the factor-label method. However, in Examples 3 and 4, the numbers become such that a quick mental solution is not readily achieved by most students.

EXAMPLE 3

MOLE WEIGHTS YIELD WEIGHT RELATIONSHIPS

How many grams of oxygen are necessary to react with an excess of hydrogen to produce 270 g of water?

Solution From the balanced equation

$$2\,H_2 + O_2 \longrightarrow 2\,H_2O$$

the mole ratio between oxygen and water is immediately evident and is 1 mole of oxygen molecules per 2 moles of water molecules, or

$$\frac{\text{1 mole oxygen molecules}}{\text{2 moles water molecules}}$$

This mole ratio can be changed into a weight ratio because the mole weight can be easily calculated from the atomic weights involved. One molecule of oxygen (O_2) weighs 32 amu (16 amu for each oxygen atom). Therefore, a mole of oxygen molecules weighs 32 g. Similarly, 2 moles of water weigh 36 g [2(16 + 1 + 1)]. Therefore, the weight ratio is:

$$\frac{1\ \cancel{\text{mole oxygen molecules}} \times \dfrac{32\text{ g oxygen}}{\cancel{\text{mole oxygen molecules}}}}{2\ \cancel{\text{moles water molecules}} \times \dfrac{18\text{ g water}}{\cancel{\text{mole water molecules}}}}$$

or

$$\frac{32\text{ g oxygen}}{36\text{ g water}}$$

This weight relationship is exactly the conversion factor needed to answer the original question:

$$?\text{ grams oxygen} = 270\ \cancel{\text{g water}} \times \frac{32\text{ g oxygen}}{36\ \cancel{\text{g water}}}$$

$$= 240\text{ g oxygen}$$

Note: It should be observed that a weight relationship could be established between any two of the three pure substances involved in the reaction, regardless of whether they are reactants or products.

EXAMPLE 4

How many molecules of water are produced in the decomposition of eight molecules of table sugar? The unbalanced equation is as follows:

$$C_{12}H_{22}O_{11} \longrightarrow C + H_2O$$

Solution Balance the equation

$$C_{12}H_{22}O_{11} \longrightarrow 12\ C + 11\ H_2O$$

$$?\text{ molecules of water} = 8\ \cancel{\text{molecules of sugar}} \times \frac{11\text{ molecules water}}{1\ \cancel{\text{molecule sugar}}}$$

$$= 88\text{ molecules of water}$$

EXAMPLE 5

How many grams of mercuric oxide are necessary to produce 50 g of oxygen? Mercuric oxide decomposes as follows:

$$2\ HgO \longrightarrow 2\ Hg + O_2$$

Solution

Weight of 2 moles of HgO = 2(201 + 16) = 2(217) = 434 g

Weight of 1 mole of O_2 = 2(16) = 32 g

$$?\text{ g HgO} = 50\ \cancel{\text{g oxygen}} \times \frac{434\text{ g mercuric oxide}}{32\ \cancel{\text{g oxygen}}}$$

$$= 678\text{ g mercuric oxide}$$

EXAMPLE 6

How many pounds of mercuric oxide are necessary to produce 50 lb of oxygen by the reaction:

$$2\ HgO \longrightarrow 2\ Hg + O_2$$

Solution Note that the problem is the same as Example 5 except for the units of chemicals. Also note that the conversion factor of Example 5

$$\frac{434\text{ g mercuric oxide}}{32\text{ g oxygen}}$$

can be converted to any other units desired:

$$\frac{434\ \cancel{g}\text{ mercuric oxide} \times \dfrac{1\text{ lb}}{454\ \cancel{g}}}{32\ \cancel{g}\text{ oxygen} \times \dfrac{1\text{ lb}}{454\ \cancel{g}}} = \frac{434\text{ lb mercuric oxide}}{32\text{ lb oxygen}}$$

It is evident that the ratio, 434/32, expresses the ratio between weights of mercuric oxide and oxygen in this reaction regardless of the units employed.

$$?\text{ pounds mercuric oxide} = 50\ \cancel{\text{lb oxygen}} \times \frac{434\text{ lb mercuric oxide}}{32\ \cancel{\text{lb oxygen}}}$$

$$= 678\text{ lb of mercuric oxide}$$

PROBLEMS

1. What weight of oxygen is necessary to burn 28 g of methane (CH_4)? The equation is:

   $$CH_4 + 2\ O_2 \longrightarrow CO_2 + 2\ H_2O$$

   Ans. 112 g oxygen

2. Potassium chlorate ($KClO_3$) releases oxygen when heated according to the equation:

   $$2\ KClO_3 \longrightarrow 2\ KCl + 3\ O_2$$

   What weight of potassium chlorate is necessary to produce 1.43 g of oxygen? *Ans.* 3.65 g $KClO_3$

3. Fe_3O_4 is a magnetic oxide of iron. What weight of this oxide can be produced from 150 g of iron? *Ans.* 207 g oxide

4. Steam reacts with hot carbon to produce a fuel called water gas; it is a mixture of carbon monoxide and hydrogen. The equation is:

   $$H_2O + C \longrightarrow CO + H_2$$

   What weight of carbon is necessary to produce 10 g of hydrogen by this reaction? *Ans.* 60 g carbon

5. Iron oxide (Fe_2O_3) can be reduced to metallic iron by heating it with carbon.

   $$2\ Fe_2O_3 + 3\ C \longrightarrow 4\ Fe + 3\ CO_2$$

   How many tons of carbon would be necessary to reduce 5 tons of the iron oxide in this reaction? *Ans.* 0.56 ton carbon

6. How many grams of hydrogen are necessary to reduce 1 lb (454 g) of lead oxide (PbO) by the reaction:

$$PbO + H_2 \longrightarrow Pb + H_2O$$

Ans. 3.91 g hydrogen

7. Hydrogen can be produced by the reaction of iron with steam.

$$4\ H_2O + 3\ Fe \longrightarrow 4\ H_2 + Fe_3O_4$$

What weight of iron would be needed to produce 0.5 lb of hydrogen?

Ans. 10.5 lb iron

8. Tin ore, containing SnO_2, can be reduced to tin by heating with carbon.

$$SnO_2 + C \longrightarrow Sn + CO_2$$

How many tons of tin can be produced from 100 tons of SnO_2?

Ans. 79 tons tin

Answers to Self-Test Questions

Chapter 1

Self-Test 1

1. observed experimental facts
2. the same
3. serendipity
4. silicon
5. chemophobia
6. Clean Air Act
7. heart attacks

Chapter 2

Self-Test 2A

1. dirt, wood, dusty air, salt water, etc.
2. rain water, gold, quartz, diamond, etc.
3. false
4. energy
5. solution
6. one or more pure substances
7. 109
8. false
9. filtration, distillation, recrystallization, chromatography
10. elemental composition
11. **(a)** oxygen, **(b)** iron, **(c)** hydrogen
12. student recall

Self-Test 2B

1. **(a)** oxygen, an oxygen atom, or 1 mole of oxygen atoms
   (b) two oxygen molecules each containing two atoms, or two moles of oxygen molecules
   (c) a molecule of methane or 1 mole of methane molecules
   (d) reacts to form or yields
   (e) a water molecule or 1 mole of water molecules
   (f) two water molecules or 2 moles of water molecules
2. the element, an atom of the element, or 1 mole of the elemental atoms
3. the elements present and the relative number of each type of atom
4. coefficient
5. **(a)** 6,6,1,6 (The 1 is understood.)
   (b) 6
   (c) 6
   (d) 44, 180
   (e) 264
6. 136 kcal
7. **(a)** 2,1,2
   (b) 1,2,1
   (c) 4,3,2
8. curiosity, profit (or comfort)
9. easier to use
10. two dozen, 2 moles of hydrogen atoms
11. dynamic (Reactants and products are interchanging.)
12. temperature, reactant concentrations, catalysts

Chapter 3

Self-Test 3A

1. b
2. d
3. b
4. 10 protons, 10 electrons, 11 neutrons
5. a
6. b
7. c
8. a
9. hydrogen-3, hydrogen-2
10. atomic number, mass number
11. c

Self-Test 3B

1. R, R, T, N, I
2. Mendeleev
3. atomic number
4. periodic
5. ions
6. ionic
7. 1
8. CaI_2
9. Cl^-
10. valence
11. losing
12. gaining
13. Rb^+, S^{2-} Rb_2S
    Ca^{2+}, O^{2-} CaO
    Mg^{2+}, P^{3-} Mg_3P_2

Self-Test 3C

1. covalent
2. 3
3. 2
4. 4
5. 6
6. tetrahedral
7. d
8. triple
9. hydrogen bonding

Chapter 4

Self-Test 4A

1. solvent
2. solute
3. electrolyte
4. the proton
5. 0.5
6. donates, accepts
7. neutralization
8. pH 2
9. pH 1
10. neutral
11. bases

Self-Test 4B

1. combustion
2. CO_2
3. loss
4. CO
5. reduction
6. b
7. coke, a form of carbon
8. electrical
9. hydrogen and oxygen
10. aluminum
11. false
12. c
13. corrosion
14. galvanizing

Chapter 5

Self-Test 5A

1. kinetic
2. potential
3. coal, petroleum, and natural gas
4. oxygen, water
5. entropy
6. required
7. natural gas
8. quantity
9. 20% to 25% efficiency
10. hydrogen and carbon
11. producing electricity
12. carbon monoxide, hydrogen
13. coal
14. natural gas

Self-Test 5B

1. electricity
2. about 38%
3. energy from trash that does not have to occupy landfills, possible air pollution and heavy metals scattered in the environment
4. fission
5. fission
6. less than
7. released energy
8. nuclear fission reaction
9. nonfissionable
10. false
11. 100%
12. arsenic, boron

Chapter 6

Self-Test 6A

1. graphite
2. diamond
3. tetrahedral
4. structural
5. ethene (ethylene)
6. double
7. ethyne (acetylene)
8. $—C_2H_5$

Self-Test 6B

1. benzene
2. **(a)** alcohol **(b)** carboxylic acid **(c)** aldehyde **(d)** ketone
3. **(a)** C_2H_6, C_2H_5OH, CH_3CHO, CH_3COOH, $(C_2H_5)_2O$, $C_2H_5NH_2$
   (b) ethane, ethyl alcohol, acetaldehyde, acetic acid, diethyl ether, ethyl amine
   (c) $—C_2H_5$ or ethyl group in ethanol, diethyl ether, ethyl amine
   $—CH_3$ or methyl group in acetaldehyde and acetic acid
4. **(a)** diethyl ether **(b)** ethanol **(c)** acetic acid **(d)** acetone
5. 12 (6 C and 6 H)

Chapter 7

Self-Test 7A

1. fractional distillation
2. higher
3. catalytic re-forming
4. petroleum
5. cracking
6. MTBE (methyl tertiary-butyl ether)
7. ethanol
8. c

Self-Test 7B

1. 42%
2. acetaldehyde
3. denatured
4. rubbing (isopropyl)
5. acetic acid (vinegar)
6. acetic acid
7. formaldehyde
8. methanol
9. New Zealand
10. alcohols
11. esters
12. alkaloids, caffeine
13. methanol

Chapter 8

Self-Test 8A

1. monomers
2. thermoplastic
3. a. 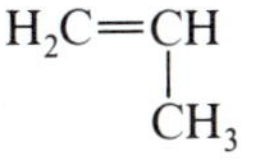 $H_2C{=}CH{-}CH_3$

   b. 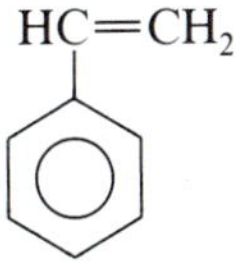 $HC{=}CH_2$ (with benzene ring attached to CH)

   c. $F_2C{=}CF_2$

   d. $H_2C{=}CHCl$
4. a. 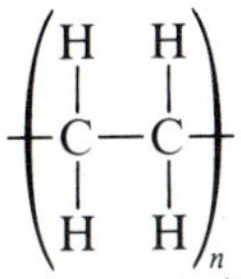 $\left(-\text{CH}_2-\text{CH}_2- \right)_n$ (H H / C—C / H H)

   b. 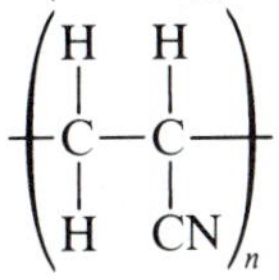 $\left(-\text{CH}_2-\text{CH}(\text{CN})- \right)_n$ (H H / C—C / H CN)
5. isoprene

Self-Test 8B

1. condensation
2. H_2O
3. 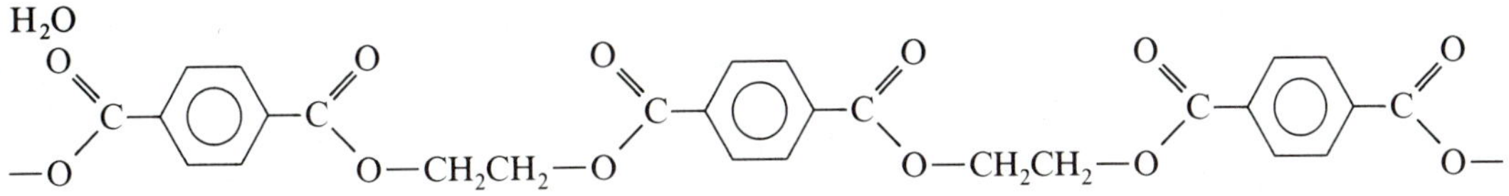 —O—C(=O)—C₆H₄—C(=O)—O—CH₂CH₂—O—C(=O)—C₆H₄—C(=O)—O—CH₂CH₂—O—C(=O)—C₆H₄—C(=O)—O—

   $-O-\overset{O}{\overset{\|}{C}}-C_6H_4-\overset{O}{\overset{\|}{C}}-O-CH_2CH_2-O-\overset{O}{\overset{\|}{C}}-C_6H_4-\overset{O}{\overset{\|}{C}}-O-CH_2CH_2-O-\overset{O}{\overset{\|}{C}}-C_6H_4-\overset{O}{\overset{\|}{C}}-O-$
4. condensation
5. polyester
6. thermosetting
7. 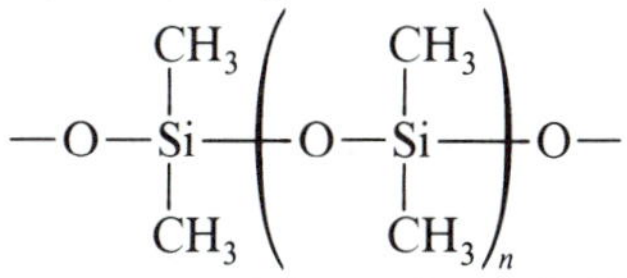 $-O-Si(CH_3)_2-\left(O-Si(CH_3)_2\right)_n-O-$
8. O (oxygen); it is a polymer held together by a network of Si—O bonds
9. 4

Chapter 9

Self-Test 9A

1. 4
2. rotation of polarized light
3. glucose, fructose
4. monosaccharides
5. glucose
6. hydrogen bonding
7. glucose
8. glucose
9. cellulose
10. glycerol and fatty acids
11. the absence or presence of double bonds. Saturated fats have only single C—C bonds, whereas unsaturated fats have one or more C=C double bonds.

Self-Test 9B

1. amino acids
2. 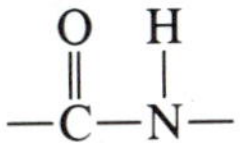 $-\overset{O}{\overset{\|}{C}}-\overset{H}{\overset{|}{N}}-$
3. 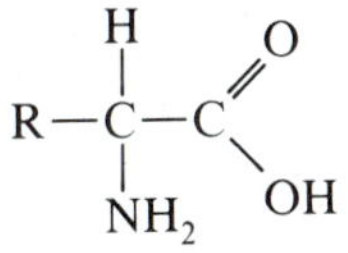 $R-CH(NH_2)-C(=O)OH$
4. glycine
5. **(a)** 27

   (b) 6
6. key, lock
7. hydrogen

Self-Test 9C

1. CO_2 and H_2O; energy
2. ATP
3. ADP, Energy
4. hydrolysis
5. DNA
6. ribose, deoxyribose
7. a sugar (ribose or deoxyribose), phosphoric acid, and a nitrogenous base
8. double helix
9. hydrogen bonds
10. tRNA (transfer-RNA)
11. A and T or U; G and C; T and A; U and A
12. 3 billion

Chapter 10

Self-Test 10A

1. dehydration, hydrolysis
2. oxidizing
3. hemoglobin, oxygen
4. false
5. CN^-, cytochrome oxidase, oxygen
6. c
7. heavy metal poisons, complex
8. true
9. water, food, paint

Self-Test 10B

1. neurotoxins
2. synapse
3. acetylcholine
4. d
5. b
6. b

Self-Test 10C

1. teratogen
2. mutagen
3. b
4. caffeine, ozone
5. mutagenicity
6. carcinogen
7. benzene—gasoline
   mineral oil—sun tanning oils
   sucrose—table sugar (not a carcinogen)
   asbestos—auto brake linings
   arsenic—pesticides
8. maximum tolerated dose
9. true

Chapter 11

Self-Test 11A

1. 70
2. oceans
3. metal ions, pesticides, organic solvents
4. groundwater
5. aquifer
6. runoff

Self-Test 11B

1. biochemical oxygen demand, BOD
2. batteries, petroleum
3. solvents, pesticides
4. auto battery; lead, sulfuric acid
   paints; solvents
   motor oil; oil, metals
5. aluminum, paper, plastics, glass

Self-Test 11C

1. filtration
2. aerobic
3. anaerobic
4. chlorine
5. sodium
6. reverse osmosis

Chapter 12

Self-Test 12A

1. ppm, 10,000, 0.001
2. decrease
3. ozone
4. adsorb
5. absorb
6. warm, cool
7. coal burning
8. oxides
9. NO and O
10. O atom

Self-Test 12B

1. sulfur dioxide, carbon dioxide
2. sulfur trioxide, sulfuric acid
3. lime
4. sulfuric acid, nitric acid
5. 5.6, carbon dioxide
6. 1872
7. C—Cl bond
8. $O_2 + O \rightarrow O_3$

9. ClO
10. C—Br
11. Antarctica
12. ultraviolet

Self-Test 12C

1. clear-cutting, automobiles, fossil fuel burning to produce electricity
2. photosynthesis, dissolving in oceans
3. carbon dioxide, ozone, water vapor, carbon dioxide
4. true
5. 350 ppm
6. Texas
7. toluene
8. smoking
9. soil and rocks under the home

Chapter 13

Self-Test 13A

1. clays, silts, sandy soils, loams
2. sour
3. acidic, basic
4. size of soil particles, chemical composition of soils
5. a trivalent ion like Fe^{3+} (hydrolyzes more than Na^{+})
6. humus
7. nitrogen, phosphorus, potassium
8. calcium, magnesium, sulfur
9. element
10. true

Self-Test 13B

1. false
2. nitrogen, phosphorus as P_2O_5, potassium as K_2O
3. yes
4. gas
5. K_2CO_3
6. 33%
7. DDT
8. chlordan
9. selective herbicide
10. 2,4-D

Chapter 14

Self-Test 14A

1. proteins, fats, carbohydrates, vitamins, minerals, water
2. fats
3. one
4. basal metabolic rate, weight in pounds
5. fats
6. apoenzyme
7. nitrogen
8. triglycerides
9. linoleic acid
10. *cis*-fatty acids
11. source of energy, provides roughage
12. hypoglycemia
13. *trans*

Self-Test 14B

1. variety, whole, different places
2. sodium and potassium
3. greater than 1
4. kidney
5. iodine
6. false
7. fat, water
8. provitamin
9. antioxidant
10. vitamin B_6 (pyridoxine)
11. coenzymes, growth and energy production
12. synergistic, potentiation
13. drying, salting
14. false
15. more easily oxidized
16. complexing agent (sequestrant)
17. false
18. flavor enhancer
19. generally recognized as safe
20. alitame

Chapter 15

Self-Test 15A

1. generic
2. antibiotic
3. cholesterol, triglycerides
4. faster
5. relaxing
6. LDLs
7. true
8. DNA
9. antimetabolites
10. leukemia
11. TPA (tissue plasminogen activator)
12. true
13. heart disease

Self-Test 15B

1. folic acids
2. cell walls

3. histamine
4. testosterone
5. progesterone and estradiol
6. muscle building
7. acne, baldness, changes in sexual desire, liver damage, liver cancer
8. insulin

Self-Test 15C

1. neurotransmitters
2. dopamine
3. heroin
4. enkephalins
5. cocaine
6. heroin
7. AIDS
8. phosphate
9. T cells
10. designer drug
11. norepinephrine
12. acetylcholine
13. methamphetamine

Chapter 16

Self-Test 16A

1. Food and Drug Administration
2. keratin, true
3. cystine
4. 4
5. disulfide bonds
6. melanin (black), red pigment
7. greater than 7
8. increases
9. c

Self-Test 16B

1. emollient
2. moisturizer
3. hydrogen bonding
4. an eosin dye, or lake
5. astringent
6. black color
7. melanin
8. ultraviolet light absorber
9. 4
10. aluminum, astringents

Self-Test 16C

1. detergent
2. saponification
3. soap
4. air
5. lanolin
6. oxidizes
7. sodium hypochlorite

Index/Glossary

Note: d following a page number indicates a definition; i indicates an illustration or figure; s indicates a structure; and t indicates a table. Glossary terms, printed in boldface, are defined here as well as in the text.

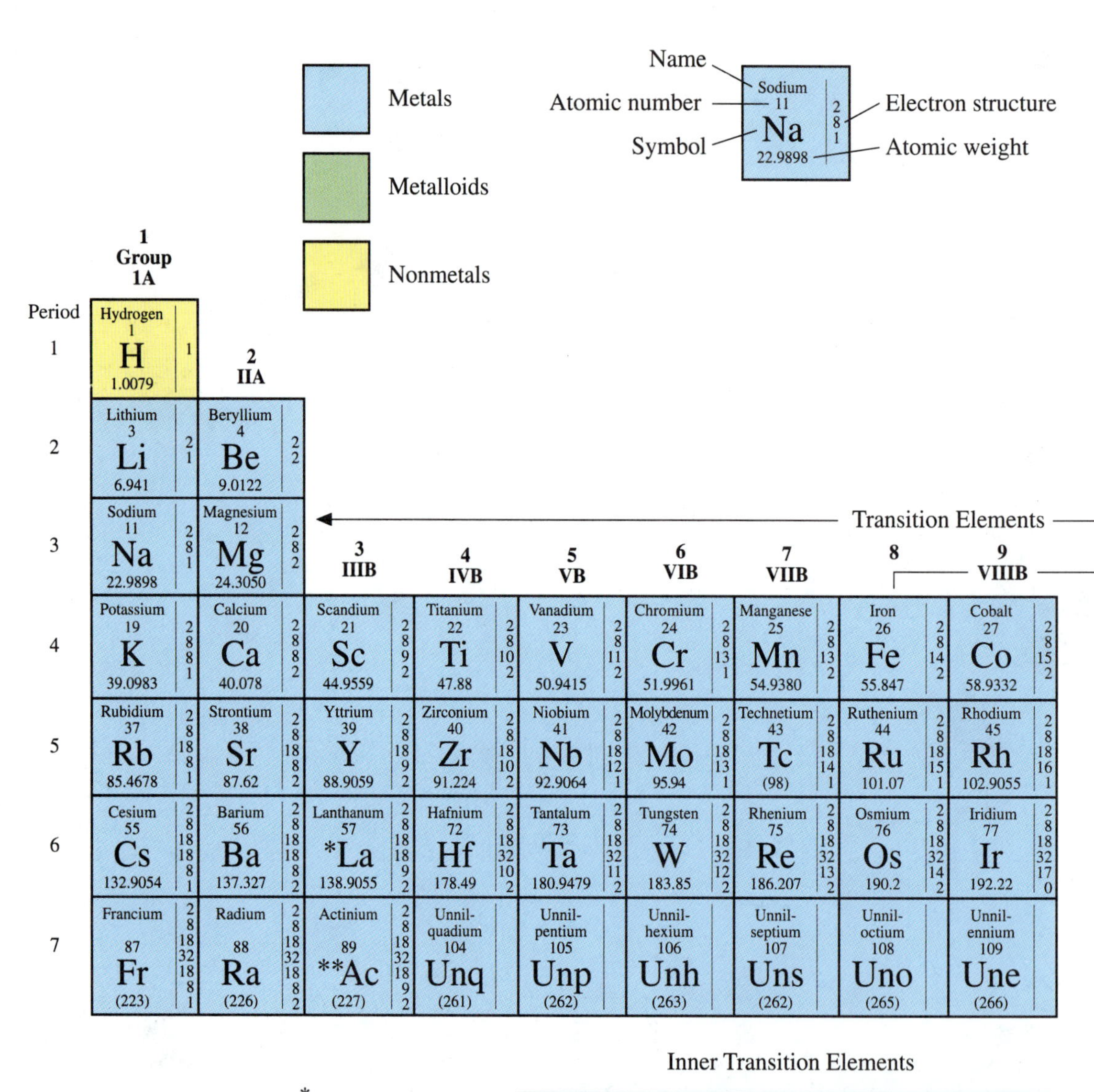

Inner Transition Elements

* Lanthanide Series 6	Cerium 58 Ce 140.115 (2 8 18 20 8 2)	Praseo-dymium 59 Pr 140.9076 (2 8 18 21 8 2)	Neodymium 60 Nd 144.24 (2 8 18 22 8 2)	Promethium 61 Pm (145) (2 8 18 23 8 2)	Samarium 62 Sm 150.36 (2 8 18 24 8 2)
** Actinide Series 7	Thorium 90 Th 232.0381 (2 8 18 32 18 10 2)	Protac-tinium 91 Pa 231.0359 (2 8 18 32 20 9 2)	Uranium 92 U 238.0289 (2 8 18 32 21 9 2)	Neptunium 93 Np (237) (2 8 18 32 22 9 2)	Plutonium 94 Pu (244) (2 8 18 32 23 9 2)

Note: Atomic masses are 1987 IUPAC values (up to four decimal places).